AF323873

CHINESE EXPLORED ASIA
LONG BEFORE THE EUROPEANS

An Ancient World Map Tells All

CHINESE EXPLORED ASIA

LONG BEFORE THE EUROPEANS

An Ancient World Map Tells All

Sheng-Wei Wang

China-US Friendship Exchange, USA

World Scientific

NEW JERSEY · LONDON · SINGAPORE · BEIJING · SHANGHAI · HONG KONG · TAIPEI · CHENNAI · TOKYO

Published by

World Scientific Publishing Co. Pte. Ltd. .

5 Toh Tuck Link, Singapore 596224

USA office: 27 Warren Street, Suite 401-402, Hackensack, NJ 07601

UK office: 57 Shelton Street, Covent Garden, London WC2H 9HE

Library of Congress Cataloging-in-Publication Data
Names: Wang, Sheng-Wei author
Title: Chinese explored Asia long before the Europeans : an ancient world map tells all /
    Sheng-Wei Wang, China-US Friendship Exchange, USA.
Description: Hackensack, NJ : World Scientific, [2026] | Includes bibliographical references and index.
Identifiers: LCCN 2025006816 | ISBN 9789819810963 hardcover |
    ISBN 9789819810970 ebook | ISBN 9789819810987 ebook other
Subjects: LCSH: Discoveries in geography--Chinese |
    Cartography--Asia--History | World maps--Early works to 1800 |
    Explorers--China | Physical geography--Asia | Asia--Discovery and
    exploration--Chinese | China--History--Ming dynasty, 1368-1644
Classification: LCC G322 .W34 2025 | DDC 910.951--dc23/eng/20250610
LC record available at https://lccn.loc.gov/2025006816

British Library Cataloguing-in-Publication Data
A catalogue record for this book is available from the British Library.

ISBN 978-981-98-1096-3 (hardcover)
ISBN 978-981-98-1097-0 (ebook for institutions)
ISBN 978-981-98-1098-7 (ebook for individuals)

For any available supplementary material, please visit
https://www.worldscientific.com/worldscibooks/10.1142/14249#t=suppl

Desk Editors: Aanand Jayaraman/Yulin Jiang

Typeset by Stallion Press
Email: enquiries@stallionpress.com

About the Author

Sheng-Wei Wang is an independent scholar and writer who founded the China–US Friendship Exchange, Inc. in 2006. She is passionate about uncovering our shared history of maritime globalization.

She was born in Taiwan, lived in the US and Germany, and currently resides in Hong Kong. She has a B.S. in Chemistry from Tsing Hua University (Hsinchu, Taiwan) and a Ph.D. in Chemical Physics from the University of Southern California. She was awarded the Alexander von Humboldt Postdoctoral Fellowship to work under the guidance of Professor Gerhard Ertl (2007 Nobel Prize in Chemistry). She also did research at the California Institute of Technology and the Stanford Linear Accelerator Centre. Later she was a staff scientist at the Lawrence Berkeley National Laboratory.

In 2013, she started research on pre-Columbian Chinese global maritime exploration. She has since single-authored two English books on this subject: *The Last Journey of the San Bao Eunuch, Admiral Zheng He* (2019), and *Chinese Global Exploration in the Pre-Columbian Era: Evidence from an Ancient World Map* (2023). This is her third single-authored English book on related topics. In 2022, she founded WorldDiscovery.net to share her work and to promote an inclusive and open dialogue about Chinese maritime exploration since Antiquity.

In 2023, she was interviewed on her books by the Hong Kong International Business Channel (HKIBC), and in early 2024 by the English-language newspaper *South China Morning Post* (SCMP). Chinese versions of the SCMP interview reports were soon republished by the *Reference News* (参考消息) and the *Journal of Qinghai University* (青海大学学报). In the same year, on 11 July — China's National Maritime Day — she was awarded the honor of "Zheng He Cultural Promotion Ambassador" by the Jiangsu Provincial Zheng He Research Society and others in Nanjing, China, for her contributions to Zheng He research. In 2025, the renowned scholar Bruce Cornet praises her book as follows: "This book presents a groundbreaking reinterpretation of the world exploration history. Dr. Wang's meticulous research opens new windows to the Pre-Columbian era, challenging Eurocentric perspectives." 5 January 2025, *The Economist* Book

Review. She is interviewed by the Radio Television Hong Kong (RTHK) on 15 May 2025 to talk about "UN Inscription of Ming Navigator Zheng He".

She has been invited to give presentations to academics and the public worldwide to discuss how the globalized world is influenced by the common history of mankind. This new book provides abundant evidence showing that the Chinese most likely have conducted peaceful global maritime exploration before the Europeans. She hopes that this book will further stimulate China's maritime development and enables the readers to rediscover the true contributions the Chinese made to the world's maritime civilization.

Acknowledgment

The author is grateful for Professor Emeritus Michel A. Van Hove's critical reading of the content, constructive suggestions, and technical assistance in preparing the figures in published form for this book.

Chinese Nomenclature

In this book, all the Chinese characters in the main text and in the Appendix are expressed first by the Chinese traditional characters followed by the Chinese simplified characters, separated by a slash "/". If both characters are the same, that expression will not be repeated.

Contents

Introduction

In the Introduction of my previous book, titled *Chinese Global Exploration in the Pre-Columbian Era: Evidence from an Ancient World Map,* I point out the following issue:[1]

In school, we all learn that Christopher Columbus discovered America, and that Europeans ushered in the Age of Discovery. However, even the most rigorous scholars admit the existence of a very puzzling phenomenon on the ancient Western maps: many supposedly "unknown" areas were already mapped out before the Age of Discovery. To resolve this enigma requires considering the role played by China — then the world's strongest maritime power.

But how do we go about it? I have decided in that previous book (denoted here as "Book 1") and the present book (denoted here as "Book 2") to thoroughly analyse an important ancient world map — Kunyu Wanguo Quantu (坤輿萬國全圖/坤舆万国全图)[2] or Complete Geographical Map of All the Kingdoms of the World, written with Chinese characters — abbreviated here as KWQ, to seek the truth. The map was published in 1602 by Italian Jesuit priest Matteo Ricci in China,[3] and for over 400 years since its publication, it has been regarded as strictly of European origin and based on maps which Ricci brought with him to China in 1582 of the Ming Dynasty. In my books, I want to find out: (1) whether the KWQ map also has "many supposedly 'unknown' areas that were mapped out before the Age of Discovery"; and (2) what was the true origin and era of the KWQ.

These are two very challenging issues for researchers to tackle, especially because the Ming emperors imposed the *Haijin* (海禁) — or sea ban — soon after the Treasure Fleets led by Admiral Zheng He (鄭和/郑和; 1371–1433/1435) completed their seventh (and last) voyage in the 1430s. The emperors stopped all future voyages of the Treasure Fleets and destroyed the nautical maps (although a few might have escaped this ill fate) and precious documents gathered by the Chinese mariners during their voyages, resulting in the later Chinese losing their geographical knowledge of the world. By contrast, China's retreat from the world stage provided a great opportunity for the European maritime powers to quickly fill in this power void and rise on the world's stage. Their authorities, navigators, and map makers who were able to access and copy the Chinese maps could have used the Chinese maps and documents for guidance to "rediscover" many supposedly "unknown" areas, hence creating all the confusion.

If this is true, then a thorough analysis of this famous world map — Kunyu Wanguo Quantu — may be the right starting point for me to extract some useful information to show: (1) how different the content of the KWQ is from the major contemporary European maps in the sixteenth century; (2) how the difference enables me to determine the true origin of the KWQ; and (3) what methodology I can devise for extracting the eras of the different regions from the KWQ.

To achieve these three goals, I found that in addition to the conventional cartographic approach, a non-conventional methodology must be used in the analysis.

Within the conventional cartographic approach, I have painstakingly compared all the geographical items depicted on the KWQ with those on the selected contemporary European maps in the sixteenth century to check the extent of their overlap. This task has been applied in Book 1 to over 550 geographical items and more than 40 annotations depicted in different regions outside of Asia on the KWQ. Readers can refer to Book 1 for the details of such comparisons. In this book, I have likewise listed specifically the geographical names and annotations which only appear on the KWQ, and those which are of Chinese origin (as best as I could). These lists provide abundant evidence for showing the Chinese origin of the KWQ.

To extract the eras of the different regions depicted on the KWQ, I realise that every geographical item depicted on any map must carry a "time label" or "era" associated with its respective history (for example the founding and extinction dates of a kingdom or the founding date of a city and so on), and when many geographical items appear in the same region on the same map at the same time, it means that their eras overlap, and the smallest (narrowest) overlapping era can be regarded as the era of that region on that map or the last era of that region being explored before the cartographers depict these data on their maps.

Most researchers in the past either did not perform or lacked a thorough and comprehensive investigation of the histories of all the geographical items on the KWQ to obtain these "time labels" or "eras". Hence, they were not able to deduce the eras of the different regions depicted on the KWQ. Because very often these eras were associated with a kingdom's founding and extinction, the final era extracted also represented the political era (landscape) of that region on the map.

Using both the conventional cartographic approach and my "time label" methodology for extracting the eras, I have obtained the geographical and historical information of "all" the relevant items depicted on the KWQ, region by region, in Book 1 and Book 2. From the results, I have determined the origin of each portion on the map and the era when that region was last explored before the relevant geographical data were depicted on the map. I emphasize "all" to avoid obtaining erroneous results based on incomplete data as some of the researchers have done.

At the end of my analyses, it became very clear who was in that region earlier: the Chinese or the Europeans.

Analysing the annotations depicted on the KWQ can also give valuable information. For example, the single annotation for Japan can be analysed to give the era of Japan on the KWQ as 897 A.D., which was in the Tang Dynasty. Similarly, on the KWQ, there are also annotations in the region of Africa which show unique characteristics of Chinese culture. On the other hand, the major contemporary European maps in the sixteenth century analysed in my two books include very few or even no annotations.

In summary, my main findings in Book 1 for the non-Asian region on the KWQ are as follows: (1) the KWQ is a map of Chinese origin; (2) contrary to general perception, Zheng He's mariners reached far beyond the east coast of Africa; (3) Europe on the KWQ reveals its political landscape over four centuries before Ricci's time; (4) the maps of Australia, New Zealand, Antarctica and the Americas contain Chinese geographical data that were updated by Zheng He's mariners during their sixth voyage in the 1420s (an era long before Christopher Columbus landed in the "New World" in 1492); and (5) the date of the geographical data in the African part of the map is 1433, which was during the time of the Ming mariners' seventh (and last) voyage.

In the present Book 2, I have extended my analyses to the whole region of Asia depicted on the KWQ. After analysing 522 geographical items and 80 annotations, and comparing the information obtained with those on the selected contemporary European maps in the sixteenth century, I have extracted the eras of the Asian regions on the KWQ as follows: (1) Japan in 897 A.D.; (2) Joseon in 1433; (3) China in 1433; (4) Southeast Asia in 1432–1433; (5) Indian Subcontinent and its eastern periphery in 1432–1433; (6) West and Central Asia in 1433; and (7) Asia to the North of China in 1428–1433.

Since all these eras are long before the first Europeans' arrival in Asia in the "Age of Discovery", the Asian portion of the KWQ must be of Chinese origin. Moreover, in the different regions of Asia, there is a high percentage of geographical items which are depicted only on the KWQ, but not on any of the major sixteenth century European maps. These percentages are over 90% in Japan and the Korean Peninsula, c. 37% in the Southeast Asian region, c. 41% in the Indian Subcontinent and its eastern periphery region, c. 23% in the West and Central Asian region, and c. 55% in Asia to the north of China. These data show that the KWQ can not be a direct or adapted copy of the European maps. Such findings add an additional layer of evidence to support the Chinese origin of the Asian portion of the KWQ.

In the specific case of China, Matteo Ricci explicitly confessed that he had used Chinese records for the China portion of the KWQ:

大明聲名文物之盛自十五度至四十二度皆是其餘四海朝貢之國甚多此總圖略載嶽瀆省道大略餘詳統志省志不能殫述

My translation and comments (between square brackets) are shown below:

The fame and rich cultural relics of the Great Ming Dynasty are shown [on this map] from fifteen degrees to forty-two degrees [in latitude]. Many countries from all over the world came to send in tributes. This comprehensive map briefly contains the outline of the mountains, rivers, and provincial roads and structures. The rest of the detailed information in the general records and provincial records cannot be stated [or depicted] here in detail.

The last sentence of this annotation states that the China portion on the KWQ can only state or depict some general information, and the more detailed information from the Chinese local general records and provincial records cannot be covered on this map.

In conclusion, my systematic and exhaustive research can serve as a foundation for future China studies by students and experts in this field and it offers new insights into the history of world maritime exploration. The two books can also serve as an extensive database of more than one thousand carefully curated geographical and historical facts pertaining to large parts of the world in the Pre-Columbian era.

Endnotes

[1]Sheng-Wei Wang. "Introduction." *Chinese Global Exploration in the Pre-Columbian Era: Evidence from an Ancient World Map*, Singapore: World Scientific Publishing Co., 2023, pp. ix–xi.
[2]*Library of Congress.* "Kun yu wan guo quan tu". www.loc.gov, Library of Congress, 2025, loc.gov/item/2010585650.
[3]Ronnie Po-Chia Hsia. *Matteo Ricci and the Catholic Mission to China, 1583–1610: A Short History with Documents (Passages: Key Moments in History).* Indianapolis, USA: Hackett Publishing Company, Inc., 2016.

Chapter 1

An Ancient World Map Depicts Chinese Exploration of Japan in 897 A.D., Long Before the First Europeans Landed in that Region in 1542/1543

Abstract

This chapter reports that an ancient world map — Kunyu Wanguo Quantu (坤輿萬國全圖/坤輿万国全图) or Complete Geographical Map of All the Kingdoms of the World published by Matteo Ricci in 1602 in China — depicts Japan in 897 A.D. of the Chinese Tang Dynasty, almost 650 years before the first Europeans visited Japan. A detailed analysis of all the geographical items depicted on the Japan portion of this map, and Japan's surrounding islands and seas shows that this portion of the map is of Chinese origin, instead of a direct or adapted copy of the major European maps in the sixteenth century.

Keywords: China, Europeans, Japan, Kunyu Wanguo Quantu, Matteo Ricci, Tang Dynasty

1. Introduction

For over four hundred years, the ancient world map — Kunyu Wanguo Quantu (坤輿萬國全圖/坤輿万国全图)[1] (abbreviated here as KWQ) — written with Chinese characters and having latitudinal and longitudinal lines, has been generally regarded as a European map. The map was published in 1602 by Matteo Ricci (an Italian Jesuit priest and one of the founding figures of the Jesuit China missions) and was believed to be based on the European maps which Ricci brought with him to China in 1582.[2] In my previous book titled *Chinese Global Exploration in the Pre-Columbian Era: Evidence from an Ancient World Map*,[3] I have provided thorough analyses and concrete evidence to show the eras from which the geographical information on several major portions of the KWQ were obtained. These eras are as follows: (1) Chinese explored Cape Breton Island on Canada's Atlantic Coast long before the Europeans; (2) geographical information on Europe was provided by Chinese explorers in the Southern Song Dynasty (南宋; 1127–1279); (3) information on Mo Wa La Ni Jia (墨瓦蠟泥加/墨瓦蜡泥加; referring to the regions of Australia, New Zealand, *Tierra del Fuego*/Land of Fire in South America and Antarctica) was obtained by Ming (1368–1644) mariners during their sixth voyage in the 1420s to the Western Ocean (referring to the Indian

Ocean and beyond); (4) information on Africa was obtained by the Ming mariners in 1433 during their seventh (and last) voyage to the Western Ocean; and (5) information on the Americas was also obtained by the Ming mariners during their sixth voyage in the 1420s to the Western Ocean. These mariners belonged to the Treasure Fleets commanded by Admiral Zheng He (鄭和/郑和; 1371–1433/1435) in the early fifteenth century. The findings show that these portions of the KWQ most likely used Chinese maps as source maps (these source maps no longer exist), instead of copying from other contemporary European maps.

In this chapter, I extend my analysis to the map of Japan on KWQ, abbreviated here as the Japan-KWQ. As in my previous book, the task is to find the origin of this part of the map and, by analysing the annotations and examining the locations and histories of every geographical item depicted on the Japan-KWQ, to extract the political era of Japan revealed by the map. That era was the era when the information on the Japan portion of the map was obtained.

I shall begin by first detailing the Europeans' knowledge of Japan, and then reviewing the Chinese knowledge of Japan.

2. Japan was Basically Unknown in Europe, Until the Europeans Read Marco Polo's Travel Book

Figure 1.1(a) shows a map made by a Greek scholar in c. 194 B.C., which contains no Japan, but includes China and India. Figure 1.1(b) shows a map titled *Tabula Rogeriana* drawn by an Arabian scholar in 1154, whose knowledge of mapmaking and worldview were more advanced than those of Europe at the time. The map still does not have an island that could be identified as Japan.

Japan was basically unknown to the Europeans in Antiquity. Only in the seventh century, did Isodore of Seville (c. 560–636), who was widely regarded in the nineteenth century as "the last scholar of the ancient world",[6] claim that the Garden of Eden was in far-eastern Asia. Although

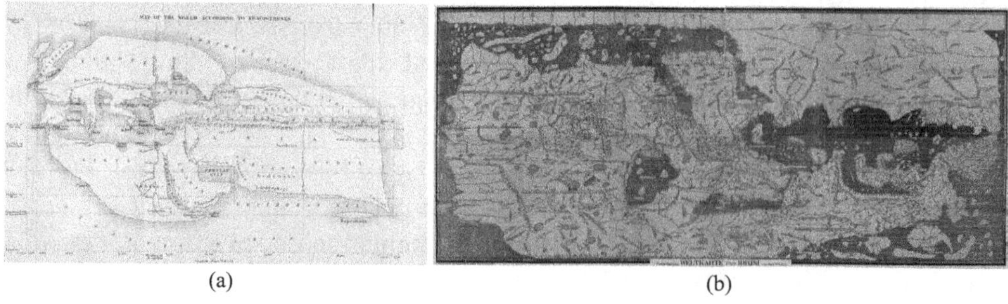

(a)   (b)

Fig. 1.1.   (a) Nineteenth-century reconstruction of Eratosthenes' map in c. 194 B.C. of the whole world known to the Greeks (public domain);[4] Japan is not on the map, but China and India are. (b) *Tabula Rogeriana* (public domain)[5] made in 1154 by al-Idrisi, an Arabian scholar. The map shows the whole world known to the Arabians at the time and it was regarded as the most accurate world map for the next three centuries. However, it shows no island that could be identified as Japan.

several different locations of the Garden of Eden were "discovered" and debated, but the route to it was sealed off by God with a flaming sword. His claim seems to have influenced map makers in the next few hundred years in searching for this mysterious Garden of Eden. Finally, the oldest map of Japan was drawn in the eleventh century by Mahmud of Kashgar (Mahmud al-Kashgari, a Kara-Khanid scholar; 1029–1101)[7] on his world map, as shown in Fig. 1.2.

This circular map shows the four cardinal directions and some seas and rivers. Although the map has many errors, the eastern regions show both China[8] and Japan (being an island in the eastern part of Asia).[9] Japan is in the black half-circle at the top (the top is East); China is in the "hook" hanging from that half-circle.[10]

Apart from Isidore of Seville's claim and Mahmud al-Kashgari's Oldest Map of Japan (outside China), Japan was basically unknown in Europe, since it was very closed off to the rest of the world at the time. The situation changed only until the Europeans read about Marco Polo's travel book.[12] Marco Polo (1254–1324) was a Venetian and the most famous traveler of his time. His voyages started in 1271, the first year of China's Yuan Dynasty (1271–1368) and lasted nearly a quarter of a century. The book records his traveling through Asia in the late thirteenth and early fourteenth centuries and describes Japan in detail. Japan was called Zipangu or Cipangu in his book. Since he served the Kublai Khan on numerous diplomatic missions, he could know about Japan from the court of Kublai Khan. But it is unlikely that he visited Japan.

Hence, when we see in Fig. 1.3(a) that on the 1411–1415 De Virga World Map there is a large island in the Indian Ocean labeled "Caparu sive Java magna", it possibly combines Marco Polo's descriptions of Java and Japan.[13] Similarly, the Fra Mauro World Map drawn in the 1450s also

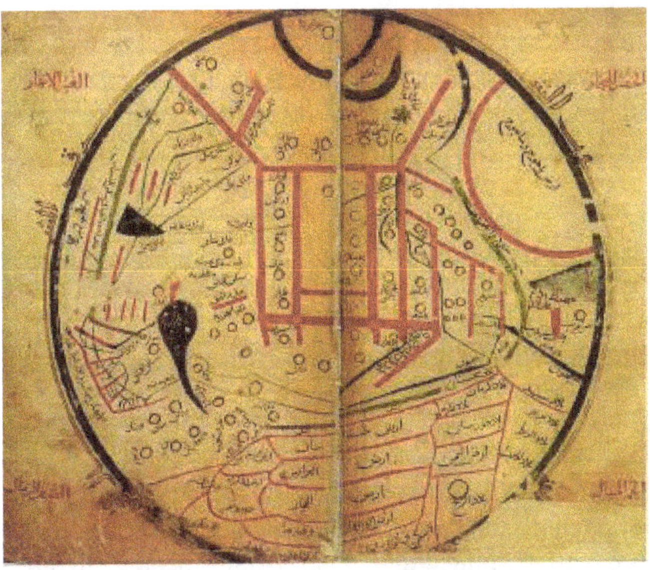

Fig. 1.2.   Map of the eleventh century from Mahmud al-Kashgari (public domain),[11] which contains the Oldest Map of Japan (outside China).

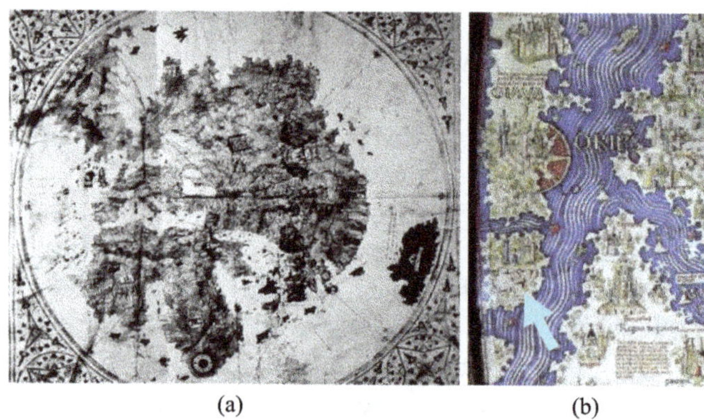

(a)   (b)

Fig. 1.3. (a) The De Virga World Map (1411–1415) (public domain).[14] (b) Depicts Cipangu (Japan) extracted from the Fra Mauro World Map (c. 1450) (public domain),[15] pointed out by the arrow.

depicts a part of Japan, probably Kyushu. It appears south of the island of Java, see Fig. 1.3(b), denoted "Isola de Cimpagu (Island of Cimpagu)" which is a misspelling of Cipangu. Such a depiction is also based on Marco Polo's description.

Then in the late fifteenth century, we have the Erdapfel which is a terrestrial globe produced by Martin Behaim from 1490 to 1492, and the oldest surviving terrestrial globe. But on this globe, Cipangu (Japan) is oversized and well south of its true position, as Marco Polo had described in his memoirs. This was the last known globe made according to the theories of Ptolemy (150 A.D.) before the discovery of America.[16]

The first map of Japan printed in a Western manuscript and drawn from the written account of Marco Polo is the 1528 simple diagram drawn by Benedetto Bordone.[17] This predates the first European visit to Japan; hence, this map bears no resemblance to how Japan really looks. Instead, the map looks more like a drawing of only the main island of Honshu, not of the entire Japanese archipelago.[18]

Finally, around the middle of the sixteenth century, Portuguese traders António Mota and Francisco Zeimoto (and possibly a third named António Peixoto),[19] landed on Tanegashima Island (種子島/种子岛) at the southern tip of the Japanese archipelago after their boat was blown off course near Japan in 1543 (some sources say 1542).[20] This was the first time Europeans came in direct contact with Japan, followed by Spanish missionaries in 1549,[21] and Dutch and English traders in the 1600s.[22]

Only in c. 1571 did Portuguese cartographers like Fernão Vaz Dourado (c. 1520–c. 1580) start to draw better maps of the then-known parts of Japan. These maps give higher accuracy in depicting lands and continents. However, on Fernão Vaz Dourado's map,[23] the shape of the largest island Honshu is grossly distorted and Japan's second largest island Hokkaido (formerly known as Ezo, Yezo, Yeso, or Yesso)[24] is omitted (since the island had not been visited by the Europeans before 1589).[25] In the 1589–1590 trans-Pacific crossing from Macau to Acapulco, João da Gama (c. 1540–after 1591) — a Portuguese explorer and colonial administrator in the Far East and the grandson of Vasco da Gama (c. 1460–1524) — sailed from Macau to the northeast and rounded the northern end of Japan.[26] He landed at a place which has since been called Ezo (Yezo),[27] later renamed the island of Hokkaido in 1869. He finally arrived at Acapulco in Mexico in March 1590. João da Gama possibly died in Spain or Portugal after 1591 or 1592. No European cartographer knew how to draw the island of Hokkaido accurately even after João da Gama returned to Spain or Portugal around 1590–1591/1592.

About 25 years later, the 1617 Portuguese map of Japan by Christopher Blancus and Inacio Moreira[28] became the most accurate representation of Japan then, because Moreira visited Japan himself at least twice at the end of the sixteenth century and did not rely on second-hand accounts as the previous European cartographers did. Hence, he was able to render the size and shape of the Japanese islands more precisely than anyone in Europe had done before. However, only a small southern portion of Hokkaido is rendered on this 1617 map. The map shows that the Europeans had rapidly begun to accumulate knowledge of Asian geography. *But all these came after Matteo Ricci published the KWQ in 1602 and his death in 1610; hence, the 1617 Portuguese map could not be the source map of the Japan-KWQ.*

3. China Has Been Japan's Neighbour Since Antiquity; the Japan-KWQ Is of Chinese Origin and Depicts Japan in 897 A.D.

Contrary to Europeans' unfamiliarity with Japan until around the middle of the sixteenth century, China has been Japan's neighbour since ancient days. The China–Japan historical connection can be summarized by the following two respects.

3.1 *Japan–China relations*

There is a Chinese legend: the first Chinese Emperor — Qin Shi Huang (秦始皇; 259–210 B.C.) — sent hundreds of young men and women to Japan to search for medicines of immortality; they never returned, perhaps for fear of the consequences of failure, while they reached Japan and colonised it.[29]

According to Chinese historical records, the first mention of the Japanese archipelago appears in the Chinese book titled *Hou Han Shu* (後漢書/后汉书) or *The Book of the Later Han* written in 398–445 A.D. It states that in the year 57 A.D., the Emperor of the Han Dynasty (202 B.C.–9 A.D., 25–220 A.D.) gave a golden seal to Wa (Japan). The King of Na's gold seal (漢委奴國王印/漢委奴国王印) was discovered in northern Kyushu in the eighteenth century.[30] Here the "King of Na" refers to the "King of Na State of Wa (Japan)" which was a vassal state of the Han Dynasty. This record shows that China's direct diplomatic relationship with Japan goes back to times in the Later Han/Eastern Han Dynasty (25 A.D.–220 A.D.), and Japan began to adopt many Chinese traditions.

Then, during the Sui Dynasty (隋朝; 581–618) and the Tang Dynasty (618–907, with an interregnum between 690 and 705), Japan wanted to learn from China how to establish itself as a sovereign nation in Northeast Asia. The Japanese Imperial Court sent many students on nineteen imperial missions to China.

China-Japan economic and trade exchanges continued in the Song Dynasty (960–1279),[31] but interruptions occurred in the Yuan Dynasty (1271–1368).

In 1274 and 1281, the Kamakura shogunate withstood two Mongol invasions launched by Kublai Khan of the Yuan Dynasty.[32] The economic and trade exchanges between Japan and the Yuan Dynasty were interrupted during this period.

In the Ming Dynasty, Wokou (倭寇), which literally translates into "Japanese pirates", were pirates of varying origins (also including Chinese, Korean, and Portuguese), who raided the coastlines of China. The Imperial Court of Ming decreed that Ningbo (寧波/宁波) was the only place where China-Japan relations could take place.[33]

The above short summary shows that Japanese missions to Tang China (遣唐使) represent Japanese earnest efforts to learn from the Chinese civilization in the seventh, eighth and ninth centuries. The nature of their contacts also evolved gradually from political and ceremonial acknowledgment to cultural exchanges and commercial ties in the Song Dynasty but were interrupted in the Yuan Dynasty and became restrictive in the Ming Dynasty.

3.2 *Japan on ancient Chinese maps*

Concerning the Chinese maps which show Japan, the Huayi Tu (華夷圖/华夷图) or Map of China and the Barbarian Countries, engraved as a stone stele and dated 1136 (some source says 1137; or even as early as 1034), may be the earliest surviving map of China that relates China with other foreign states (see Fig. 1.4).[34] It has very accurate latitudes and longitudes for cities in the interior part of China, which may indicate the use of spherical trigonometry projection.[35]

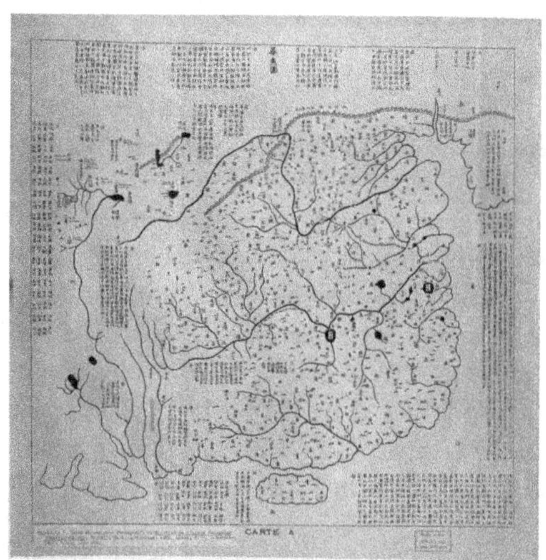

Fig. 1.4.   This map shows the 1136 Huayi Tu (華夷圖/华夷图) with annotations or Map of China and Barbaric Countries with annotations (public domain).[36]

The Huayi Tu covers China during the Jin/Great Jin Dynasty (金朝 or 大金; 1115–1234) and the Southern Song Dynasty (南宋; 1127–1279; the Han-led Song Dynasty was based in southern China). The former is also called the "Jurchen Jin Dynasty" ("女真金朝"), because members of the ruling Wanyan clan were of Jurchen (女真) descent and the Jin emperors referred to their state as *Zhongguo* (中國/中国; China). They were conquered in 1234 by the Mongols and the Southern Song Dynasty.

The Huayi Tu is based on the Hainei Huayi Tu (海內華夷圖/海内华夷图) or Map of China and the Barbarian Countries within the Seas. The map was produced in 801 A.D. by Jia Dan (賈耽/贾耽; 730–805 A.D.) who was a Chinese cartographer, military general, and politician during the Tang Dynasty. The Hainei Huayi Tu was presented to Emperor Dezong of Tang (唐德宗) in 801.[37] But this map is lost.

The texts arranged around the edges of the Huayi Tu were taken from the Hainei Huayi Tu; they include brief descriptions of Korea, Japan and other foreign states, as well as lists of the states to the south and west of China. These descriptions show that China already had a good knowledge about Japan in the Tang Dynasty due to their close contacts.

The next surviving ancient Chinese map showing Japan is the c. 1389 Da Ming Hunyi Tu (大明混一圖/大明混一图) or Composite Map of the Ming Empire in the early Ming Dynasty (see Fig. 1.5; however, the exact date of creation remains unknown). It is an extensive Chinese map measuring 386 cm × 456 cm in size. The original texts were written in Classical Chinese, but on the surviving copy Manchu labels were later superimposed. This is because the map was created sometime during the Ming Dynasty (some sources say the map was based on the Yuan Dynasty information) and then handed over to the new rulers of China, the Qing (清; 1644–1912).

Fig. 1.5.   The Da Ming Hunyi Tu (大明混一圖/大明混一图) or Composite Map of the Ming Empire (public domain)[39] dated around 1389 in the early Ming Dynasty; the map stretches eastward to Korea and central Japan.

Nurhaci (努爾哈赤/努尔哈赤), unified Jurchen clans (known later as Manchus) and founded the Later Jin Dynasty (1616–1636) in 1616, renouncing the Ming overlordship. His son Hong Taiji (皇太極/皇太极) was declared Emperor of the Great Qing in 1636.[38] The dynasty seized control of Beijing in 1644, which is considered the start of the dynasty's rule.

The map depicts Eurasia, placing China in the center and stretching northward to Mongolia, southward to Java, eastward to Korea and central Japan, and westward to Europe and Africa.

In summary, from the above analyses, Japan is shown on ancient Chinese maps at least from the Tang Dynasty, and the geographical locations of Japan on these maps are far more realistic than those on the European and Islamic maps. The reason is simple: Chinese, not Europeans or Arabs, had landed on the Japanese archipelago since Antiquity.

The above knowledge helps to build a sound basis for the analysis in the following section where the Japan-KWQ will be compared with the major European maps of the sixteenth century to determine the Japan-KWQ's origin and the political era revealed by the map.

4. The Japan-KWQ is of Chinese Origin, not a Direct or Adapted Copy of the Major European Maps

Japan consists of four main islands: Hokkaido (北海道; the northernmost and second largest main island); Honshu (本州; the largest and most populous island); Shikoku (四國/四国; the smallest main island); and Kyushu (九州; the westernmost main island).

Figure 1.6(a) shows the Japan-KWQ — the map of Japan and its surrounding areas — extracted from the KWQ; (b) shows the Japan-Mercator (1569) extracted from the 1569 World Map by Gerardus Mercator; (c) shows the Japan-Ortelius (1570) extracted from the 1570 World Map by Abraham Ortelius; and (d) shows the Japan-Plancius extracted from the 1594 World Map by Petrus Plancius.

Figure 1.6(a) depicts the four main islands of Japan in reasonably good agreement with the Japan archipelago on a modern map. However, in Figs. 1.6(b) and 1.6(c), we can see that all the main islands of Japan are distorted into relatively large and small fragments, and Hokkaido becomes an island chain made of many small islands. Although Fig. 1.6(d) gives the relatively best representation of the three main islands of Japan, it completely omits Hokkaido. The three European maps in Figs. 1.6(b)–1.6(d) were drawn based on Marco Polo's description because although Portuguese traders had landed on a Japanese island in 1542/1543, the four main islands had not been fully surveyed when these three European cartographers drew their maps. Specifically, in Section 2, I have mentioned the 1588–1590 Macau-Acapulco trans-Pacific crossing made by João da Gama and his landing in 1589 at a place which has since been called Ezo (Yezo), today's island of Hokkaido. He returned to Portugal or Spain in 1590 when the Japan-Mercator (1569) and the Japan-Ortelius (1570) were already published. Hence, neither Mercator nor Ortelius could have the realistic knowledge needed to draw Hokkaido. The Japan-Plancius (1594) does not depict Hokkaido, even though 1594 was a few years after João da Gama returned to Europe.

Figure 1.7(a) is the very detailed 1570 Map of Tartary from *Theatrum Orbis Terrarium*, (meaning, "Theatre of the lands of the World") by Abraham Ortelius. It shows that the shapes of the three

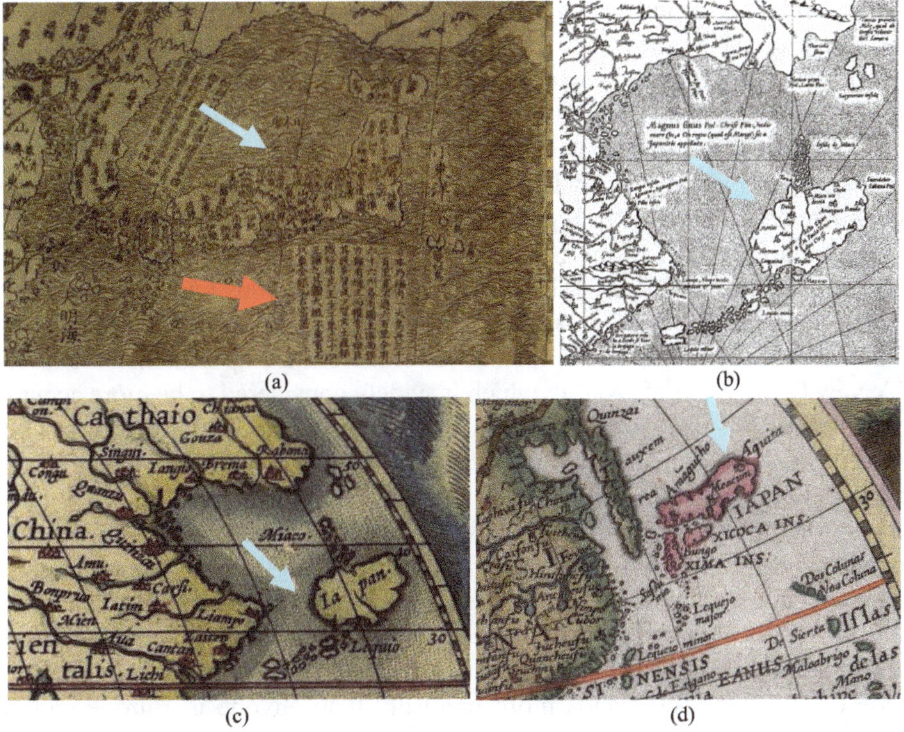

Fig. 1.6.   It shows Japan's main islands pointed out by a blue arrow in each of the sub-figures: (a) is the Japan-KWQ extracted from the KWQ (public domain);[40] the red arrow points to the only annotation on the Japan-KWQ, which will be discussed in Section 6 to show Japan's political era revealed by this map. (b) the Japan-Mercator (1569) extracted from the 1569 World Map by Gerardus Mercator (public domain);[41] (c) the Japan-Ortelius (1570) extracted from the 1570 World Map by Abraham Ortelius (public domain);[42] and (d) the Japan-Plancius (1594) extracted from the 1594 World Map by Petrus Plancius (public domain).[43]

Fig. 1.7.   (a) The 1570 Map of Tartary from *Theatrum Orbis Terrarium* by Abraham Ortelius (public domain);[44] (b) the 1589 *Descriptio Maris Pacifici* (Description of the Pacific Ocean), also by Abraham Ortelius (public domain);[45] Japan is depicted at the upper left corner of this map.

Japanese main islands are somewhat improved, and more details are added, but Hokkaido is still omitted; however, Fig. 1.7(b) shows the 1589 *Maris Pacifici* (*Descriptio Maris Pacifici*, meaning "Description of the Pacific Ocean") also by Abraham Ortelius. This map contains all four main islands of Japan; some details of this map may have been influenced by the 1568 description of Japan in a manuscript written by Fernão Vaz Dourado (c. 1520–c. 1580; a Portuguese cartographer), rather than by a map. However, Hokkaido is depicted as a huge island to the north of the island of Honshu, two smaller islands, Kyushu and Shikoku are nearby. The year 1568 was 21 years before João da Gama first reached the island of Hokkaido. It remains a mystery: how did Fernão Vaz Dourado know of the existence of Hokkaido? How did Ortelius know Hokkaido in 1570?

We notice that, in the first edition of Abraham Ortelius' *Theatrum Orbis Terrarum* (meaning, "Theatre of the lands of the World") published in 1570, there is a map named *Asiae Nova Descriptio* (meaning, "A New Description of Asia")[46] which is widely considered to be more accurate and complete than earlier maps of Asia. *Asiae Nova Descriptio* was informed by the travels of Marco Polo, however, the shape of Japan is distorted.[47] There are also other errors in depicting many geographical items in Asia on this map. Hence, I have decided not to use it as Japan-Ortelius (1570); instead, I have extracted Japan-Ortelius (1570) from his 1570 World Map to compare it with the Japan-KWQ.

As detailed in Section 2, only in 1617 did the two Portuguese — Christopher Blancus and Inacio Moreira — engrave the most accurate European representations of Japan of that time. As mentioned earlier, this is because, unlike the previous European maps which relied on second-hand accounts, Moreira explored Japan at least twice at the end of the sixteenth century to render the size and shape of the Japanese islands more precisely than anyone in Europe had done before. However, only the southern part of Hokkaido is rendered on this map, and the map was engraved after Matteo Ricci published the KWQ in 1602 and his passing in 6010. Hence, it could not be the source map of the Japan-KWQ.

Finally, when we compare the four maps in Figs. 1.6(a)–1.6(d) with Japan on a modern map shown in Fig. 1.8, it becomes very clear that the Japan-KWQ in Fig. 1.6(a) gives far better representations of Japan's four main islands in terms of their shapes, sizes, and relative geographical positions than the three European maps.

Why is this the case? The reason is apparent: although Portuguese traders had landed on a Japanese island in 1542/1543, the European explorers had not reached all the main islands of Japan before 1589, and the European cartographers still drew maps of Japan based on Marco Polo's descriptions, as I have explained in Sections 2 and 4 (this section). But Marco Polo is unlikely to have visited Japan. Hence, the more accurate Japan-KWQ cannot be a direct or adapted copy of the European maps of less accuracy.

But Matteo Ricci was in China when he published the Japan-KWQ, and Chinese knew Japan well and drew maps of Japan since Antiquity. So, did Ricci use Chinese maps as his sources to draw the Japan-KWQ? There are strong reasons to support this viewpoint. The European maps he brought with him could not be his source maps in making the Japan-KWQ, since the Japan-KWQ is more accurate than the major sixteenth-century European maps. The Europeans did not have sufficient knowledge about Japan and were not able to draw realistic maps of Japan before 1617.

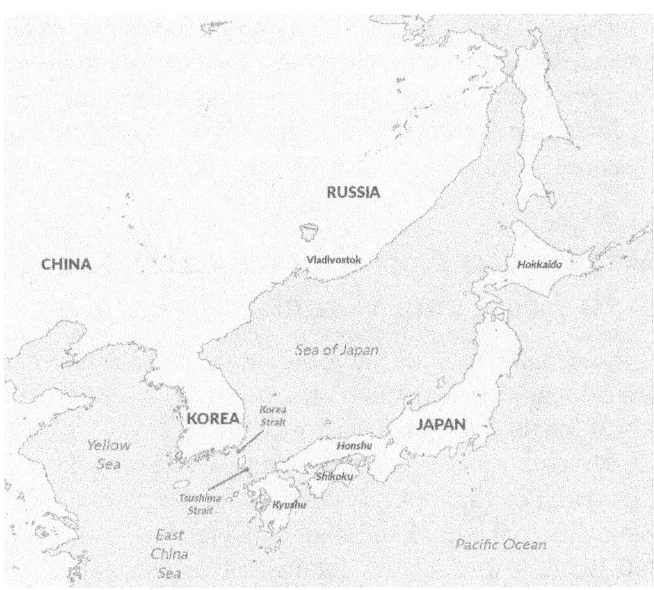

Fig. 1.8.   The map shows Japan's four main islands, Hokkaido, Honshu, Shikoku, and Kyushu, and their surroundings on a modern map (public domain).[48]

Could Matteo Ricci's source maps be any of the Japanese maps? The answer is also "not likely", because early Japanese maps were not accurate even when evaluated by Western standards. For example, it is said that the first maps of Japan (these maps no longer exist) were drawn in 646 A.D.[49] During the Shōmu (聖武天皇/圣武天皇; 701–756) reign, maps known as *Gyōki-zu* (行基圖/行基図), named for the high priest Gyōki Bosatsu (668–749) who immigrated to Japan from Korea in the eighth century, were developed. Among them, the maps derived from early surveys conducted in 646, 738, and 796, show the following: the northeasterly extent of Japan is near the island of Sado, the westerly extent is Kyushu and the southerly extent is the tip of Shikoku. This sort of geographical rendering indicates a gross difference from a modern map, besides the fact that Kyushu stretches much further south than Shikoku, and Sado is more northerly than northeasterly. The *Gyōki-zu* style of maps formed the dominant framework in Japanese cartography until the late medieval (1300–1500 A.D.) and Edo (1603–1867) periods.

In addition, Japan was a closed society for many years, and they also had a long-lasting indifference to exploration; the theory of the Earth as a sphere is thought to have arrived with Francis Xavier (1506–1552; a Spanish Navarrese Catholic missionary and saint who was a co-founder of the Society of Jesus)[50] when he came to Japan in c. 1550. Besides, in the feudal society, the ordinary Japanese citizens were forbidden to travel and the Japanese government in Edo (Tokyo) had no interest in accurate map-making because maps could be used by enemies to gain military advantage.[51]

China–Japan relationships started since Antiquity and Chinese cartographers such as Jia Dan had been able to produce maps of Japan during the Tang Dynasty, which is many hundreds of years before the Europeans first arrived in Japan. These facts lend strong support to the Chinese origin of the Japan-KWQ. Moreover, the accuracy of the Japan-KWQ further supports its Chinese origin as detailed in the next section.

5. The Japan-KWQ Depicts Correctly the Extreme Points of Japan and Covers 82% of Its Legal and Administrative Units

First geographically, a close examination of the locations of Japan's extreme geographic points on the Japan-KWQ and their comparisons with those on a modern map give the following results: (1) on a modern map, the maximal longitudinal span (east-west) is 16.2° (145.8°–129.6°) and the maximal latitudinal span (north-south) of Japan's four main islands is 14.5° (45.5°–31°). They are compared with c. 20.2° (142°–121.8°) and c. 13.3° (43 *du*–29.5 *du*, and multiplied by 0.9856 for converting the KWQ latitudinal reading *du* to the modern longitudinal reading °; this is because the KWQ latitudinal degree 1 *du* = 360/365.25 = 0.9856°; but for qualitative discussions, *du* will be replaced by °, namely, 1 *du* ≈ 1°, since the error is only about 1.4% and within the error of our visual reading),[52] respectively, on the Japan-KWQ. The error in the longitudinal span on the Japan-KWQ is an overestimation by about 4° (c. 20.2°–16.2°), and the error in the latitudinal span on the Japan-KWQ is about 1.2° (c. 13.3°–14.5°). Table 1.1 compares the coordinates themselves for the extreme points in the four cardinal directions on the Japan-KWQ and on a modern map. Their errors in latitudinal

Table 1.1. Comparing the most extreme points of Japan's four main islands on a modern map with those on the Japan-KWQ.

The most extreme points of Japan's four main islands on a modern map	Modern coordinates	Extreme points on the Japan-KWQ	The Japan-KWQ coordinates relative to the Equator and Greenwich Meridian (the KWQ latitudinal reading *du* is already converted into the modern reading °)[53]
Easternmost Point	Cape Nosappu, Hokkaido	43.4°N, 145.8°E	In Hokkaido	41.4°N, 142.0°E
Westernmost Point	Cape Kōzakihana, Kyushu	33.2°N, 129.6°E	In Kyushu	31.7°N, 121.8°E
Southernmost Point	Cape Sata, Kyushu	31.0°N 130.7°E	In Kyushu	29.1°N, 124.5°E
Northernmost Point	Cape Sōya, Hokkaido	45.5°N, 141.9°E	In Hokkaido	42.4°N, 135.5°E

coordinates are no more than c. 3.1° (the largest for the Northernmost Point comparison), but their errors in longitudinal coordinates can be as large as c. 7.8° for the Westernmost Point comparison.

The Japan-Mercator (1569), the Japan-Ortelius (1570), and the Japan-Plancius (1594) depict Japan's four main islands very poorly (they are based on Marco Polo's fictitious descriptions) or even omit Hokkaido; hence, it is not meaningful to compare their most extreme points of Japan's four main islands with those on the Japan-KWQ and a modern map.

We next turn our attention to the legal and administrative units and their geographies of Japan on the Japan-KWQ: the three major European maps do not give the administrative units — *Gokishichidō* — referring to the "five provinces and seven circuits (五畿七道)" of Japan (68 units in total) at the time. On the contrary, the Japan-KWQ depicts 56 (a little over 82%) out of the total 68 such units, although there exist two major mistakes on this map: (1) the Northern Land Circuit/Hokurikudō (北陸道/北陆道) is erroneously depicted in Hokkaido, and only five of the seven provinces are depicted; and (2) some of the provinces in the region of the Eastern Sea Circuit/Tōkaidō are erroneously placed in the region of the Eastern Mountains Circuit/Tōsandō (東山道/东山道) and vice versa. There are similar errors such as the omissions of ten more provinces on other parts of the Japan-KWQ, totaling twelve provinces missing, which will be further discussed in the Appendix; the detailed analyses of the names and geographical locations of these circuits and the provinces under their rule are also given in the Appendix at the end of this chapter. The following section explains this legal and governmental system in ancient Japan and the political era revealed by the Japan-KWQ.

6. The Japan-KWQ Depicts Japan in 897 A.D., Almost 650 Years Before the Europeans' First Arrival

On the Japan-KWQ, there is only one annotation. It is pointed at by the red arrow in Fig. 1.6(a). The annotation gives a general description of the country of Japan and reveals its political system and the era of the map:

日本乃海內一大島長三千二百里寬不過六百里今有六十六州各有國主俗尚強力雖有總王而權常在強臣其民多習武少習文土產銀鐵好漆其王生子年三十以王讓之其國不重寶石只重金銀及古窯器

My translation and comments (between square brackets) are as follows:

Japan is a large archipelago in the sea. It is 3,200 *li* long [one *li* is about ½ km] and no more than 600 *li* wide. At present, it has 66 states [州 or provinces] and each has its own ruler. Their culture respects strength. Although there is a chief ruler, the country's power is often held in the hands of the strong ministers. Most of their people practice martial arts, only few study art and literature. The local productions are silver, iron, and good quality lacquer. Their chief ruler [Emperor Uda (宇多天皇); 867–931 A.D.] has sons. When he [Uda] was 30 years old [this was in 897 A.D.], he abdicated in favour of his eldest son [Prince Atsuhito who would later come to be known as

Emperor Daigo (醍醐天皇) and reigned for 33 years from 897 to 930 A.D.]. The country's people attach no importance to precious stones, but care only about gold, silver, and ancient kiln wares.

A close examination of the above annotation will be made in four subsections as follows.

6.1 *The Japan-KWQ gives a reasonable estimate of the dimension of the Japanese archipelago*

Firstly, the four main (largest) islands of Japan are Hokkaido, Honshu, Shikoku, and Kyushu. Since a Chinese *li* is 0.5 km, 3,200 *li* is c. 1,600 km/994 mi and 600 *li* is c. 300 km/186 mi. On a modern map, to make a rough estimate of the total length of the four main islands, Shikoku can be ignored from the calculation due to its smallest size and special geographical location.

The total length of Japan can be roughly estimated by summing the lengths of the remaining three islands as follows: (1) the largest island, Honshu, forms a northeast-southwest arc extending about 1,300 km/810 miles and varies greatly in width from 50 to 230 km/31 to 143 mi);[54] (2) between Wakkanai in the far north and Hakodate near the southern end of Hokkaido, the straight-line distance is 413 km/255 mi,[55] which can be approximated as the length of Hokkaido, and the island's width can be approximated as the distance from Sapporo to Kushiro, which is c. 349 km/217 mi; and (3) the Kyushu Expressway, roughly 346 km/215 mi long, spans the length of Kyushu, and about half of that length, 173 km/108 mi, can be added to the length of Honshu as part of the length of Japan; the width of Kyushu does not exceed c. 300 km/186 mi. Adding the three lengths together (1,300 + 413 + 173 = 1,886 km) can give a crude estimate of the total length of Japan's four main islands as 1,886 km/1,172 mi, to be compared with c. 1600 km/994 mi given by the annotation on the Japan-KWQ. Knowing how simplistic and crude the model calculation is, an overestimate of c. 18% in depicting the length of the Japanese archipelago is quite reasonable. It is harder to estimate the width of Hokkaido because it has a very complicated shape and the distance from Sapporo to Kushiro is c. 349 km/217 mi. However, 349 km/217 mi is just a little over 16% wider than the width (300 km/186 mi) of the Japanese archipelago given by the annotation. The widths of Honshu and Kyushu are no more than 300 km/186 mi, in consistency with the description given by the same annotation.

Hence, we can conclude that the Japan-KWQ gives a reasonable estimate of the real dimension of the Japanese archipelago.

6.2 *The Japan-KWQ depicts Japan's centralised legal and governmental system — Gokishichidō — borrowed from China*

In the previously mentioned annotation, the "66 states [or provinces; 州]" refers to the five provinces near the capital region and the 61 provinces mainly in the regions of the three main islands. The three main islands were Honshu, Kyushu, and Shikoku. Although there were Japanese settlers who had ruled the southern tip of Hokkaido Island (the old name was Ezo, renamed Hokkaido in 1869),[56] this was from the sixteenth century and Hokkaido was then considered *foreign territory* inhabited by the indigenous people of the island, known as the Ainu people.[57] The 66 provinces were under the rule of the seven "circuits". Here, "circuit" means "dō" or "道"; each "circuit" was a region which contained provinces of its own; running through each of the seven regions was a road or dō (道) of the same name, connecting the imperial capital with all the provincial capitals along its route.

But why 66, not 68 provinces? The reason is as follows: although the two island provinces — Iki Province (壱岐國/壱岐国) and Tsushima Province (對馬國/对马国) — were also under the centralised legal and governmental system in Japan, and should make the total number of provinces 68, they were often not counted as one of the 66 provinces. Hence, when counting the provinces on Kyushu (九州, literally, "nine provinces"), only nine provinces were included, instead of eleven, after excluding the Iki Province (壱岐國/壱岐国) and the Tsushima Province (對馬國/对马国). There were four provinces on Shikoku (四國/四国, literally "four provinces"), 51 provinces on Honshu including the five provinces near the capital region, and two more island provinces (not on Honshu), making the total 66.

"Circuit" is part of the *Gokishichidō* (meaning, "five provinces and seven circuits"; 五畿七道) system which was the name for Japan's ancient administrative units under the *Ritsuryō* (律令) system (the historical law system in Japan, based on the philosophies of Confucianism and Chinese Legalism). *Gokishichidō* was organised during the Asuka period (飛鳥時代/飞鸟时代; 538–710 A.D.) as part of a legal and governmental system.[58] The purpose of *Gokishichidō* was to strengthen Japan's centralization to protect the results of its sociopolitical reforms. In 645, the Taika Era Reforms (大化改新) were the first signs of implementation of the system.[59]

However, a detailed examination in the Appendix of all the provinces on the Japan-KWQ finds that only 51 provinces under the rule of the seven circuits are depicted on the map, making the total count of the provinces as 56 (51 + 5) including the five provinces near the capital region. Nevertheless, the number of provinces depicted on the Japan-KWQ is still over 82% of the total number of 68 (66 plus the two island provinces) under the *Ritsuryō* system.

As stated, *Gokishichidō* consisted of five provinces in the Kinai (畿内) or capital region, and seven dō or circuits away from the capital region. Although these units did not survive as administrative structures beyond the Muromachi period (室町時代/室町时代; 1336–1573), they did remain important geographical entities until the nineteenth century.[60]

The five Kinai provinces were: (1) Yamato Province (大和國/大和国; see label 34 in Fig. 1.9; now Nara Prefecture); (2) Yamashiro Province (山城國/山城国; see label 35 in Fig. 1.9; now the southern part of Kyoto Prefecture, including the city of Kyoto); (3) Kawachi Province (河內國/河内国; see label 36 in Fig. 1.9; now the southeastern part of Osaka Prefecture); (4) Settsu Province (摂津國/摂津国; see label 37 in Fig. 1.9; now the northern part of Osaka Prefecture, including the city of Osaka, and parts of Hyōgo Prefecture); and (5) Izumi Province (和泉國/和泉国; see label 38 in Fig. 1.9; now the southern part of Osaka Prefecture). In the Appendix, I show that the Japan-KWQ does depict the five Kinai provinces around the capital region, but their geographical locations do not match entirely those shown in Fig. 1.9.

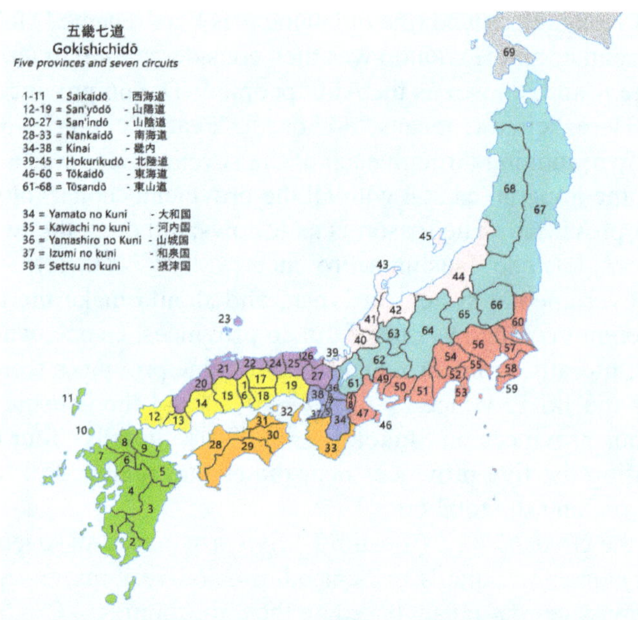

Fig. 1.9.   Map of the *Gokishichidō* ("five provinces and seven circuits or dō"; 五畿七道; *Source*: Artanisen, under CC BY-SA 4.0, https://commons.wikimedia.org/wiki/File:Gokishichido_Seven_Circuits_Japan_Map.png).[67] *Gokishichidō* was the name of ancient administrative units in Japan. In 645, the Taika Era Reforms (大化改新) were the first signs of implementation of the system.[68] Kinai (畿内; capital region) was the name for the five ancient provinces around the capital, and dō (道) was the name for the seven administrative areas stretching away from the Kinai region in different directions. Running through each of the seven areas was a road/dō (道) of the same name, connecting the imperial capital with all the provincial capitals along its route.

The seven dō or circuits were: (1) Western Sea Circuit/Saikaidō (eleven provinces; see labels 1–11 in Fig. 1.9; 西海道; it was the entire island of Kyushu plus the two island provinces — Iki Province (壱岐國/壱岐国) and Tsushima Province (對馬國/对马国); (2) Light (Southern) Mountains Circuit/San'yōdō (eight provinces; see labels 12–19 in Fig. 1.9; 山陽道/山阳道; in the western part of Honshu and along the northern side of the Seto Inland Sea);[61] (3) Dark (Northern) Mountains Circuit/San'indo (eight provinces; see labels 20–27 in Fig. 1.9; 山陰道/山阴道; in the western part of Honshu and along the Sea of Japan coast, plus the island province — Oki Province);[62] (4) Southern Sea Circuit/Nankaidō (six provinces; see labels 28–33 in Fig. 1.9; 南海道; the entire Shikoku, the southern portion of the Kii Peninsula and the island province Awaji lying between Honshu and Shikoku);[63] (5) Northern Land Circuit/Hokurikudō (seven provinces; see labels 39–45 in Fig. 1.9; 北陸道/北陆道; approximately located in the central part of Honshu and along the northeast direction of the Sea of Japan coast);[64] (6) Eastern Sea Circuit/Tōkaidō (fifteen provinces; see labels 46–60 in Fig. 1.9; 東海道/东海道; approximately located in the central part of Honshu and along Japan's Pacific coast);[65] and (7) Eastern Mountains Circuit/Tosandō (eight provinces; see labels 61–68 in Fig. 1.9 東山道/东山道; located in the central and northeastern parts of Honshu, along the northeast direction through the Japanese Alps and along Japan's Pacific coast).[66] Detailed descriptions of these provinces on the Japan-KWQ can be found in the Appendix.

In the Appendix, all the administrative units on the Japan-KWQ are examined and some of the major findings are reported here: (1) both Ōsumi Province (大隅國/大隅国; Item 2) and Hyūga Province (日向國/日向国; Item 3) are missing from the Western Sea Circuit/Saikaidō (西海道); (2) Tanba Province (丹波; Item 27) is missing from the Dark (Northern) Mountains Circuit/San'indō; (3) Sanuki Province (讚岐/赞岐; Item 31) is missing from the Southern Sea Circuit/Nankaidō; (4) Echizen (越前; Item 39) and Wakasa (若狭; Item 40) Provinces are missing from the Northern Land Circuit/Hokurikudō (北陸道/北陆道); (5) Shima (志摩; Item 46), Ise (伊勢/伊势; Item 47), Iga (伊賀/伊贺; Item 48), Owari (尾張/尾张; Item 49), Kai (甲斐; Item 54), and Sagami (相模國/相模国; Item 55) are missing from the Eastern Sea Circuit/Tokaido (東海道/东海道). The total number of missing provinces on the Japan-KWQ is twelve, making the total count of the provinces as 56 (51 + 5), the correct number should be 68 under the *Ritsuryō* system (律令制).

6.3 *The Japan-KWQ reflects the political era of Japan in 897 A.D.*

The political era revealed by the Japan-KWQ can be extracted from the text "其王生子年三十以王讓之" in the annotation (after adding the punctuation) as:

其王生子，年三十，以王讓之。

My translation is as follows:

Their chief ruler has sons. When he was 30 years old, he abdicated[69] in favour of his eldest son.

In the Japanese history, from the first emperor (who reigned 660–585 BC) to the 107th emperor (who reigned 1586–1611; overlapping with the year 1602 when Matteo Ricci published the KWQ), only Japan's 59th Emperor Uda (宇多天皇; he was born on 10 June in the year of 867 A.D. and died on 3 September in the year of 931 A.D.; reigned from 17 September 887 to 4 August 897)[70] fits the description of the above annotation. On 4 August, 897, after he became 30 years old, he abdicated in favour of his eldest son — Prince Atsuhito — who then was only twelve years old, to become Japan's 60th Emperor Daigo (醍醐天皇); he lived from 885 to 930 and reigned from 897 to 930.

In the Chinese language, it is incorrect to interpret "年三十 (was 30 years old)" as "reigned for 30 years (在位三十年)". Furthermore, if "年三十 (was 30 years old)" is interpreted as "reigning for 30 years" and giving up the throne to the eldest son, then none of the 107 emperors of Japan (from the first emperor to the 107th emperor) would fit the description of this annotation, including Emperor Go-Komatsu (后小松天皇). According to the pre-Meiji (Meiji era started from 1868) scholars, Go-Komatsu's reign as the legitimate emperor was only from 1392 to 1412, a total of 20 years,[71] since from 1382 to 1392 he was the emperor of the Northern Dynasty, not the whole Japan.

This annotation reveals the political era of Japan in 897 when the Japanese imperial power was changing hands to a young emperor — Emperor Daigo — at age twelve.[72] This era overlapped with the Chinese late Tang Dynasty (618–907, with an interregnum between 690 and 705). Recall that between 607 and 838, Japan sent nineteen missions to China to learn from the Chinese culture and civilization.[73] After the abolition of the mission to the Tang Dynasty in 894, Japan began to develop its own national style and culture. The Chinese cartographer(s) depicted Japan in 897 to record this important political change in Japan, and more than 700 years later, the map became the source map for Matteo Ricci to draw and publish the Japan-KWQ in 1602.

In 905, a set of laws and regulations – the *Engishiki* (延喜式; *Procedures of the Engi Era* is a Japanese book about laws and customs) – was compiled under the order of Emperor Daigo. The major part of the writing was completed in 927[74] and includes detailed regulations on government systems such as the *Gokishichidō* system and etiquette, which have become important documents for the study of ancient Japanese history. After several revisions, the work was used as a basis for reform starting in 967; the *Gokishichidō* system was abolished in 1871 during the Meiji era (1868–1912).[75]

The year 897 precedes by almost 650 years the first European arrival in Japan in 1543 (some sources say 1542). Several years later, in 1549, Japan chose to end its recognition of the Ming-led worldview and order, and to cancel any further tribute missions[76] the country had with China during the Muromachi period (c. 1336–1573).

6.4 *The Japan-KWQ correctly describes Japan's ceramic products developed since Antiquity*

We come to the final part of the annotation – 其國不重寶石只重金銀及古窯器 (The country's people attach no importance to precious stones, but only care about gold, silver, and ancient kiln ware) – in which the three Chinese characters "古窯器" at the end of the sentence merit special attention.

According to the Chinese encyclopedia titled *Shuowen Jiezi* (說文解字/说文解字)[77] or *Explaining Simple and Analysing Compound Characters,* written in the East Han Dynasty (25–220 A.D.), 窯 has the following meaning:

窯，(说文解字): "烧瓦灶也"。

My translation and comment (between square brackets) are as follows:

Yao (窯) or Kiln, in *Shuowen Jiezi* (說文解字/说文解字) or *Explaining Simple and Analysing Compound Characters,* refers to Shao Wa Zao (燒瓦灶/烧瓦灶; the "tile- or brick-making stove") [The bricks and tiles fired in the kiln can be collectively referred to as kiln wares (窯器)].

Japan has been rich in ceramics (common examples are earthenware, porcelain, and brick) since ancient times, and the well-known "Six Ancient Kilns (六古窯/六古窑)" in Japan have inherited the technology from the early Jomon period (繩文時代/绳文时代; c. 6,000–300 B.C.).[78] The description made in the annotation on the Japan-KWQ is consistent with the history of Japanese ceramics. None of the three sixteenth-century European maps document this well-known Japanese trait developed since Antiquity. The annotation on the Japan-KWQ adds an extra layer of evidence to support that the Japan-KWQ is of Chinese origin, not a copy or adapted copy of the three major European maps.

7. Conclusions

In summary, in the Introduction, I state that the purpose of this chapter is to analyse the map of Japan on the KWQ, abbreviated here as the Japan-KWQ, to determine the origin of this map and its era. By analysing the annotation and examining the location and history of every geographical item depicted on the Japan-KWQ, I have extracted the political era of Japan revealed by the map; this era was also the era or date when these places were last explored.

Based on the analyses presented in Sections 1–6, the following conclusions can be reached: (1) the Japan-KWQ is of Chinese origin, not a copy or adapted copy of the three major European maps, because it is a much more accurate map than the European maps. (2) China-Japan interactions started centuries before Christ was born, but Europeans' first arrival in Japan occurred as late as in the sixteenth century. Chinese cartographer Jia Dan (賈耽/贾耽;730–805 A.D.) already drew a map with texts describing Korea, Japan and China's other foreign states in 801 A.D. and presented it to Emperor Dezong of Tang (唐德宗); (3) the Japan-KWQ gives a fairly accurate geographical depiction of Japan in terms of its shape, size, and the most extreme points in four cardinal directions, although the map mislocated some of the circuits and omitted twelve provinces under the rule of their circuits; these circuits belonged to Japan's centralised legal and governmental system — *Gokishichidō* (meaning, "five provinces and seven circuits"; 五畿七道) — borrowed in the seventh century from China; (4) the Japanese culture respected strength; (5) the main Japanese local productions were silver, iron and good quality lacquer; (6) the famous Japanese ceramics inherited the technology developed in ancient days; and (7) the Japan-KWQ reveals Japan in 897 A.D., soon after Japan abolished in 894 its missions to the Tang Dynasty and began to develop its own national style and culture to become a sovereign nation in Asia. In the year 897, the Japanese imperial power changed hands from Emperor Uda to his eldest son, twelve-year-old Emperor Daigo, who reigned for the next 33 years. The Chinese cartographer(s) depicted Japan in 897 to record this important political change in Japan, and more than 700 years later, the map became the source map for Matteo Ricci to draw and publish the portion of Japan on the KWQ in 1602.

Appendix

The geographical items on the Japan-KWQ are grouped in the following table into two categories: (1) those that are incorporated into the *Gokishichidō* ("five provinces and seven circuits"; 五畿七道) system are listed according to the order on a modern map shown in Fig. 1.9; and (2) those that are the seas and islands surrounding the four main islands of Japan. Moreover, all the geographical items are numbered in sequence in the table. The provinces that should be depicted but are not shown on the Japan-KWQ will also be listed and pointed out. For the "five provinces" near the capital region, a "**" sign is placed in front of the number, and for the "seven circuits", a "*" sign is placed in front of the number. My comments are given in the last column of the table. On the Japan-KWQ, most geographical items are written in Chinese traditional characters. Since many readers use the Chinese simplified characters, I include both: all the Chinese characters in the main text and in the Appendix are expressed first with the Chinese traditional characters, then followed by the Chinese simplified characters, separated by a slash "/". When both characters are the same, that expression is presented only once. The Chinese "pinyin" spelling of each item is given right after each item is introduced. Here I would like to point out that Kanji (漢字/汉字; meaning, "Han characters" or Chinese traditional characters) are the logographic Chinese characters taken from the Chinese script used in the writing of Japanese (from the fifth century to the present). They were made a major part of the Japanese writing system during the time of Old Japanese and are still used, along with the subsequently derived syllabic scripts of Hiragana (from c. 800 A.D. to the present) and Katakana (from c. 800 A.D. to the present). Hence, many of the geographical items depicted on the Japan-KWQ can be understood by the Japanese readers.

Item #	Name	Pinyin	Etymology	History, Geography and My Comments
			Category I: The *Gokishichidō* ("five provinces and seven circuits or dō"; 五畿七道) system revealed on the Japan-KWQ	
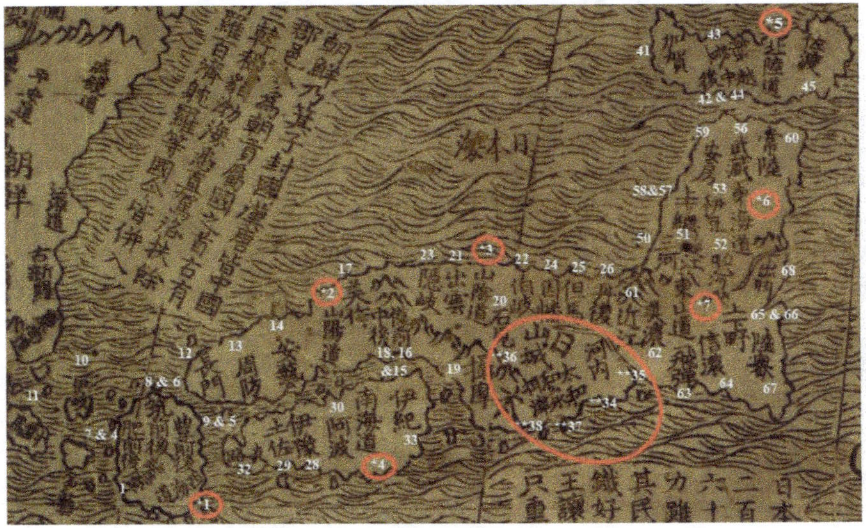				**Fig. A1.1.** This figure is extracted from Fig. 1.6(a) (public domain).
*1	西海道	Xi Hai Dao	Western Sea Circuit/Saikaidō (the entire island of Kyushu plus the two island provinces – Iki Province (壱岐國/壱岐国) and Tsushima Province (對馬國/对马国)[79]	Saikaidō (西海道, literally, "western sea circuit" or 'western sea region') was one of the seven circuits in the *Gokishichidō* ("five provinces and seven circuits"; 五畿七道) system. It consisted of Kyushu and surrounding islands. The following were the eleven provinces under its rule, as labeled in Fig. 1.9 on a modern map. However, Ōsumi Province (大隅國/大隅国; Item 2) and Hyūga Province (日向國/日向国; Item 3) are missing on the Japan-KWQ: 1. Satsuma Province (薩摩國/萨摩国):[80] in 702, Changgen Province (transliteration of 唱更國/唱更国) took the opportunity to separate from Hyūga Province (日向國/日向国); in 704, Changgen Province (transliteration of 唱更國/唱更国) was

(*Continued*)

Item #	Name	Pinyin	Etymology	History, Geography and My Comments
				renamed "Sama Province (薩麻國/薩麻国)"; and after the middle of the eighth century, it was again renamed "Satsuma Province (薩摩國/萨摩国)".[81] 2. Ōsumi Province (大隅國/大隅国):[82] in the *Gu Shi Ji* (古事記/古事记) or *Kojiki*/*Records of Ancient Matters* which is usually considered to be the oldest extant literary work in Japan and semi-historical accounts down to 641[83] and the *Ri Ben Shu Ji* (日本書紀/日本书纪) or *Nihon Shoki*/*Chronicles of Japan* which is the second-oldest book of classical Japanese history,[84] Hyūga (日向國/日向国) is called Kumaso Province (熊曽國/熊曽国); in 713, Ōsumi Province (大隅国) was separated from the land of Hyūga (日向國/日向国).[85] 3. Hyūga Province (日向國/日向国):[86] earlier it was called Kumaso Province (熊曾國/熊曾国); in 713, the land of Hyūga was administratively separated from Ōsumi Province (大隅國/大隅国).[87] 4. Higo Province (肥後國/肥后国):[88] in 696, the ancient Hi Province (火國/火国) was separated into Hizen Province (肥前國/肥前国) and Higo Province (肥後國/肥后国).[89] 5. Bungo Province (豐後國/丰后国):[90] at the end of the seventh century, Toyo Province (豐國/丰国) was split into Buzen (豐前國/丰前国; literally, "the front of Toyo") and Bungo (豐後國/丰后国; "the back of Toyo").[91] 6. Chikugo Province (筑後國/筑后国):[92] in 698, the ancient Tsukushi Province (筑紫國/筑紫国) was split into Chikuzen Province (筑前國/筑前国) and Chikugo Province (筑後國/筑后国).[93] 7. Hizen Province (肥前國/肥前国):[94] the ancient Hi Province (火國/火国) was separated into Hizen Province (肥前國/肥前国) and Higo Province

(肥後國/肥后国). The name Hizen Province (肥前國/肥前国) appears from 696 A.D. in the early chronicle *Xu Ri Ben Ji* (続日本紀/続日本纪) or *Shoku Nihongi*/*Chronicle of Japan Continued* which is an imperially commissioned Japanese history text written in 797 A.D.[95]

8. Chikuzen Province (筑前國/筑前国):[96] in 698, the ancient Tsukushi Province (筑紫國/筑紫国) was split into Chikuzen Province (筑前國/筑前国) and Chikugo Province/筑後國/筑后国.[97]

9. Buzen Province (豐前國/丰前国):[98] at the end of the seventh century, Toyo Province (豐國/丰国) was split into Buzen (豐前國/丰前国; literally, "the front of Toyo") and Bungo (豐後國/丰后国; "the back of Toyo").[99]

10. Iki Province (壱岐國/壱岐国):[100] the Iki Islands have been inhabited since the Japanese Paleolithic era and were organised as Iki Province under the *Ritsuryō* reforms in the latter half of the seventh century.[101]

11. Tsushima Province (對馬國/对马国):[102] it was on Tsushima Island and possibly recognised in the fifth century as a province. Under the *Ritsuryō* system, Tsushima formally became a province.[103]

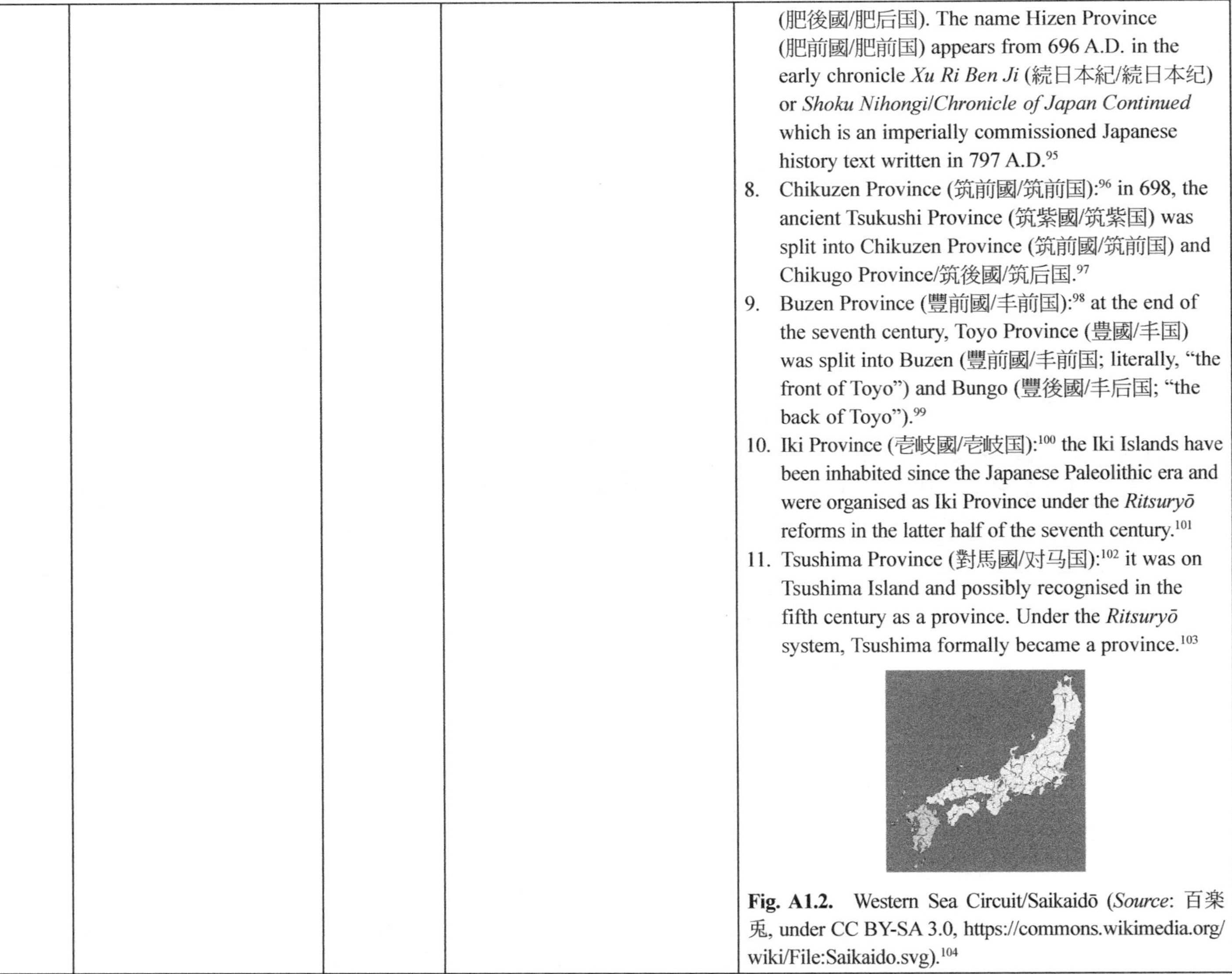

Fig. A1.2. Western Sea Circuit/Saikaidō (*Source*: 百楽兎, under CC BY-SA 3.0, https://commons.wikimedia.org/wiki/File:Saikaido.svg).[104]

(*Continued*)

Item #	Name	Pinyin	Etymology	History, Geography and My Comments
1	薩摩	Sa Mo Guo	Satsuma Province	Satsuma (薩摩/萨摩) was one of the eleven provinces under the rule of the Western Sea Circuit/Saikaidō. **Fig. A1.3.** Satsuma Province (*Source*: Ash_Crow, under CC BY-SA 3.0, https://commons.wikimedia.org/wiki/File:Provinces_of_Japan-Satsuma.svg).[105]
2	大隅	Da Yu	Ōsumi Province	Not depicted on the Japan-KWQ.
3	日向	Ri Xiang	Hyūga Province	Not depicted on the Japan-KWQ.
7&4	肥前後	Fei Qian Hou	Hizen and Higo Provinces	Hizen (肥前; Item 7) and Higo (肥後/肥后; Item 4) were two of the eleven provinces under the rule of the Western Sea Circuit/Saikaidō. (a)　　(b) **Fig. A1.4.** (a) Hizen Province (item 7; *Source*: Ash_Crow, under CC BY-SA 3.0, https://commons.wikimedia.org/wiki/File:Provinces_of_Japan-Hizen.svg);[106] (b) Higo Province (Item 4; *Source*: Ash_Crow, under CC BY-SA 3.0, https://commons.wikimedia.org/wiki/File:Provinces_of_Japan-Higo.svg).[107]

| 9&5 | 豐前後 | Feng Qian Hou | Buzen and Bungo Provinces | Buzen (豐前/丰前: Item 9) and Bungo (豐後/丰后; Item 5) were two of the eleven provinces under the rule of the Western Sea Circuit/Saikaidō.
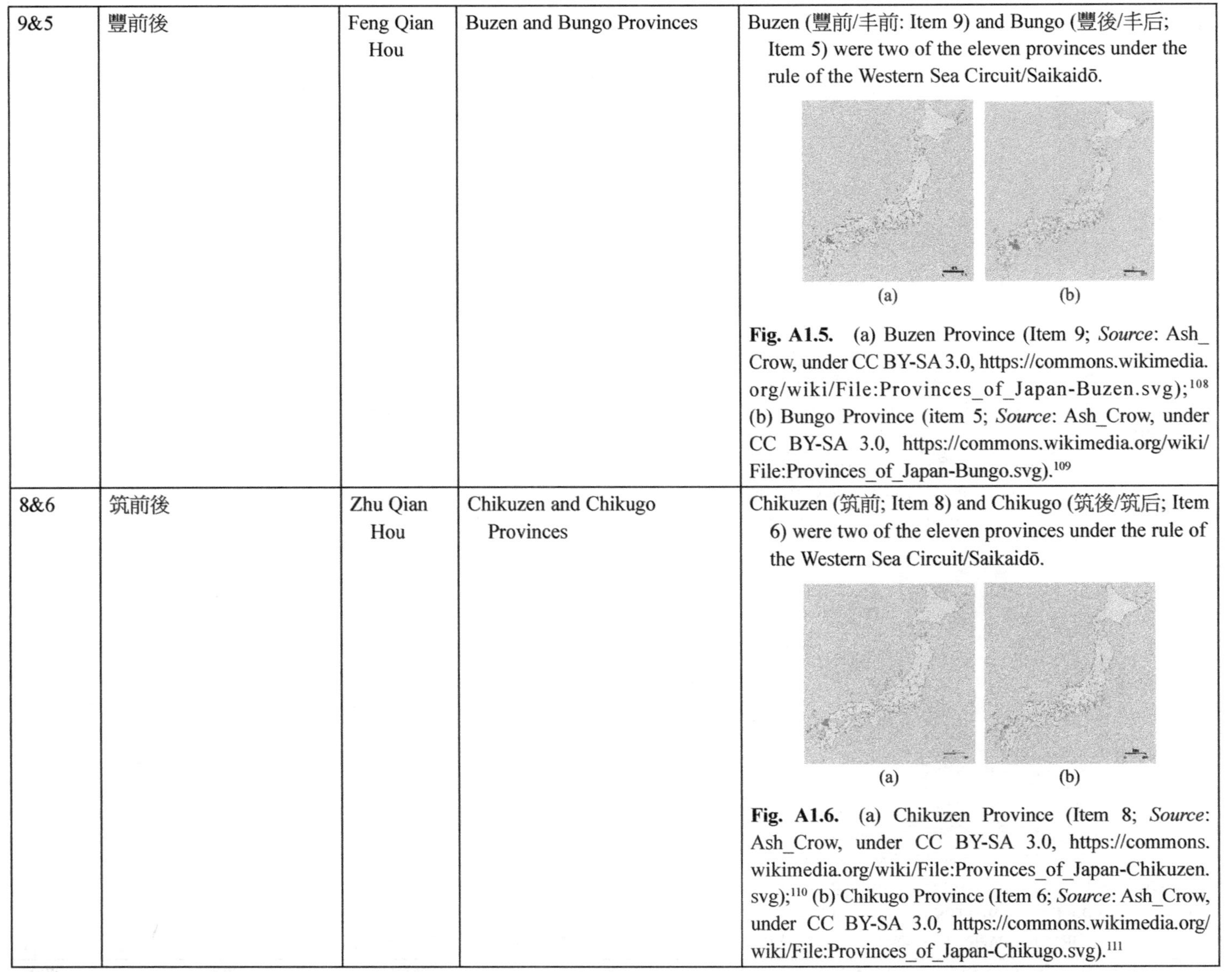
Fig. A1.5. (a) Buzen Province (Item 9; *Source*: Ash_Crow, under CC BY-SA 3.0, https://commons.wikimedia.org/wiki/File:Provinces_of_Japan-Buzen.svg);[108] (b) Bungo Province (item 5; *Source*: Ash_Crow, under CC BY-SA 3.0, https://commons.wikimedia.org/wiki/File:Provinces_of_Japan-Bungo.svg).[109] |
| 8&6 | 筑前後 | Zhu Qian Hou | Chikuzen and Chikugo Provinces | Chikuzen (筑前; Item 8) and Chikugo (筑後/筑后; Item 6) were two of the eleven provinces under the rule of the Western Sea Circuit/Saikaidō.
Fig. A1.6. (a) Chikuzen Province (Item 8; *Source*: Ash_Crow, under CC BY-SA 3.0, https://commons.wikimedia.org/wiki/File:Provinces_of_Japan-Chikuzen.svg);[110] (b) Chikugo Province (Item 6; *Source*: Ash_Crow, under CC BY-SA 3.0, https://commons.wikimedia.org/wiki/File:Provinces_of_Japan-Chikugo.svg).[111] |

(*Continued*)

Item #	Name	Pinyin	Etymology	History, Geography and My Comments
10	壹岐/壱岐	Yi Qi Guo	Iki Province	Iki (壱岐) was one of the eleven provinces under the rule of the Western Sea Circuit/Saikaidō. Its territory was in the present Iki Island (壱岐島/壱岐岛). On the Japan-KWQ, Iki is written as 壹岐. Fig. A1.7. Iki Province (small dot pointed at by the arrow; *Source*: 百楽兎, under CC BY-SA 3.0, https://commons.wikimedia.org/wiki/File:Old_Japan_Iki.svg).[112]
11	對馬	Dui Ma Guo	Tsushima Province	Tsushima (對馬/对马) was one of the eleven provinces under the rule of the Western Sea Circuit/Saikaidō. Its territory was in the present Tsushima Island (對馬島/对马岛). Fig. A1.8. Tsushima Province (*Source*: 百楽兎, under CC BY-SA 3.0, https://commons.wikimedia.org/wiki/File:Provinces_of_Japan-Tsushima.svg).[113]

Remarks on Western Sea Circuit/Saikaidō (西海道)**:** Both Ōsumi Province (大隅國/大隅国; Item 2) and Hyūga Province (日向國/日向国; Item 3) are missing from the Western Sea Circuit/Saikaidō (西海道) on the Japan-KWQ, making the total number of provinces under the rule of the Western Sea Circuit/Saikaidō nine instead of eleven. Because in 713, Ōsumi Province (大隅國/大隅国) was separated from the land of Hyūga (日向國/日向国), it is not surprising to see that when Ōsumi Province (大隅國/大隅国) is omitted from the map, Hyūga Province (日向國/日向国) is also omitted. However, Satsuma Province (薩摩國/萨摩国) was separated from Hyūga Province (日向國/日向国) in the earlier year 702, so, it is not omitted on the Japan-KWQ. The rest of the nine provinces are depicted on the Japan-KWQ at reasonably correct geographical locations.

| *2 | 山陽道 | Shan Yang Dao | Light (Southern) Mountains Circuit/San'yōdō[114] was located in the western part of Honshu and along the northern side of the Seto Inland Sea (瀬戸内海); this name derives from the idea that the southern side is the "sunny (陽/阳)" side of the central mountain chain running through Honshu; while the northern side was the "shady" side and named as Dark (Northern) Mountains Circuit/San'indō (山陰道/山阴道) which is to be discussed in Item *3. | Light (Southern) Mountains Circuit/San'yōdō (山陽道/山阳道). It was one of the seven circuits in the *Gokishichidō* ("five provinces and seven circuits"; 五畿七道) system. The following were the eight provinces under its rule as labelled in Fig. 1.9 on a modern map:

12. Nagato Province (長門國/长门国):[115] its old name was Xue Men Guo (transliteration of 穴門國/穴门国), then changed to Nagato Province (長門國/长门国) in c. 665.[116]
13. Suō Province (周防國/周防国):[117] its old name was Zhou Fang (transliteration of 周芳), then changed to Suō (周防) near the end of the seventh century.[118] After the Taika Era Reforms (大化改新 or 大化革新; took place in 645), it became an independent province.
14. Aki Province (安藝國/安艺国):[119] it was an old province of Japan; Emperor Shōmu (reigned from 724 to 749) ordered two official temples (one for male Buddhist priests and one for nuns) to be founded in Aki Province.
15. Bingo Province (備後國/备后国):[120] it was established in 689 from Ji Bei Province (transliteration of 吉備國/吉备国)[121] after the Taika Era Reforms (大化改新 or 大化革新; took place in 645).
16. Bitchū Province (備中國/备中国):[122] it was established in 689 from Ji Bei Province (吉備國/吉备国)[123] after the Taika Era Reforms (大化改新 or 大化革新; took place in 645). |

(Continued)

Item #	Name	Pinyin	Etymology	History, Geography and My Comments
				17. Mimasaka Province (美作國/美作国);[124] in 713 a few districts of Bizen Province (備前國/备前国) were separated into a new province called Mimasaka Province (美作國/美作国).[125] 18. Bizen Province (備前國/备前国):[126] it was established in 689 from Ji Bei Province (吉備國/吉备国)[127] after the Taika Era Reforms (大化改新 or 大化革新; took place in 645). 19. Harima Province (播磨國/播磨国):[128] it was established in the seventh century. In 713, Harima Province was already recorded in the ancient Japanese book *Feng Tu Ji* (風土記/风土记) or *Fudoki/Description of regional climate, culture, etc.* **Fig. A1.9.** Light (Southern) Mountains Circuit San'yōdō (*Source*: 百楽兎, under CC BY-SA 3.0, https://commons.wikimedia.org/wiki/File:Sanyodo.svg).[129]
12	長門	Chang Men	Nagato Province	Nagato (長門/长门) was one of the eight provinces under the rule of the Light (Southern) Mountains Circuit/San'yōdō (山陽道/山阳道).

				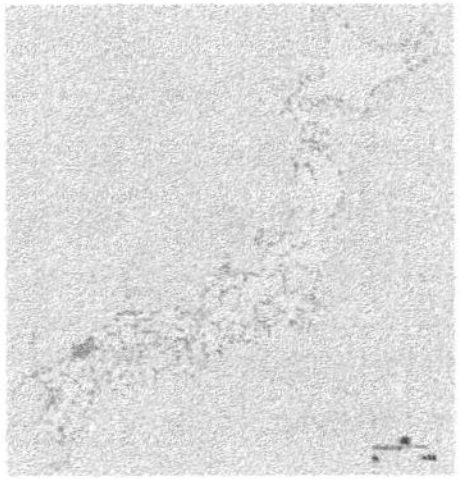 Fig. A1.10. Nagato Province (*Source*: Ash_Crow, under CC BY-SA 3.0, https://commons.wikimedia.org/wiki/File:Provinces_of_Japan-Nagato.svg).[130]
13	周防	Zhou Fang	Suō Province	Suō (周防) was one of the eight provinces under the rule of the Light (Southern) Mountains Circuit/San'yōdō (山陽道/山阳道). 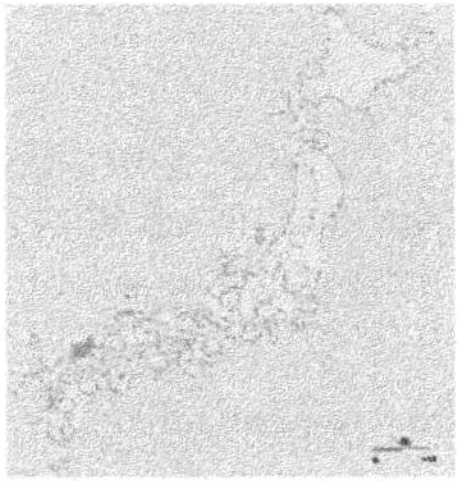 Fig. A1.11. Suo Province (*Source*: Ash_Crow, under CC BY-SA 3.0, https://commons.wikimedia.org/wiki/File:Provinces_of_Japan-Suo.svg).[131]
14	安藝	An Yi	Aki Province	Aki (安藝/安艺) was one of the eight provinces under the rule of the Light (Southern) Mountains Circuit/San'yōdō (山陽道/山阳道).

(*Continued*)

Item #	Name	Pinyin	Etymology	History, Geography and My Comments
				Fig. A1.12. Aki Province (*Source*: Ash_Crow, under CC BY-SA 3.0, https://commons.wikimedia.org/wiki/File:Provinces_of_Japan-Aki.svg).[132]
17	美作	Mei Zuo	Mimasaka Province	Mimasaka (美作) was one of the eight provinces under the rule of the Light (Southern) Mountains Circuit/San'yōdō (山陽道/山阳道). **Fig. A1.13.** Mimasaka Province (*Source*: Ash_Crow, under CC BY-SA 3.0, https://commons.wikimedia.org/wiki/File:Provinces_of_Japan-Mimasaka.svg).[133]
15, 16 &18	備前中後	Bei Qian Zhong Hou	Bizen, Bitchu and Bingo Provinces	Bizen (備前/备前; Item 18), Bitchu (備中/备中; Item 16) and Bingo (備後/备后; Item 15) were three of the eight provinces under the rule of the Light (Southern) Mountains Circuit/San'yōdō (山陽道/山阳道).

19	攝摩 (播磨)	She Mo (Bo Mo)	Harima Province	

Fig. A1.14. (a) Bizen Province (*Source*: Ash_Crow, under CC BY-SA 3.0, https://commons.wikimedia.org/wiki/File:Provinces_of_Japan-Bizen.svg);[134] (b) Bitchu Province (*Source*: Ash_Crow, under CC BY-SA 3.0, https://commons.wikimedia.org/wiki/File:Provinces_of_Japan-Bitchu.svg);[135] and (c) Bingo Province (*Source*: Ash_Crow, under CC BY-SA 3.0, https://commons.wikimedia.org/wiki/File:Provinces_of_Japan-Bingo.svg).[136]

Harima (攝摩/摄摩) was one of the eight provinces under the rule of the Light (Southern) Mountains Circuit/San'yōdō (山陽道/山阳道). On the Japan-KWQ, 播磨 is written as 攝摩.

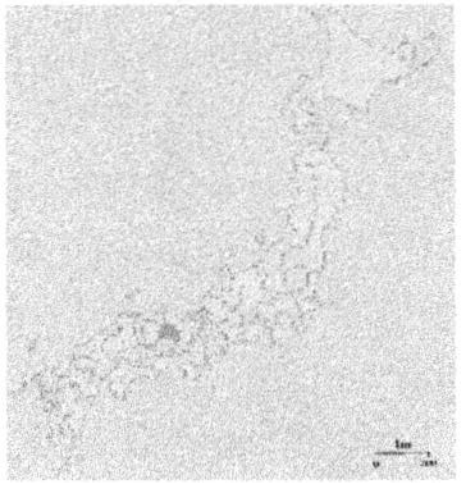

Fig. A1.15. Harima Province (*Source*: Ash_Crow, under CC BY-SA 3.0, https://commons.wikimedia.org/wiki/File:Provinces_of_Japan-Harima.svg).[137]

Remark on Light (Southern) Mountains Circuit/San'yōdō: (山陽道/山阳道)**:** The Japan-KWQ depicts reasonably well all the eight provinces under the rule of the Light (Southern) Mountains Circuit/San'yōdō as shown in Fig. 1.9 on a modern map, excepting that the location of Harima (攝摩/摄摩 or 播磨) is misplaced on the Kii Peninsula, being not at its true location, and Mimasaka (美作) is too far to the north leaving no space to accommodate the provinces ruled by the Dark (Northern) Mountains Circuit/San'indō (山陰道/山阴道). This latter case will be further discussed below.

Item #	Name	Pinyin	Etymology	History, Geography and My Comments
*3	山陰道	Shan Yin Dao	Dark (Northern) Mountains Circuit/San'indō (located in the western part of Honshu and along the Sea of Japan coast, plus the island province – Oki Province).[138] This name derives from the idea that the northern side is the "shady (陰/阴)" side of the central mountain chain running through Honshu.	Dark (Northern) mountains circuit/San'indō (山陰道/山阴道; literally, "the shaded side of a mountain") was one of the seven circuits in the *Gokishichidō* ("five provinces and seven circuits"; 五畿七道) system. The following were the eight provinces under its rule: 20. Iwami Province (石見國/石见国):[139] it was established in the seventh century.[140] 21. Izumo Province (出雲國/出云国):[141] it was an independent dynasty that existed before the arrival of the Yamato kingship (ancient Izumo).[142] Scholars believe that the place belonged to Tou Ma Guo (transliteration of 投馬國/投马国) but became a province in the *Ritsuryō* system. 22. Hōki Province (伯耆國/伯耆国):[143] ancient records in 698 and 712 show names closely related to Hōki;[144] after the Taika Era Reforms (大化改新 or 大化革新; took place in 645), the place became an independent province. 23. Oki Province (隱岐國/隐岐国):[145] it was organised as a province under the *Ritsuryō* system in the latter part of the seventh century after the Taika Era Reforms in 645.[146] 24. Inaba Province (因幡國/因幡国):[147] it was the ancient Dao Yu Guo (transliteration of 稻羽國/稻羽国);[148] it changed name to Inaba Province (因幡國/因幡国) after the Taika Era Reforms. 25. Tajima Province (但馬國/但马国):[149] the area was once under the control of the ancient Tanba Kingdom (丹波國/丹波国) which was later divided into Tajima, Tango and Tanba. However, since the name "Tajima" appears in the *Ri Ben Shu Jii* (日本書紀/日本书纪) or *Nihon Shoki/Chronicles of Japan* in an entry dated 675 A.D., this division occurred before the formalization of the Japanese

				provinces (the formalization occurred after the Taika Reforms of the year 645). 26. Tango Province (丹後國/丹后国)[150]: it was separated from Tanba or Tamba Province (丹波國/丹波国) in 713.[151] 27. Tanba Province (丹波國/丹波国)[152]: it broke off from Tango in 713. 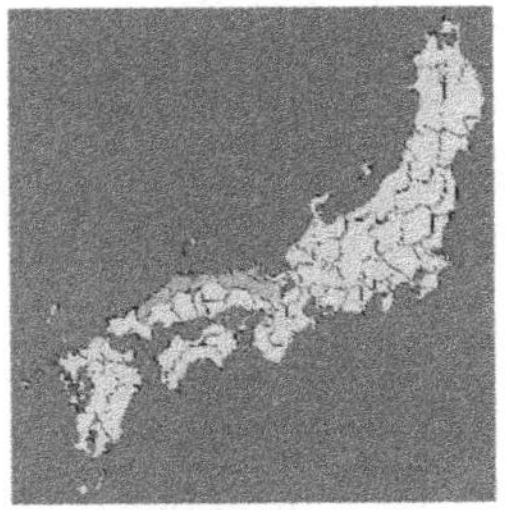 **Fig. A1.16.** Dark (Northern) Mountains Circuit/San'indō (*Source*: 百楽兎, under CC BY-SA 3.0, https://commons.wikimedia.org/wiki/File:Sanindo.svg).[153]
20	石見	Shi Jian	Iwami Province	Iwami (石見/石见) was one of the eight provinces under the rule of the Dark (Northern) Mountains Circuit/San'indō (山陰道/山阴道). On the Japan-KWQ, Iwami is mistakenly placed to the southeast of Izomo (出雲/出云); it should be to its west. 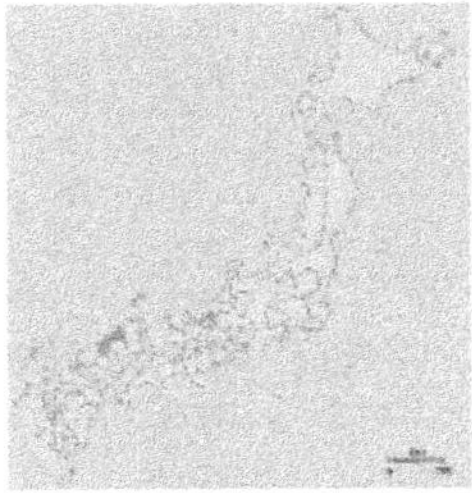 **Fig. A1.17.** Iwami Province (*Source*: Ash_Crow, under CC BY-SA 3.0, https://commons.wikimedia.org/wiki/File:Provinces_of_Japan-Iwami.svg).[154]

(*Continued*)

Item #	Name	Pinyin	Etymology	History, Geography and My Comments
21	出雲	Cu Yun Guo	Izumo Province	Izumo (出雲/出云) was one of the eight provinces under the rule of the Dark (Northern) Mountains CirIcuit/San'indō (山陰道/山阴道). **Fig. A1.18.** Izumo Province (*Source*: Ash_Crow, under CC BY-SA 3.0, https://commons.wikimedia.org/wiki/File:Provinces_of_Japan-Izumo.svg).[155]
22	伯歧 (耆)	Bo Qi Guo	Hōki Province	Hōki (伯耆 or 伯歧) was one of the eight provinces under the rule of the Dark (Northern) Mountains Circuit/San'indō (山陰道/山阴道). **Fig. A1.19.** Hōki Province (*Source*: Ash_Crow, under CC BY-SA 3.0, https://commons.wikimedia.org/wiki/File:Provinces_of_Japan-Hoki.svg).[156]
23	隱岐	Yin Qi	Oki Province	Oki (隱岐/隐岐) was one of the eight provinces under the rule of the Dark (Northern) Mountains Circuit/San'indō (山陰道/山阴道).

				Oki (隱岐/隐岐) was a province of Japan which consisted of the Oki Islands in the Sea of Japan; it was located off the coast of the Izumo and Hōki Provinces. But the Japan-KWQ erroneously depicts Oki (隱岐/隐岐) as a coastal region to the west of Izomo (出雲/出云) Province. 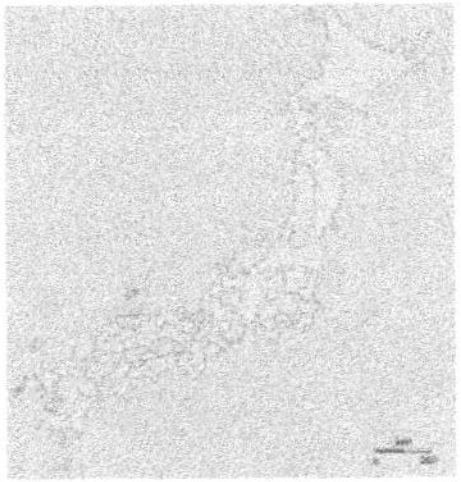 **Fig. A1.20.** Oki Province (*Source*: Ash_Crow, under CC BY-SA 3.0, https://commons.wikimedia.org/wiki/File:Provinces_of_Japan-Oki.svg).[157]
24	因幡	Yin Fan Guo	Inaba Province	Inaba (因幡) was one of the eight provinces under the rule of the Dark (Northern) Mountains Circuit/San'indō (山陰道/山阴道). 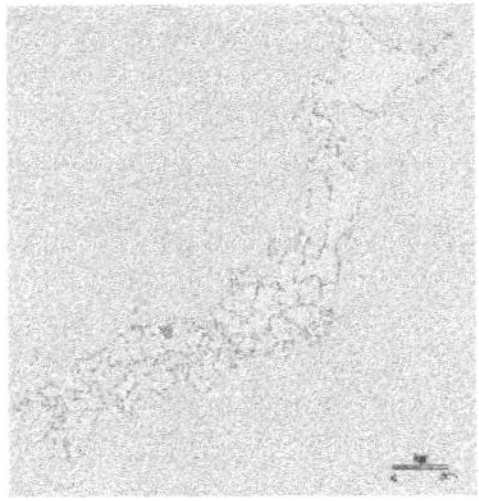 **Fig. A1.21.** Inaba Province (*Source*: Ash_Crow, under CC BY-SA 3.0, https://commons.wikimedia.org/wiki/File:Provinces_of_Japan-Inaba.svg).[158]
25	但馬	Dan Ma	Tajima Province	Tajima (但馬/但马) was one of the eight provinces under the rule of the Dark (Northern) Mountains Circuit/San'indō (山陰道/山阴道).

(Continued)

Item #	Name	Pinyin	Etymology	History, Geography and My Comments
				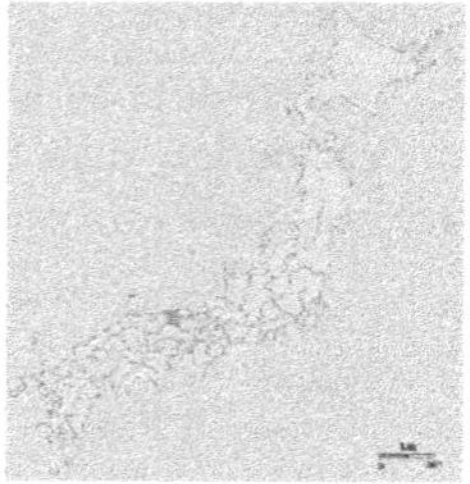 Fig. A1.22. Tajima Province (*Source*: Ash_Crow, under CC BY-SA 3.0, https://commons.wikimedia.org/wiki/File:Provinces_of_Japan-Tajima.svg).[159]
26	丹後	Dan Hou Guo	Tango Province	Tango (丹後/丹后) was one of the eight provinces under the rule of the Dark (Northern) Mountains Circuit/San'indō (山陰道). 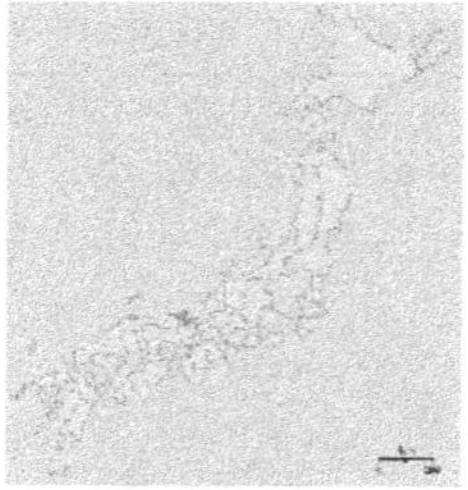 Fig. A1.23. Tango Province (*Source*: Ash_Crow, under CC BY-SA 3.0, https://commons.wikimedia.org/wiki/File:Provinces_of_Japan-Tango.svg).[160]
27	丹波	Da Bo	Tanba Province	Not depicted on the Japan-KWQ.

Remark on the Dark (Northern) Mountains Circuit/San'indō: The Japan-KWQ, by mistake, does not depict Tanba (丹波; Item 27) Province as one of the eight provinces ruled by the Dark (Northern) Mountains Circuit/San'indō. Furthermore, some of the provinces are not correctly depicted to the north of the Light (Southern) Mountains Circuit/San'yōdō (山陽道/山阳道), for example: Oki (隱岐/隐岐) Province consisted of the Oki Islands in the Sea of Japan and should be located off the coast of the Izumo (出雲/出云; Item 21) and Hōki (伯耆; Item 22) Provinces, but the Japan-KWQ erroneously depicts Oki (隱岐/隐岐) Province as a coastal region to the west of Izomo (出雲/出云) Province. In addition, Iwami (石見/石见; Item 20) is mistakenly placed to the southeast of Izomo (出雲/出云; Item 21); it should be to its west.

| *4 | 南海道 | Nan Hai Dao | Southern Sea Circuit/Nankaid-ō (the entire Shikoku, the southern portion of the Kii Peninsula and the island province Awaji which lies between Honshu and Shikoku)[161] | Nankaidō (南海道, literally, "southern sea circuit" or "southern sea region") was one of the seven circuits in the *Gokishichidō* ("five provinces and seven circuits"; 五畿七道) system. The following were the six provinces under its rule as shown in Fig. 1.9 on a modern map. Among them, Awaji (淡路) was an island province off Shikoku (Japan's smallest main island) and Kii Province (紀伊國/纪伊国) was on Honshu (Japan's largest main island). Hence, there were only four provinces on Honshu (四國/四国, literally "four provinces"):

28. Iyo Province (伊予國/伊予国):[162] it was formed by the *Ritsuryo* reforms. In 645, the Taika Era Reforms were the first signs of implementation of the system.[163]
29. Tosa Province (土佐國/土佐国):[164] it was formed by the *Ritsuryo* reforms; the name "Tosa" appears in the *Ri Ben Shu Jii* (日本書紀/日本书纪) or *Nihon Shoki/Chronicles of Japan* in an entry dated March 675 A.D.
30. Awa Province (阿波國/阿波国):[165] in 713, the Liguo Province (栗國/栗国) formally changed its name to Awa Province (阿波國/阿波国).[166]
31. Sanuki Province (讚岐國/赞岐国):[167] it was formed by the *Ritsuryo* reforms.
32. Awaji Province (淡路國/淡路国):[168] it was founded in the seventh century and changed its name to Awaji Province (淡路國/淡路国) after the *Ritsuryo* reforms;[169] Awaji Province was an island province between Kii Province (紀伊國/纪伊国; item 33) and Awa Province (阿波國/阿波国; Item 30), see Fig. 1.9. |

(*Continued*)

Item #	Name	Pinyin	Etymology	History, Geography and My Comments
				33. Kii Province (紀伊國/纪伊国):[170] it was formerly known as "The Country of Wood (木之國/木之国)" and changed to Kii Province (紀伊國/纪伊国) in 713.[171] **Fig. A1.24.** Southern Sea Circuit/Nankaidō (*Source*: 百楽兎, under CC BY-SA 3.0, https://commons.wikimedia.org/wiki/File:Nankaido.svg).[172]
28	伊豫(伊予)	Yi Yu (Yi Yu)	Iyo Province	Iyo (伊予) was one of the six provinces under the rule of the Southern Sea Circuit/Nankaidō. On the Japan-KWQ, 伊予 is written as 伊豫. They have similar pronunciation. **Fig. A1.25.** Iyo Province (*Source*: Ash_Crow, under CC BY-SA 3.0, https://commons.wikimedia.org/wiki/File:Provinces_of_Japan-Iyo.svg).[173]

29	土佐	Tu Zho Guo	Tosa Province	Tosa (土佐) was one of the six provinces under the rule of the Southern Sea Circuit/Nankaidō. **Fig. A1.26.** Tosa Province (*Source*: Ash_Crow, under CC BY-SA 3.0, https://commons.wikimedia.org/wiki/File:Provinces_of_Japan-Tosa.svg).[174]
30	阿波	A Bo Guo	Awa Province	Awa (阿波) was one of the six provinces under the rule of the Southern Sea Circuit/Nankaidō. **Fig. A1.27.** Awa Province (*Source*: Ash_Crow, under CC BY-SA 3.0, https://commons.wikimedia.org/wiki/File:Provinces_of_Japan-Awa_(Tokushima).svg).[175]
31	讚岐	Zan Qi	Sanuki Province	Not depicted on the Japan-KWQ.
32	淡路	Dian Lu Guo	Awaji Province	Awaji (淡路) was one of the six provinces under the rule of the Southern Sea Circuit/Nankaidō. But the Japan-KWQ mistakenly depicts it in the westernmost region of Shikoku (四國/四国) Island; it should be depicted between Kii (紀伊; Item 33) Province and Awa (阿波; Item 30) Province.

(*Continued*)

Item #	Name	Pinyin	Etymology	History, Geography and My Comments
				Fig. A1.28. Awaji Province (*Source*: Ash_Crow, under CC BY-SA 3.0, https://commons.wikimedia.org/wiki/ File:Provinces_of_Japan-Awaji.svg).[176]
33	伊紀[紀伊]	Yi Ji [Ji Yi]	Kii Province	Kii (紀伊) was one of the six provinces under the rule of the Southern Sea Circuit/Nankaidō. It should be on Honshu (本州) Island, but on the Japan-KWQ, it is mistakenly depicted on Shikoku (四國/四国) Island. Also, 紀伊 is written as 伊紀 on the Japan-KWQ. Fig. A1.29. Kii Province (*Source*: Ash_Crow, under CC BY-SA 3.0, https://commons.wikimedia.org/wiki/ File:Provinces_of_Japan-Kii.svg).[177]

Remark on the Southern Sea Circuit/Nankaidō: The Japan-KWQ, by mistake, does not depict Sanuki (讚岐/赞岐; Item 31) as one of the six provinces ruled by the Southern Sea Circuit/Nankaidō. In addition, Awaji (淡路; Item 32), an island in the eastern part of the Seto Inland Sea between the islands of Honshu and Shikoku, is erroneously depicted as the westernmost region of Shikoku.

| ** 34 | 太和
[大和] | Da He | Yamato Province[178] | Yamato (大和), also called Washū (和州) was one of the five provinces in and around the imperial capital. This area was defined as "Kinai (畿内)" in the *Gokishichidō* ("five provinces and seven circuits"; 五畿七道) system. 大 means "great" or "big" and 和 is "Wa" or "倭", meaning "dwarf, pygmy"; together the name was written as "大倭". Due to its offensive connotation, in 738, the name was revised to the more desirable characters "大養徳/大养德"; in 747, it was changed back to "大倭"; and the final revision into "大和" was made in c. 757.[179] From this history, we know that the Japan-KWQ must be a map drawn after c.757.
The Japan-KWQ writes 大和 as太和, since "太 also means "great".
Fig. A1.30. Yamato Province (*Source*: Ash_Crow, under CC BY-SA 3.0, https://commons.wikimedia.org/wiki/File:Provinces_of_Japan-Yamato.svg).[180] |
| ** 35 | 河内 | He Nei | Kawachi Province[181] | Kawachi (河内) was one of the five provinces in and around the imperial capital in the *Gokishichidō* ("five provinces and seven circuits"; 五畿七道) system. Kawachi Province was established in the seventh century.
Under Dōkyō's (道鏡/道镜, 700–772) administration,[182] *Yuge-no-Miya* (由義宮/由义宫) was established in Kawachi Province as the "Western Capital (西京; *Nishi-no-Miyako*)". |

(*Continued*)

Item #	Name	Pinyin	Etymology	History, Geography and My Comments
				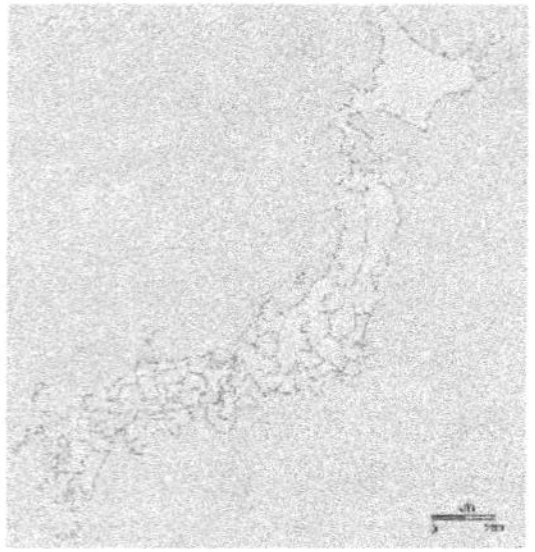 **Fig. A1.31.** Kawachi Province (*Source*: Ash_Crow, under CC BY-SA 3.0, https://commons.wikimedia.org/wiki/File:Provinces_of_Japan-Kawachi.svg).[183]
** 36	山城	Shan Cheng	Yamashiro Province[184]	Yamashiro (山城) was one of the five provinces in and around the imperial capital in the *Gokishichidō* ("five provinces and seven circuits"; 五畿七道) system. "Yamashiro" was formerly written with the characters meaning "mountain" (山) and "era" (代) in the seventh century. In 794, Emperor Kanmu made his new capital utilise the surroundings as natural fortification, and the character for "shiro" was finally changed to "castle" (山城國/山城国).[185] *The name indicates that the Japan-KWQ is a map drawn later than 794.* 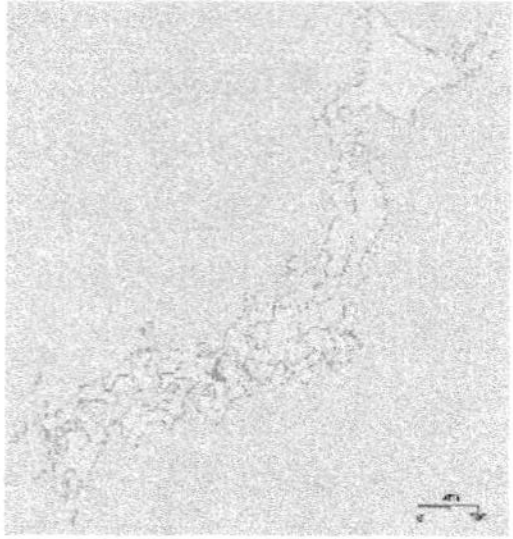 **Fig. A1.32.** Yamashiro Province (*Source*: Ash_Crow, under CC BY-SA 3.0, https://commons.wikimedia.org/wiki/File:Provinces_of_Japan-Yamashiro.svg).[186]

** 37	和泉	He Quan	Izumi Province[187]	Izumi (和泉) was one of the five provinces in and around the imperial capital in the *Gokishichidō* ("five provinces and seven circuits"; 五畿七道) system. In 716, the Izumi and Hine Districts, and the Ōtori District were separated from Kawachi (河內國/河内国; **35), and the three districts were made into a province named *Izumi-gen* (和泉監). The name "Izumi" means "spring" (泉), but is written with two characters, the character for "peace" (和) being prepended due to an imperial edict in 713.[188] In 740, Izumi was abolished and merged back into Kawachi Province. However, in 757, it was re-established with a normal province designation *kuni* (國/国). **Fig. A1.33.** Izumi Province (*Source*: Ash_Crow, under CC BY-SA 3.0, https://commons.wikimedia.org/wiki/File:Provinces_of_Japan-Izumi.svg).[189]
** 38	攝津	She Jin	Settsu Province[190]	Settsu (攝津/摄津) was one of the five provinces in and around the imperial capital in the *Gokishichidō* ("five provinces and seven circuits"; 五畿七道) system. It included the important harbour city Osaka.

(Continued)

Item #	Name	Pinyin	Etymology	History, Geography and My Comments
				Fig. A1.34. Settsu Province (*Source*: Ash_Crow, under CC BY-SA 3.0, https://commons.wikimedia.org/wiki/File:Provinces_of_Japan-Settsu.svg).[191]

Remark on the five provinces in and around the imperial capital in the *Gokishichidō* system: On the map of the Japan-KWQ, all the five provinces in the area surrounding the capital (Kinai; 畿内) are depicted (see the five provinces inside the large red circle in Fig. A1.1). But their relative locations are incorrect (see Fig. 1.9 for their correct relative geographical locations on a modern map); the imperial capital was first - Heijō-kyō at Nara (平城京; 710–740 and 745–784), then Heian-kyō at Kyoto (平安京; 794–1868 with an interruption in 1180).

Item #	Name	Pinyin	Etymology	History, Geography and My Comments
*5	北陆道	Bei Lu Dao	Northern Land Circuit/ Hokuriku-dō (approximately located in the central part of Honshu and along the northeast direction of the Sea of Japan coast).[192] The name literally means "North Land Way" which refers to a series of roads that connected the capitals of each of the provinces that made up the region.	As explained in Section 6.2 of the main text, *Gokishichidō* was the name of *Ritsuryō* system (律令制; the legal and governmental system) in Japan. The idea was initiated during the Asuka period (538–710 A.D.). It was borrowed from the Chinese Tang Dynasty and consisted of "five provinces and seven circuits (五畿七道)" as administrative units. The Northern Land Circuit/Hokurikudō (北陸道/北陆道) was one of the seven circuits. When the *Gokishichidō* system was initially established after the Taika Era Reforms (in the year 645), the Northern Land Circuit/Hokurikudō consisted of just two provinces: Wakasa (若狹國/若狹国) and Koshi (越國/越国). During the reign of Emperor Temmu (天武天皇; c. 631–686), Koshi was divided into three regions: Echizen, Etchū and Echigo (越前國/越前国、越中國/越中国、越後國/越後国), and Sado Island (an island off the coast of Niigata in the Chubu region of Japan) was added as a fifth province (Sado Province; 佐渡國/佐渡国). Later, Noto (能登)

and Kaga (加賀/加贺) were carved out of Echizen (越前國/越前国) to form seven provinces in total. The system's full implementation started from the Nara period (奈良時代/奈良时代; c. 710–794). The following were the seven provinces as shown in Fig. 1.9 on a modern map:

39. Echizen Province (越前國/越前国):[193] during 689–692, Koshi Province (高志國/高志国; the old name of 越國/越国) established in the seventh century was divided into three separate provinces:[194] Echizen (越前國/越前国), Etchū (越中國/越中国), and Echigo (越後國/越后国).

40. Wakasa Province (若狹國/若狭国):[195] Wakasa existed as a political entity before the *Ritsuryō* system and the implementation of the Taihō Code (大寶律令/大宝律令) in the Nara period (c.710–794). Wakasa Province was formally established around 701 A.D. with the creation of the *Ritsuryō* provincial system.

41. Kaga Province (加賀國/加贺国):[196] in 823 A.D., the two eastern districts, Kaga and Enuma, were separated from Echizen Province (越前國/越前国) to form Kaga Province; it was the last province to be created under the *Ritsuryō* system.[197]

42. Etchū Province (越中國/越中国):[198] see Item 39 for details.

43. Noto Province (能登國/能登国):[199] in 718 A.D., four districts of Echizen Province (越前國/越前国) were split off into Noto Province. However, in the year 741, the province was abolished, and merged into Etchū Province. Noto Province was subsequently re-established in 757.[200]

44. Echigo Province (越後國/越后国):[201] see Item 39 for details.

(*Continued*)

Item #	Name	Pinyin	Etymology	History, Geography and My Comments
				45. Sado Province (佐渡國/佐渡国):[202] the province was on Sado Island. It was part of the ancient Koshi Province (高志國/高志国; the old name of 越國/越国) established in the seventh century; after 743, it was part of the Echigo Province (越後國/越后国) but split off from Echigo Province in 752.[203] On the Japan-KWQ, the entire Northern Land Circuit/Hokurikudō (北陸道/北陆道) is erroneously placed in Hokkaido; besides, only five of the seven provinces: Kaga Province (加賀國/加贺国; Item 41), Etchū Province (越中國/越中国; Item 42), Noto Province (能登國/能登国; Item 43), Echigo Province (越後國/越后国; Item 44) and Sado Province (佐渡國/佐渡国; Item 45) are included in the Northern Land Circuit/Hokurikudō; Echizen Province (越前國/越前国; Item 39) and Wakasa Province (若狭國/若狭国; Item 40) are missing on the Japan-KWQ. **Fig. A1.35.** Northern Land Circuit/Hokurikudō (*Source*: 百楽兎, under CC BY-SA 3.0, https://commons.wikimedia.org/wiki/File:Hokurikudo.svg).[204]
39	越前	Yue Qian	Echizen Province	Not depicted on the Japan-KWQ.
40	若狭	Ruo Xia	Wakasa Province	Not depicted on the Japan-KWQ.
41	加贺	Jia He	Kaga Province	Kaga (加賀/加贺) was one of the seven provinces under the rule of the Northern Land Circuit/Hokurikudō. It was in Honshu.

				On the Japan-KWQ, Kaga (加賀/加贺) is erroneously depicted in Hokkaido. 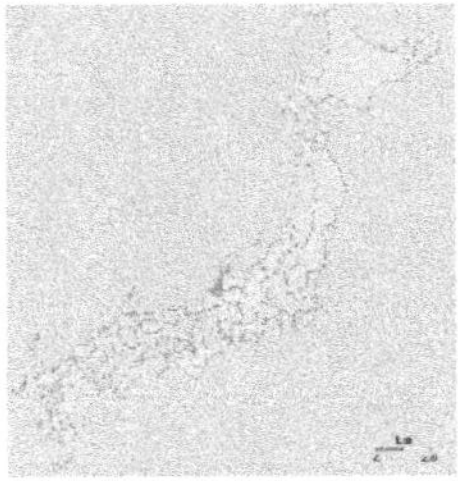 **Fig. A1.36.** Kaga Province (*Source*: Ash_Crow, under CC BY-SA 3.0, https://commons.wikimedia.org/wiki/File:Provinces_of_Japan-Kaga.svg).[205]
42 & 44	越中後	Yue Zhong Hou	Etchū and Echigo Provinces	越中後/越中后 refers to Etchū (越中; Item 42) and Echigo (越後/越后; Item 44) Provinces. They were two of the seven provinces under the rule of the Northern Land Circuit/Hokurikudō. They were in Honshu. But on the Japan-KWQ, Etchū and Echigo are erroneously depicted in Hokkaido. 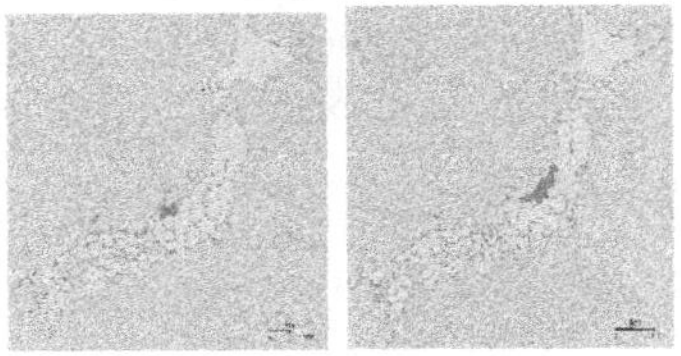 (a)　(b) **Fig. A1.37.** (a) Etchū Province (*Source*: Ash_Crow, under CC BY-SA 3.0, https://commons.wikimedia.org/wiki/File:Provinces_of_Japan-Etchu.svg);[206] (b) Echigo Province (*Source*: Ash_Crow, under CC BY-SA 3.0, https://commons.wikimedia.org/wiki/File:Provinces_of_Japan-Echigo.svg).[207]

(Continued)

Item #	Name	Pinyin	Etymology	History, Geography and My Comments
43	能登	Neng Deng	Noto Province	Noto (能登) was one of the seven provinces under the rule of the Northern land circuit/Hokurikudō. It was on the Noto Peninsula in Honshu. On the Japan-KWQ, Noto (能登) is erroneously depicted in Hokkaido. **Fig. A1.38.** Noto Province (*Source*: Ash_Crow, under CC BY-SA 3.0, https://commons.wikimedia.org/wiki/File:Provinces_of_Japan-Noto.svg).[208]
45	佐渡	Zuo Du	Sado Province	Sado (佐渡) was one of the seven provinces under the rule of the Northern land circuit/Hokurikudō. It was on Sado Island (佐渡島/佐渡岛) in Honshu. On the Japan-KWQ, Sado (佐渡) is erroneously depicted in Hokkaido. **Fig. A1.39.** Sado Province (*Source*: Ash_Crow, under CC BY-SA 3.0, https://commons.wikimedia.org/wiki/File:Provinces_of_Japan-Sado.svg).[209]

Remark on the **Northern Land Circuit/Hokurikudō** (北陆道): On the map of the Japan-KWQ, the entire region of Northern Land Circuit/ Hokurikudō (北陆道) is erroneously placed in Hokkaido; besides, only five of the seven provinces: Kaga (加賀/加贺; Item 41), Etchu (越中; Item 42), Noto (能登; Item 43), Echigo (越後/越后; Item 44) and Sado (佐渡; Item 45) are depicted in the Northern Land Circuit/Hokurikudō. The other two provinces, Echizen (越前; Item 39) and Wakasa (若狭; Item 40) shown in Fig. 1.9, are missing on the Japan-KWQ.

*6	東海道	Dong Hai Dao	Eastern Sea Circuit/Tōkaidō (approximately located in the central part of Honshu and along Japan's Pacific coast)[210]	The Eastern Sea Circuit/Tōkaidō (東海道/东海道) was one of the seven circuits in the *Gokishichidō* ("five provinces and seven circuits"; 五畿七道) system.
				There were fifteen provinces under the rule of the Eastern Sea Circuit/Tōkaidō (東海道/东海道) as shown in Fig. 1.9 on a modern map. They were:
				46. Shima Province (志摩國/志摩国): it was separated from Ise Province (伊勢國/伊势国)[211] during the early eighth century;[212] it was the smallest of all provinces.
				47. Ise Province (伊勢國/伊势国):[213] it was one of the original provinces of Japan, established in the Nara period (710–794), when the former princely state of Ise was divided into Ise (伊勢國/伊势国), Iga (伊賀國/伊贺国) and Shima (志摩國/志摩国).
				48. Iga Province (伊賀國/伊贺国):[214] it was separated from Ise Province around 680 A.D. during the Asuka period (592–710).[215]
				49. Owari Province (尾張國/尾张国):[216] it is mentioned in records of the Nara period (710–794); before 704, the place was named Wei Zhi Guo (尾治國/尾治国).[217]
				50. Mikawa Province (三河國/三河国):[218] it was one of the original provinces of Japan established in the Nara period under the Taihō Code (大寶律令/大宝律令; an administrative reorganization enacted in 703).[219]
				51. Tōtōmi Province (遠江國/远江国):[220] it was one of the original provinces of Japan established in the Nara period under the Taihō Code.

(Continued)

Item #	Name	Pinyin	Etymology	History, Geography and My Comments
				52. Suruga Province (駿河國/駿河国):[221] it was also one of the original provinces of Japan established in the Nara period under the Taihō Code.
				53. Izu Province (伊豆國/伊豆国):[222] in 680, two districts were separated from the Suruga Province (駿河國/駿河国) to form the new Izu Province (伊豆國/伊豆国).[223]
				54. Kai Province (甲斐國/甲斐国):[224] it was one of the original provinces of Japan established in the Nara period under the Taihō Code.
				55. Sagami Province (相模國/相模国):[225] it was also one of the original provinces of Japan established in the seventh century.[226]
				56. Musashi Province (武藏國/武藏国):[227] Musashi Province originally belonged to the Eastern Mountains Circuit/Tōsandō; in 771 it was assigned to the Eastern Sea Circuit/Tōkaidō.[228]
				57. Shimōsa Province (下總國/下总国):[229] it was originally part of a larger territory known as Fusa Province (總國/総国), which was divided into "upper (上)" and "lower (下)" portions (i.e., Kazusa and Shimōsa) during the reign of Emperor Kōtoku (645–654).[230]
				58. Kazusa Province (上總國/上总国):[231] see explanation above.
				59. Awa Province (安房國/安房国):[232] it was elevated to the status of a full province in 718; it was merged back[233] into Kazusa in 741 but regained its independent status in 757.
				60. Hitachi Province (常陸國/常陆国):[234] was established in the seventh century; name changed to Hitachi Province (常陸國/常陆国) at the end of the seventh century.[235]

				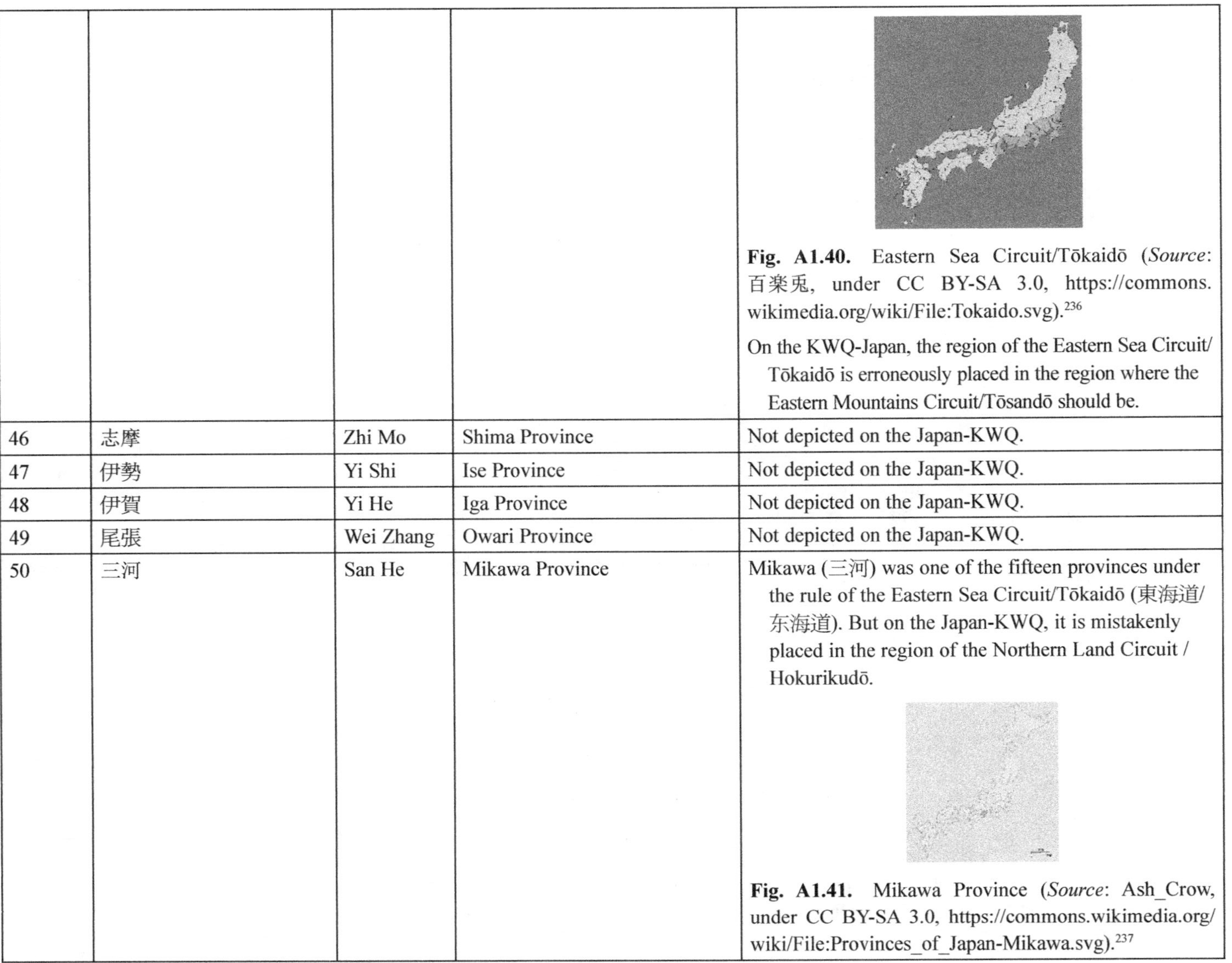 Fig. A1.40. Eastern Sea Circuit/Tōkaidō (*Source*: 百楽兎, under CC BY-SA 3.0, https://commons. wikimedia.org/wiki/File:Tokaido.svg).[236] On the KWQ-Japan, the region of the Eastern Sea Circuit/ Tōkaidō is erroneously placed in the region where the Eastern Mountains Circuit/Tōsandō should be.
46	志摩	Zhi Mo	Shima Province	Not depicted on the Japan-KWQ.
47	伊勢	Yi Shi	Ise Province	Not depicted on the Japan-KWQ.
48	伊賀	Yi He	Iga Province	Not depicted on the Japan-KWQ.
49	尾張	Wei Zhang	Owari Province	Not depicted on the Japan-KWQ.
50	三河	San He	Mikawa Province	Mikawa (三河) was one of the fifteen provinces under the rule of the Eastern Sea Circuit/Tōkaidō (東海道/ 东海道). But on the Japan-KWQ, it is mistakenly placed in the region of the Northern Land Circuit / Hokurikudō. Fig. A1.41. Mikawa Province (*Source*: Ash_Crow, under CC BY-SA 3.0, https://commons.wikimedia.org/ wiki/File:Provinces_of_Japan-Mikawa.svg).[237]

(*Continued*)

Item #	Name	Pinyin	Etymology	History, Geography and My Comments
51	遠江	Yuan Jiang	Tōtōmi Province	Tōtōmi (遠江/远江) was one of the fifteen provinces under the rule of the Eastern Sea Circuit/Tōkaidō (東海道/东海道). But on the Japan-KWQ, it is mistakenly placed in the region of the Eastern Mountains Circuit/Tōsandō. 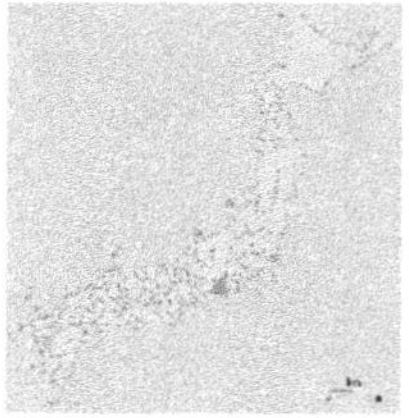 **Fig. A1.42.** Tōtōmi Province (*Source*: Ash_Crow, under CC BY-SA 3.0, https://commons.wikimedia.org/wiki/File:Provinces_of_Japan-Totomi.svg).[238]
52	駿河	Jun He	Suruga Province	Suruga (駿河/骏河) was one of the fifteen provinces under the rule of the Eastern Sea Circuit/Tōkaidō (東海道/东海道). But on the KWQ-Japan, Suruga (駿河/骏河) is mistakenly placed in the region of the Eastern Mountains Circuit/Tōsandō. 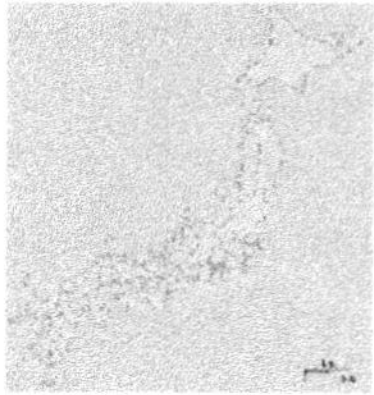 **Fig. A1.43.** Suruga Province (*Source*: Ash_Crow, under CC BY-SA 3.0, https://commons.wikimedia.org/wiki/File:Provinces_of_Japan-Suruga.svg).[239]
53	伊豆	Yi Dou	Izu Province	Izu (伊豆) was one of the fifteen provinces under the rule of the Eastern Sea Circuit/Tōkaidō (東海道/东海道). But on the Japan-KWQ, Izu (伊豆) is mistakenly placed in the region of the Eastern Mountains Circuit/Tōsandō.

				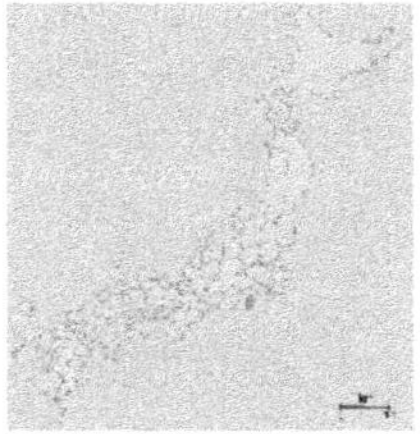 Fig. A1.44. Izu Province (*Source*: Ash_Crow, under CC BY-SA 3.0, https://commons.wikimedia.org/wiki/File:Provinces_of_Japan-Izu.svg).[240]
54	甲斐	Jia Fei	Kai Province	Not depicted on the Japan-KWQ.
55	相模	Xiang Mo	Sagami Province	Not depicted on the Japan-KWQ.
56	武藏	Wu Cang	Musashi Province	Musashi (武藏) was one of the fifteen provinces under the rule of the Eastern Sea Circuit/Tōkaidō (東海道/东海道) after 771; however, before the year 771, Musashi was one of the provinces of the Eastern Mountain Circuit/Tōkaidō. Since the annotation given in Section 6.2 of the main text reveals that the Japan-KWQ depicts Japan in 897 A.D., not before 771, hence, on the Japan-KWQ, Musashi (武藏) is mistakenly placed in the region of the Eastern Mountains Circuit/Tōsandō. 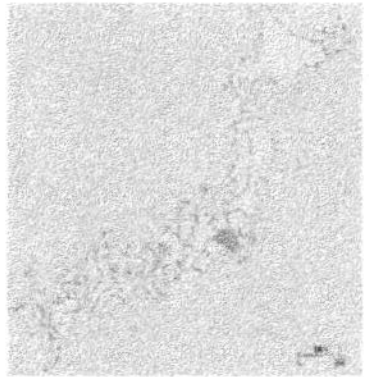 Fig. A1.45. Hitachi Province (*Source*: Ash_Crow, under CC BY-SA 3.0, https://commons.wikimedia.org/wiki/File:Provinces_of_Japan-Musashi.svg).[241]
58 & 57	上下總	Zhang Xia Zong	Kazusa and Shimōsa Provinces	Kazusa (上總/上总; Item 58) and Shimōsa (下總/下总; Item 57), shortened as 上下總 on the Japan-KWQ,

(*Continued*)

Item #	Name	Pinyin	Etymology	History, Geography and My Comments
				were two of the fifteen provinces under the rule of the Eastern Sea Circuit/Tōkaidō (東海道/东海道). But on the Japan-KWQ, they are mistakenly placed in the region of the Eastern Mountains Circuit/Tōsandō. Fig. A1.46. (a) Kazusa Province (*Source*: Ash_Crow, under CC BY-SA 3.0, https://commons.wikimedia.org/wiki/File:Provinces_of_Japan-Kazusa.svg);[242] (b) Shimosa Province (*Source*: Ash_Crow, under CC BY-SA 3.0, https://commons.wikimedia.org/wiki/File:Provinces_of_Japan-Shimosa.svg).[243]
59	安房	An Fang	Awa Province in today's Chiba Prefecture	Awa (安房) in today's Chiba Prefecture was one of the fifteen provinces under the rule of the Eastern Sea Circuit/Tōkaidō (東海道/东海道). But on the Japan-KWQ, Awa (安房) is mistakenly placed in the region of the Eastern Mountains Circuit/Tōsandō. Fig. A1.47. Awa Province (*Source*: Ash_Crow, under CC BY-SA 3.0, https://commons.wikimedia.org/wiki/File:Provinces_of_Japan-Awa_(Chiba).svg).[244]

| 60 | 常陸 | Chang Lu | Hitachi Province | Hitachi (常陸/常陆) was one of the fifteen provinces under the rule of the Eastern Sea Circuit/Tōkaidō (東海道/东海道). But on the Japan-KWQ, Hitachi (常陸/常陆) is mistakenly placed in the region of the Eastern Mountains Circuit/Tōsandō.

Fig. A1.48. Hitachi Province (*Source*: Ash_Crow, under CC BY-SA 3.0, https://commons.wikimedia.org/wiki/File:Provinces_of_Japan-Hitachi.svg).[245] |

Remark on the Eastern Sea Circuit/Tōkaidō (東海道/东海道)**:** Six provinces are missing on the Japan-KWQ. They are Shima (志摩; Item 46), Ise (伊勢/伊势; Item 47), Iga (伊賀/伊贺; Item 48), Owari (尾張/尾张; Item 49), Kai (甲斐; Item 54), and Sagami (相模; Item 55). In addition, Mikawa (三河; Item 50) is mistakenly placed in the region of the Northern Land Circuit /Hokurikudō, and the remaining eight provinces: Tōtōmi (遠江/远江; Item 51), Suruga (駿河/骏河; Item 52), Izu (伊豆; Item 53), Musashi (武藏; Item 56), Shimōsa (下總/下总; Item 57), Kazusa (上總/上总; Item 58), Awa (安房; Item 59) and Hitachi (常陸/常陆; Item 60) are erroneously placed in the region where the Eastern Mountains Circuit/Tōsandō should be.

| *7 | 東山道 | Dong Shan Dao | Eastern Mountains Circuit/Tōsandō (located in the central and northeastern parts of Honshu, along the northeast direction through the Japanese Alps and along Japan's Pacific coast)[246] | The Eastern Mountains Circuit/Tōsandō, literally, "eastern mountain circuit" or "eastern mountain region", is a Japanese geographical term. It means both an ancient division of the country and the main road running through it. The Eastern Mountain Circuit/Tōkaidō (東山道/东山道) was one of the seven circuits in the Gokishichidō ("five provinces and seven circuits"; 五畿七道) system. However, on the Japan-KWQ, the Eastern Mountains Circuit/Tōsandō is mostly depicted by mistake in the region where the Eastern Sea Circuit/Tōkaidō should be. There were eight provinces under the rule of the Eastern Mountain Circuit/Tōsandō (東山道/东山道). They are shown in Fig. 1.9 on a modern map: |

(*Continued*)

Item #	Name	Pinyin	Etymology	History, Geography and My Comments
				61. Ōmi Province (近江國/近江国):[247] the name of the province was set to "Ōmi" after the enactment and enforcement of the Taihō Code in 703 A.D. and the decree of 713 A.D.[248] 62. Mino Province (美濃國/美浓国):[249] the use of the *kanji* "美濃" is found in the *Gu Shi Ji* (古事記)/(古事记) or *Kojiki* (*Records of Ancient Matters*)[250] and became prevalent after the eighth century.[251] 63. Hida Province (飛驒國/飞驒国):[252] it existed as a political entity before the *Ritsuryō* system and the implementation of the Taihō Code of the Nara period. Its name was changed to Hida Province (飛驒國/飞驒国) in 702.[253] 64. Shinano Province (信濃國/信浓国):[254] after the Taika Era Reforms (大化改新 or 大化革新; took place in 645), it was established as a province.[255] 65. Kōzuke Province (上野國/上野国):[256] in 713, with the standardization of province names into two *kanji* characters (which are Chinese characters taken from the Chinese script, and used in the writing of Japanese), the ancient land of Keno or Kenu (毛野) became Kamitsuke (上野; later it changed to modern Kōzuke) and Shimozuke (下野).[257] 66. Shimotsuke Province (下野國/下野国):[258] see explanation in Item 65. 67. Mutsu Province (陸奧國/陆奥国): it was located near the edge of the Eastern Mountains Circuit/Tōsandō; it separated in the seventh century from Hitachi Province (常陸國/常陆国) to become Mutsu Province (陸奧國/陆奥国), and later became the largest province as it expanded northward.[259]

				68. Dewa Province (出羽國/出羽国):[260] it was promoted to the status of a province in 712 A.D.[261] **Fig. A1.49.** Eastern Mountains Circuit/Tōsandō (*Source*: 百楽兎, under CC BY-SA 3.0, https://commons.wikimedia.org/wiki/File:Tosando.svg).[262]
61	近江	Jin Jiang	Ōmi Province	Ōmi (近江) was one of the eight provinces under the rule of the Eastern Mountains Circuit/Tōsandō (東山道/东山道). However, on the Japan-KWQ, Ōmi (近江) is depicted too far away from the Kii Peninsula. **Fig. A1.50.** Ōmi Province (*Source*: Ash_Crow, under CC BY-SA 3.0, https://commons.wikimedia.org/wiki/File:Provinces_of_Japan-Omi.svg).[263]
62	美濃	Mei Nong	Mino Province	Mino (美濃/美浓) was one of the eight provinces under the rule of the Eastern Mountains Circuit/Tōsandō (東山道/东山道). On the Japan-KWQ, it is depicted roughly where it should be.

(*Continued*)

Item #	Name	Pinyin	Etymology	History, Geography and My Comments
				Fig. A1.51. Mino Province (*Source*: Ash_Crow, under CC BY-SA 3.0, File:Provinces_of_Japan-Mino.svg).[264]
63	飛彈[驒]	Fei Dan [Tuo]	Hida Province	Hida (飛騨/飞騨) was one of the eight provinces under the rule of the Eastern Mountains Circuit/Tōsandō (東山道/东山道). However, on the Japan-KWQ, it is translated into "飛彈" and mistakenly placed in the region of the Eastern Sea Circuit/Tōkaidō. Fig. A1.52. Hida Province (*Source*: Ash_Crow, under CC BY-SA 3.0, File:Provinces_of_Japan-Hida.svg).[265]
64	信濃	Xin Nong	Shinano Province	Shinano (信濃/信浓) was one of the eight provinces under the rule of the Eastern Mountains Circuit/Tōsandō (東山道/东山道). However, on the Japan-KWQ, it is mistakenly placed in the region of the Eastern Sea Circuit/Tōkaidō.

				Fig. A1.53. Shinano Province (*Source:* Ash_Crow, under CC BY-SA 3.0, https://commons.wikimedia.org/wiki/File:Provinces_of_Japan-Shinano.svg).[266]
65 & 66	上下野	Shang Xia Ye	Kōzuke and Shimotsuke Provinces	Kōzuke (上野) and Shimotsuke (下野) were two of the eight provinces under the rule of the Eastern Mountains Circuit/Tōsandō (東山道/东山道). However, on the KWQ-Japan, they are mistakenly depicted in the region of the Eastern Sea Circuit/Tōkaidō. **Fig. A1.54.** (a) Kōzuke Province (*Source:* Ash_Crow, under CC BY-SA 3.0, https://commons.wikimedia.org/wiki/File:Provinces_of_Japan-Kozuke.svg);[267] (b) Shimotsuke Province (*Source:* Ash_Crow, under CC BY-SA 3.0, File:Provinces_of_Japan-Shimotsuke.svg).[268]
67	陸奧	Lu Ao Guo	Mutsu Province	Mutsu (陸奧/陆奥) was one of the eight provinces under the rule of the Eastern Mountains Circuit/Tōsandō (東山道/东山道). However, on the KWQ-Japan, it is mistakenly placed in the region of the Eastern Sea Circuit/Tōkaidō.

(Continued)

Item #	Name	Pinyin	Etymology	History, Geography and My Comments
				Fig. A1.55. Mutsu Province (*Source*: Ash_Crow, under CC BY-SA 3.0, https://commons.wikimedia.org/wiki/File:Provinces_of_Japan-Mutsu.svg).[269]
68	出羽	Chu Yu	Dewa Province	Dewa (出羽) was one of the eight provinces under the rule of the Eastern Mountains Circuit/Tōsandō (東山道/东山道). However, on the KWQ-Japan, it is mistakenly depicted in the region of the Eastern Sea Circuit/Tōkaidō facing the Pacific Ocean. **Fig. A1.56.** Dewa Province (*Source*: Ash_Crow, under CC BY-SA 3.0, https://commons.wikimedia.org/wiki/File:Provinces_of_Japan-Dewa.svg).[270]

Remark on the Eastern Mountains Circuit/Tōsandō (東山道/东山道): The Japan-KWQ erroneously depicts most of the eight provinces under the rule of the Eastern Mountains Circuit/Tōsandō (東山道/东山道) in the region where the Eastern Sea Circuit/Tōkaidō (東海道/东海道) should be. The only exception is Mino (美濃/美浓) which is depicted roughly where it was.

Category II: Islands and seas surrounding the four main islands of Japan

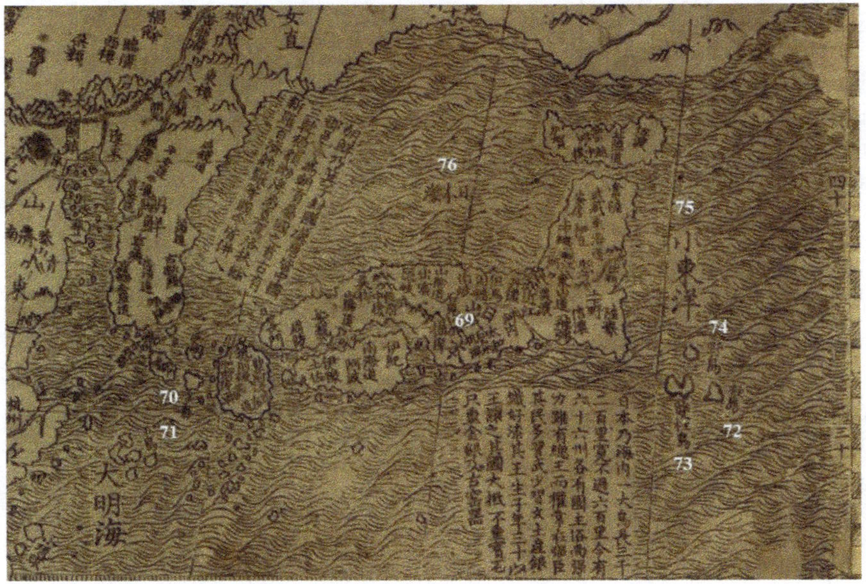

Fig. A1.57.   The map of the Japan-KWQ extracted from the KWQ (public domain).

| 69 | 日本 | | Ri Ben | The Japanese names for Japan are Nihon and Nippon. They are both written in Japanese using the *kanji* "日本". | Japan (日本) is an island country in East Asia. The ancient Japan had four main islands: Hokkaido, Honshu (the "mainland"), Shikoku, and Kyushu as discussed earlier.

The origin of the name Japan is not certain in the Japanese historical records. Historians say the Japanese called their country Yamato ("和" or "倭") in its early history. But the Chinese book *Xin Tangshu: Japan* (新唐書・日本傳/新唐书・日本传) or *New Book of Tang: Japan* records the following:[271]

咸亨元年, 遣使賀平高麗。後稍習夏音, 惡倭名, 更號日本。使者自言, 因近日所出, 以爲名。 |

(*Continued*)

Item #	Name	Pinyin	Etymology	History, Geography and My Comments
				My translation and comments (between square brackets) are as follows: In the first year [670 A.D.] of the Xianheng (咸亨) period [of Emperor Gaozong of Tang (唐高宗)], Japan sent envoys to congratulate the Tang Dynasty on the pacification of Goguryeo [高句麗/高句丽; 37 B.C.–668 A.D.; it was a Korean kingdom located in the northern and central parts of the Korean Peninsula and the southern and central parts of Northeast China]. Later, after they learned a little Chinese, they disliked the name "Yamato (和; it is 'Wa' or '倭')" [their country's name at the time], so they changed it to "日本 (Japan)". The envoys claimed that this was because their country's location was close to where the sun rose [日 means "sun" and 本 "source or origin"]. The earliest archaeological discovery of the name "日本 (Japan)" was an epitaph made in 678[272] and unearthed in 2011 in Xian, China. Hence, it can be said that the term "日本 (Japan)" was used no later than the late seventh century. *This is a solid proof of Tang-Japan contacts.*
70	多藝[多禰]	Duo Yi/Duo Mi	Tane Province; it only existed between 702 and 824 (Tane was annexed to Ōsumi Province).[273] On the Japan-KWQ, 多禰 is written as 多藝, perhaps due to their close pronunciation.	According to *Xu Ri Ben Ji* (続日本紀/続日本纪) or *Shoku Nihong/Chronicle of Japan Continued* (an imperially commissioned Japanese history text completed in 797; it came out after *Nihon Shoki/Chronicles of Japan*), Tane (多禰/多祢) Province was established in 702 and became one of the then twelve provinces under the rule of the Western Sea Circuit/Saikaidō; but it was annexed to Ōsumi Province in 824. Its territory was mainly in the present Tanegashima Island (種子島/种子岛) and Yakushima Island (屋久島/屋久岛), both belonging to the Ōsumi Islands (大隅諸島/大隅诸岛).

Tanegashima is traditionally known as the site of the introduction of European firearms to Japan in 1543 (some reports say 1542).[274] The two Portuguese traders, António Mota and Francisco Zeimoto (and possibly a third named António Peixoto), who had the first contact with Japan, should be recognised as the first Europeans to introduce firearms to the Japanese.

Yakushima Island (屋久島/屋久岛) was first mentioned in written documents of the Chinese Sui Dynasty in the sixth century, and often mentioned in the diaries of travellers during the Tang Dynasty (618–907; with an interregnum between 690 and 705). But contact with Europeans on Yakushima only began with the arrival of an Italian missionary called Giovanni Batista Sidotti in 1708. The Japanese politician and scholar Arai Hakuseki published the *Xi Yang Ji Wen* (西洋紀聞/西洋纪闻) or *Seiyō Kibun*/*Chronicle of the West* based on his conversations with Sidotti.[275]

Since the Old Japan Tane Province was incorporated into Ōsumi (大隅; Item 2) Province in 824, the total number of the provinces ruled under the Western Sea Circuit/Saikaidō was reduced to eleven.

In the main text of this chapter, the annotation reveals the political era of Japan in 897 A.D. By that time the Old Japan Tane (多禰/多祢; Item 70) Province no longer existed, but the islands which used to belong to Tane Province still existed and is depicted on the Japan-KWQ. The Japan-KWQ keeps the name of Tane (多禰/多祢) when these islands were already incorporated into Ōsumi (大隅; Item 2) Province after 824, but the cartographer omitted Ōsumi (大隅; Item 2) Province from the Japan-KWQ. Recall that Hyūga Province (日向國/日向国) broke off from Ōsumi Province (大隅國/大隅国) in 713. Since Ōsumi Province (大隅國/大隅国) is omitted

(*Continued*)

Item #	Name	Pinyin	Etymology	History, Geography and My Comments
				from the Japan-KWQ, Hyūga Province (日向國/日向国) is also omitted from the Japan-KWQ. (a)　　　　　(b) **Fig. A1.58.** (a) The Old Japan Tane Province (bottom left islands in red colour; *Source*: This file is lacking author information., under CC BY-SA 2.0, https://commons.wikimedia.org/wiki/File:Old_Japan_Tane.svg);[276] (b) Ōsumi Province (*Source*: Ash_Crow, under CC BY-SA 3.0, https://commons.wikimedia.org/wiki/File:Provinces_of_Japan-Osumi.svg).[277]
71	五島	Wu Dao	Gotō Islands[278]	Gotō Islands (五島列島/五岛列岛; literally: "five-island archipelago")[279] are Japanese islands in the East China Sea, off the western coast of Kyushu. Portuguese introduced Christianity to the islands in the late sixteenth century[280] and became the important historical roots of Christianity in Japan within the islands. However, neither Japan-Mercator (1569) nor Japan-Ortelius (1570) nor Japan-Plancius (1594) depicts Gotō Islands.
72	一柱島	Zhu Dao	Izu Ōshima[281]	On the Japan-KWQ, 一柱島/一柱岛 (31.0°N, 146.5°E; the KWQ latitudinal reading 31.5 *du* is already converted into the modern latitudinal reading 31.0° after multiplying it by 0.9856) is today's Izu Ōshima (伊豆大島/伊豆大岛; 34.7°N, 139.4°E). The island is an inhabited volcanic island in the Izu archipelago in the Philippine Sea, off the coast of Honshu.

				Izu Ōshima is the largest island in the Izu archipelago or the Izu Islands (伊豆諸島/伊豆诸岛).[282] The "Seven Islands of Izu" (伊豆七島/伊豆七岛) are the seven inhabited islands of the Izu Islands: from north to south they are Izu Ōshima, To-shima, Nii-jima, Kōzu-shima, Miyake-jima, Hachijō-kojima, and Mikura-jima, see Fig. A1.59. Neither Japan-Mercator (1569) nor Japan-Ortelius (1570) nor Japan-Plancius (1594) depicts Izu Ōshima Island. 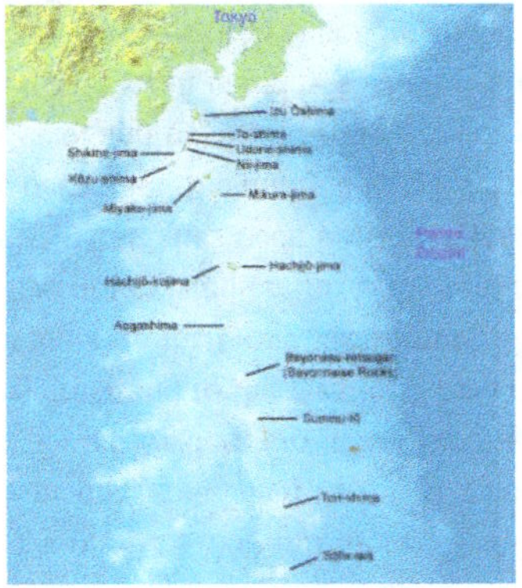 **Fig. A1.59.** Map of the Izu Islands; Izu Ōshima Island is at the top (public domain).[283]
73	雙柱島	Shuang Zhu Dao	Nii-jima	雙柱島/双柱岛 (Item 73) on the Japan-KWQ may be today's Nii-jima[284] (新島/新岛; see Fig. A1.59). The island is made of eight rhyolitic lava domes in two groups at the northern and southern ends of the island, separated by a low, flat isthmus, thus fitting the description of 雙柱/双柱, meaning "double pillars". Neither the Japan-Mercator (1569) nor the Japan-Ortelius (1570) nor the Japan-Plancius (1594) depicts Nii-jima.

(Continued)

Item #	Name	Pinyin	Etymology	History, Geography and My Comments
74	野島	Ye Dao	Ye Island	There is no island to the north of Izu Ōshima in Fig. A1.59, since Izu Ōshima, the largest of the islands, is also the northernmost among the islands in the Izu archipelago. Hence, Ye Island cannot be identified on a modern map. Neither the Japan-Mercator (1569) nor the Japan-Ortelius (1570) nor the Japan-Plancius (1594) depicts Ye Island (野島/野岛).
75	小東洋	Xiao Dong Yang	Small Eastern Ocean; it is today's Pacific Ocean.	小 means "small", 東 "east", and 洋 "ocean". 小東洋 means "Small Eastern Ocean" which is today's North Pacific Ocean; the Philippine Sea is a section of the western North Pacific Ocean, lying east and north of the Philippines. As a contrast, there is 大西洋 (Great Western Ocean) near Europe on the KWQ introduced by Matteo Ricci. 大 means "great" and 西 "west"; hence, 大西洋means "Great Western Ocean", meaning that the ocean was great and to the west of China or the ocean belonged to the Great Westerners. This term, 大西洋, had not existed on the Chinese maps before 1602. Only "西洋" has been used by the Chinese to mean the ocean and the many countries to the west (Indian Ocean and beyond) of their country. The Great Western Ocean is today's Atlantic Ocean. Neither Japan-Mercator (1569) nor Japan-Ortelius (1570) nor Japan-Plancius (1594) depicts the Small Eastern Ocean.
76	日本海	Ri Ben Hai	Sea of Japan; originally Jīng Hai (鯨海/鲸海, literally "Whale Sea") in China.[285] The sea area was also described as "the sea of Mangi", "the sea of Cin" or "the sea of China" on Western maps produced before 1602, the year when Matteo Ricci published the KWQ.[286]	日本 means "Japan". 日本海 refers to today's "Sea of Japan", see Fig. 1.8. The Japanese archipelago separates the Sea of Japan from the Pacific Ocean. The Sea of Japan is the marginal sea between the Japanese archipelago, Sakhalin, the Korean Peninsula, and the mainland of the Russian Far East. During the Northern and Southern Dynasties, the ethnic groups on the Korean Peninsula called the sea to the east of the Korean Peninsula the East Sea (Donghae; 東

海/东海). In 1602, Matteo Ricci was the first person to use the name "日本海 (Sea of Japan)". From then on, the Chinese place name – Whale Sea (鯨海/鯨海) or "the sea of Mangi" or "the sea of Cin" or "the sea of China" – that typified the Eastern world was no longer used. However, in ancient days, even the Japanese themselves used "佐渡外海 ('Sado Island Offshore' or 'Outer Sea of Sado Island')" to represent this sea area instead of the "Sea of Japan"; and the part of the sea area close to the northern part of the Tsushima Strait on the Korean Peninsula was called the "Sea of Korea", "East Sea" or "Sea of Choson" by Koreans.

It is apparent that in 1602 Matteo Ricci changed the Chinese name of this sea and used "日本海 (Sea of Japan)" to depict it.

Neither the Japan-Mercator (1569) nor the Japan-Ortelius (1570) nor the Japan-Plancius (1594) gives a name to this marginal sea between the Japanese archipelago, Sakhalin, the Korean Peninsula, and the mainland of the Russian Far East.

Remark on islands and seas surrounding the four main islands of Japan: Other than 野島/野岛 (Ye Island; Item 74; to the north of the Izu Ōshima), which cannot be identified on a modern map, the Japan-KWQ depicts major islands and seas surrounding the main islands of Japan correctly. Matteo Ricci was the first person to use the name "日本海 (Sea of Japan)" and from then on, the old Chinese name for this sea – Whale Sea (鯨海) or "the sea of Mangi" or "the sea of Cin" or "the sea of China" that typified the Eastern world – was no longer in use. In modern days, Sea of Japan is the dominant term used in English for this sea, and the name in most European languages is equivalent; but this sea is sometimes still called by different names in surrounding countries.

Endnotes

[1]Library of Congress. "Kun yu wan guo quan tu." www.loc.gov, Library of Congress, 2025, loc.gov/item/2010585650.

[2]Ronnie Po-Chia Hsia. *Matteo Ricci and the Catholic Mission to China, 1583–1610: A Short History with Documents (Passages: Key Moments in History)*. Indianapolis, USA: Hackett Publishing Company, Inc., 2016.

[3]Sheng-Wei Wang. *Chinese Global Exploration in the Pre-Columbian Era: Evidence from an Ancient World map*. Singapore: World Scientific, 2023.

[4]"File:Mappa di Eratostene.jpg." *Wikimedia Commons: The Free Media Repository*, Wikimedia Foundation, 16 Mar. 2023, upload.wikimedia.org/wikipedia/commons/e/e8/Mappa_di_Eratostene.jpg.

[5]"File:TabulaRogeriana upside-down.jpg." *Wikimedia Commons*, 4 Apr. 2023, upload.wikimedia.org/wikipedia/commons/a/a1/TabulaRogeriana_upside-down.jpg.

[6]Charles F. Montalembert. *Les Moines d'Occident depuis Saint Benoît jusqu'à Saint Bernard [The Monks of the West from Saint Benedict to Saint Bernard]*. Paris, France: J. Lecoffre, 1860.

[7]B. Atalay, *Divânu Lügati't-Türk Tercumesi, Vol. 1*, Ankara, Turkey: Turk Tarih Kurumu Basimevi (Turkish Historical Society Publishing House), 1940, p. 30.

[8]Sevim Tekeli. "The Oldest Map of Japan Drawn by Mahmud of Kashgar." https://muslimheritage.com/, Muslim Heritage, 3 Jan. 2007, https://muslimheritage.com/the-oldest-map-of-japan-drawn-by-mahmud-of-kashgar/#:~:text=Although%20the%20Japanese%20map%20was%20included%20for%20the,by%20Mahmud%20of%20Kashgar%20in%20early%2011th%20century.

[9]B. Atalay. *Op. cit.*

[10]Dan. "Mahmud al-Kashgari." www.danstopicals.com, Dan's topical stamps, 2025, https://www.danstopicals.com/kashgari.htm.

[11]"File:Kashgari map.jpg." *Wikimedia Commons*, 30 May 2022, upload.wikimedia.org/wikipedia/commons/b/bc/Kashgari_map.jpg.

[12]Marco Polo. Ronald Latham, trans. *The Travels of Marco Polo*. London, UK: Folio Society, 1958; Marco Polo. Nigel Cliff, ed., trans., and Intro. Coralie Bickford-Smith, lllust. *The Travels*. London, UK: Penguin Classics, 2015.

[13]Evelyn Edson. *The World Map, 1300–1492 (Illustrated edition)*, Maryland, USA: the Johns Hopkins University Press, 2007, p. 88.

[14]"File:DeVirgaDetail.jpg." *Wikimedia Commons*, 12 Dec. 2014, upload.wikimedia.org/wikipedia/commons/7/7b/DeVirgaDetail.jpg.

[15]"File:FramauroJava.jpg." *Wikimedia Common*, 13 May 2023, upload.wikimedia.org/wikipedia/commons/1/1d/FramauroJava.jpg.

[16]R. V. Tooley, *Maps and Map Makers*, London, UK: Bonanza Books, 1952, p. 25.

[17]Anna Jamieson. "Discover Japan Through These 6 Antique Maps." https://japanobjects.com/, Japan Objects, 2025, japanobjects.com/features/antique-maps.

[18]*Ibid.*

[19]Olof G. Lidin. *Tanegashima-The Arrival of Europe in Japan*. Copenhagen, Denmark: Nordic Institute of Asian Studies, 2002.

[20]Michael Cooper, *et al. The Southern Barbarians: The First Europeans in Japan*. Tokyo, Japan: Kodansha International/Sophia University, 1971.

[21]Diego Pacheco. "Xavier and Tanegashima." *Monumenta Nipponica*, vol. 29, no. 4, 1974, pp. 477–480. doi:10.2307/2383897.

[22]Giles Milton. *Samurai William: The Englishman Who Opened Japan*. London, UK: Penguin Books, 2003; William de Lange. *Pars Japonica: The First Dutch Expedition to Reach the Shores of Japan...Brought by the English Pilot Will Adams, Hero of Shogun*. Warren, CT, USA: Floating World Editions, 2006.

[23]"File:Atlas de Fernao Vaz Dourado (Asia).jpg." *Wikimedia Commons*, 23 Oct. 2021, commons.wikimedia. org/wiki/File:Atlas_de_Fernao_Vaz_Dourado_(Asia).jpg.

[24]Japan's governance was limited to Oshima Peninsula – the southernmost part of Hokkaido – until the seventeenth century; in 1869, following the Meiji Restoration, Ezo was annexed by Japan under on-going colonial practices, and renamed Hokkaido.

[25]Fr. Manuel Teixeira. "Count Moric Benyovszky: A Hungarian Cruzoe in Asia." *Asian Studies: Journal of Critical Perspectives on Asia*, vol. 4, no. 1, 1966, pp. 127–134.

[26]*Ibid.*

[27]*Ibid.*

[28]Anna Jamieson. *Op. cit.*

[29]Ong Siew Chey. *China Condensed: 5000 Years of History & Culture*, Singapore: Marshall Cavendish International Asia Pte Ltd, 2011, p. 17.

[30]Museum Staff. "About the Fukuoka City Museum." http://museum.city.fukuoka.jp/, The Fukuoka City Museum, 2025, museum.city.fukuoka.jp/en/exhibition.html.

[31]Ping Bu and Shinichi Kitaoka, eds. *The History of China-Japan Relations: From Ancient World to Modern International Order*. London, UK: Palgrave Macmillan, 2023.

[32]National Archives of Japan. "Joint project celebrating the 50th anniversary of Japan-Mongolia diplomatic relations. Relations between Japan and Mongolia in the 13th century." https://web.archive.org/, National Archives of Japan, 2025, www.archives.go.jp/about/activity/international/jp_mn50/english/ch01.html.

[33]Jacques Gernet. *A history of Chinese civilization (2nd, illustrated, revised ed.)*. Cambridge, UK: Cambridge University Press, 1996, p. 420.

[34]Morris Rossabi, ed. *Eurasian Influences on Yuan China*, Singapore: ISEAS Publishing, 2013, p. 131.

[35]Charles H. Hapgood. "Chapter 5: The ancient maps of the East and West." *Maps of the Ancient Sea Kings: Evidence of Advanced Civilization in the Ice Age*, Kempton, IL, USA: Adventures Unlimited Press, 1997, pp. 135–147.

[36]"File:華夷圖附注.jpg." *Wikimedia Commons*, 26 Mar. 2023, upload.wikimedia.org/wikipedia/commons/a/ac/華夷圖附注.jpg.

[37]Leo Bagrow. *History of Cartography*, New Jersey, USA: Transaction Publishers, 1963, p. 199.

[38]Mark Gamsa. *Manchuria – A Concise History*, London, UK: Bloomsbury, 2020, p. 155.

[39]"File:Da-ming-hun-yi-tu.jpg." *Wikimedia Commons*, 17 Oct. 2016, upload.wikimedia.org/wikipedia/commons/c/cd/Da-ming-hun-yi-tu.jpg.

[40]"File:Kunyu Wanguo Quantu by Matteo Ricci Plate 1-3.jpg." *Wikimedia Commons*, 23 Oct. 2020, commons.wikimedia.org/wiki/File:Kunyu_Wanguo_Quantu_by_Matteo_Ricci_Plate_1-3.jpg.

[41]"File:Mercator 1569 world map composite.jpg." *Wikimedia Commons*, 26 Nov. 2016, upload.wikimedia. org/wikipedia/commons/4/4b/Mercator_1569_world_map_composite.jpg.

[42]"File:OrteliusWorldMap1570.jpg." *Wikimedia Commons*, 12 July 2022, upload.wikimedia.org/wikipedia/commons/e/e2/OrteliusWorldMap1570.jpg.

[43]"File:1594 double hemisphere world map by Petrus Plancius.jpg." *Wikimedia Commons*, 2 Sept. 2022, upload.wikimedia.org/wikipedia/commons/0/0d/1594_double_hemisphere_world_map_by_Petrus_Plancius.jpg.

[44]"File:Tartary from Theatrum orbis terrarum, by Abraham Ortelius.jpg." *Wikimedia Commons*, 15 Feb. 2023, upload.wikimedia.org/wikipedia/commons/8/85/Tartary_from_Theatrum_orbis_terrarum%2C_by_Abraham_Ortelius.jpg.

[45]"File:Ortelius – Maris Pacifici 1589.jpg." *Wikimedia Commons*, 8 July 2012, upload.wikimedia.org/wikipedia/commons/4/4e/Ortelius_-_Maris_Pacifici_1589.jpg.

[46]"File:CEM-09-Asiae-Nova-Descriptio-China-2510.jpg." *Wikimedia Commons*, 16 June 2022, upload.wikimedia.org/wikipedia/commons/6/6a/CEM-09-Asiae-Nova-Descriptio-China-2510.jpg.

[47]Abraham Ortelius. *A New Description of Asia*. (1579 Map). Retrieved from the Library of Congress, www.loc.gov/item/2021668733/.

[48]"File:TsuShima Strait.png." *Wikimedia Commons*, 2 Oct. 2022, upload.wikimedia.org/wikipedia/commons/a/ab/TsuShima_Strait.png.

[49]M. Ramming. "The Evolution of Cartography in Japan." *Imago Mundi*, vol. 2, 1937, pp. 17–21. JSTOR, http://www.jstor.org/stable/1149827.

[50]Ian Johnson. *The Miracles of Francis Xavier*. Winnipeg, MB, Canada: Good News Fellowship Ministries, 2013.

[51]F. J. Manasek. *Mercator's World: The Magazine of Maps, Atlases, Globes, and Charts (Volume 2, No. 1)*. Piscataway, NJ, USA: Aster Publishing, 1997.

[52]Sheng-Wei Wang. *Chinese Global Exploration in the Pre-Columbian Era: Evidence from an Ancient World Map*, Singapore: World Scientific Publishing Co., pp. 5–6, 239–240.

[53]*Ibid*.

[54]Aljos Farjon and Denis Filer. *An Atlas of the World's Conifers: An Analysis of Their Distribution, Biogeography, Diversity and Conservation Status*, Leiden, Netherlands: Brill, 2013, p. 268.

[55]Distance calculator can be found at https://www.distance.to/Wakkanai/Hakodate/45.77630674696859, 142.7217974317134.

[56]Pia M. Jolliffe. "Forced Labour in Imperial Japan's First Colony: Hokkaidō." *The Asia-Pacific Journal*, vol. 18, iss. 20, no, 6, 2020. https://apjjf.org/2020/20/jolliffe.

[57]Philip Seaton. "Japanese Empire in Hokkaido." https://oxfordre.com, Oxford Research Encyclopedia of Asian History, 2017, oxfordre.com/asianhistory/display/10.1093/acrefore/9780190277727.001.0001/acrefore-9780190277727-e-76.

[58]Louis-Frédéric Nussbaum. Käthe Roth, trans. *Japan Encyclopedia*, Cambridge, MA, USA: Harvard University Press, 2002, p. 255. https://books.google.com.hk/books?id=p2QnPijAEmEC&pg=PA255&redir_esc=y#v=onepage&q&f=false.

[59]K. Asakawa. *The early institutional life of Japan: a study in the reform of 645 A.D.* London, UK: Forgotten Books, 2018, p. 324.

[60]Isaac Titsingh. *Annales des empereurs du Japon*. Charleston, South Carolina, USA: Nabu Press, 2011, p. 57, at Google Books. https://books.google.com.hk/books?id=18oNAAAAIAAJ&pg=PA57&redir_esc=y#v=onepage&q&f=false.

[61]Isaac Titsingh. *Op. cit.*, p. 65, at Google Books. https://books.google.com.hk/books?id=18oNAAAAIAAJ&pg=PA65&redir_esc=y#v=onepage&q&f=false.

[62]*Ibid*.

[63]Isaac Titsingh. *Op. cit.*, pp. 65–66, at Google Books. https://books.google.com.hk/books?id=18oNAAAAIAAJ&pg=PA65&redir_esc=y#v=onepage&q&f=false.

[64]Isaac Titsingh. *Op. cit.*, p. 66, at Google Books.
[65]Isaac Titsingh. *Op. cit.*, p. 57, at Google Books.
[66]*Ibid.*
[67]"File:Gokishichido Seven Circuits Japan Map.png." *Wikimedia Commons*, 7 July 2022, upload.wikimedia.org/wikipedia/commons/e/e6/Gokishichido_Seven_Circuits_Japan_Map.png
[68]K. Asakawa. *Op. cit.*
[69]Louis-Frédéric Nussbaum. Käthe Roth, trans. *Op. cit.*
[70]Japanese dictionary (日本国語大辞典). "宇多天皇." https://kotobank.jp/, 精選版日本国語大辞典 「宇多天皇」の意, 2025, kotobank.jp/word/宇多天皇-34727.
[71]Isaac Titsingh. *Annales des empereurs du japon*, Paris, France: The Oriental Translation Fund of Great Britain and Ireland, 1834, pp. 317–327. An electronic version can be found at https://books.google.com.hk/books?id=18oNAAAAIAAJ&pg=PA25&redir_esc=y#v=onepage&q&f=false.
[72]Richard A. B. Ponsonby-Fane. *The Imperial House of Japan*, Kyoto, Japan: Ponsonby Memorial Society, 1959, pp. 68–69.
[73]Joshua A. Fogel. *Articulating the Sinosphere: Sino-Japanese Relations in Space and Time.* Cambridge, MA, USA: Harvard University Press, 2009, pp. 102-107; Michael Hoffman. "Cultures Combined in the Mists of Time: Origins of the China-Japan relationship." *The Asia Pacific Journal*, vol. 4, no. 2, 16 Feb. 2006. https://apjjf.org/Michael-Hoffman/1911/article.html.
[74]Louis-Frédéric Nussbaum. Käthe Roth, trans. *Op. cit.*, p. 178, at Google Books.
[75]W. G. Beasley. *The Meiji Restoration.* Stanford, USA: Stanford University Press, 1972.
[76]Christopher Howe. *The Origins of Japanese Trade Supremacy: Development and Technology in Asia from 1540 to the Pacific War*, Chicago, US: University of Chicago Press, 1996, p. 337.
[77]This encyclopedia on Chinese history and literature can be found on the Internet at http://www.chinaknowledge.de/Literature/Science/shuowenjiezi.html.
[78]Ministry of Foreign Affairs of Japan. "The Thousand-Year Legacy and Evolution of the 'Six Ancient Kilns'." https://web-japan.org/, Web Japan, 2025, web-japan.org/trends/11_food/jfd202112_six-ancient-kilns.html.
[79]William E. Deal. *Handbook to Life in Medieval and Early Modern Japan.* Oxford, UK: Oxford University Press, 2005, p. 83, at Google Books. https://books.google.com.hk/books?id=i0ni1NmbYe0C&pg=PA83&redir_esc=y#v=onepage&q&f=false.
[80]Louis-Frédéric Nussbaum. Käthe Roth, trans. *Op. cit*, p. 829, at Google Books.
[81]Dao Ye Teng Hua (稻叶藤花). "日本令制国的名称由来是 (Origin of the names of the Ryōseikoku/Provinces of Japan)?" www.zhihu.com, Zhi Hu, 2025, zhihu.com/question/40789961.
[82]Louis-Frédéric Nussbaum. Käthe Roth, trans. *Op. cit.*, p. 762, at Google Books.
[83]Stephania Burke. "Hero's Journey for the Yamato Takeru Myth in Kojiki (712)." http://orias.berkeley.edu/, The Office of Resources for International and Area Studies (ORIAS), a unit of International and Area Studies (IAS) at the University of California, Berkeley, 2025, https://web.archive.org/web/20121005004012/http://orias.berkeley.edu/hero/yamato/plot_yamato.html.
[84]W. G. Aston. *Nihongi: Chronicles of Japan from the Earliest Times to A.D. 697 (translation edition)*, North Clarendon, VT, USA: Tuttle Publishing, 2011, p. xv.
[85]Dao Ye Teng Hua (稻叶藤花). *Op. cit.*
[86]Louis-Frédéric Nussbaum. Käthe Roth, trans. *Op. cit.*, p. 365, at Google Books. https://books.google.com.hk/books?id=p2QnPijAEmEC&pg=PA365&redir_esc=y#v=onepage&q&f=false.
[87]Dao Ye Teng Hua (稻叶藤花). *Op. cit.*
[88]Louis-Frédéric Nussbaum. Käthe Roth, trans. *Op. cit.*, p. 310, at Google Books.

[89]Dao Ye Teng Hua (稻叶藤花). *Op. cit.*

[90]Louis-Frédéric Nussbaum. Käthe Roth, trans. *Op. cit.,* p. 90, at Google Books.

[91]Dao Ye Teng Hua (稻叶藤花). *Op. cit.*

[92]Louis-Frédéric Nussbaum. Käthe Roth, trans. *Op. cit.,* p. 113, at Google Books. https://books.google.com. hk/books?id=p2QnPijAEmEC&pg=PA113&redir_esc=y#v=onepage&q&f=false.

[93]Dao Ye Teng Hua (稻叶藤花). *Op. cit.*

[94]Louis-Frédéric Nussbaum. Käthe Roth, trans. *Op. cit.,* 338, at Google Books. https://books.google.com.hk/ books?id=p2QnPijAEmEC&pg=PA338&redir_esc=y#v=onepage&q&f=false.

[95]Ross Bender. "Performative Loci of the Imperial Edicts in Nara Japan, 749-70." *Oral tradition*, vol. 24, no. 1, 2009, pp. 249–268.

[96]Louis-Frédéric Nussbaum. Käthe Roth, trans. *Op. cit.,* p. 114, at Google Books.

[97]Dao Ye Teng Hua (稻叶藤花). *Op. cit.*

[98]Louis-Frédéric Nussbaum. Käthe Roth, trans. *Op. cit.,* p. 96, at Google Books.

[99]Dao Ye Teng Hua (稻叶藤花). *Op. cit.*

[100]Louis-Frédéric Nussbaum. Käthe Roth, trans. *Op. cit.,* p. 379, at Google Books. https://books.google.com. hk/books?id=p2QnPijAEmEC&pg=PA379&redir_esc=y#v=onepage&q&f=false.

[101]Dao Ye Teng Hua (稻叶藤花). *Op. cit.*

[102]Louis-Frédéric Nussbaum. Käthe Roth, trans. *Op. cit.,* p. 780, at Google Books.

[103]Dao Ye Teng Hua (稻叶藤花).

[104]"File:Saikaido.svg." *Wikimedia Commons*, 15 Aug. 2021, upload.wikimedia.org/wikipedia/commons/ thumb/1/10/Saikaido.svg/2015px-Saikaido.svg.png.

[105]"File:Provinces of Japan-Satsuma.svg." *Wikimedia Commons*, 16 Mar. 2021, upload.wikimedia.org/ wikipedia/commons/thumb/b/b9/Provinces_of_Japan-Satsuma.svg/1960px-Provinces_of_Japan-Satsuma. svg.png.

[106]"File:Provinces of Japan-Hizen.svg." *Wikimedia Commons*, 17 Sept. 2020, upload.wikimedia.org/ wikipedia/commons/thumb/6/66/Provinces_of_Japan-Hizen.svg/1960px-Provinces_of_Japan-Hizen.svg.png.

[107]"File:Provinces of Japan-Higo.svg." *Wikimedia Commons*, 17 Sept. 2020, upload.wikimedia.org/wikipedia/ commons/thumb/d/d9/Provinces_of_Japan-Higo.svg/1960px-Provinces_of_Japan-Higo.svg.png.

[108]"File:Provinces of Japan-Buzen.svg." *Wikimedia Commons*, 7 Sept. 2020, upload.wikimedia.org/wikipedia/ commons/thumb/6/6d/Provinces_of_Japan-Buzen.svg/1960px-Provinces_of_Japan-Buzen.svg.png.

[109]"File:Provinces of Japan-Bungo.svg." *Wikimedia Commons*, 17 Sept. 2020, upload.wikimedia.org/ wikipedia/commons/thumb/a/ad/Provinces_of_Japan-Bungo.svg/1960px-Provinces_of_Japan-Bungo.svg.png.

[110]"File:Provinces of Japan-Chikuzen.svg." *Wikimedia Commons*, 19 Sept. 2020, upload.wikimedia.org/ wikipedia/commons/thumb/a/a6/Provinces_of_Japan-Chikuzen.svg/1960px-Provinces_of_Japan-Chikuzen. svg.png.

[111]"File:Provinces of Japan-Chikugo.svg." *Wikimedia Commons*, 16 Sept. 2020, upload.wikimedia.org/ wikipedia/commons/thumb/c/cc/Provinces_of_Japan-Chikugo.svg/1960px-Provinces_of_Japan-Chikugo. svg.png.

[112]"File:Old Japan Iki.svg." *Wikimedia Commons*, 17 Oct. 2020, upload.wikimedia.org/wikipedia/commons/ thumb/b/bb/Old_Japan_Iki.svg/2020px-Old_Japan_Iki.svg.png.

[113]"File:Provinces of Japan-Tsushima.svg." *Wikimedia Commons*, 14 Sept. 2020, upload.wikimedia.org/ wikipedia/commons/thumb/2/2f/Provinces_of_Japan-Tsushima.svg/1960px-Provinces_of_Japan-Tsushima. svg.png.

[114]Isaac Titsingh. *Op. cit.,* p. 65. https://books.google.com.hk/books?id=18oNAAAAIAAJ&pg=PA65&redir_esc=y#v=onepage&q&f=false.

[115]Louis-Frédéric Nussbaum. Käthe Roth, trans. *Op. cit.*, p. 684, at Google Books.
[116]Dao Ye Teng Hua (稻叶藤花). *Op. cit.*
[117]Louis-Frédéric Nussbaum. Käthe Roth, trans. *Op. cit.*, p. 916, at Google Books.
[118]Dao Ye Teng Hua (稻叶藤花). *Op. cit.*
[119]Louis-Frédéric Nussbaum. Käthe Roth, trans. *Op. cit.*, p. 18, at Google Books.
[120]Louis-Frédéric Nussbaum. Käthe Roth, trans. *Op. cit.*, p. 76, at Google Books.
[121]Dao Ye Teng Hua (稻叶藤花). *Op. cit.*
[122]Louis-Frédéric Nussbaum. Käthe Roth, trans. *Op. cit.*, p. 77, at Google Books.
[123]Dao Ye Teng Hua (稻叶藤花). *Op. cit.*
[124]Louis-Frédéric Nussbaum. Käthe Roth, trans. *Op. cit.*, p. 631, at Google Books.
[125]Dao Ye Teng Hua (稻叶藤花). *Op. cit.*
[126]Louis-Frédéric Nussbaum. Käthe Roth, trans. *Op. cit.*, p. 78, at Google Books.
[127]Dao Ye Teng Hua (稻叶藤花). *Op. cit.*
[128]Louis-Frédéric Nussbaum. Käthe Roth, trans. *Op. cit.*, p. 290, at Google Books.
[129]"File:Sanyodo.svg." *Wikimedia Commons*, 15 Aug. 2021, upload.wikimedia.org/wikipedia/commons/thumb/8/86/Sanyodo.svg/2015px-Sanyodo.svg.png.
[130]"File:Provinces of Japan-Nagato.svg." *Wikimedia Commons*, 17 Sept. 2020, upload.wikimedia.org/wikipedia/commons/thumb/8/87/Provinces_of_Japan-Nagato.svg/1960px-Provinces_of_Japan-Nagato.svg.png.
[131]"File:Provinces of Japan-Suo.svg." *Wikimedia Commons*, 8 Sept. 2020, upload.wikimedia.org/wikipedia/commons/thumb/1/1f/Provinces_of_Japan-Suo.svg/1960px-Provinces_of_Japan-Suo.svg.png.
[132]"File:Provinces of Japan-Aki.svg." *Wikimedia Commons*, 10 Sept. 2020, upload.wikimedia.org/wikipedia/commons/thumb/0/0f/Provinces_of_Japan-Aki.svg/1960px-Provinces_of_Japan-Aki.svg.png.
[133]"File:Provinces of Japan-Mimasaka.svg." *Wikimedia Commons*, 27 July 2021, upload.wikimedia.org/wikipedia/commons/thumb/d/d6/Provinces_of_Japan-Mimasaka.svg/1960px-Provinces_of_Japan-Mimasaka.svg.png.
[134]"File:Provinces of Japan-Bizen.svg." *Wikimedia Commons*, 12 Sept. 2020, upload.wikimedia.org/wikipedia/commons/thumb/f/f7/Provinces_of_Japan-Bizen.svg/1960px-Provinces_of_Japan-Bizen.svg.png.
[135]"File:Provinces of Japan-Bitchu.svg." *Wikimedia Commons*, 9 Sept. 2020, upload.wikimedia.org/wikipedia/commons/thumb/f/f4/Provinces_of_Japan-Bitchu.svg/760px-Provinces_of_Japan-Bitchu.svg.png.
[136]"File:Provinces of Japan-Bingo.svg." *Wikimedia Commons*, 14 Sept. 2020, upload.wikimedia.org/wikipedia/commons/thumb/a/af/Provinces_of_Japan-Bingo.svg/1960px-Provinces_of_Japan-Bingo.svg.png.
[137]"File:Provinces of Japan-Harima.svg." *Wikimedia Commons*, 14 Sept. 2020, upload.wikimedia.org/wikipedia/commons/thumb/1/15/Provinces_of_Japan-Harima.svg/760px-Provinces_of_Japan-Harima.svg.png.
[138]Isaac Titsingh. *Op. cit.,* p. 65, at Google Books.
[139]Louis-Frédéric Nussbaum. Käthe Roth, trans. *Op. cit.*, p. 408, at Google Books. https://books.google.com.hk/books?id=p2QnPijAEmEC&pg=PA408&redir_esc=y#v=onepage&q&f=false.
[140]Dao Ye Teng Hua (稻叶藤花). *Op. cit.*
[141]Louis-Frédéric Nussbaum. Käthe Roth, trans. *Op. cit.*, pp. 385 and 412, at Google Books. https://books.google.com.hk/books?id=p2QnPijAEmEC&pg=PA412&redir_esc=y#v=onepage&q&f=false.
[142]Dao Ye Teng Hua (稻叶藤花).
[143]Louis-Frédéric Nussbaum. Käthe Roth, trans. *Op. cit.*, p. 411, at Google Books. https://books.google.com.hk/books?id=p2QnPijAEmEC&pg=PA411&redir_esc=y#v=onepage&q&f=false.
[144]Dao Ye Teng Hua (稻叶藤花). *Op. cit.*

[145]Louis-Frédéric Nussbaum. Käthe Roth, trans. *Op. cit.*, p. 62, at Google Books.

[146]Dao Ye Teng Hua (稻叶藤花). *Op. cit.*

[147]Louis-Frédéric Nussbaum. Käthe Roth, trans. *Op. cit.*, pp. 385 and 411, at Google Books.

[148]Dao Ye Teng Hua (稻叶藤花). *Op. cit.*

[149]Louis-Frédéric Nussbaum. Käthe Roth, trans. *Op. cit.*, p. 411, at Google Books. https://books.google.com.hk/books?id=p2QnPijAEmEC&pg=PA411&redir_esc=y#v=onepage&q&f=false.

[150]*Ibid.*

[151]Dao Ye Teng Hua (稻叶藤花). *Op. cit.*

[152]Louis-Frédéric Nussbaum. Käthe Roth, trans. *Op. cit.*, p. 943, at Google Books.

[153]"File:Sanindo.svg." *Wikimedia Commons*, 15 Aug. 2021, upload.wikimedia.org/wikipedia/commons/thumb/c/c0/Sanindo.svg/2015px-Sanindo.svg.png.

[154]"File:Provinces of Japan-Iwami.svg." *Wikimedia Commons*, 6 Sept. 2020, upload.wikimedia.org/wikipedia/commons/thumb/9/94/Provinces_of_Japan-Iwami.svg/1960px-Provinces_of_Japan-Iwami.svg.png.

[155]"File:Provinces of Japan-Izumo.svg." *Wikimedia Commons*, 9 Sept. 2020, upload.wikimedia.org/wikipedia/commons/thumb/9/97/Provinces_of_Japan-Izumo.svg/1960px-Provinces_of_Japan-Izumo.svg.png.

[156]"File:Provinces of Japan-Hoki.svg." *Wikimedia Commons*, 9 Aug. 2020, upload.wikimedia.org/wikipedia/commons/thumb/d/da/Provinces_of_Japan-Hoki.svg/1960px-Provinces_of_Japan-Hoki.svg.png.

[157]"File:Provinces of Japan-Oki.svg." *Wikimedia Commons*, 19 Sept. 2020, upload.wikimedia.org/wikipedia/commons/thumb/5/55/Provinces_of_Japan-Oki.svg/1960px-Provinces_of_Japan-Oki.svg.png.

[158]"File:Provinces of Japan-Inaba.svg." *Wikimedia Commons*, 16 March 2021, upload.wikimedia.org/wikipedia/commons/thumb/3/39/Provinces_of_Japan-Inaba.svg/1960px-Provinces_of_Japan-Inaba.svg.png.

[159]"File:Provinces of Japan-Tajima.svg." *Wikimedia Commons*, 15 Sept. 2020, upload.wikimedia.org/wikipedia/commons/thumb/d/d2/Provinces_of_Japan-Tajima.svg/1960px-Provinces_of_Japan-Tajima.svg.png.

[160]"File:Provinces of Japan-Tango.svg." *Wikimedia Commons*, 19 Sept. 2020, upload.wikimedia.org/wikipedia/commons/thumb/f/f3/Provinces_of_Japan-Tango.svg/1960px-Provinces_of_Japan-Tango.svg.png.

[161]Isaac Titsingh. *Op. cit.*, pp. 65–66, at Google Books. https://books.google.com.hk/books?id=18oNAAAAIAAJ&pg=PA65&redir_esc=y#v=onepage&q&f=false.

[162]Louis-Frédéric Nussbaum. Käthe Roth, trans. *Op. cit.*, p. 988, at Google Books.

[163]K. Asakawa. *Op. cit.*, p. 324.

[164]Louis-Frédéric Nussbaum. Käthe Roth, trans. *Op. cit.*, p. 988, at Google Books.

[165]Louis-Frédéric Nussbaum. Käthe Roth, trans. *Op. cit.*, p. 62, at Google Books.

[166]Dao Ye Teng Hua (稻叶藤花). *Op. cit.*

[167]Louis-Frédéric Nussbaum. Käthe Roth, trans. *Op. cit.*, p. 988, at Google Books.

[168]Louis-Frédéric Nussbaum. Käthe Roth, trans. *Op. cit.*, p. 61, at Google Books.

[169]Dao Ye Teng Hua (稻叶藤花). *Op. cit.*

[170]Louis-Frédéric Nussbaum. Käthe Roth, trans. *Op. cit.*, p. 515, at Google Books.

[171]Dao Ye Teng Hua (稻叶藤花). *Op. cit.*

[172]"File:Nankaido.svg." *Wikimedia Commons*, 9 Dec. 2022, upload.wikimedia.org/wikipedia/commons/thumb/8/8e/Nankaido.svg/2015px-Nankaido.svg.png

[173]"File:Provinces of Japan-Iyo.svg." *Wikimedia Commons*, 12 Sept. 2020, upload.wikimedia.org/wikipedia/commons/thumb/4/4c/Provinces_of_Japan-Iyo.svg/1960px-Provinces_of_Japan-Iyo.svg.png.

[174]"File:Provinces of Japan-Tosa.svg." *Wikimedia Commons*, 13 Sept. 2020, upload.wikimedia.org/wikipedia/commons/thumb/8/87/Provinces_of_Japan-Tosa.svg/1960px-Provinces_of_Japan-Tosa.svg.png.

[175]"File:Provinces of Japan-Awa (Tokushima).svg." *Wikimedia Commons*, 11 Sept. 2020, upload.wikimedia. org/wikipedia/commons/thumb/5/51/Provinces_of_Japan-Awa_(Tokushima).svg/1960px-Provinces_of_ Japan-Awa_(Tokushima).svg.png.

[176]"File:Provinces of Japan-Awaji.svg." *Wikimedia Commons*, 12 Sept. 2020, upload.wikimedia.org/ wikipedia/commons/thumb/2/23/Provinces_of_Japan-Awaji.svg/1960px-Provinces_of_Japan-Awaji.svg.png.

[177]"File:Provinces of Japan-Kii.svg." *Wikimedia Commons*, 7 Sept. 2020, upload.wikimedia.org/wikipedia/ commons/thumb/1/10/Provinces_of_Japan-Kii.svg/1960px-Provinces_of_Japan-Kii.svg.png.

[178]Louis-Frédéric Nussbaum. Käthe Roth, trans. *Op. cit.*, p. 1046, at Google Books.

[179]Dao Ye Teng Hua (稻叶藤花). *Op. cit.*

[180]"File:Provinces of Japan-Yamato.svg." *Wikimedia Commons*, 8 Aug. 2021, upload.wikimedia.org/wikipedia/ commons/thumb/8/86/Provinces_of_Japan-Yamato.svg/1960px-Provinces_of_Japan-Yamato.svg.png.

[181]Louis-Frédéric Nussbaum. Käthe Roth, trans. *Op. cit.*, p. 496, at Google Books.

[182]Donald H. Shively and William H. McCullough. *The Cambridge History of Japan*, Cambridge, UK: Cambridge University Press, 1999, p. 453, at Google Books. https://books.google.com.hk/books?id=eiTW WfoyuyAC&pg=PA453&dq=&redir_esc=y#v=onepage&q&f=false.

[183]"File:Provinces of Japan-Kawachi.svg." *Wikimedia Commons*, 13 Sept. 2020, upload.wikimedia.org/ wikipedia/commons/thumb/6/63/Provinces_of_Japan-Kawachi.svg/760px-Provinces_of_Japan-Kawachi. svg.png.

[184]Louis-Frédéric Nussbaum. Käthe Roth, trans. *Op. cit.*, p. 1045, at Google Books.

[185]Dao Ye Teng Hua (稻叶藤花). *Op. cit.*

[186]"File:Provinces of Japan-Yamashiro.svg." *Wikimedia Commons*, 12 Sept. 2020, upload.wikimedia.org/ wikipedia/commons/thumb/3/3e/Provinces_of_Japan-Yamashiro.svg/760px-Provinces_of_Japan- Yamashiro.svg.png.

[187]Louis-Frédéric Nussbaum. Käthe Roth, trans. *Op. cit.*, p. 411, at Google Books. https://books.google.com. hk/books?id=p2QnPijAEmEC&pg=PA411&redir_esc=y#v=onepage&q&f=false.

[188]Dao Ye Teng Hua (稻叶藤花). *Op. cit.*

[189]"File:Provinces of Japan-Izumi.svg." *Wikimedia Commons*, 19 Sept. 2020, upload.wikimedia.org/ wikipedia/commons/thumb/a/a9/Provinces_of_Japan-Izumi.svg/1960px-Provinces_of_Japan-Izumi.svg.png.

[190]Louis-Frédéric Nussbaum. Käthe Roth, trans. *Op. cit.*, p. 846, at Google Books.

[191]"File:Provinces of Japan-Settsu.svg." *Wikimedia Commons*, 12 Sept. 2020, upload.wikimedia.org/ wikipedia/commons/thumb/4/42/Provinces_of_Japan-Settsu.svg/1960px-Provinces_of_Japan-Settsu.svg.png.

[192]Isaac Titsingh. *Op. cit.*, p. 66, at Google Books. https://books.google.com.hk/books?id=18oNAAAAIAA J&pg=PA66&redir_esc=y#v=onepage&q&f=false.

[193]Louis-Frédéric Nussbaum. Käthe Roth, trans. *Op. cit.*, p. 165, at Google Books.

[194]Dao Ye Teng Hua (稻叶藤花). *Op. cit.*

[195]Louis-Frédéric Nussbaum. Käthe Roth, trans. *Op. cit.*, p. 1025, at Google Books.

[196]Louis-Frédéric Nussbaum. Käthe Roth, trans. *Op. cit.*, p. 445, at Google Books. https://books.google.com. hk/books?id=p2QnPijAEmEC&pg=PA445&redir_esc=y#v=onepage&q&f=false.

[197]Dao Ye Teng Hua (稻叶藤花). *Op. cit.*

[198]Louis-Frédéric Nussbaum. Käthe Roth, trans. *Op. cit.*, p. 728, at Google Books.

[199]*Ibid.*

[200]Dao Ye Teng Hua (稻叶藤花).

[201]Louis-Frédéric Nussbaum. Käthe Roth, trans. *Op. cit.*, p. 164, at Google Books.

[202]Louis-Frédéric Nussbaum. Käthe Roth, trans. *Op. cit.*, p. 803, at Google Books.

[203]Dao Ye Teng Hua (稻叶藤花). *Op. cit.*

[204]"File:Hokurikudo.svg." *Wikimedia Commons*, 15 Aug. 2021, upload.wikimedia.org/wikipedia/commons/thumb/9/9c/Hokurikudo.svg/2015px-Hokurikudo.svg.png.

[205]"File:Provinces of Japan-Kaga.svg." *Wikimedia Commons*, 14 Sept. 2020, upload.wikimedia.org/wikipedia/commons/thumb/b/b5/Provinces_of_Japan-Kaga.svg/1960px-Provinces_of_Japan-Kaga.svg.png.

[206]"File:Provinces of Japan-Etchu.svg." *Wikimedia Commons*, 9 Sept. 2020, upload.wikimedia.org/wikipedia/commons/thumb/6/63/Provinces_of_Japan-Etchu.svg/1960px-Provinces_of_Japan-Etchu.svg.png.

[207]"File:Provinces of Japan-Echigo.svg." *Wikimedia Commons*, 11 Sept. 2020, upload.wikimedia.org/wikipedia/commons/thumb/0/0e/Provinces_of_Japan-Echigo.svg/1960px-Provinces_of_Japan-Echigo.svg.png.

[208]"File:Provinces of Japan-Noto.svg." *Wikimedia Commons*, 18 Sept. 2020, upload.wikimedia.org/wikipedia/commons/thumb/a/a8/Provinces_of_Japan-Noto.svg/1960px-Provinces_of_Japan-Noto.svg.png.

[209]"File:Provinces of Japan-Sado.svg." *Wikimedia Commons*, 17 Sept. 2020, upload.wikimedia.org/wikipedia/commons/thumb/9/9a/Provinces_of_Japan-Sado.svg/1960px-Provinces_of_Japan-Sado.svg.png.

[210]Isaac Titsingh. *Op. cit.*, p. 57. https://books.google.com.hk/books?id=18oNAAAAIAAJ&pg=PA57&redir_esc=y#v=onepage&q&f=false.

[211]Louis-Frédéric Nussbaum. *Op. cit.*, p. 857.

[212]Dao Ye Teng Hua (稻叶藤花). *Op. cit.*

[213]Louis-Frédéric Nussbaum. Käthe Roth, trans. *Op. cit.*, p. 395, at Google Books. https://books.google.com.hk/books?id=p2QnPijAEmEC&pg=PA395&redir_esc=y#v=onepage&q&f=false.

[214]Louis-Frédéric Nussbaum. Käthe Roth, trans. *Op. cit.*, p. 373, at Google Books. https://books.google.com.hk/books?id=p2QnPijAEmEC&pg=PA373&redir_esc=y#v=onepage&q&f=false.

[215]Dao Ye Teng Hua (稻叶藤花). *Op. cit.*

[216]Louis-Frédéric Nussbaum. Käthe Roth, trans. *Op. cit.*, p. 629, at Google Books.

[217]Dao Ye Teng Hua (稻叶藤花). *Op. cit.*

[218]Louis-Frédéric Nussbaum. Käthe Roth, trans. *Op. cit.*, p. 629, at Google Books.

[219]Louis-Frédéric Nussbaum. Käthe Roth, trans. *Op. cit.*, p. 924, at Google Books.

[220]Louis-Frédéric Nussbaum. Käthe Roth, trans. *Op. cit.*, p. 990, at Google Books.

[221]Louis-Frédéric Nussbaum. Käthe Roth, trans. *Op. cit.*, p. 916, at Google Books.

[222]Louis-Frédéric Nussbaum. Käthe Roth, trans. *Op. cit.*, p. 411, at Google Books. https://books.google.com.hk/books?id=p2QnPijAEmEC&pg=PA411&redir_esc=y#v=onepage&q&f=false.

[223]Dao Ye Teng Hua (稻叶藤花). *Op. cit.*

[224]Louis-Frédéric Nussbaum. Käthe Roth, trans. *Op. cit.*, p. 448, at Google Books. https://books.google.com.hk/books?id=p2QnPijAEmEC&pg=PA448&dq=&redir_esc=y#v=onepage&q&f=false.

[225]Louis-Frédéric Nussbaum. Käthe Roth, trans. *Op. cit.*, pp. 466–467, at Google Books.

[226]Dao Ye Teng Hua (稻叶藤花). *Op. cit.*

[227]Louis-Frédéric Nussbaum. Käthe Roth, trans. *Op. cit.*, pp. 669–671, at Google Books.

[228]Dao Ye Teng Hua (稻叶藤花). *Op. cit.*

[229]Louis-Frédéric Nussbaum. Käthe Roth, trans. *Op. cit.*, p. 862, at Google Books.

[230]Dao Ye Teng Hua (稻叶藤花). *Op. cit.*

[231]Louis-Frédéric Nussbaum. Käthe Roth, trans. *Op. cit.*, p. 502, at Google Books.

[232]Louis-Frédéric Nussbaum. Käthe Roth, trans. *Op. cit.*, p. 62, at Google Books.

[233]Dao Ye Teng Hua (稻叶藤花). *Op. cit.*

[234]Louis-Frédéric Nussbaum. Käthe Roth, trans. *Op. cit.*, p. 336, at Google Books.

[235]Dao Ye Teng Hua (稻叶藤花). *Op. cit.*

[236]"File:Tokaido.svg." *Wikimedia Commons*, 30 Dec. 2021, upload.wikimedia.org/wikipedia/commons/thumb/0/05/Tokaido.svg/2015px-Tokaido.svg.png.

237"File:Provinces of Japan-Mikawa.svg." *Wikimedia Commons*, 8 Sept. 2020, upload.wikimedia.org/wikipedia/commons/thumb/e/e1/Provinces_of_Japan-Mikawa.svg/1960px-Provinces_of_Japan-Mikawa.svg.png.

238"File:Provinces of Japan-Totomi.svg." *Wikimedia Commons*, 13 Sept. 2020, upload.wikimedia.org/wikipedia/commons/thumb/0/08/Provinces_of_Japan-Totomi.svg/1960px-Provinces_of_Japan-Totomi.svg.png.

239"File:Provinces of Japan-Suruga.svg." *Wikimedia Commons*, 9 Sept. 2020, upload.wikimedia.org/wikipedia/commons/thumb/5/54/Provinces_of_Japan-Suruga.svg/760px-Provinces_of_Japan-Suruga.svg.png.

240"File:Provinces of Japan-Izu.svg." *Wikimedia Commons*, 18 Sept. 2020, upload.wikimedia.org/wikipedia/commons/thumb/4/49/Provinces_of_Japan-Izu.svg/1960px-Provinces_of_Japan-Izu.svg.png.

241"File:Provinces of Japan-Musashi.svg." *Wikimedia Commons*, 16 Sept. 2020, upload.wikimedia.org/wikipedia/commons/thumb/e/e0/Provinces_of_Japan-Musashi.svg/1960px-Provinces_of_Japan-Musashi.svg.png.

242"File:Provinces of Japan-Kazusa.svg." *Wikimedia Commons*, 19 Sept. 2020, upload.wikimedia.org/wikipedia/commons/thumb/7/77/Provinces_of_Japan-Kazusa.svg/1960px-Provinces_of_Japan-Kazusa.svg.png.

243"File:Provinces of Japan-Shimosa.svg." *Wikimedia Commons*, 6 Sept. 2020, upload.wikimedia.org/wikipedia/commons/thumb/b/bd/Provinces_of_Japan-Shimosa.svg/1960px-Provinces_of_Japan-Shimosa.svg.png.

244"File:Provinces of Japan-Awa (Chiba).svg." *Wikimedia Commons*, 6 Aug. 2021, upload.wikimedia.org/wikipedia/commons/thumb/2/25/Provinces_of_Japan-Awa_(Chiba).svg/1960px-Provinces_of_Japan-Awa_(Chiba).svg.png.

245"File:Provinces of Japan-Hitachi.svg." *Wikimedia Commons*, 18 Sept. 2020, upload.wikimedia.org/wikipedia/commons/thumb/8/80/Provinces_of_Japan-Hitachi.svg/1960px-Provinces_of_Japan-Hitachi.svg.png.

246Isaac Titsingh. *Op. cit.*, p. 57, at Google Books. https://books.google.com.hk/books?id=18oNAAAAIAAJ&pg=PA57&redir_esc=y#v=onepage&q&f=false.

247Louis-Frédéric Nussbaum. Käthe Roth, trans. *Op. cit.*, p. 750, at Google Books.

248Dao Ye Teng Hua (稻叶藤花). *Op. cit.*

249Louis-Frédéric Nussbaum. Käthe Roth, trans. *Op. cit.*

250Jaroslav Průšek and Zbigniew Słupski, eds. *Dictionary of Oriental Literatures 1: East Asia.* Oxfordshire, England, UK: Routledge, 2006.

251Dao Ye Teng Hua (稻叶藤花). *Op. cit.*

252Louis-Frédéric Nussbaum. Käthe Roth, trans. *Op. cit.*, p. 307, at Google Books.

253Dao Ye Teng Hua (稻叶藤花). *Op. cit.*

254Louis-Frédéric Nussbaum. Käthe Roth, trans. *Op. cit.*, p. 863, at Google Books.

255Dao Ye Teng Hua (稻叶藤花). *Op. cit.*

256Louis-Frédéric Nussbaum. Käthe Roth, trans. *Op. cit.*, p. 990, at Google Books.

257Dao Ye Teng Hua (稻叶藤花). *Op. cit.*

258Louis-Frédéric Nussbaum. Käthe Roth, trans. *Op. cit.*

259Dao Ye Teng Hua (稻叶藤花). *Op. cit.*

260Louis-Frédéric Nussbaum. Käthe Roth, trans. *Op. cit.*

261Dao Ye Teng Hua (稻叶藤花). *Op. cit.*

262"File:Tosando.svg." *Wikimedia Commons*, 15 Aug. 2021, upload.wikimedia.org/wikipedia/commons/thumb/3/30/Tosando.svg/2015px-Tosando.svg.png.

[263]"File:Provinces of Japan-Omi.svg." *Wikimedia Commons*, 9 Sept. 2020, upload.wikimedia.org/wikipedia/commons/thumb/9/95/Provinces_of_Japan-Omi.svg/1960px-Provinces_of_Japan-Omi.svg.png.

[264]"File:Provinces of Japan-Mino.svg." *Wikimedia Commons*, 15 Sept. 2020, upload.wikimedia.org/wikipedia/commons/thumb/8/88/Provinces_of_Japan-Mino.svg/760px-Provinces_of_Japan-Mino.svg.png.

[265]"File:Provinces of Japan-Hida.svg." *Wikimedia Commons*, 7 Sept. 2020, upload.wikimedia.org/wikipedia/commons/thumb/f/fd/Provinces_of_Japan-Hida.svg/1960px-Provinces_of_Japan-Hida.svg.png.

[266]"File:Provinces of Japan-Shinano.svg." Wikimedia Commons, 8 Sept. 2020, upload.wikimedia.org/wikipedia/commons/thumb/3/3c/Provinces_of_Japan-Shinano.svg/1960px-Provinces_of_Japan-Shinano.svg.png.

[267]"File:Provinces of Japan-Kozuke.svg." *Wikimedia Commons*, 14 Sept. 2020, upload.wikimedia.org/wikipedia/commons/thumb/9/9a/Provinces_of_Japan-Kozuke.svg/1960px-Provinces_of_Japan-Kozuke.svg.png.

[268]"File:Provinces of Japan-Shimotsuke.svg." *Wikimedia Commons*, 17 Sept. 2020, upload.wikimedia.org/wikipedia/commons/thumb/c/c2/Provinces_of_Japan-Shimotsuke.svg/1960px-Provinces_of_Japan-Shimotsuke.svg.png.

[269]"File:Provinces of Japan-Mutsu.svg." *Wikimedia Commons*, 7 Sept. 2020, upload.wikimedia.org/wikipedia/commons/thumb/b/bd/Provinces_of_Japan-Mutsu.svg/1960px-Provinces_of_Japan-Mutsu.svg.png.

[270]"File:Provinces of Japan-Dewa.svg." *Wikimedia Commons*, 16 Sept. 2020, upload.wikimedia.org/wikipedia/commons/thumb/9/9b/Provinces_of_Japan-Dewa.svg/1960px-Provinces_of_Japan-Dewa.svg.png.

[271]Ouyang Xiu (欧阳修), *et al.* "Biography of Dongyi (东夷传) in Chapter 220 (卷二二〇): The 145th Biography (列传第一四五)." *Xin Tangshu* (新唐书 • 东夷传) or *New Book of Tang: Japan*. An Internet version can be found at https://ctext.org/wiki.pl?if=gb&chapter=272608&remap=gb.

[272]Wang Lianlong (王连龙). "Research on Baekje People Mi Jun's Epitaph 〈百济人（祢军墓志）考论)." *Frontiers in Social Sciences*, vol. 7, 2011, pp. 123–129 ((社会科学战线）2011年第7期，第123–129页).

[273]Louis-Frédéric Nussbaum. Käthe Roth, trans. *Op. cit.*, p. 447, at Google Books. https://books.google.com.hk/books?id=p2QnPijAEmEC&pg=PA447&redir_esc=y#v=onepage&q&f=false.

[274]George Sansom. *A History of Japan*, 1334–1615, Stanford, CA, USA: Stanford University Press, 1961, p. 263.

[275]Jason Josephson. *The Invention of Religion in Japan*, Chicago, IL, USA: University of Chicago Press. 2012, pp. 43–44.

[276]"File:Old Japan Tane.svg." *Wikimedia Commons*, 26 Sept. 2020, upload.wikimedia.org/wikipedia/commons/thumb/b/b9/Old_Japan_Tane.svg/2015px-Old_Japan_Tane.svg.png.

[277]"File:Provinces of Japan-Osumi.svg." *Wikimedia Commons*, 16 Mar. 2021, upload.wikimedia.org/wikipedia/commons/thumb/e/ee/Provinces_of_Japan-Osumi.svg/1960px-Provinces_of_Japan-Osumi.svg.png.

[278]Teikoku Shoin Co. *Teikoku's Complete Atlas of Japan*. Tokyo, Japan: Teikoku-Shoin Co. Ltd. Editorial Department, 1973.

[279]*Ibid.*

[280]Jesús López-Gay. "In Remembrance of the Historian Juan G. Ruiz-de-Medina, S.J." *Bulletin of Portuguese – Japanese Studies* (*Universidade Nova de Lisboa Lisboa, Portugal*), núm. 1, december, 2000, pp. 145–149.

[281]Louis-Frédéric Nussbaum. Käthe Roth, trans. *Op. cit.*, p. 412, at Google Books. https://books.google.com.hk/books?id=p2QnPijAEmEC&pg=PA412&redir_esc=y#v=onepage&q&f=false.

[282]*Ibid.*
[283]"File:Map of Izu Islands.png." *Wikimedia Commons*, 21 July 2022, upload.wikimedia.org/wikipedia/commons/e/e8/Map_of_Izu_Islands.png.
[284]Louis-Frédéric Nussbaum. Käthe Roth, trans. *Op. cit.*, p. 412, at Google Books. https://books.google.com.hk/books?id=p2QnPijAEmEC&pg=PA412&redir_esc=y#v=onepage&q&f=false.
[285]Liu Yingsheng (刘迎胜). "Study on Jingchuan and Jinghai: The Sea of Japan in Ancient East Asian Atlases – A Chinese Perspective on the Dispute over the Name of the Sea of Japan/East Sea between South Korea and Japan (鲸川与鲸海小考: 古代东亚图籍中的日本海 – 韩日有关日本海/东海名称争议的中国视角)." *Yuan Shǐ Jí Mín Zu Yu Bian Jiang Yan Jiu Ji Kan* (元史及民族与边疆研究集刊) or *Collective Publications on Yuan History, and Ethnic and Frontier Studies, vol. 32.* Shanghai, China: Shanghai Ancient Books Publishing House (上海古籍出版社), January 2017. https://www.sohu.com/a/208445626_632473.
[286]The Sasakawa Peace Foundation. "The History of the Name of the Sea of Japan." www.spf.org, Ocean Policy Research Institute, No. 55, Nov. 20, 2002, https://www.spf.org/en/opri/newsletter/55_1.html?full=55_1.

Chapter 2

An Ancient World Map Depicts the Chinese Exploration of Korea in 1433, Long Before the First European Man Landed There in 1582

Abstract

This chapter reports that an ancient world map — Kunyu Wanguo Quantu (坤輿萬國全圖/坤輿万国全图) or Complete Geographical Map of All the Kingdoms of the World published by Matteo Ricci in 1602 in China — depicts Korea in 1433, long before the first Europeans landed on Korea. An insightful analysis of the annotation, locations, and histories of the fifteen geographical items depicted on the Korean portion of this map shows that the map is of Chinese origin, not a direct or adapted copy of the major European maps before 1602.

Keywords: China, Europeans, Korea, Kunyu Wanguo Quantu, Matteo Ricci

1. Introduction

For over four hundred years, the ancient world map — Kunyu Wanguo Quantu (坤輿萬國全圖/坤輿万国全图)[1] abbreviated here as KWQ — written with Chinese characters and having latitudinal and longitudinal lines, has been generally regarded as a European map. The map was published in 1602 by Matteo Ricci (an Italian Jesuit priest and one of the founding figures of the Jesuit China missions) and was believed to be based on the European maps that Ricci brought with him to China in 1582.[2] In my previous book titled *Chinese Global Exploration in the Pre-Columbian Era: Evidence from an Ancient World Map*,[3] I conducted thorough analyses and presented concrete evidence to show the eras of the geographical information that I deduced on several major portions of the KWQ. My findings are as follows: (1) The Chinese explored Cape Breton Island on Canada's Atlantic Coast long before the Europeans; (2) information on both Mo Wa La Ni Jia (墨瓦蠟泥加/墨瓦蜡泥加; referring to the regions of Australia, New Zealand, *Tierra del Fuego*/Land of Fire in South America, and Antarctica) and the Americas was obtained by the Ming (1368–1644) mariners in the 1420s during their sixth voyage to the Western Ocean (referring to the Indian

Ocean and beyond); (3) information on Europe was obtained by the Chinese in the Southern Song Dynasty (南宋; 1127–1279); and (4) information on Africa was obtained by the Ming mariners in 1433 during their seventh (and last) voyage to the Western Ocean. Here, the mariners I mentioned in (2) and (4) belonged to the Treasure Fleets commanded by Admiral Zheng He (鄭和/郑和; 1371–1433/1435) in the early fifteenth century. My findings also show that all these portions of the KWQ most likely used Chinese maps as source maps (these source maps no longer exist), instead of copying from other contemporary European maps.

In Chapter 1 of the present book, I have extended my analysis to the map of Japan on the KWQ (abbreviated as Japan-KWQ) and discovered that the Japan-KWQ reveals Japan in 897 A.D. during the Chinese Tang Dynasty (618–907, with an interregnum between 690 and 705), almost 650 years before the first Europeans visited Japan.

All the above-mentioned findings, representing my earnest effort in seeking the true maritime world history, have produced fruitful results which merit attention. Hence, it would be very interesting to carry out similar analysis of the Korean portion of the KWQ, abbreviated as Korean-KWQ to find its origin and era. Such a task involves analysing the annotation and examining every geographical item (fifteen in total) depicted on the Korean-KWQ and their histories. But before doing so, I shall review the Europeans' knowledge of Korea, then the Chinese knowledge of Korea, to facilitate the later in-depth discussions in this chapter.

2. The First Mention of Korea — Cauli (or Kauli) — By Marco Polo, Did Not Spark the Europeans' Interest; The Recorded European Visits to Korea Start in the Late Sixteenth Century

Korea was first mentioned to the Europeans by Marco Polo (c. 1254–1324) in his travel book[4] around 1300, in which the term "Cauli" (or Kauli) was used to describe what was believed to be a portion of the Korean Peninsula. "Cauli" was the Italian transliteration of Goryeo (高麗/高丽; 918–1392), named in homage to the earlier northern Goguryeo Kingdom (高句麗/高句丽; 37 B.C.–668 A.D.). Goryeo was a Korean state founded in 918 during Korea's national division called the Later Three Kingdoms Period (889–935 A.D.; some sources say 936 A.D.). The new kingdom unified and ruled the Korean Peninsula until 1392.[5] However, Marco Polo's book gives very limited and vague information on Korea; thus, it did not spark the Europeans' interest in exploration of Korea.

Later, when the mid-fifteenth century ushered in the European Renaissance and the "Age of Discovery", the Europeans' primary interests of discovery and exploration also turned to the Orient. In 1488, Portuguese mariner and explorer Bartolomeu Dias (c. 1450–1500)[6] became the first European navigator to round the Cape of Good Hope of Africa and established a sea route between Europe and Asia. By the end of the fifteenth century, in 1497, another Portuguese explorer Vasco da Gama (c. 1460–1524)[7] traveled south, making three trips around the Cape of Good Hope to India.

Although by the middle of the sixteenth century the Portuguese missionaries and traders were well established in both China and Japan, they never had any direct contact with Korea; nor did the Spanish or the Dutch ever establish diplomatic relations with Korea before the twentieth century.

By the time the first Europeans truly learned about the existence of Korea, it was nearly five decades after Vasco da Gama rounded the Cape of Good Hope to reach India, and this time too it was the Portuguese.[8] That was because from 1543 the Portuguese had a trade post on the Island of Hirado (平戶島/平户岛; in today's Nagasaki Prefecture, Japan) where they got to know that in the northwestern direction from that island and passing Tsushima Island (對馬島/对马岛) of Japan, there was a nation which they called Couray.[9]

It took about four more decades before the first European finally entered Korea: he was a man referred to as "Pingni" or "Mari" by Korean sources. He landed with some Chinese on Jeju Island (South Korea's largest island) in 1582. But he was immediately deported to China.[10]

A few years later, it was Jan Huyghen van Linschoten (1563–1611)[11] who got to know about Korea from a Dutch sailor Dirck Gerritszoon Pomp, alias Dirck China (1544–c. 1608; the first known Dutchman to visit China and Japan).[12] He went to sea in 1584 aboard the Portuguese vessel "Santa Cruz" and arrived in Nagasaki (in today's Nagasaki Prefecture, Kyushu, Japan) in 1585. Dirck gave oral information to Jan Huyghen van Linschoten about the existence of Korea. He published his *Reys-gheschrift* (travel notes) in 1595. His original writing was in Dutch and its translation in English is as follows:[13]

> ... so stretches the coast [from Japan] again to the north, recedes after that inward, northwest ward, to which Coast those from Japan trade with the Nation which is called Cooray, from which I have good, comprehensive and true information, as well as from the navigation to this Country, from the pilots, who investigated the situation there and sailed there.

In *Itinerario* published one year later, he gives further information on Korea. His original writing was in Dutch and its English translation is as follows:[14]

> A little above Japan, on 34 and 35 degrees, not far from the coast of China, is another big island, called Insula de Core, from which until now, there is no certainty concerning size, people, nor what trade there is.

The above-mentioned two translated paragraphs tell us that in the mid-1580s, the Europeans knew of Korea as "Cooray" or as a big island called "Insula de Core" by the Portuguese. However, Korea is a peninsula that extends from c. 34°N to 43°N, and is not "a big island". Besides, the Korean mainland being "a little above Japan" is not correct either because the Korean mainland is about 180–200 km/112–124 mi to the northwest of the Japanese mainland (Honshu and Kyushu Islands).

It is true that little was known to the Europeans about the country's mainland because Jan Huyghen van Linschoten wrote that "until now, there is no certainty concerning size, people, nor what trade there is."

Moreover, when the first European explorers — the Portuguese — sighted Jeju Island (濟州島/济州岛), they called it *Ilha de Ladrones* ("Island of Thieves").[15] The island was also known to Europeans as Quelpart, a name that came from the first Dutch ship, the Quelpart, to spot the island. The ship was led by Hendrik Hamel in 1653.[16]

You may wonder why the Europeans thought that Korea was an island in those days. The answer will be proposed in Section 3. For the time being, I shall continue to elaborate on the European contact with Korea after reviewing Jan Huyghen van Linschoten's information.

Starting in 1590, the Annual Letters from Jesuit missionaries in the Far East began describing Korea in more detail, initially from indirect sources and later via news from Father Gregorio de Céspedes (1550–1611; a Spanish Jesuit priest).[17] He visited Korea in late 1593[18] (he arrived in Busan on 27 December 1593)[19] and was invited by one of the three leading generals of the Japanese invasion army (the invasion caused the Imjin War; an initial invasion in 1592, a brief truce in 1596, and a second invasion in 1597; the conflict ended in 1598 with the withdrawal of Japanese forces).[20] He stayed in Korea until April 1594. After that time, the geographical questions regarding the relationship of Korea to Japan, China, and the Tartary were beginning to be answered. Here, "Tartary" is a generic name used by the Europeans up until the middle of the nineteenth century for a vast part of Asia bounded by the Caspian Sea, the Ural Mountains, the Pacific Ocean, and the northern borders of China, India, and Persia, when this region was largely unknown to European geographers. This vast region will be discussed in Chapter 7 of this book.

In the seventeenth century, there are two known cases — in 1627 and 1653 — of Westerners visiting Korea and staying there involuntarily. These Westerners were Dutch.

In 1627, a Dutch sailor, likely the first Dutch visitor to Korea, Jan Janse de Weltevree (1595–year of death unknown), and his company landed on the shores of Jeolla Province (*Jeolla-do*; 全羅道/全罗道; one of the historical eight provinces of Korea during the Joseon Kingdom in present-day southwestern Korea; see Section 6 of this chapter) due to rough seas encountered en route to Japan.[21] Weltevree permanently settled in Korea,[22] but the two other captives were killed in 1636 during a raid of the Manchu when they fought in the Korean army.[23]

In 1653, Dutchman Hendrik Hamel and 35 others were on board their ship, the Sparrow Hawk, sailing from Taiwan toward Nagasaki. The ship met strong and deadly Jeju winds that forced it onto the rocks of the island's south shore. They were caught and brought to today's Seoul (漢城/汉城; capital of the Joseon Kingdom) and then served in the Joseon (朝鮮/朝鲜; 1392–1897; Joseon was the last dynastic kingdom of Korea) military under Weltevree's supervision. Thirteen years later, in 1666, eight of them escaped, returned to Holland, and told their stories. Hamel wrote a famous book on his adventure.[24] It was the first book about Korea from within.

Hendrik Hamel's book changed the Europeans' entire perception of Korea. But it was too late to have any impact on Matteo Ricci's making of the Korea-KWQ because he had passed away 54 years earlier in 1610.

This review of European contact with Korea tells us that little about Korea was known to the Europeans until they read Hendrik Hamel's book after 1666, which records his and his fellow mariners' life on the mainland of Korea. With this historical backdrop, we should not expect to see any realistic maps of Korea drawn by European cartographers before 1602 to serve as the source maps for Matteo Ricci to draw the Korea-KWQ. In the next section, I will make clear that major maps of Korea drawn by European cartographers before 1602 show that this is precisely the case.

3. Major Maps of Korea Made by Europeans Before 1602

From Section 2, we know that the Europeans did not enter the Korean mainland before the late sixteenth century. This was long after Marco Polo made an account of the country in his travel book[25] around 1300, which was widely circulated in Europe and became the principal source of European knowledge of Asia. In comparison with his description of Japan, Marco Polo's writing on Korea was very limited and vague; hence, it did not motivate the Europeans to explore the place right away.

However, surprisingly, on the Fra Mauro map drawn in c. 1450s (some sources say c. 1460), as shown in Fig. 2.1, the Asian part shows not only the Arabian Peninsula, Persia, the Indian sub-continent, Burma, and China but also the Korean Peninsula.

It remains a mystery how Fra Mauro (c. 1400–1464) was able to draw Korea correctly as a peninsula as early as around the mid-fifteenth century, more than a century before the first

Fig. 2.1.   The Fra Mauro map (inverted, North is at the top) shows the Korean Peninsula (see endnote 26; public domain). Japan may be on the rightmost edge, with the Korean Peninsula to its west and pointing southward. But as explained in Chapter 1 of this book, the map also depicts a part of Japan, probably Kyushu, south of the island of Java, see Fig. 1.3(b), denoted "Isola de Cimpagu", lit. "Island of Cimpagu", which is a misspelling of Cipangu, an interpretation based on Marco Polo's description of Japan.

Europeans landed in Korea, unless he copied the map of Korea from the Chinese or a Korean source.

Such a viewpoint is supported by the following strong arguments: (1) the Chinese have known of Korea since Antiquity; not surprisingly, the Chinese 1389 Da Ming Hunyi Tu (大明混一圖/大明混一图)[27] or Composite Map of the Ming Empire, as shown in Fig. 2.10(a), stretches eastward to Korea. (2) The 1402 Korean Hun Yi Jiang Li Li Dai Guo Du Zhi Tu (混一疆理歷代國都之圖/混一疆理历代国都之图)[28] or Honil Gangni Yeokdae Gukdo Ji Do/Map of Integrated Lands and Regions of Historical Countries and Capitals (of China), completed by the Korean scholars Kwon Kun and Yi Hoe, shows the complete Korean Peninsula. The map is often abbreviated as Kangnido.[29] It is the oldest extant Korean world map[30] and the most familiar example of the known-world maps. The map is based on Chinese maps and has additional input from western sources derived from Islamic scholarship in the Mongol Empire.[31] Both Da Ming Hunyi Tu and Kangnido were revised after their production, making their original form uncertain. However, the surviving copies of the *Kangnido can be used to infer the original content of the Chinese map, implying its source map is most likely a Chinese map.* (3) Fra Mauro (c. 1400–1464)[32] traveled extensively in his youth as a merchant and a soldier. Later, he became a monk and lived near the island of Murano near Venice, where he was employed as an accountant and professional cartographer, making maps for the rulers of the main seafaring nations of the time, such as Venice and Portugal (Prince Henry the Navigator was most interested in European maritime discoveries and maritime expansion). Mauro would frequently consult with merchants of the city after they returned from overseas voyages and get maps from them, including those from Asia. In particular, Niccolò de' Conti (c. 1395–1469)[33] — an Italian merchant, explorer, and writer — traveled to Asia during the early fifteenth century and was one of the sources for the Fra Mauro map. The map surprisingly indicates that there was a sea route from Europe around Africa to India a few decades before Bartolomeu Dias and Vasco da Gama rounded the Cape of Good Hope to reach Asia.

Niccolò de' Conti's travels occurred around the same time and in the same places as the Chinese expeditions of Admiral Zheng He (鄭和/郑和; 1371–1433/1435). His travel accounts are consistent with those of the Chinese writers who were on Zheng He's ships like Ma Huan (馬歡/马欢; c. 1380–1460) and Fei Xin (費信/费信; c. 1385/1388–after 1436). Ma Huan was a Chinese voyager and translator and accompanied Admiral Zheng He on the fourth, fifth, and seventh of his seven voyages to the Western Ocean. In 1451, Ma Huan completed the final version of his book titled *Yingya Shenglan* (瀛涯勝覽/瀛涯胜览)[34] or *The Overall Survey of the Ocean's Shores*, which records the countries visited by the Ming Treasure Ships. Fei Xin was a member of the military forces of Zheng He's Treasure Fleet. He accompanied Admiral Zheng He on the first four of his seven voyages to the Western Ocean. Fei Xin wrote a book titled *Xingcha Shenglan* (星槎勝覽/星槎胜览)[35] or *Description of the Starry Raft* (Preface dated 1436) to describe the foreign places that the Chinese mariners had seen.

It is very likely that Niccolò de' Conti passed Chinese maps or geographical information gathered by Zheng He's mariners to Fra Mauro and his collaborators for them to draw the Fra Mauro map in which Korea is correctly presented as a peninsula.

Now, we can examine the Europeans' perception of the shape of ancient Korea before Matteo Ricci published the KWQ at the beginning of the seventeenth century.

3.1 *Korean Peninsula perceived as an island*

Other than the surprisingly correct depiction of the Korean Peninsula on the Fra Mauro map around the middle of the fifteenth century, European maps of the sixteenth century often wrongfully presented the Korean Peninsula as an island. It was only from the late sixteenth century onward, through missionaries and merchants who were sojourning in Asia, that place names related to the Korean Peninsula started to appear on European world maps.

Now, we come to the point of discussing why Korea came to be displayed as an island by the Europeans.

One explanation is that, interestingly, ancient Koreans themselves depicted Korea as "almost" an island,[36] with only one mountain — Mount Peaktu (白頭山/白头山; Chinese Changbai Mountain; 長白山/长白山) — connecting Korea to the mainland. Two rivers — the Yalu River (鴨綠江/鸭绿江; Amrok River or Amnok River) and the Tumen River (圖們江/图们江) — flowing from that mountain were drawn so wide that the Western cartographers took Korea to be an island.

The other explanation is that, as shown in Fig. 2.2(a), the Yalu River forms the western portion of the modern boundary between China and North Korea; its estuary empties into Korea Bay in

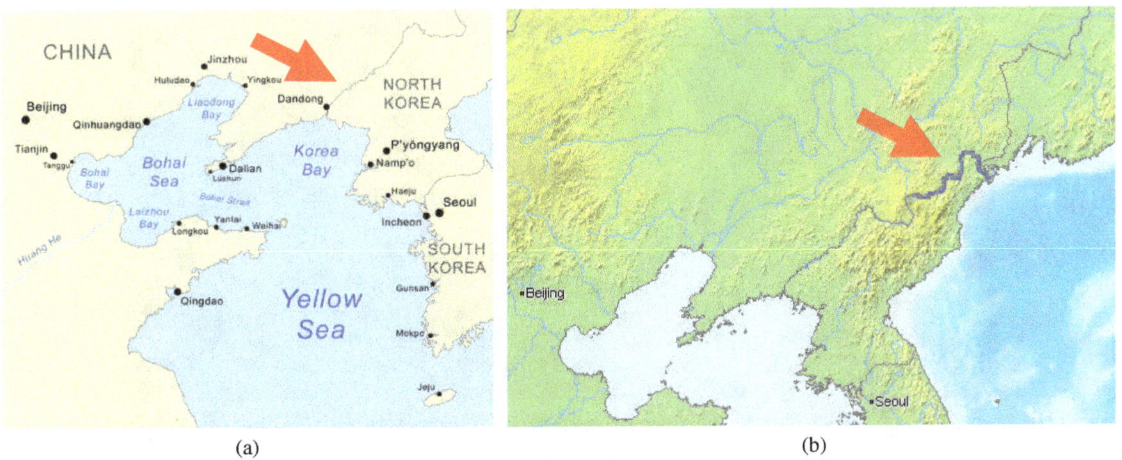

(a)   (b)

Fig. 2.2.   (a) shows the map of the Bohai Sea (渤海) and surrounding bays (see endnote 37; *Source*: Kmusser, under CC BY-SA 2.5, https://commons.wikimedia.org/wiki/File:Bohaiseamap2.png); the Yalu River is indicated by a red arrow. (b) shows the location of the Tumen River (see endnote 38; public domain), which is also indicated by a red arrow.

the Yellow Sea between Dandong (丹東/丹东; 40.0°N, 124.4°E) in China and Sinuiju (新義州市/新义州市; 40.1°N, 124.3°E) in North Korea. As shown in Fig. 2.2(b), the Tumen River forms the border on the eastern side (at the China–North Korea–Russia tri-border area) and flows into the Sea of Japan. It is also possible that mariners having navigated partway up these rivers concluded that the two waterways joined to form a strait that reached from sea to sea. If so, they had not proceeded far enough inland to observe that the region of Mount Peaktu is the source of both rivers. Notice that the Tumen River starts a bit east of the 2,744-m Mount Peaktu peak; on Google maps, a tributary of the Tumen starts close to the peak.

Here are two examples that treat Korea as an island before 1602:

(1) Luis Teixeira completed a new map of Japan in 1592 with Korea appearing as a large, carrot-shaped island,[39] as shown in Fig. 2.3. In that year, Teixeira sent his manuscript to Abraham Ortelius (1527–1598) who published it in his edition *Theatrum Orbis Terrarum* ("Theatre of the Lands of the World") in 1595. The map became popular among the European map

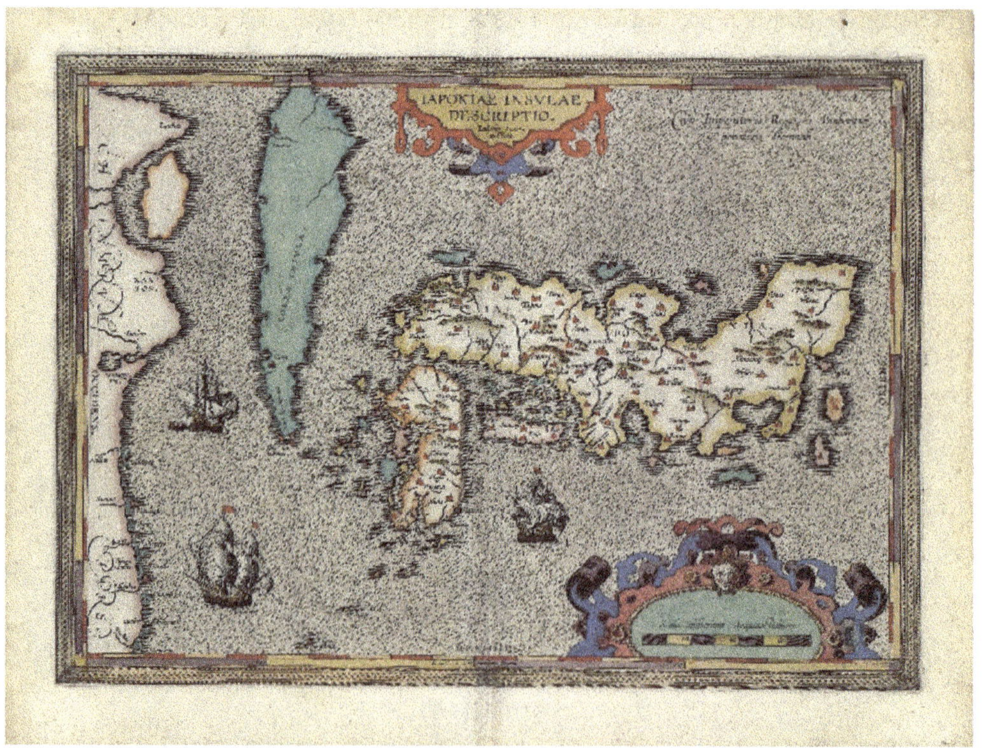

Fig. 2.3. Luis Teixeira's map of Japan in 1592 with Korea appearing as a large, carrot-shaped island denoted as COREA INSVLA (see endnote 41; public domain).

publishers. This is the first Western map on which only the Korean Peninsula and the Japanese Islands are drawn. On the map, Korea is drawn as a long island and marked as COREA INSVLA (Island of Korea); in the inland sections, the names of *Cory, Punta dos Ladrones*, and *Tauxem* are written, along with the name *Ilhas dos Ladrones* on Jeju Island.

(2) On the map of Southeast Asia which was included in the 1595 *Reys-gheschrift van de Navigatien der Portugaloysers in Orienten* (*Travel Accounts of Portuguese Navigation in the Orient*),[40] Korea is marked as "ILHA DE COREA" (lit. "Island of Korea") in the shape of a circle, and "*Costa de Conray*" (lit. "Coast of Conray") and "I. dos Ladrones" (lit. "Island of Thieves") are written as well. In addition, a small island marked as *Corea* (Korea) is drawn on the upper part of the map.

3.2 *Korean Peninsula perceived as a peninsula*

The following are examples of European maps in the late sixteenth century showing Korea as a peninsula:

(1) Figure 2.4(a) shows the 1571 Universal Atlas of Fernão Vaz Dourado (c. 1520–c. 1580; a Portuguese marine cartographer). It includes the Korean Peninsula, China, and Japan. Korea appears in the shape of a sweet potato on the top right corner and Japan as a shrimp-shaped island nearby. "Costa de Conrai", lit. "Coast of Conrai", is inscribed across the Korean Peninsula and the Chinese continent: "Conrai" refers to Goryeo (高麗/高丽; 918–1392). It was the first Western map to write a Korean country name on the Korean Peninsula.

(2) Figure 2.4(b) shows the 1594 world map created by the Dutch-Flemish astronomer, cartographer, and clergyman Petrus Plancius (1552–1622), who was one of the founders of the Dutch East India Company, for which he drew over 100 maps. The map is known to be the oldest existing map in Europe to refer to Korea with the marking "Corea". There are two earlier maps, Fig. 2.4(c) drawn by Bartholomeu Velho (date of birth unknown–1568) in 1562 and Fig. 2.4(d) drawn by Abraham Ortelius in 1589, which also show the Korean Peninsula. But they gave no specific name to the peninsula.

From the above-provided figures and discussion, we know that the European cartographers before the seventeenth century drew Korea as an island, as a poorly shaped peninsula, or as a peninsula labeled with the wrong name. Since none of their maps of Korea have the quality of the Korea-KWQ to be presented in Section 5 of this chapter, how could these European maps be the source maps of the Korea-KWQ?

Hence, we must focus our attention on the Chinese maps which most likely were the source maps of the Korea-KWQ. But before doing so, we must first comprehend the deeply intertwined Sino-Korean relations since Antiquity. This will be done in the following section.

Fig. 2.4.   (a) shows the Atlas Portugués de 1571 by Fernão Vaz Dourado (see endnote 42; *Source*: Cardeña2, under CC BY-SA 3.0, https://commons.wikimedia.org/wiki/File:Atlas_de_Fernao_Vaz_Dourado_(Asia).jpg). The Korean Peninsula appears in the shape of a sweet potato on the top right corner and Japan as a shrimp-shaped island nearby; it was the first Western map to write a Korean country name on the Korean Peninsula. (b) extracted from the 1594 world map created by Petrus Plancius (see endnote 43; public domain); it is known to be the oldest existing map in Europe to refer to Korea with the marking "Corea". (c) extracted (and strongly magnified) from the figure of the heavenly bodies — Illuminated illustration of the Ptolemaic geocentric conception of the Universe (see endnote 44; public domain) — by Portuguese cosmographer and cartographer Bartolomeu Velho from his work *Cosmographia*, made in France in 1568 (Bibliothèque nationale de France, Paris). (d) shows the 1589 Maris Pacifici (*Descriptio Maris Pacifici*, lit. "Description of the Pacific Ocean") by Abraham Ortelius from his *Theatrum Orbis Terrarum* (see endnote 45; public domain); it shows the Korean Peninsula at the upper left corner labeled as C. *de Liampo*, and the four main islands of Japan are depicted next to it.

4. China and Korea have Shared a Border and History Since Antiquity; Korea has Appeared on Ancient Chinese Maps Since the Tang Dynasty

While the first direct European contact with Korea occurred in the late sixteenth century, relations between the Chinese and Koreans started in Antiquity. Chinese dynasties shared borders and had peace and waged wars with ancient Korea; they even established local administrative units in Korea several times in history. The history of Korea and its intertwined relations with the Chinese dynasties can be briefly reviewed in the following.

4.1. *The history of Korea and Sino-Korean relations*

Archaeologists believe that the ancestors of today's Koreans came from Mongolia and Siberia. According to the 1281 *San Guo Yi Shi* (三國遺事/三国遗事) or *Samguk Yusa/Memorabilia of the Three Kingdoms*,[46] Dangun or Tangun (檀君), also known as Dangun Wanggeom (檀君王儉/檀君王俭), was the legendary founder and God-king of Gojoseon[47] (古朝鮮/古朝鲜; Old Joseon; the first Korean kingdom; 2333–108 B.C.). He ruled the northern part of the Korean Peninsula and the region around the present-day Liaoning Province in Northeast China. A legend says that in c. 1100 B.C., a Chinese sage named Jizi (箕子or Gija) and his followers were in Gojoseon and they taught the locals agriculture and many other things, including the proper ceremonies. Although there have been several interpretations of Gojoseon and Gija Joseon, as well as debates regarding Gija Joseon's existence,[48] they are not the main subjects of this chapter and will not be discussed further. Nevertheless, it is certain that Gojoseon existed by the fourth century B.C., conducted trade and waged war with two northeastern Chinese states, Zhao (趙/赵) and Yan (燕), in the Chinese Warring Period (戰國時期/战国时期; 403–221 B.C.), and shared a border with Yan.[49]

Then, in 108 B.C., during the Chinese Western Han Dynasty (西漢/西汉; 206 B.C.–9 A.D.), Gojoseon in the north of the Korean Peninsula was overthrown by Chinese armies.[50]

The political map of Korea from 108 B.C. to the late fifteenth century can be shown graphically in Figs. 2.5–2.8 with the following explanations.

Figure 2.5(a) shows Korea in 108 B.C., before Gojoseon was overthrown by the Chinese armies. The political map includes the northern and southern regions of the Korean Peninsula and Manchuria.

In the northern region were Buyeo (扶餘/扶余), Goguryeo (古朝鮮/古朝鲜), Okjeo (沃沮), Dongye (東濊/东濊), and other minor statelets. China installed four commanderies (郡邑)[51] in the former Gojoseon territory, but three of them fell quickly to Korean resistance.

In the southern region of the Korean Peninsula were the little-known Korean state of Jin State (辰國/辰国) and a minor statelet Ye (濊).

Figure 2.5(b) shows Korea in 1 A.D.

In the northern region, Goguryeo[52] (高句麗/高句丽; it was founded in 37 B.C. but was mentioned in Chinese records as early as 75 B.C.) emerged. The kingdom gradually conquered and absorbed all its neighbours and destroyed the last Chinese commandery in 313 A.D.

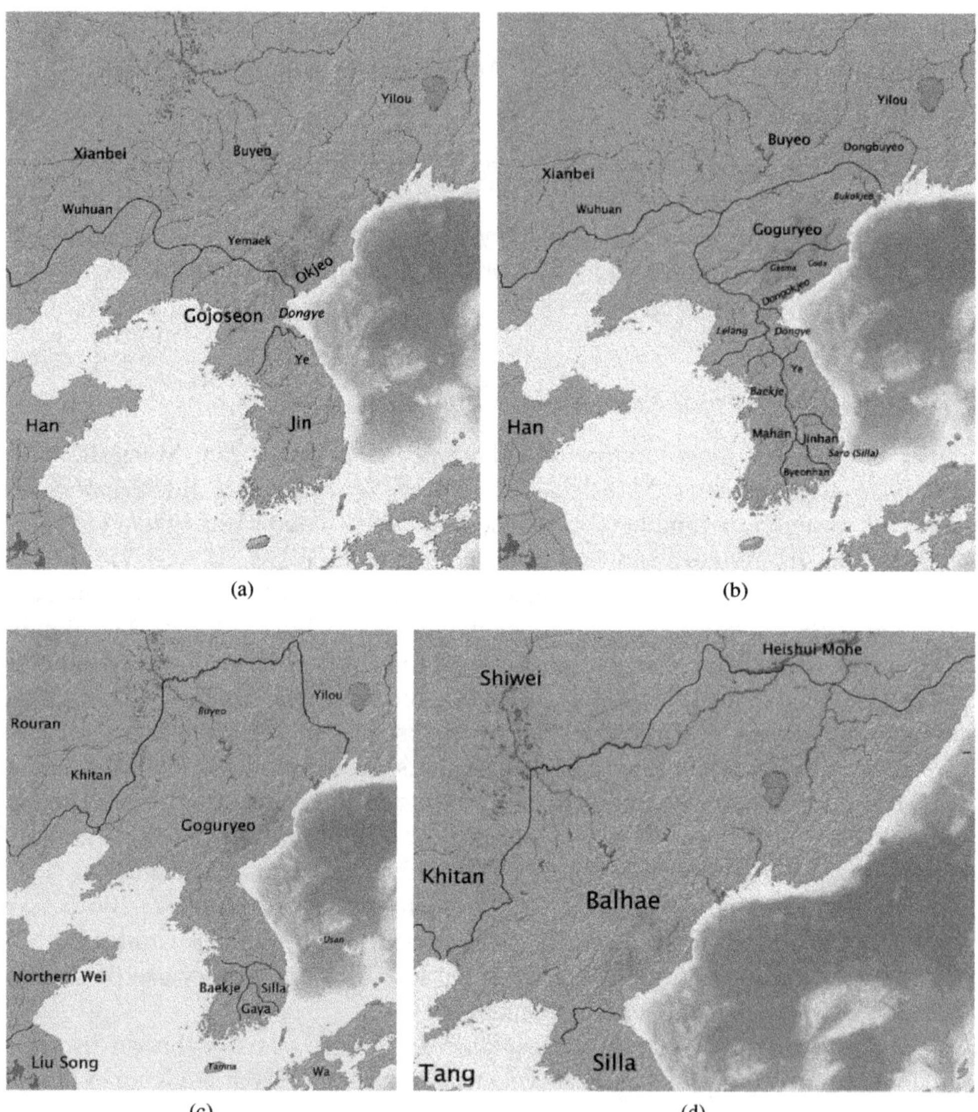

Fig. 2.5. (a) Korea in 108 B.C. (see endnote 62; *Source*: Historiographer at English Wikipedia, under CC BY-SA 3.0, https://commons.wikimedia.org/wiki/File:History_of_Korea-108_BC.png). (b) The Proto-Three Kingdoms, c. 1 A.D. (see endnote 63; *Source*: Historiographer at English Wikipedia, under CC BY-SA 3.0, https://commons.wikimedia.org/wiki/File:History_of_Korea-001.png). (c) Korea in 476, the moment of greatest territorial expansion of Goguryeo (see endnote 64; *Source*: Myself, under CC BY-SA 3.0, https://commons.wikimedia.org/wiki/File:History_of_Korea-476.png) during the Three Kingdoms Period. (d) Balhae and Unified Silla in 830 A.D. (see endnote 65; public domain).

In the central and southern region, there emerged the loose confederacies of Jinhan (辰韓/辰韩; it was a loose confederacy of chiefdoms), Byeonhan (弁韓/弁韩), and Mahan (馬韓/马韩), or collectively Samhan (三韓/三韩)[53] during the Proto-Three Kingdoms Period (原三國時代/原三国时代; started around the first century B.C.). This period was the early part of the Three Kingdoms of Goguryeo, Baekje, and Silla: It was after the fall of Gojoseon but before the maturation of the Three Kingdoms into the full-fledged kingdoms around 313;[54] it is also called the Samhan Period (三韓時代/三韩时代). During the Proto-Three Kingdoms Period, numerous states sprang up from the former territories of Gojoseon, as shown in Fig. 2.5(b).

The Samhan Period was important in Korea's history because Baekje (百濟/百济; 18 B.C.–660 A.D.)[55] was founded in 18 B.C. in Mahan's territory and began to slowly overtake it; Silla (新羅/新罗; Old Silla: 57 B.C.–668 A.D. and Unified Silla: 668–935) was founded by the unification of six chiefdoms within Jinhan; and Byeonhan (弁韓/弁韩) — existed from around the beginning of the Common Era to the fourth century — was absorbed into the later Gaya (加倻 or 駕洛/驾洛) Confederacy (42–562),[56] which in turn was annexed by Silla.[57]

Figure 2.5(c) shows Korea in 476 in the Three Kingdoms Period (三國時代/三国时代; 57 B.C.–668 A.D.) when the three kingdoms — Goguryeo, Baekje, and Silla — coexisted.[58]

The three kingdoms occupied the entire Korean Peninsula and roughly half of Manchuria (located mostly in present-day China), along with smaller parts of present-day Russia. But there existed the Gaya Confederacy (42–562) in the southern region of the Korean Peninsula and relatively large states like Buyeo (扶餘/扶余; c. second century B.C.–494 A.D; it was annexed by Goguryeo in 494) to the north of the Peninsula and also occupied the other parts of Manchuria in modern China.[59]

Figure 2.5(c) shows that among the three kingdoms, Baekje and Silla dominated the southern half of the Korean Peninsula and Tamna (耽羅/耽罗; today's Jeju Island), whereas the vast Goguryeo Kingdom controlled the Liaodong Peninsula, Manchuria, and the northern half of the Korean Peninsula. Goguryeo (高句麗/高句丽) was later known as Goryeo (高麗/高丽; in 918, Goryeo was founded as the successor to Goguryeo and inherited its name). Historically, Goguryeo (37 B.C.–668 A.D.), Later Goguryeo (901–918; renamed to Taebong in 911), and Goryeo (918–1392) all used the name "Goryeo", from which the modern name Korea is derived.

According to the mainstream Korean historians, during the Tang Dynasty (618–907, with an interregnum between 690 and 705), Silla allied with China and unified the Korean Peninsula in 668 for the first time in Korean history to start the era of Unified Silla (668–935). After the fall of Baekje and Goguryeo, the Tang Dynasty established a short-lived military government to administer parts of the Korean Peninsula. However, a Silla–Tang War (c. 670–676 A.D.) broke out, resulting in Silla (joined by Goguryeo and Baekje loyalists) expelling the Protectorate armies from the peninsula in 676.[60]

Figure 2.5(d) shows that Unified Silla/Late Silla coexisted with Balhae (渤海; 698–926) in 830 A.D. during the Northern and Southern States Period (南北國時代/南北国时代; 698–926). Balhae was a multi-ethnic kingdom whose land extended to what is today's Northeast China, the Korean Peninsula, and the Russian Far East.[61]

Figure 2.6 shows that Unified Silla (668–935) was eventually divided into the Later Three Kingdoms (後三國/后三国; 889–936): Later Baekje (892–936), Later Goguryeo (901–911; renamed Taebong 泰封 in 911; 911–918), and Late Silla (Unified Silla; 668–935).

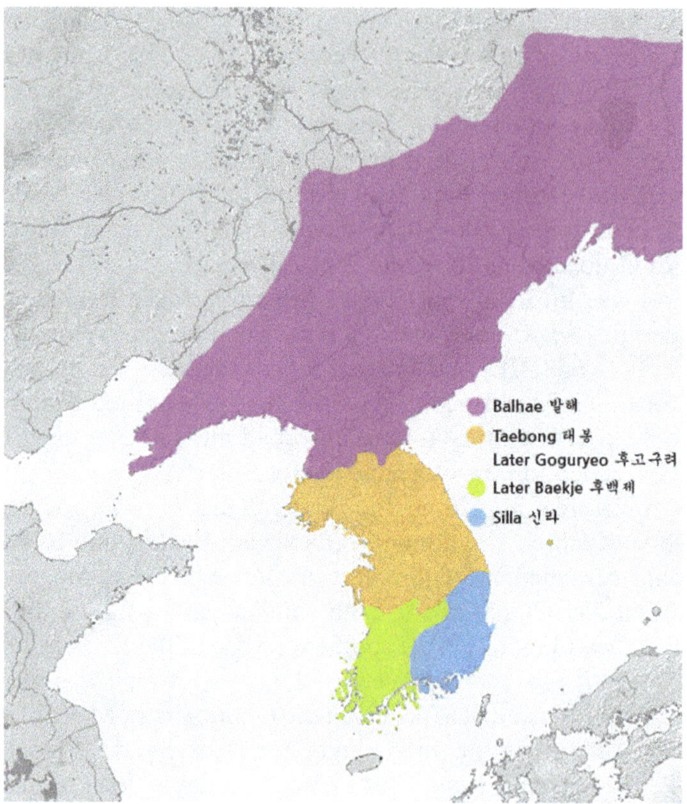

Fig. 2.6. Later Three Kingdoms: Later Baekje, Later Goguryeo (changed name to Taebong in 911), and Late Silla (Unified Silla) coexisted with Balhae (see endnote 67; *Source*: KJS615, under CC BY-SA 3.0, https://commons. wikimedia.org/wiki/File:History_of_Korea-Later_three_Kingdoms_Period-915_CE.gif) during the Northern and Southern States Period (698–926).

Late Silla was ultimately annexed in 935 by the new Goguryeo revivalist state of Goryeo (高麗/高丽; 918–1392)[66] whose founder Taejo of Goryeo (高麗太祖/高丽太祖; ruled from 918 to 943) was recognised as the *de facto* ruler of Korea by the Tang Dynasty of China in 932 A.D. and achieved unification of the Later Three Kingdoms in 936 (some sources say 935).

Figure 2.7(a) shows the Mongol-led Yuan Dynasty in 1294, and Goryeo is marked as a vassal.

Goryeo was under the rule of the Mongol Empire/Great Mongol Nation (1206–1368) and the Mongol-led Yuan Dynasty (Great Yuan; 大元; 1271–1368), which ruled China from 1271 to 1368 and the Korean Peninsula from c. 1270 to 1356.[68]

As early as 1258, the Mongols annexed the northern areas of the Korean Peninsula[69] and incorporated them into their empire as the Ssangseong Prefecture (雙城摠管府/双城摠管府; an administrative division of the Yuan Dynasty; 1258–1356) and the Dongnyeong Prefecture (東寧府/东宁府; 1259–1290; an administrative division of the Yuan Dynasty). In 1290, the Dongnyeong Prefecture was abolished, and its jurisdiction was transferred to Goryeo. When the Yuan Dynasty weakened in 1356, Goryeo also recovered the Ssangseong Prefecture.

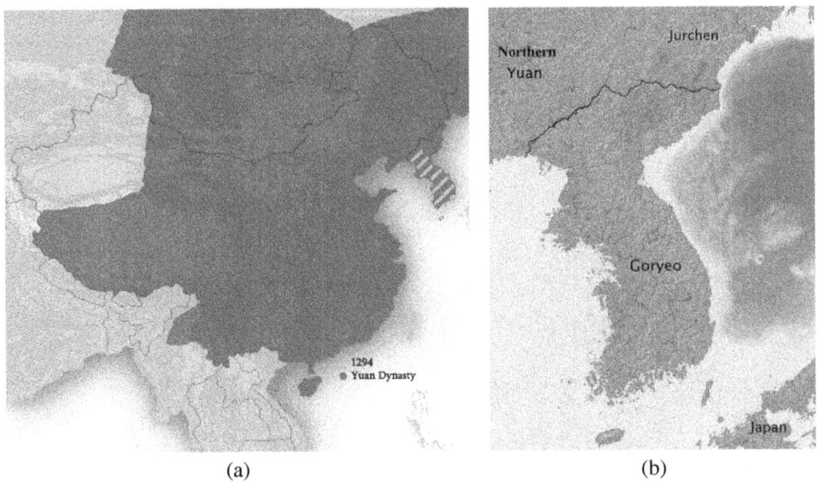

(a)   (b)

Fig. 2.7. (a) shows the Yuen Dynasty in 1294; Goryeo is marked as a vassal (see endnote 72; *Source*: Yuen_ Dynasty_1294.png: Ian Kiu, under CC BY-SA 3.0, https://commons.wikimedia.org/wiki/File:Yuen_Dynasty_1294_-_ Goryeo_as_vassal.png). (b) shows Goryeo in 1389 during the early Ming Dynasty (see endnote 73; *Source*: Historyclassproffesor, under CC BY-SA 3.0, https://commons.wikimedia.org/wiki/File:Koryo_map.png).

Figure 2.7(b) shows Goryeo in 1389 during the early Ming Dynasty.

In 1368, following the defeat of the Yuan forces by the Han people-led Ming Dynasty (明朝 or Great Ming; 大明; 1368–1644), Ming became the ruling dynasty of China. The Genghisid rulers (Genghis Khan; c.1162–1227; the founder and first khagan of the Mongol Empire) retreated to the Mongolian Plateau as the Northern Yuan Dynasty (北元; 1368–1635) until they surrendered to the Later Jin Dynasty (1616–1636; 後金/后金; a Jurchen-led royal dynasty of China in Manchuria, later evolved into the Qing Dynasty).[70]

By the late fourteenth century, Goryeo became weak due to years of war with the Mongols. In 1388, the pro-Yuan faction in the royal court in Goryeo refused the Ming's demand that territories of the former Ssangseong Prefectures (雙城摠管府/双城摠管府) be handed over to Ming China. Goryeo General Yi Seong-gye (李成桂; 1335–1408)[71] was chosen to lead the attack against Ming China. However, he revolted, ascended the throne himself, and renamed the new kingdom Joseon (朝鮮/朝鲜; 1392–1897), a tribute to the ancient Korean state of Gojoseon (古朝鮮/古朝鲜).

Figure 2.8 shows Joseon and its surrounding countries and external areas. The map shows united Mongol tribes and the Jurchen land in the later part of the fifteenth century.

In 1418, Sejong the Great (朝鮮世宗/朝鲜世宗; 1397–1450; the fourth ruler of the Joseon Dynasty) ascended the throne. In 1433, Sejong sent his prominent general to the north to destroy the Jurchens (女真人), who later became the Manchus (滿族/满族) living in Manchuria. The military campaign successfully captured several fortresses and established four counties[74] and six commanderies[75] to safeguard the people from the Jurchens. The kingdom's northernmost borders were expanded to the natural boundaries at the banks of the Amrok and Tuman rivers through the subjugation of the Jurchens.

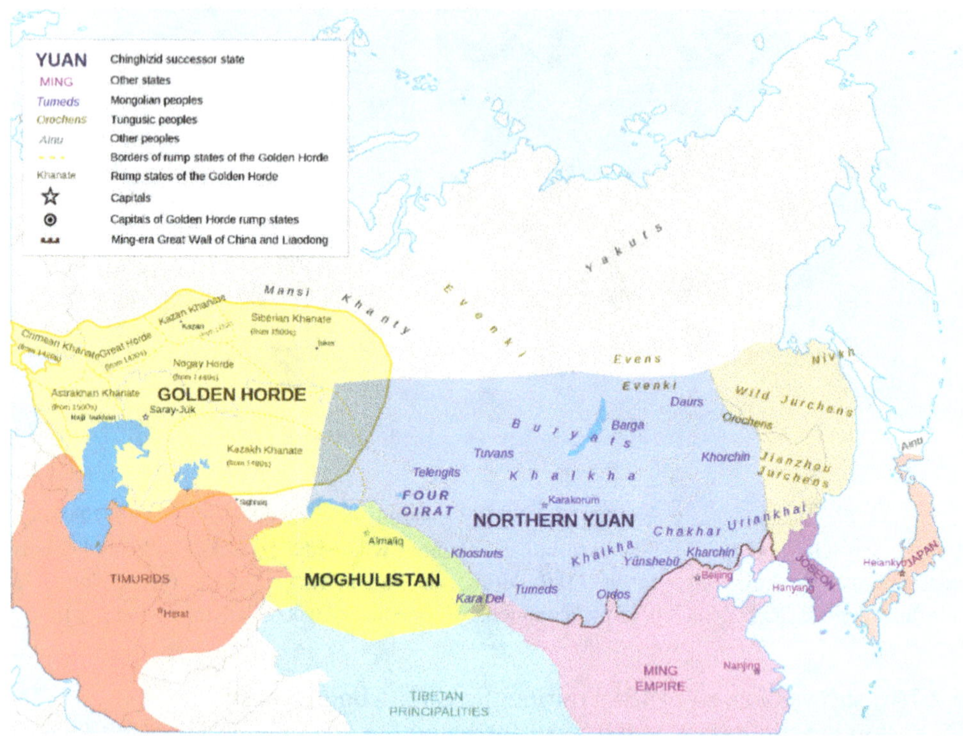

Fig. 2.8. This modern map shows the political condition in the later part of the fifteenth century when the Mongol tribes of Moghulistan were united (see endnote 77; *Source*: Elvonudinium, under CC BY-SA 3.0, https://commons.wikimedia.org/wiki/File:Northern_Yuan_and_Golden_Horde.svg). To the west, south, and north of the Joseon Kingdom are the Golden Horde (1242–1502), Timurid Empire (1370–1507), Moghulistan (Eastern Chagatai Khanate; 1347–1570 or 1347–1705; a detailed discussion is provided in Chapter 7 of this book), Northern Yuan (1368–1635), Ming Empire (1368–1644), Joseon (1392–1897), and the areas occupied by Wild Jurchens (野人女真), Jianzhou Jurchens (建州女真), and area occupied by the Yakuts (雅庫特人/雅庫特人) in Siberia.

The issue of controlling the Jurchens was a point of contention between Joseon Korea and the early Ming.

The Ming Dynasty distinguished three different groups of Jurchens:[76] the Wild Jurchens (野人女真) of what became Russian Manchuria, the Haixi Jurchens (海西女真) of modern Heilongjiang Province, and the Jianzhou Jurchens (建州女真) of modern Jilin Province.

During the Yongle (永樂/永乐; Emperor Yongle reigned from 1402 to 1424) and Xuande (宣德; Emperor Xuande reigned from 1425 to 1435) years, Chinese commanderies were established in association with the tribal military units which were ruled by the hereditary tribal leaders. In the Yongle Period, 178 Guard Battalions were set up in Manchuria. But none of these are depicted on the Jurchen lands in Fig. 2.8, indicating that in the later part of the fifteenth century, the Ming Empire seems to no longer have control of the Jurchen lands.

From Figs. 2.5–2.8, the northern borders of Korea have shown continuous changes resulting from the changing political conditions of these northern territories. Such changes can provide important clues for determining the era of the Korea-KWQ in Section 6 of this chapter.

In summary, China and Korea have had long and intertwined political relations since Antiquity. Moreover, as early as the middle of the fourth century, the Korean dynasties had already established a tributary relationship with China. According to historical records, Goguryeo (173 tribute missions),[78] Baekje (45 tribute missions),[79] Silla (19 tribute missions),[80] Unified Silla (63 tribute missions in the eighth century),[81] Goryeo (envoy missions),[82] and Joseon (826 tribute missions)[83] all had tributary relations with Chinese dynasties. For more than 1,500 years afterward, this relationship remained in place until it was officially terminated with the Treaty of Shimonoseki[84] (a treaty signed in 1895 between the Empire of Japan and Qing China, ending the First Sino-Japanese War). In this treaty, China recognised the independence of Korea and ceded Taiwan, the Pescadores, and the Liaodong Peninsula in Manchuria to Japan.

The Chinese people had been on the Korean land more than 2,000 years before the first Europeans' arrival because, in 108 B.C., the Chinese armies of the Western Han Dynasty overthrew Gojoseon in the north of the Korean Peninsula. Hence, we should not be surprised that the Korean Peninsula was accurately drawn on ancient Chinese maps long before European cartographers made such efforts on their maps.

The following section will show when and how the ancient Chinese cartographers started to depict Korea on their maps.

4.2. *Korea on ancient Chinese maps*

Concerning the Chinese maps that show Korea, Huayi Tu (華夷圖/华夷图) or Map of China and the Barbarian Countries, engraved on a stone stele and dated 1136 (some sources say 1137, or even as early as 1034), may be the earliest surviving map. It relates China with other foreign states as shown in Fig. 2.9. The map has very accurate latitudes and longitudes for cities in the interior part of China, which may indicate the use of spherical trigonometry projection.[85]

Huayi Tu covers China during the Jin/Great Jin Dynasty (金朝 or 大金; 1115–1234), sometimes called the "Jurchen Dynasty" or the "Jurchen Jin" because members of the ruling Wanyan (完顏/完颜) clan were of Jurchen descent. The Jin emperors referred to their state as China (Zhongguo; 中國/中国).[87] Later, they were conquered in 1234 by the Mongols and the Han-led Southern Song Dynasty (南宋; 1127–1279) based in southern China. Huayi Tu is based on Hainei Huayi Tu (海內華夷圖/海内华夷图) or Map of China and the Barbarian Countries within the Seas produced in 801 A.D. by Jia Dan (賈耽/贾耽). Jia Dan was a Chinese cartographer, military general, and politician during the Tang Dynasty (618–907, with an interregnum between 690 and 705). Hainei Huayi Tu was presented to Emperor Dezong of Tang (唐德宗; reigned from 779 to 805) in 801.[88] However, the map is lost.

The texts arranged around the edges of Huayi Tu were taken from Hainei Huayi Tu; they include brief descriptions of Korea, Japan, and other foreign states as well as lists of the states to the south and west of China. This is not surprising because China–Korea contact started in Antiquity, as detailed in Section 4.1.

The next surviving ancient Chinese map showing Korea is the c. 1389 Da Ming Hunyi Tu (大明混一圖/大明混一图) or Composite Map of the Ming Empire as shown in Fig. 2.10(a).

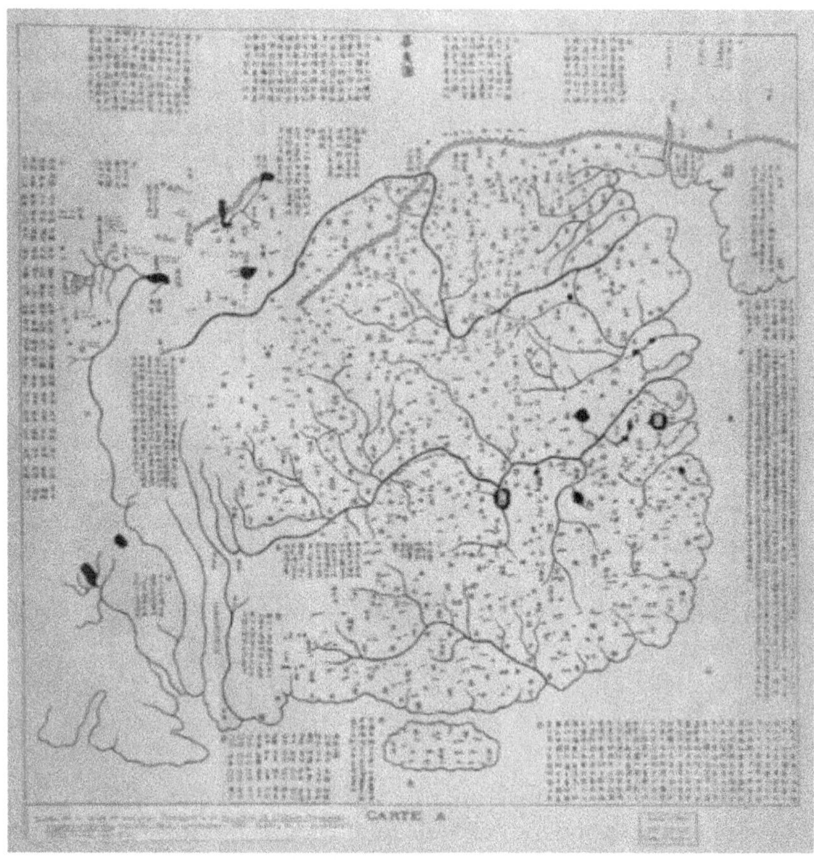

Fig. 2.9. This map shows the 1136 Huayi Tu (華夷圖/华夷图) or Map of China and Barbaric Countries (see endnote 86; public domain).

However, the exact date of its creation remains unknown. It is an extensive Chinese map measuring 386 cm × 456 cm in size. The original text was written in Classical Chinese, but Manchu labels were later superimposed on the surviving copy. This is because the map was created sometime during the Ming Dynasty (some sources say the map was based on information from the Yuan Dynasty) and then handed over to the new rulers of China, the Qing (Great Qing; 清朝 or 大清; 1636–1912).

Figure 2.10(b) shows the oldest extant Korean world map — Hun Yi Jiang Li Li Dai Guo Du Zhi Tu (混一疆理歷代國都之圖/混一疆理历代国都之图) or Honil Gangni Yeokdae Gukdo Ji Do/Map of Integrated Lands and Regions of Historical Countries and Capitals (of China), often abbreviated as Kangnido — completed by the Korean scholars Kwon Kun and Yi Hoe in 1402, during the Joseon Dynasty.[89] Both the Da Ming Hunyi Tu (大明混一圖/大明混一图) or Composite Map of the Ming Empire and the Kangnido were revised after their production, making their original form uncertain. However, the surviving copies of the Kangnido can still be used to infer the original content of the Chinese map, showing the map's Chinese origin.

(a)   (b)

Fig. 2.10.   (a) shows the Da Ming Hunyi Tu (大明混一圖/大明混一图) or Composite Map of the Ming Empire (see endnote 90; public domain) from around 1389 in the early Ming Dynasty; the map stretches eastward to show the Korean Peninsula and Japan. (b) shows the 1402 Kangnido map (see endnote 91; public domain).

From the provided analyses, we know that Korea was on ancient Chinese maps at least from the period of the Tang Dynasty, and the geographical shape of Korea as a peninsula, not an island, matches the modern maps well. Hence, these Chinese maps show a far more realistic Korea than the European maps. The reason is simple: The Chinese, not the Europeans, had landed on the Korean Peninsula in ancient times.

In summary, the knowledge in Sections 4.1 and 4.2 helps to build a sound basis for the analysis in the next section when the Korea-KWQ is compared with the major European maps of the sixteenth century to determine its origin and political era.

5. The Korea-KWQ is of Chinese Origin, not a Direct or Adapted Copy of the Major European Maps from the Sixteenth Century

In this section, the Korea-KWQ will be compared with the Korean portion of three major sixteenth-century European world maps to see their differences. These four maps of Korea are shown in Figs. 2.11(a)–2.11(d).

5.1. *The Korea-KWQ is a far superior map in presenting Korea*

The Korea-Mercator (1569) and the Korea-Ortelius (1570) omitted the Korean Peninsula entirely, while the Korea-Plancius (1594) depicts the Korean Peninsula like a double-blazed long knife, although it has "Corea" written across the whole land. Only the Korea-KWQ gives a realistic shape of the Korean Peninsula and labels the peninsula as "朝鮮/朝鲜", lit. "Joseon".

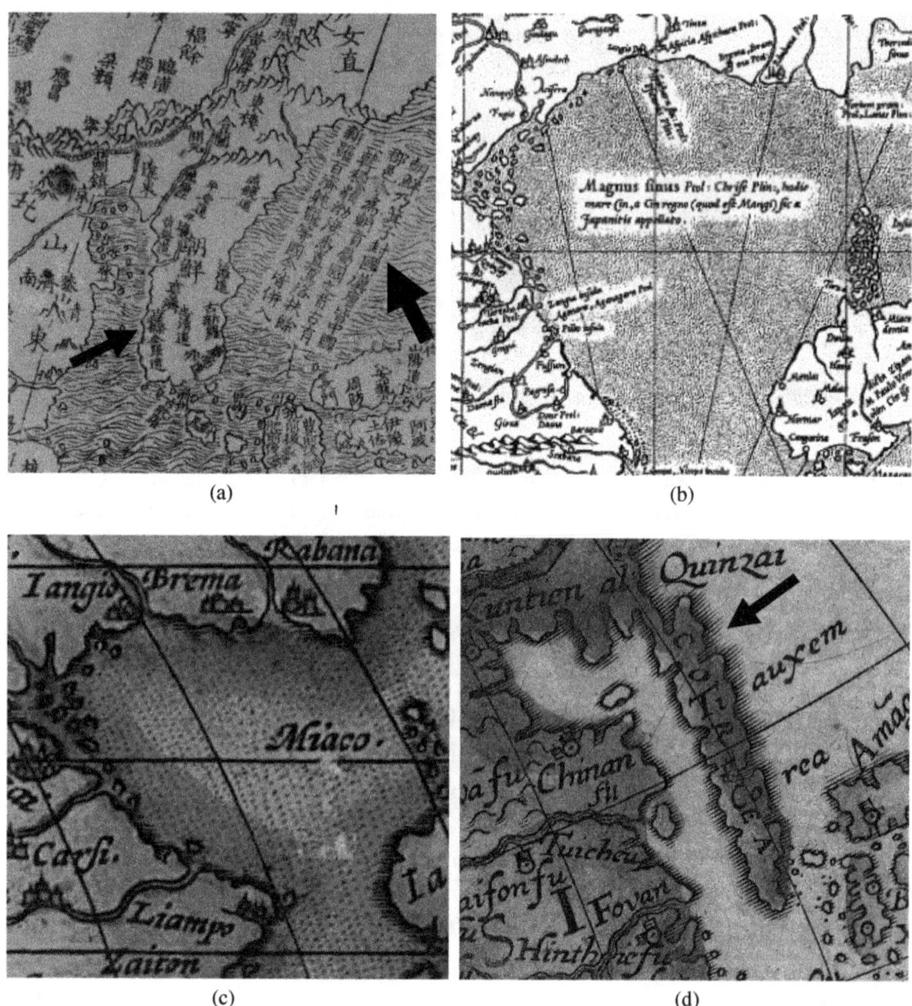

Fig. 2.11. (a) shows the Korea-KWQ extracted from the KWQ (see endnote 92; public domain); the small arrow indicates the Korean Peninsula and the large arrow indicated the annotation which will be thoroughly discussed in Section 6 to reveal Korea's political era during the time of the Korea-KWQ; (b) shows the Korea-Mercator (1569) extracted from the 1569 World Map by Gerardus Mercator (see endnote 93; public domain); (c) shows the Korea-Ortelius (1570) extracted from the 1570 World Map by Abraham Ortelius (see endnote 94; public domain); and (d) shows the Korea-Plancius (1594) extracted from the 1594 World Map by Petrus Plancius (see endnote 95; public domain); the arrow indicates the Korean Peninsula.

The 1589 Maris Pacifici (*Descriptio Maris Pacifici*, lit. "Description of the Pacific Ocean") by Abraham Ortelius does show the Korean Peninsula, but the name is depicted as C. *de Liampo*, instead of "Corea". Since "C." is the abbreviation of "cape" in many European languages, and a cape is smaller and narrower than a peninsula, C. *de Liampo* does not give a good representation of the Korean Peninsula in comparison with the Korea-KWQ. Hence, the Korea-KWQ cannot be a direct or adapted copy of the C. *de Liampo.*

The Korea-KWQ can only be of Chinese origin because Chinese cartographers knew how to draw maps of Korea no later than the Tang Dynasty. Matteo Ricci was in China; he must have used Chinese maps as his source maps to draw the Korea-KWQ.

5.2. *The Korea-KWQ correctly depicts the major geographical points in comparison with those on a modern map*

On a modern map, from west to east, the Yalu River, the Mount Paektu, and the Tumen River divide the two countries of China and Korea. Besides the general geographical comparison in Section 5.1, which displays the superiority of the Korea-KWQ to the contemporary European maps: a close examination of several major geographical points on the KWQ-Korea and their comparisons with those on a modern map shows the following: (1) The largest error in latitudes is 1.9° (34.2°N vs. 32.3°N) which occurs at the southernmost point at Ttangkkeut in Haenam-gun in today's South Korea and (2) the largest error in longitudes is 10.9° (125.4°E vs. 114.5°E) which occurs at the westernmost point at the mouth of the Taedong River in today's North Korea.

Table 2.1 gives the detailed coordinates of these comparisons at several key geographical points on the Korean Mainland. The latitudinal accuracy is impressive, and the c. 10° longitudinal errors are not surprising for the early fifteenth-century Chinese explorers. I have observed and

Table 2.1. Comparing the key geographical points of the Korean Mainland on a modern map with those on the Korea-KWQ.

The key geographical points of the Korean Mainland on a modern map		Modern coordinates	Coordinates in Greenwich meridian of similar points on the Korea-KWQ; all the KWQ latitudinal data have been multiplied by 0.9856 to compare with modern latitudinal data
Southernmost point	Ttangkkeut in Haenam-gun (South Korea)	34.2°N, 126.3°E	32.3°N, 117.5°E
Westernmost point	Sinŭiju in North P'yŏngan Province (Norh Korea)	40.1°N, 124.4°E	38.3°N, 114.0°E
Estuary of the Yalu River	Between Dandong (China) and Sinuiju (North Korea)	Between 40.0°N, 124.4°E and 40.1°N, 124.3°E.	Between 38.2°N, 114.2°E and 38.3°N, 114.0°E.
Mouth of the Tumen River	Sea of Japan (North Korea)	42.3°N 130.7°E	41.9°N 120.5°E
Mouth of the Taedong River	Korea Bay in Yellow Sea (North Korea)	38.7°N, 125.4°E	37.0°N, 114.5°E
Major city	Pyongyang (North Korea)	39.0°N 125.8°E	38.4°N, 116.5°E

discussed the order of magnitude of longitudinal errors made by Zheng He's mariners when they explored the Americas in Chapter 5 of my previous Book titled *Chinese Global Exploration in the Pre-Columbian Era: Evidence from an Ancient World Map*.[96]

Based on the coordinates listed in Table 2.1, we know that the Korea-KWQ depicts the key geographical points on the Korean Mainland with an accuracy that none of the contemporary European maps could match. How could the Korea-KWQ be a direct or adapted copy of any of the three major sixteenth century European maps? The Korea-KWQ cannot be a copy or adapted copy of any Korean maps, either, because the Korean cartographers were using Chinese, Japanese, and European maps as their source maps to draw their world maps and the maps of Korea.[97] Matteo Ricci was in China when he published the KWQ and this section details that the Chinese have known Korea since Antiquity and Chinese cartographers drew the Korean Peninsula no later than the Tang Dynasty. If the Korea-KWQ is not of Chinese origin, who else could have drawn this superior map? From whom did Matteo Ricci get all his accurate geographical information on the Korea-KWQ?

6. The Korea-KWQ Depicts Korea in 1433, Long Before the Arrival of the First Europeans

To determine the political era revealed by the Korea-KWQ, the following three aspects must be taken into account:

6.1. *The annotation on the Korea-KWQ reveals a Korea no earlier than 1404*

This annotation is written next to the Korean Peninsula in Fig. 2.11(a) and this finding is based on the following detailed analysis. The annotation is as follows:

朝鮮乃箕子封國漢唐皆中國郡邑今爲朝貢屬國之首古有三韓穢貊渤海悉直駕洛扶餘新羅百濟耽羅等國今皆併入

My English translation and comments (between square brackets) are as follows:

Gojoseon/Old Joseon [古朝鮮/古朝鲜; 2333–108 B.C.] was founded by Jizi [Gija; 箕子; a member of the Chinese Shang Dynasty royal house, who ruled Gija Joseon or 箕氏朝鮮/箕氏朝鲜 from c. 1100–194 B.C.; the addition of Go or 古, lit. "ancient" or "old", is used in historiography to distinguish the kingdom from the Joseon Dynasty founded in 1392 A.D.]. During the Han and Tang dynasties, the kingdom was a Chinese commandery. Today, Joseon [the Joseon Dynasty founded in 1392 A.D.] ranks top of China's tributary countries. In ancient days, there were kingdoms called Samhan [三韓/三韩], Yemaek [穢貊/秽貊], Balhae [渤海], Sil-jick Joo [悉直], Gaya [加倻 or 駕洛/驾洛], Buyeo [扶餘/扶余], Silla [新羅/新罗], Baekje [百濟/百济], and Tamna [耽羅/耽罗], etc. Now they are all incorporated into one country [referring to the Joseon Kingdom founded in 1392 A.D.].

The above-stated Samhan (三韓/三韩 or Three Han) has been explained in Section 4.1. It refers to the collective name of the Byeonhan (弁韓/弁韩), Jinhan (辰韓/辰韩), and Mahan

(馬韓/马韩) confederacies that emerged in the first century B.C. during the Proto-Three Kingdoms of Korea. They were in the central and southern regions of the Korean Peninsula as shown in Fig. 2.5(b). The Samhan confederacies eventually merged and developed into Baekje (百濟/百济; 18 B.C.–660 A.D.), Gaya (加倻 or 駕洛/驾洛; 42–562), and Silla (57 B.C.–935 A.D.) as shown in Fig. 2.5(c). Gaya was a confederation that ruled central-southern Korea during the Three Kingdoms Period (57 B.C.–668 A.D.). Gaya was the fourth of the "Three" Kingdoms (Baekje, Silla, and Goguryeo) before they (the Gaya nations) were absorbed by Silla in 562.[98]

During the fifth century, Goguryeo (37 B.C.–668 A.D.) reached its height of power and pressured Baekje and Silla. Moreover, during the Chinese Sui and Tang Dynasties, Goguryeo was constantly at war with the Sui and Tang militaries and its national strength was weakened. Finally, according to mainstream Korean historians, in 668, Goguryeo was destroyed by the coalition forces of the Tang Dynasty and Silla.[99] Also, the joint forces of the Tang Dynasty and Silla attacked Baekje earlier in 660 A.D., and Baekje perished. However, Silla was ultimately annexed in 935 A.D. by the new Goguryeo revivalist state of Goryeo.[100]

Yemaek (穢貊/秽貊) was an ancient ethnic group from Manchuria and Korea as shown in Fig. 2.5(a). This group was a precursor[101] of the Korean Goguryeo Kingdom. Yemaek became extinct around the fifth century.

Balhae or Bohai (渤海; 698 A.D.–926 A.D.) was a multi-ethnic kingdom whose land extends to what is today's Northeast China, the Korean Peninsula, and the Russian Far East as shown in Fig. 2.5(d). Balhae was conquered by the Khitan-led Liao Dynasty in 926.[102] The Mongols called northern China "Khitan", a name that later referred to China in general.

Sil-jick Joo (悉直; today's Samcheok, which is a city in Gangwon-do, South Korea) was a small kingdom. It was annexed to Silla in 102 A.D.[103] and annexed to Goguryeo in 468.[104]

The Gaya (加倻 or 駕洛/驾洛; 42–562) Confederation[105] consisted of many small city-states and was the fourth of the "Three" Kingdoms; it was absorbed by Silla in 562.[106]

Buyeo (扶餘/扶余; c. second century B.C.–494 A.D.)[107] was an ancient kingdom, located in southern Manchuria and the northern Korean Peninsula. It was annexed by Goguryeo in 494.

Silla (新羅/新罗; 57 B.C.–935 A.D.) was one of the ancient Three Kingdoms of Korea, as explained earlier. It was handed over to the Goryeo in 935.

Baekje (百濟/百济; 18 B.C.–660 A.D.) was one of the ancient Three Kingdoms of Korea, as explained earlier. It was conquered by Silla and the Chinese Tang Dynasty in 660.

Tamna (耽羅/耽罗; since Antiquity–1404 A.D.) was a state in Jeju Island. It existed from ancient times until it was absorbed by the Joseon Kingdom (朝鮮/朝鲜; 1392–1897) in 1404, following a long period of being a tributary state or autonomous administrative region of various Korean kingdoms. According to *San Guo Shi Ji* (三國史記/三国史记) or *Samguk Sagi/History of the Three Kingdoms*, Tamna in 476 entered a tributary relationship with Baekje; as Baekje waned, Tamna turned to Silla. Tamna briefly reclaimed its independence after the fall of Silla in 935. However, it was subjugated by the Goryeo Dynasty in 938 and officially annexed in 1105. Nevertheless, the kingdom maintained local autonomy until 1404, when Taejong of Joseon unified Korea and brought the Tamna Kingdom to an end.

Joseon (朝鮮/朝鲜; 1392–1897),[108] officially the Great Joseon (大朝鮮國/大朝鲜国), was the last dynastic kingdom of Korea. It was replaced by the Korean Empire (大韓帝國/大韩帝国; 1897–1910) in 1897. The empire stood until Japan's annexation of Korea in August 1910.

In summary, the annotation states that all the above-mentioned kingdoms were incorporated into one country [the Joseon Dynasty founded in 1392 A.D.] at the time of the drawing of the map. Since the Tamna Kingdom was the last country incorporated into Joseon in 1404, we can be sure that *the Korea revealed by the Korea-KWQ must date no earlier than 1404.* This is further supported by the annotation which states that "Today, Joseon [the Joseon Kingdom founded in 1392 A.D.] ranks top of China's tributary countries" because relations between Ming China and Joseon Korea improved dramatically. During the era of the Ming Emperor Hongwu (洪武帝; 1328–1398; reigned from 1368 to 1398) in the early Ming Dynasty when it overthrew the Mongol Yuan Dynasty, the royal court in Goryeo split into two conflicting factions, one supporting the Ming and the other standing by the Yuan. The relationship between Ming and Joseon became much more amicable and mutually profitable during the reigns of Emperor Yongle (1360–1424; reigned 1402–1424) and Emperor Xuande (1399–1435; reigned 1425–1435), and Joseon Korea ranked top of China's tributary countries.

6.2. *The northern border of the Korea-KWQ reveals Korea from 1433*

The Korean Taejo of Joseon (朝鮮太祖/朝鲜太祖; 1335–1408; reigned 1392–1398) declared the name of the kingdom to be Joseon with the endorsement by the neighbouring Ming Dynasty's Emperor Hongwu. The name was a tribute to the ancient Korean state of Gojoseon.

Then, in 1418, during the period of the Ming Yongle Emperor, Sejong the Great (世宗大王; reigned 1418–1450) ascended the throne. On the northern border, Sejong wanted protection against the Jurchens (女真人),[109] who lived in the Hamgil Province (later named Hamgyong in 1509) under the rule of Ming China. Later they became the Manchus (滿族/满族), living in Manchuria. In 1433, during the era of Emperor Xuande (1399–1435; reigned 1425–1435), Sejong sent his prominent general to the north to destroy the Jurchens. The military campaign successfully pushed north, captured several fortresses, and established four counties[110] and six commanderies.[111] Sejong of Joseon succeeded in making the Tumen River the border between the Jurchens and Joseon. Sejong restored Korean territory, reaching roughly the present-day border between North Korea and China: Such a border runs from the estuary of the Yalu River (鴨綠江/鸭绿江; Amrok River) in Korea Bay in the west to the Tumen River (圖們江/图们江) in the east (near today's tripoint where the China–Russia border and the North Korea–Russia border meet).[112]

The above-mentioned borders with China and Jurchen are evidenced on the Korea-KWQ shown in Fig. 2.11(a) and 2.12(a). In these two figures, the territory of Jurchen (女真 or 女直 or 女貞) is depicted to the north and northeast of the Tumen River, and several Chinese place names are depicted not only to the west and north of the Yalu River, but also on the Jurchen land (a clear example is the Nurgan Regional Military Commission). *From these territorial boundaries, we can confirm that the Korea-KWQ reveals the northernmost borders established by Sejong the Great starting from 1433.* Since 1433 is later than 1404, the present finding is consistent with the earlier confirmation that the political era revealed by the Korea-KWQ cannot be earlier than 1404.

The next question is whether the Korea-KWQ shows Korea in 1433 or after 1433.

First, recall that the entire KWQ has been deduced in my previous book titled *Chinese Global Exploration in the Pre-Columbian Era: Evidence from an Ancient World Map* as a map drawn in

1433; this means that the Korea-KWQ cannot be a map after 1433. Since the Jurchen-Joseon border depicted on the Korea-KWQ was established from 1433, not before 1433; this will determine the era of the Korea-KWQ as 1433.

Next, since Jurchen was then under the rule of the Ming Dynasty, and since the Ming emperor established the Nurgan Regional Military Commission (奴兒干都指揮使司/奴儿干都指挥使司) on the Jurchen land, this border condition would also determine the era of the China portion (including the Jurchen land) of the KWQ as 1433. On the Korea-KWQ, there is an annotation to the east and northeast of the Jurchen land as indicated by a smaller arrow in Fig. 2.12(a). This annotation is written next to the place name "奴兒干":

奴兒干都司皆女直地元為胡里改今設一百十四衛所其分地不詳

My translation and comments (between square brackets) are as follows:

The Nurgan Regional Military Commission [奴兒干都指揮使司] is located on the Jurchen land. In the Yuan Dynasty, the region was occupied by the Huligai tribes. Now the Jurchen tribes are divided into 114 Guard Battalions [in the Ming Dynasty, the Jurchen became vassals to the Ming emperors]. But their detailed ownerships of the lands are unknown.

During the Yuan Dynasty, the Taowen, Huligai, and Wodolian Jurchen tribes[113] lived in Heilongjiang (黑龍江/黑龙江; a province in northeast China) around Yilan (依蘭/依兰) when it was part of Liaoyang Province (遼陽等處行中書省/辽阳等处行中书省) and governed as a circuit (道; a historical political division of China). These tribes became the Jianzhou Jurchens in the Ming Dynasty.

In 1409, during the era of Emperor Yongle (who reigned from 1402 to 1424), the Ming government created the Nurgan Regional Military Commission (奴兒干都指揮使司/奴儿干都指挥使司; 1409–1435),[114] abbreviated as Nurgan Command Post (奴兒干都司/奴儿干都司). It was a Chinese seat or administrative unit established on Jurchen land. The outpost Nurgan (奴兒干/奴儿干) was at Telin (特林; modern Tyr, Russia), near the mouth of the Amur River (Heilongjiang; 黑龍江/黑龙江, lit. "Black Dragon River"), about 100 km/62 mi from the sea. The Jurchens came under the nominal administration of the Nurgan Command Post.

The Ming official's last inspection of the Nurgan Command Post was in 1433 by Eunuch Yishiha (a Jurchen eunuch of the Ming Dynasty of China; fl. 1409–1451), and the Nurgan Regional Military Commission was formally abolished in 1435.[115]

The Yongning Temple Stele (永寧寺碑/永宁寺碑) was erected by Eunuch Yishiha in 1413 in the Nurgan outpost during the Ming Yongle Era. The inscriptions on the stele are in four languages: Chinese, Jurchen, Mongol, and Tibetan. They give a detailed record of the Ming court's efforts to assert suzerainty over the Jurchens and commemorate the founding of the Yongning Temple (永寧寺/永宁寺; lit. "The Temple of Eternal Peace", dedicated to the Goddess of Mercy, Guanyin).[116] The location of the temple is in the village of Tyr near Nikolayevsk-on-Amur in today's Russia. A stele with a monolingual Chinese inscription commemorating the rebuilding of the temple by Yishiha was erected in 1433. Both monuments are now held at the Arsenyev Museum in Vladivostok. Figure 2.12(b) shows a sketch of the 1413 Yongning Temple Stele.

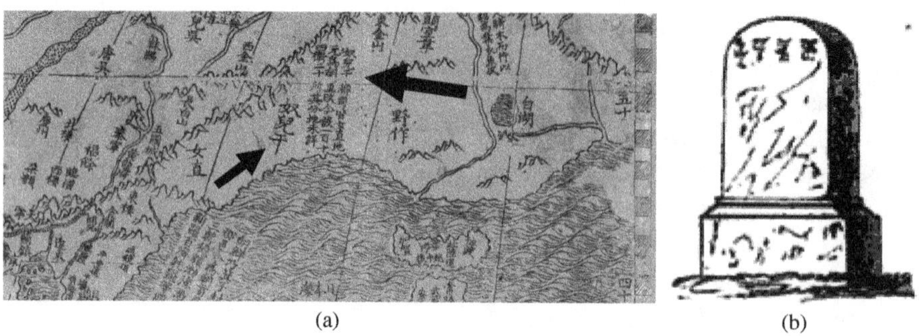

Fig. 2.12.   (a) shows the location of Nurgan City (indicated by a small arrow) and the annotation (indicated by the large arrow); this map has been extracted from the KWQ (public domain). (b) sketch of the 1413 Yongning Temple Stele (see endnote 117; public domain) from *The Russians on the Amur (1861)* by Ernst Georg Ravenstein (1834–1913).

The above gives detailed history that the Nurgan Regional Military Commission (奴兒干都指 揮使司/奴儿干都指挥使司) depicted on the Jurchen land was last inspected by the Ming envoy in 1433. Although the Commission existed in 1409–1435, the years 1434–1435 are after the entire KWQ was drawn and the years 1409–1432 are before the Joseon-Jurchen border was established. Hence, both the Korea-KWQ and the China portion (including the Jurchen land) of the KWQ are of the era 1433.

The Ming rule of Manchuria began with its conquest of the region in the late 1380s after the fall of the Mongol-led Yuan Dynasty and reached its peak in the early fifteenth century with the establishment of the Nurgan Regional Military Commission. Chinese Guard Battalions (衛所/ 卫所) established in association with the tribal military units which were ruled by the hereditary tribal leaders. "Guard Battalion" was an administrative division at the regional level. In the Yongle period, 178 Guard Battalions were set up in Manchuria. However, the annotation states a far lower number of 114. This is because the Ming power waned considerably in Manchuria after 1424 (Emperor Yongle died in 1424), and the Nurgan Regional Military Commission was abolished in 1435. Hence, it is not surprising that the number of Guard Battalions dropped from 178 in the Yongle Era (1402–1424) to 114 in the Xuande Era (1425–1435). The year 1433 falls in the Xuande Era.

Starting in the 1580s, the Jianzhou Jurchen chieftain Nurhaci (努爾哈赤/努尔哈赤; 1559– 1626; as a chieftain, he was nominally a Ming vassal) began to take control of most of Manchuria over the next several decades (compare Figs. 2.13 and 2.14 to see the Ming's retreat from the Jurchen land from the early fifteenth to the late sixteenth century). By the early seventeenth century, in 1616, Nurhaci established the Later Jin Dynasty (後金/后金; 1616–1636) of China. The Qing Dynasty established by his son Hong Taiji (皇太極/皇太极; 1592–1643) would eventually conquer the Ming and take control of Southern China.

Figure 2.14(a) clearly shows that by the year 1580, the Nurgan Regional Military Commission was long gone and erased from the map, while the Jurchens became more powerful and were in full control of their own land. They were renamed Manchus in 1635 by Hong Taiji. Figure 2.14(b) shows the Japanese invasion of Korea (1592–1598). Notice that in the northern Korean Peninsula,

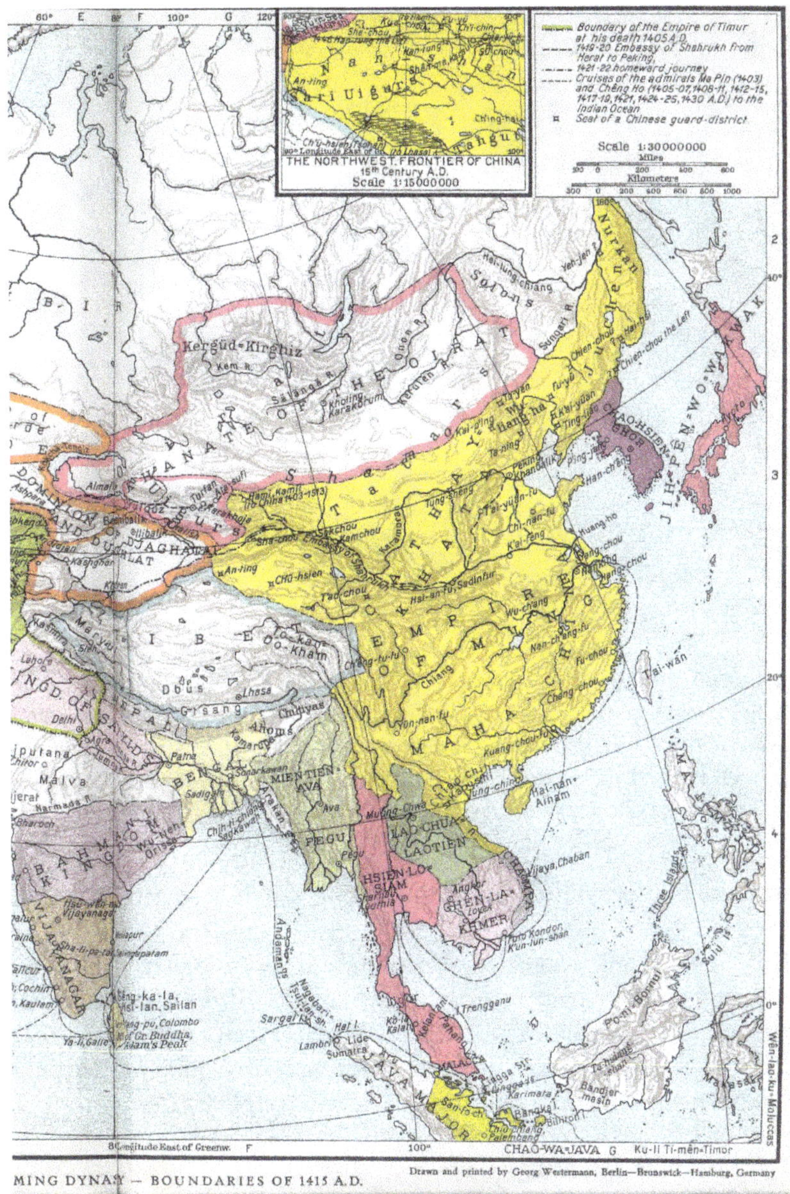

Fig. 2.13. The Ming Empire (in yellow) under the Yongle Emperor in 1415 (see endnote 118; public domain); the Nurgan Regional Military Commission and Jurchen are in the upper right corner.

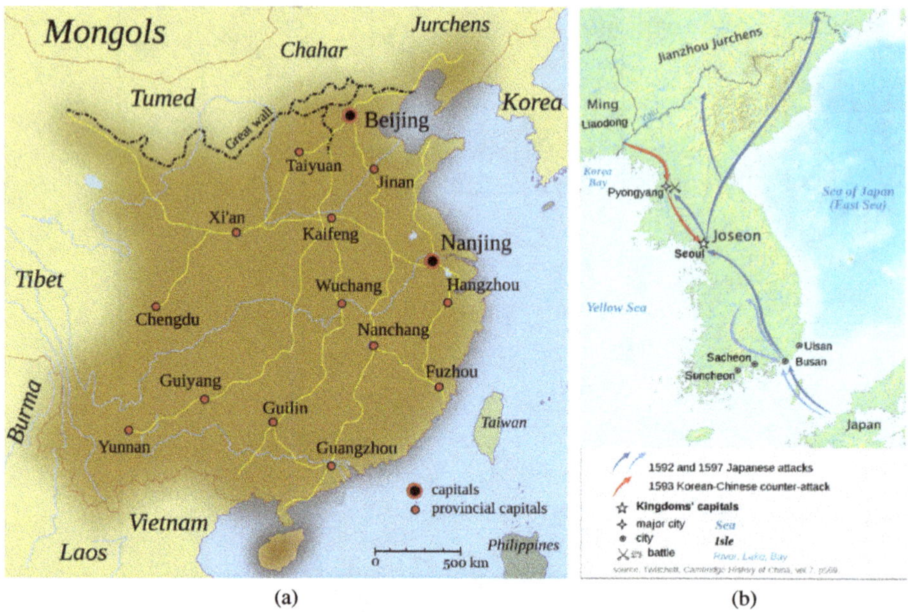

Fig. 2.14. (a) shows Ming China and its surrounding areas around 1580 (see endnote 119; *Source*: Říše Ming.png: Michal Klajban, under CC BY-SA 3.0, https://commons.wikimedia.org/wiki/File:Ming_Empire_cca_1580_(en).svg), revealing that Ming China lost most of the Jurchen land it had ruled in the early fifteenth century. (b) shows Joseon and its neighbouring countries from 1592 to 1597 during the Japanese invasion of Korea (see endnote 120; *Source*: Yug, under CC CC0 1.0 Universal Public Domain Dedication, https://commons.wikimedia.org/wiki/File:History_of_Korea-1592-1597.svg), depicting that Jianzhen Jurchens occupied most of the regions north of the Yalu and Tumen Rivers.

Joseon shared its northern border with the Jianzhou Jurchens, while sharing its western border with Ming Liaodong.

If the Korea-KWQ was really drawn by Matteo Ricci in 1602 based on European maps, it should show Korea's northern border as it is depicted in Fig. 2.14(b), instead of what is depicted in Fig. 2.13. The presence of the Nurgan Regional Military Commission and Nurgan City, as well as Ming China's control of the vast Jurchen land north of the Yalu and Tumen Rivers, is strong evidence of the Korea-KWQ showing the political conditions of Korea and China as they were in 1433.

6.3. *The provincial structure depicted on the Korea-KWQ is consistent with what Korea should have been in 1433, except for the Hamgyong Province*

During most of the Joseon Dynasty, the country was organised into eight provinces[121] (see Fig. 2.15. These provinces are listed in Table 2.2: (1) Chungcheong Province (*Chungcheon-do*; 忠清道; it was formed in 1356 during the Goryeo Dynasty); it is mistakenly depicted on the Korea-KWQ as landlocked; (2) Gangwon Province (*Gangwon-do*; 江原道; it was formed in 1395 during the Joseon Dynasty); (3) Gyeonggi Province (*Gyeonggi-do*; 京畿道; it has been a politically

Fig. 2.15.   Provinces of Joseon till 1896 with English labels (see endnote 122; *Source*: Kanguole, under CC BY-SA 4.0, https://commons.wikimedia.org/wiki/File:Eight_provinces_of_Korea-en.svg).

Table 2.2.   Eight provinces in Joseon until 1896

Province #	Name	Description
1	Chungcheong Province (*Chungcheon-do*; 忠清道)	It was formed in 1356 during the Goryeo Dynasty; it is mistakenly depicted on the Korea-KWQ as landlocked.
2	Gangwon Province (*Gangwon-do*; 江原道)	It was formed in 1395 during the Joseon Dynasty.
3	Gyeonggi Province (*Gyeonggi-do*; 京畿道)	It has been a politically important area since 18 B.C., during the Three Kingdoms Period); over time, Seoul in Gyeonggi Province was the capital of various Korean states, including Baekje, Joseon, the Korean Empire, Goryeo (as a secondary capital), and presently South Korea.
4	Gyeongsang Province (*Gyeongsang-do*; 慶尚道/ 庆尚道)	The predecessor to Gyeongsang Province was formed during the Goryeo Dynasty; Gyeongsang acquired its current name in 1314.

(Continued)

Table 2.2.   (*Continued*)

Province #	Name	Description
5	Jeolla Province (*Jeolla-do* 全羅道/全罗道)	In 1413, when the Joseon territories were reorganised into eight *do*, the area had various names; but the original name persisted and later was shortened to simply *Jeolla-do*.
6	Hwanghae Province (*Hwanghae-do*; 黃海道)	In 1395, the province was originally organised as Punghae or *Punghae-do* (豐海道/丰海道); in 1417, the province was renamed *Hwanghae-do*.
7	Pyongan Province (*P'yŏngan-do*; 平安道)	The province was formed in 1413.
8	Hamgyong Province (*Hamgyŏng-do*; 咸鏡道/ 咸镜道)	The province was first established as Yonggil (永吉, *Yŏnggil*) in 1413; it was renamed as Hamgil (咸吉) three years later; renamed again as Yongan or *Yŏngan* (永安) in 1470; and eventually renamed as Hamgyong in 1509.

important area since 18 B.C., during the Three Kingdoms Period); over time, Seoul in Gyeonggi Province became the capital of various Korean states, including Baekje, Joseon, the Korean Empire, Goryeo (as a secondary capital), and presently South Korea; (4) Gyeongsang Province (*Gyeongsang-do;* 慶尙道/庆尙道; the predecessor to Gyeongsang Province was formed during the Goryeo Dynasty); Gyeongsang acquired its current name in 1314; (5) Jeolla Province (*Jeolla-do* 全羅道/全罗道); in 1413, when the Joseon territories were reorganised into eight *do*, the area had various names but the original name persisted, and later was shortened to simply *Jeolla-do*; (6) Hwanghae Province (*Hwanghae-do*; 黃海道); in 1395, the province was organised as Punghae or *Punghae-do* (豐海道/丰海道); in 1417, the province was renamed *Hwanghae-do*; (7) Pyongan Province (*P'yŏngan-do*; 平安道); the province was formed in 1413; and (8) Hamgyong Province (*Hamgyŏng-do*; 咸鏡道/咸镜道); the province was first established as Yonggil (永吉, *Yŏnggil*) in 1413; it was renamed Hamgil (咸吉) three years later, renamed Yongan or *Yŏngan* (永安) in 1470, and eventually renamed Hamgyong in 1509.

Among the eight provinces, Gyeonggi Province around Seoul is omitted on the Korea-KWQ. Moreover, the names of seven of the eight provinces were established in the early fifteenth century, including the latest, Hwanghae Province, in 1417. This is consistent with Korea's political condition in 1433 as revealed by the Korea-KWQ. But the name of the eighth province "Hamgyong Province" was established in 1509, not earlier or in 1433. However, the Korea-KWQ cannot be a map of the year 1509, or after 1509, because by 1509 the Ming Nurgan Regional Military Commission on the Jurchen land had long been abolished and thus should not appear on the Korea-KWQ, and yet it *does* appear. In fact, in Chapter 3 of this book, the political landscape of the whole of China on the KWQ is discussed, which is consistent with its political era of 1433, not 1509. Hence, it makes one wonder the following: *Did Matteo Ricci replace the old name of the eighth province with its 1509 new name Hamgyong to make the Korea-KWQ look "less ancient"?*

None of the maps — Korea-Mercator (1569), Korea-Ortelius (1570), Korea-Plancius (1594), and 1589 Maris Pacifici by Abraham Ortelius — correctly depicts the northern border of Joseon in

1433, nor did they annotate the Nurgan Regional Military Commission (奴兒干都指揮使司/奴儿干都指挥使司) — a Chinese administrative unit established in 1409 on the Jurchen land in Manchuria — where Eunuch Yishiha of the Ming China made a final inspection in 1433.

None of the above-mentioned European cartographers depicts the eight provinces on the Korean Peninsula.

The Ming Dynasty of China shared a long border and close relations with the Joseon Dynasty of Korea. Both dynasties also shared Confucian ideals, which due to space limitations will not be discussed in this book. Joseon had been a member of the Ming's Sino-centric East Asian regional order long before the Europeans' first arrival in Korea.

7. Conclusions

In summary, in the Introduction of this chapter, I stated that the purpose of this chapter was to analyse the depiction of Korea on an ancient world map — Kunyu Wanguo Quantu (坤輿萬國全圖/坤輿万国全图) (abbreviated as KWQ) or Complete Geographical Map of All the Kingdoms of the World published by Matteo Ricci in 1602 in China — to determine its origin and era. This Korean portion of the map is called the Korea-KWQ in this book. By analysing the annotation and examining the location and history of every geographical item depicted on the Korea-KWQ, I can extract the era of Korea as revealed by the map; this is the era or date when these places were last explored before the map was drawn.

Based on the analyses presented in Sections 1–6, the following conclusions can be reached:

(1) The Korea-KWQ is of Chinese origin, not a copy or adapted copy of the major contemporary European maps, because China–Korea interactions had been established centuries before Christ was born and before the Europeans' first arrival in Korea occurred in the late sixteenth century. Moreover, Chinese cartographer Jia Dan (賈耽/贾耽; 730–805 A.D.) had already drawn a map containing Korea in 801 A.D. and presented it to Emperor Dezong of Tang (唐德宗).

(2) The Korea-KWQ gives an accurate geographical depiction of the Korean Peninsula and its key geographical points. No other major contemporary European maps give depictions of similar accuracy.

(3) The Korea-KWQ shows Joseon's northern borders with Ming China and the Jurchens, revealing the political era of Korea in 1433 as detailed in Subsection 6.2.

(4) The China portion of the KWQ is also in the era 1433 for the following reason: the era 1433 of the Korea-KWQ is the era when the Jurchen-Joseon border was established, which is also the era of Jurchen on the map. Since China ruled Jurchen at that time and established the Nurgan Command Post on the Jurchen land, the China portion of the KWQ is also in the era 1433.

(5) The conclusion reached in point (4) will be evidenced by the rigorous analysis to be presented in Chapter 3.

(6) Only seven of the eight provinces established in the Joseon Dynasty (1392–1897) and lasting until 1895 are depicted on the Korea-KWQ. Among them, Chungcheong Province should face

the Yellow Sea but is mistakenly depicted as a landlocked province on the Korea-KWQ. The missing province is Gyeonggi Province around Seoul.

(7) The geographical location of the ancient Buyeo is depicted mistakenly on the west coast of the Korean Peninsula facing the Yellow Sea. Buyeo should be an ancient kingdom located in southern Manchuria and on the northern Korean Peninsula.

(8) The name of Hamgyong Province only existed from 1509 onward. Since we conclude in (4) that the Korea-KWQ is a map of 1433, then, the correct name of the province should be Hamgil (valid in the years 1416–1469), which was the name of the province before the name of Hamgyong was established. If Matteo Ricci and his collaborators did not change Hamgil into Hamgyong on their Chinese source map while making the Korea-KWQ, who else could have done so?

Appendix

The geographical items on the Korea-KWQ are analysed in detail in this Appendix. On the map, they are all written in Chinese traditional characters. Since many readers use the Chinese simplified characters, I have decided to use both: All the Chinese characters in the main text and in this Appendix are expressed first by the Chinese traditional characters followed by the Chinese simplified characters, separated by a slash "/". If both characters are the same, that expression will not be repeated. Moreover, all the geographical items are numbered in sequence in this Appendix. The Chinese "pinyin" of each item is given right after it is introduced. The history, geography, and my comments for each geographical item are given in the last column of the table.

Item #	Name in Chinese	Pinyin	Name in English	History, Geography, and My Comments
				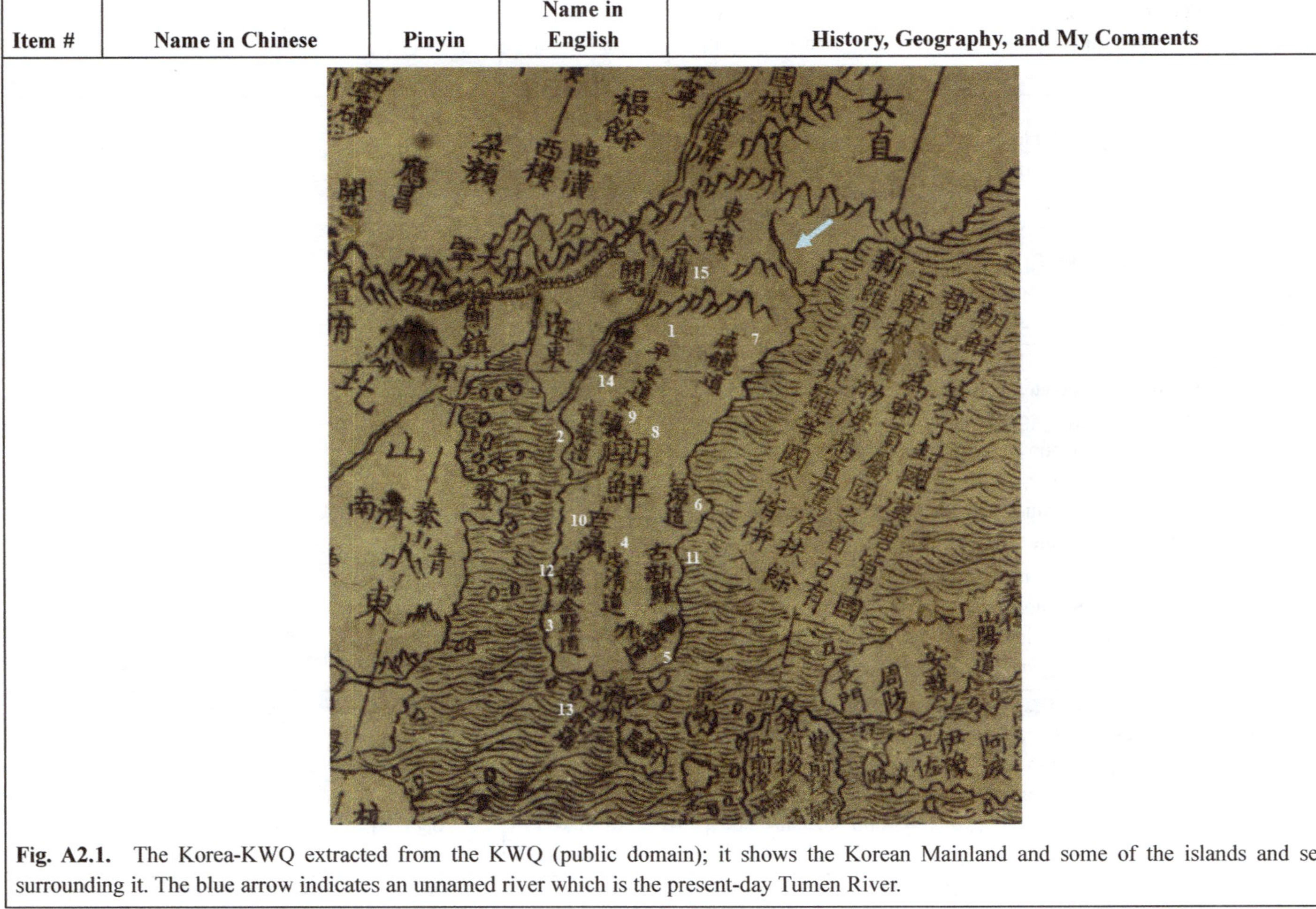

Fig. A2.1. The Korea-KWQ extracted from the KWQ (public domain); it shows the Korean Mainland and some of the islands and seas surrounding it. The blue arrow indicates an unnamed river which is the present-day Tumen River.

(*Continued*)

Item #	Name in Chinese	Pinyin	Name in English	History, Geography, and My Comments
1	平安道	Ping An Dao	Pyongan Province	Provinces (*do*; 道) have been the primary administrative division of Korea since the mid-Goryeo Dynasty in the early eleventh century. During the Joseon Dynasty, the country was redivided into eight new provinces (*do*) in 1413. Pyongan Province (*P'yŏngan-do*; 平安道) was established in 1413. Its name derived from the names of two of its principal cities, Pyeongyang (平壤) and Anju (安州).[123]
2	黃海道	Huang Hai Dao	Hwanghae Province	In 1395, the province was organised as Punghae Province (豐海道/丰海道). In 1417, the province was renamed as Hwanghae Province (黃海道). The name was derived from the names of the two principal cities of Hwangju (黃州) and Haeju (海州).[124]
3	全羅道	Quan Luo Dao	Jeolla Province	During the Samhan Period of Korean history, the area of Jeolla (全羅/全罗) was controlled by the Mahan Confederacy and the Tamna Kingdom (耽羅/耽罗) on Jeju (濟州島/济州岛). In 1413, Jeolla Province (*Jeolla-do*; 全羅道/全罗道) became one of the Eight Provinces of Korea during the Joseon Dynasty (1392–1897). It was located in the southwestern Korean Peninsula. The name was derived from the names of the principal cities of Jeonju (全州) and Naju (羅州/罗州).[125]
4	忠清道	Zhong Qing Dao	Chungcheong Province	Chungcheong Province (*Chungcheong-do*; 忠清道) was formed in 1356 during the Goryeo Dynasty. Its name was derived from the names of the principal cities of Chungju (忠州) and Cheongju (清州).[126] During the Joseon Dynasty, Chungcheong Province was one of the eight new provinces established in 1413.[127] On the Korea-KWQ, Chungcheong Province is mistakenly depicted as an inland province. The correct geographical location is shown in Fig. 2.15 facing the Yellow Sea.
5	慶尙道	Qing Shang Dao	Gyeongsang Province	Gyeongsang Province (*Gyeongsang-do*; 慶尙道/庆尚道) was one of the Eight Provinces of Korea during the Joseon Dynasty (1392–1897). Gyeongsang was in the southeast of Korea as shown in Fig. 2.15. The predecessor to Gyeongsang Province was formed during the Goryeo Dynasty (高麗/高丽; 918–1392). Gyeongsang acquired its current name in 1314. The name is derived from the names of the principal cities of Gyeongju (慶州/庆州) and Sangju (尙州).[128]

6	江原道	Jiang Yuan Dao	Gangwon Province	Gangwon Province (*Gangwon-do*; 江原道) was formed in 1395 and became one of the Eight Provinces of Korea during the Joseon Dynasty (1392–1897). The province derived its name from the names of the principal cities of Gangneung (江陵) and the provincial capital Wonju (原州).[129]
7	咸鏡道	Xian Jing Dao	Hamgyong Province	The old name of Hamgyong Province (*Hamgyŏng-do*; 咸鏡道/咸镜道) was Yonggil (永吉) Province, which was established in 1413; it was one of the Eight Provinces of Korea during the Joseon Dynasty (1392–1897). The name was changed to Hamgil (咸吉) in 1416 and again changed to Yongan (永安) in 1470. Finally, it was renamed Hamgyong Province (咸鏡道/咸镜道) in 1509 after its two principal cities, Hamhung (咸興/咸兴) and Kyongsong (鏡城/镜城).[130] Hence, in 1433, the province's name was Hamgil (咸吉). *Previously, the indigenous people living in this province were the Jurchens (女真人) under the rule of Ming China. In 1433, Sejong of Joseon sent troops there to make the Tumen River the border between Joseon and the Jurchens.* The year 1509 was during the era of the Zhengde Emperor (正德帝; 1491–1521; reigned 1505–1521), who was the eleventh emperor of the Ming Dynasty. By this era, the Nurgan Regional Military Commission (1409–1435) established by the Yongle Emperor (the third Emperor of the Ming Dynasty) had long been abolished, and Ming China had mostly lost its control over the Jurchens who became more powerful. It seems that Matteo Ricci changed the old name of Hamgil (咸吉) in 1433 to its final name Hamgyŏng (咸鏡/咸镜), which was established from 1509, to make the KWQ look "less ancient". But such a change created great inconsistencies with the political landscape revealed by Korea, the Jurchens and China on the Korea-KWQ as detailed in the main text of this chapter.

(*Continued*)

Item #	Name in Chinese	Pinyin	Name in English	History, Geography, and My Comments
8	朝鮮	Zhao Xian	Joseon	On the Korea-KWQ, Joseon (朝鮮/朝鮮) is depicted where the Gyeonggi Province (Gyeonggi-do; 京畿道; 1392–1897) should be. *Gyeonggi* means "京 (the capital)" and "畿 (the surrounding area)". Seoul was the capital of various Korean states, including Baekje, Joseon, the Korean Empire, Goryeo (as a secondary capital), and presently South Korea, and was known by other names during different Korean political eras in history. On the Korea-KWQ, Gyeonggi Province is not depicted. However, the name 朝鮮 (Joseon) is depicted near the Seoul area where the province was located.
9	平壤	Ping Rang	Pyongyang	On a modern map, Pyongyang (39.0°N, 125.8°E) is located on the Taedong River (大同江) about 109 km/68 mi upstream from its mouth at Nampo (38.7°N, 125.4°E). The river flows into Korea Bay in the Yellow Sea (黃海). On the KWQ-Korea, there is an unnamed river running through 平壤 (Pyongyang; 38.4°N, 116.5°E; the KWQ latitudinal coordinate has been converted to the modern latitudinal coordinate after multiplying it by 0.9856), which correlates well with the Taedong River. This unnamed river flows into an unnamed sea at 37.0°N, 114.5°E (the KWQ latitudinal coordinate has been converted to the modern latitudinal coordinate after multiplying it by 0.9856), which correlates closely with the coordinates of Nampo (38.7°N, 125.4°E), the mouth of the Taedong River. Again, an error around 10° is observed for the longitudes of the geographical items depicted on the Korea-KWQ. Pyongyang is one of the oldest cities in Korea.[131] It was the capital of two ancient Korean kingdoms, Gojoseon (古朝鮮/古朝鮮) and Goguryeo (高句麗/高句丽), and the secondary capital of Goryeo (高麗/高丽). Around the beginning of the Joseon Dynasty in 1394, the capital was moved to Seoul, also known as *Hanyang* and later as *Hanseong* (漢城/汉城, lit. "Fortress city on the Han River"), where it remained until the fall of the dynasty. The Seoul Capital Area contains five UNESCO World Heritage Sites and the Royal Tombs of the Joseon Dynasty.[132] But on the Korea-KWQ, the capital of the Joseon Dynasty is not depicted, but the other ancient capital Pyongyang (平壤) is depicted.

| 10 | 古百濟 | Gu Bai Ji | Ancient Baekje | Baekje or Paekche (百濟/百济; 18 B.C.–660 A.D) was a Korean kingdom located in southwestern Korea as shown in Fig. 2.5(c). It was one of the Three Kingdoms of Korea, together with Goguryeo and Silla.
Baekje was founded at Wiryeseong (慰禮城/慰礼城; present-day southern Seoul). Both Baekje and Goguryeo claimed to succeed Buyeo around the time of Gojoseon's fall.[133]
At its peak in the fourth century, Baekje controlled most of the western Korean Peninsula (Silla was its eastern neighbour), as far north as Pyongyang. It became a significant regional sea power, with political and trade relations with China and Japan.[134]
According to mainstream Korean historians, in 660, Baekje was defeated by the Tang Dynasty and Silla,[135] and ultimately submitted to Unified Silla.[136] |
| 11 | 古新羅 | Gu Xin Luo | Ancient Silla | Silla (新羅/新罗; 57 B.C.–935 A.D.) was a Korean kingdom located in the southern and central parts of the Korean Peninsula as shown in Fig. 2.5(c). Silla along with Baekje and Goguryeo formed the Three Kingdoms of Korea.
Silla began as a chiefdom in the Samhan Confederacies. According to mainstream Korean historians, Silla once allied with Sui China and then Tang China, until it annexed Geumgwan Gaya (金官伽倻) in 532, conquered Daegaya (大伽倻) in 562, and eventually conquered Baekje in 660 and Goguryeo in 668.[137] Thereafter, Unified Silla occupied most of the Korean Peninsula, while the northern part reemerged as Balhae, a successor state of Goguryeo as shown in Fig. 2.5(d).
Silla fragmented into the brief Later Three Kingdoms of Silla (Unified Silla), Later Baekje, and Taebong, and handed over power to Goryeo in 935[138] (some sources say 936). |

(*Continued*)

Item #	Name in Chinese	Pinyin	Name in English	History, Geography, and My Comments
12	古扶餘	Gu Fu Yu	Ancient Buyeo	Buyeo (扶餘/扶余; c. second century B.C.–494 A. D.) was an ancient kingdom that was centered in northern Manchuria in modern-day northeast China, as shown in Figs. 2.5(a)–2.5(c). Buyeo is a major predecessor of the Korean kingdoms of Goguryeo (37 B.C.–668 A.D.) and Baekje (18 B.C.–660 A.D.).[139] According to *Hou Han Shu* (後漢書/后汉书) or *The Book of the Later Han*, Buyeo was initially placed under the jurisdiction of the Xuantu Commandery (玄菟郡), one of the four commanderies of Han in the Western Han Dynasty (202 B.C.–9 A.D.). Later, Buyeo entered formal diplomatic relations with the Eastern Han Dynasty (25–220 A.D.) by the mid-first century A.D. as an important ally to check the Xianbei (鮮卑/鲜卑) and Goguryeo (高句麗/高句丽) threats. Buyeo was then placed under the jurisdiction of the Liaodong Commandery (遼東郡/辽东郡) of the Eastern Han. After a savage Xianbei invasion in 285, Buyeo was restored with help from the Jin Dynasty (金朝 or Great Jin; 大金; 1115–1234). However, Buyeo started to decline. A second Xianbei invasion in 346 finally destroyed the state. Only some remnants in its core region survived as vassals of Goguryeo until a final annexation in 494.[140] On the KWQ-Korea, the location of ancient Buyeo is depicted on the west coast of the Korean Peninsula (see Fig. A2.1) facing the Yellow Sea. This is a huge mistake. Figure A2.2 shows the correct geographical location of Buyeo in the third century.

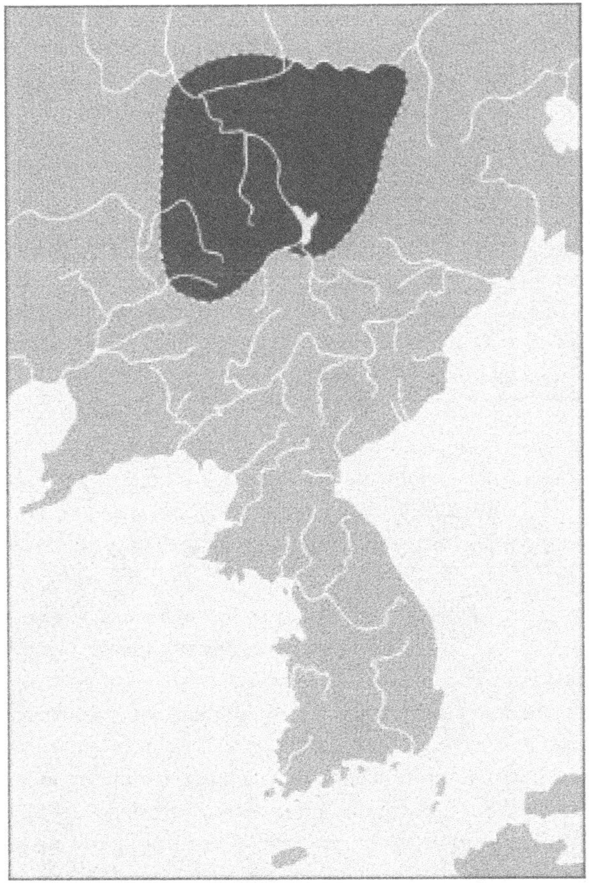

Fig. A2.2.   Map of Buyeo in the third century (see endnote 141; *Source*: Samhanin, under CC CC0 1.0 Universal Public Domain Dedication, https://commons.wikimedia.org/wiki/File:Buyeo.svg).

(Continued)

Item #	Name in Chinese	Pinyin	Name in English	History, Geography, and My Comments
13	濟州 (古耽羅)	Ji Zhou (Gu Dan Luo)	Jeju Island: The ancient Korean Tamna Kingdom once ruled Jeju Island.	Tamna (耽羅/耽罗; from Antiquity–1402 A.D.) was an ancient Korean kingdom that ruled Jeju Island. The kingdom had been a tributary state or autonomous administrative region of various Korean kingdoms. According to *San Guo Shi Ji* (三國史記/三国史记) or *Samguk Sagi/ History of the Three Kingdoms*, Tamna entered into a tributary relationship with Baekje in 476, which controlled the southwestern Korean Peninsula. As Baekje waned, Tamna turned to Silla. Near the end of the Three Kingdoms Period (57 B.C.–668 A.D.), Tamna officially subjugated itself to Silla. Tamna briefly reclaimed its independence after the fall of Silla in 935. But it was soon subjugated by the Goryeo Dynasty in 938 and officially annexed in 1105. The kingdom managed to maintain local autonomy until 1404, when Taejong of Joseon (朝鮮太宗/朝鲜太宗; 1367–1422; reigned 1400–1418) unified Korea and brought the Tamna Kingdom to an end. Based on the above-mentioned history, we know that the Korea-KWQ cannot be dated earlier than 1404.
14	鴨綠江	Ya Lu Jiang	Yalu River	The Yalu River (鴨綠江/鸭绿江) is known to Koreans as the Amrok River or Amnok River. Together with the Tumen River (圖們江/图们江) to its east, and a small portion of Mount Paektu (白頭山/白头山), the Yalu River forms the western border between North Korea and China.
15	合蘭	He Lan	Helan Prefecture	Helan Fu (合蘭府/合兰府; 府 lit. "Fu") was an administrative division at the commandery ("*jun*"; 郡縣/郡县) level, set up by the Yuan Dynasty. The seat of the government was in present-day South Hamhung, South Hamgyong Province, North Korea (北韓咸鏡南道咸鹹興南/北韩咸鏡南道咸興南). The place was annexed by Goryeo (高麗/高丽; 918–1392) near the end of the Yuan Dynasty (1271–1368).

Remark on all the geographical items depicted on the Korea-KWQ: Other than the ancient Chungcheong Province being mistakenly depicted as an inland province, ancient Buyeo being mistakenly depicted on the west coast of the Korean Peninsula to face the Yellow Sea, and Gyeonggi Province being omitted, the Korea-KWQ correctly depicts the geographical items on the Korean Mainland and its surrounding islands and seas. It seems that Matteo Ricci changed the old name of Hamgil (咸吉) in 1433 to its final name Hamgyŏng (咸鏡/咸镜), which was established from 1509, to make the KWQ look "less ancient". But such a change created severe inconsistencies with the political condition revealed by Korea and China on the Korea-KWQ and the China-KWQ as detailed in in Sections 4 and 6.

Endnotes

[1]Library of Congress. "Kun yu wan guo quan tu." www.loc.gov, Library of Congress, 2025, loc.gov/item/2010585650.

[2]Ronnie Po-Chia Hsia. *Matteo Ricci and the Catholic Mission to China, 1583–1610: A Short History with Documents (Passages: Key Moments in History)*. Indianapolis, IN, USA: Hackett Publishing Company, Inc., 2016.

[3]Sheng-Wei Wang. *Chinese Global Exploration in the Pre-Columbian Era: Evidence from an Ancient World Map*. Singapore: World Scientific Publishing Co., 2023.

[4]Marco Polo. Ronald Latham, trans. *The Travels of Marco Polo*. London, UK: Folio Society, 1958; Marco Polo. Nigel Cliff, ed., trans., and Intro. Coralie Bickford-Smith, Illust. *The Travels*. London, UK: Penguin Classics, 2015.

[5]Eugene Y. Park. *Korea: A History*. Stanford, CA, USA: Stanford University Press, 2022.

[6]Jennifer Swanson. *Bartolomeu Dias: First European Sailor to Reach the Indian Ocean (Spotlight on Explorers and Colonization)*. New York, New York, USA: Rosen Young Adult, 2017.

[7]Nigel Cliff. *The Last Crusade: The Epic Voyages of Vasco da Gama*. New York, New York, USA: Harper Perennial, 2012.

[8]Hendrick Hamel. Jean-Paul Buys, trans. *Hamel's Journal and a Description of the Kingdom of Korea 1653–1666*, 3rd revised ed. Seoul, S. Korea: Royal Asiatic Society, Korea Branch, 2011.

[9]*Ibid.*

[10]*Ibid.*

[11]Jan Huygen van Linschoten. *The Voyage of John Huyghen van Linschoten to the East Indies: The First Book, Containing His Description of the East. Volume 1.* Boston, MA, USA: Adamant Media Corporation, 2001; Henny Savenije. "Van Linschoten." *Korean Culture*, vol. 21, no. 1, 2000, pp. 4–19. http://www.cartography.henny-savenije.pe.kr/vanlinschoten.htm.

[12]Henny Savenije. *Op. cit.*

[13]*Ibid.*

[14]*Ibid.*

[15]A. E. Sokol. "The name of Quelpaert Island." *The University of Chicago Press*, vol. 38, no. 3/4, February 1948, pp. 231–235. doi:10.1086/348077. S2CID 144230819. Retrieved 8 July 2021.

[16]*Ibid.*

[17]Ralph M. Cory. "Some notes on Father Gregorio de Cespedes, Korea's first European visitor." https://www.scribd.com/, Transactions of the Royal Asiatic Society Korea Branch, 2025, https://www.scribd.com/document/205264399/SOME-NOTES-ON-FATHER-GREGORIO-DE-CESPEDES-KOREA-S-FIRST-EUROPEAN-VISITOR.

[18]Donald F. Lach. *Asia in the Making of Europe, Volume I: The Century of Discovery*, Kindle ed. Chicago, IL, USA: University of Chicago Press, 2008.

[19]*Ibid.*

[20]Samuel Hawley. *The Imjin War: Japan's Sixteenth-Century Invasion of Korea and Attempt to Conquer China*. Seoul, S. Korea: Royal Asiatic Society, Korea Branch, 2005.

[21]Hendrick Hamel. Jean-Paul Buys, trans. *Op. cit.*

[22]*Ibid.*

[23]Gari Ledyard. *The Dutch Come to Korea*. Seoul, S. Korea: Royal Asiatic Society-Korea Branch, 1971.

[24]Hendrick Hamel. Jean-Paul Buys, trans. *Op. cit.*

[25] Marco Polo. Ronald Latham, trans. *Op. cit.*

[26]"File:FraMauroDetailedMapInverted.jpg." *Wikimedia Commons: The Free Media Repository*, Wikimedia Foundation 10 Sept. 2016, upload.wikimedia.org/wikipedia/commons/8/8b/FraMauroDetailedMapInverted.jpg.

[27]"File:Da-ming-hun-yi-tu.jpg." *Wikimedia Commons*, 17 Oct. 2016, upload.wikimedia.org/wikipedia/commons/c/cd/Da-ming-hun-yi-tu.jpg.

[28]"File:KangnidoMap.jpg." *Wikipedia: The Free Enclopedia*, Wikimedia Foundation 15 Jan. 2010, en.wikipedia.org/wiki/File:KangnidoMap.jpg.

[29]Gari Ledyard. "Chapter 10: Cartography in Korea." In J. B. Harley and David Woodward, eds. *The History of Cartography, Volume Two, Book. Two: Cartography in the Traditional East and Southeast Asian Societies*, Chicago, IL, USA: The University of Chicago Press, 1994, pp. 235–345.

[30]Kenneth R. Robinson. "Choson Korea in the Ryukoku Kangnido." *Imago Mundi*, vol. 59, no. 2, June 2007, pp. 177–192, via Ingenta Connect. https://www.ingentaconnect.com/content/routledg/rimu/2007/00000059/00000002/art00002.

[31]Gari Ledyard. "Chapter 10: Cartography in Korea." *Op. cit.*

[32]A Cattaneo. *Fra Mauro's Mappa mundi and Fifteenth-Century Venice (Terrarvm Orbis)*. Turnhout, Belgium: Brepols Publishers, 2011.

[33]Michèle Gueret. *De l'Inde. Les voyages en Asie de Niccolò de' Conti*, French ed. Turnhout, Belgium: Brepols Publishers, 2004.

[34]Ma Huan. J. V. G. Mills, trans. *Ying-Yai Sheng-Lan: The Overall Survey of the Ocean's Shores (1433)*. London, UK: Hakluyt Society, 1970.

[35]The full text can be located at https://ctext.org/wiki.pl?if=gb&res=99548.

[36]Jan Huygen van Linschoten. *Op. cit.*

[37]"File:Bohaiseamap2.png." *Wikimedia Commons*, 20 Sept. 2022, upload.wikimedia.org/wikipedia/commons/0/03/Bohaiseamap2.png.

[38]"File:Location Tumen-River.png." *Wikimedia Commons*, 13 Sept. 2020, upload.wikimedia.org/wikipedia/commons/f/f4/Location_Tumen-River.png.

[39]The Sainsbury Institute. "Treasures of the Library: Maps of Japan by Luis Teixeira." https://www.sainsbury-institute.org/, Sainsbury Institute for the Study of Japanese Arts and Cultures (University of East Anglia), 2025, www.sainsbury-institute.org/e-magazine/issue-25-winter-2019/manga-at-the-british-museum-one-week-since-opening-zfjt3-bj6js/.

[40]Google Arts & Culture. "The Spread of Geographic Information on the Korean Peninsula as Seen through Ancient Maps." https://artsandculture.google.com/, National Maritime Museum of Korea, 2025, artsandculture.google.com/story/the-spread-of-geographic-information-on-the-korean-peninsula-as-seen-through-ancient-maps-korea-national-maritime-museum/0wVxy92S4WsSmQ?hl=en.

[41]"File:Map — Special Collections University of Amsterdam — OTM- HB-KZL 33.17.25.tif." *Wikimedia Commons*, 18 Mar. 2023, upload.wikimedia.org/wikipedia/commons/thumb/5/52/Map_-_Special_Collections_University_of_Amsterdam_-_OTM-_HB-KZL_33.17.25.tif/lossy-page1-6488px-Map_-_Special_Collections_University_of_Amsterdam_-_OTM-_HB-KZL_33.17.25.tif.jpg.

[42]"File:Atlas de Fernao Vaz Dourado (Asia).jpg." *Wikimedia Commons*, 23 Oct. 2021, upload.wikimedia.org/wikipedia/commons/f/f9/Atlas_de_Fernao_Vaz_Dourado_(Asia).jpg.

[43]"File:1594 double hemisphere world map by Petrus Plancius.jpg." *Wikimedia Commons*, 2 Sept. 2022, upload.wikimedia.org/wikipedia/commons/0/0d/1594_double_hemisphere_world_map_by_Petrus_Plancius.jpg.

[44]"File:Bartolomeu Velho 1568.jpg." *Wikimedia Commons*, 20 May 2013, upload.wikimedia.org/wikipedia/commons/7/7b/Bartolomeu_Velho_1568.jpg.

[45]"File:Ortelius — Maris Pacifici 1589.jpg." *Wikipedia*, 8 July 2012, upload.wikimedia.org/wikipedia/commons/4/4e/Ortelius_-_Maris_Pacifici_1589.jpg.

[46]It is a collection of legends, folktales, and historical accounts relating to the Three Kingdoms of Korea as well as to other periods and states before, during, and after the Three Kingdoms period.

[47]Northeast Asian History Foundation. *Gojoseon, Dangun, Granted: The Story Going to Find the Roots of Our Nation.* Seoul, S. Korea: Northeast Asian History Foundation, 2015.

[48]두산백과 (斗山百科全书). "기자조선 (箕子朝鮮)." https://terms.naver.com/, NAVER, 2025, terms.naver.com/entry.naver?cid=40942&docId=1071411&categoryId=33373.

[49]Gari Ledyard. "Chapter 10: Cartography in Korea." *Op. cit.*

[50]Hyŏng-sik Shin. *A Brief History of Korea.* Seoul, S. Korea: Ewha Womans University Press, 2006.

[51]*Ibid.*

[52]Gina L. Barnes. *State Formation in Korea: Emerging Elites.* Milton Park, Abingdon-on-Thames, Oxfordshire, England, UK: Routledge, 2019.

[53]Patricia Buckley Ebrey and Anne Walthall. *Pre-Modern East Asia: A Cultural, Social, and Political History, Volume I: To 1800.* Boston, MA, USA: Cengage Learning, 2013.

[54]Injae Lee, *et al.* Michael D. Shin, ed. *Korean History in Maps.* Cambridge, UK: Cambridge University Press, 2014.

[55]The Metropolitan Museum of Art. "Heilbrunn Timeline of Art History." https://web.archive.archive.org/, The Metropolitan Museum of Art, 2025, http://www.metmuseum.org/toah/ht/?period=05®ion=eak.

[56]*Ibid.*

[57]Ki-baik Lee. Edward J. Schultz and Edward W. Wagner, translators. *A New History of Korea.* Cambridge, MA, USA: Harvard University Press, 1984.

[58]Ilyon. *Samguk Yusa: Legends and History of the Three Kingdoms of Ancient Korea.* London, UK: Olympia Press, 2016.

[59]*Ibid.*

[60]Zhenping Wang. *Tang China in Multi-Polar Asia: A History of Diplomacy and War.* Honolulu, HI, USA: University of Hawaii Press, 2013.

[61]Jung Suk-bae (정석배). "발해의 북방경계에 대한 일고찰 (Study on northern borders of Balhae)." 고구려발해연구/*The Koguryo Balhae Yongu (in Korean).* 고구려발해학회/*Association of Koguryo Balhae,* vol. 54, 2016, pp. 87–125. https://www.kci.go.kr/kciportal/ci/sereArticleSearch/ciSereArtiView.kci?sereArticleSearchBean.artiId=ART002095762.

[62]"File:History of Korea-108 BC.png." *Wikimedia Commons,* 30 Apr. 2021, upload.wikimedia.org/wikipedia/commons/2/28/History_of_Korea-108_BC.png.

[63]"File:History of Korea-001.png." *Wikimedia Commons,* 27 Sept. 2020, upload.wikimedia.org/wikipedia/commons/5/5a/History_of_Korea-001.png.

[64]"File:History of Korea-476.PNG." *Wikimedia Commons,* 30 Oct. 2020, upload.wikimedia.org/wikipedia/commons/5/53/History_of_Korea-476.PNG.

[65]"File:Balhae-Territory in 830.JPG." *Wikimedia Commons,* 12 Nov. 2021, upload.wikimedia.org/wikipedia/commons/b/b6/Balhae-Territory_in_830.JPG.

[66]Kumja Paik Kim. Kaz Tsuruta, photog. *Goryeo Dynasty: Korea's Age of Enlightenment, 918-1392.* Washington, DC, USA, 2003.

[67]"File:History of Korea-Later three Kingdoms Period-915 CE.gif." *Wikimedia Commons,* 31 Oct. 2020, upload.wikimedia.org/wikipedia/commons/d/db/History_of_Korea-Later_three_Kingdoms_Period-915_CE.gif.

[68]Captivating History. *History of Korea: A Captivating Guide to Korean History, Including Events Such as the Mongol Invasions, the Split into North and South, and the Korean War.* Portland, OR, USA: Captivating History, 2020.

[69]William E. Henthorn. *Korea: The Mongol invasions.* Open Library, Legare Street Press, 2022.

[70]John Keay. *China: A History.* New York, New York, USA: Basic Books, 2011; Patricia Buckley Ebrey. *The Cambridge Illustrated History of China.* Cambridge, UK: Cambridge University Press, 2022.

[71]Byonghyon Choi, trans. *The Annals of King T'aejo: Founder of Korea's Chosŏn Dynasty*, Cambridge, MA, USA: Harvard University Press, 2014.

[72]"File:Yuen Dynasty 1294 — Goryeo as vassal.png." *Wikimedia Commons*, 18 May 2022, upload. wikimedia.org/wikipedia/commons/9/96/Yuen_Dynasty_1294_-_Goryeo_as_vassal.png.

[73]"File:Koryo map.png." *Wikimedia Commons*, 17 June 2022, upload.wikimedia.org/wikipedia/commons/c/cc/Koryo_map.png.

[74]Their locations can be found at https://web.archive.org/web/20070614134516/http://dicimg.paran.com/100_img/jpg/180/p18050200003.jpg.

[75]*Ibid.*

[76]Mao Ruizheng (茅瑞徵). *Dong Yi Kao Lue* (東夷考略) or A *Brief Study of Dongyi.* China: Tang Feng Lou (唐風樓), 1601. An electronic version can be found at https://ctext.org/wiki.pl?if=en&chapter=4521948#%E5%A5%B3%E7%9B%B4.

[77]"File:Northern Yuan and Golden Horde.svg." *Wikimedia Commons*, 4 Jan. 2023, upload.wikimedia.org/wikipedia/commons/thumb/1/14/Northern_Yuan_and_Golden_Horde.svg/2560px-Northern_Yuan_and_Golden_Horde.svg.png.

[78]Keith L. Pratt and Richard Rutt. *Korea: A Historical and Cultural Dictionary*, Milton Park, Abingdon-on-Thames, Oxfordshire, England, UK: Routledge, 1999, p. 482.

[79]*Ibid.*

[80]*Ibid.*

[81]*Ibid.*

[82]*Ibid.*

[83]David C. Kang. *East Asia Before the West: Five Centuries of Trade and Tribute*, New York City, New York, USA: Columbia University Press, 2010, p. 59.

[84]Marius B. Jansen. *Japan and China: From War to Peace, 1894–1972.* Chicago, IL, USA: Rand McNally College Publishing Company, 1975.

[85]Charles H. Hapgood. "Chapter 5: The ancient maps of the East and West." *Maps of the Ancient Sea Kings: Evidence of Advanced Civilization in the Ice Age*, Kempton, IL, USA: Adventures Unlimited Press, 1997, pp. 135–147.

[86]"File:華夷圖附注.jpg." *Wikimedia Commons*, 26 Mar. 2023, upload.wikimedia.org/wikipedia/commons/a/ac/華夷圖附注.jpg.

[87]Gang Zhao. "Reinventing China: Imperial Qing ideology and the rise of modern Chinese national identity in the early twentieth century." *Modern China*, vol. 32, no. 1, 2006, pp. 3–30. doi:10.1177/0097700405282349.

[88]Leo Agrow. *History of Cartography*, London, UK: Transaction Publishers, 1963, p. 199.

[89]Kenneth R. Robinson. "Chosŏn Korea in the Ryukoku Kangnido: Dating the oldest extant Korean map of the world (15th century)." *Imago Mundi: The International Journal for the History of Cartography*, vol. 59, no. 2, 2007, pp. 177–192. https://doi.org/10.1080/03085690701300964.

[90]"File:Da-ming-hun-yi-tu.jpg." *Wikimedia Commons*, 17 Oct. 2016, upload.wikimedia.org/wikipedia/commons/c/cd/Da-ming-hun-yi-tu.jpg.

[91]"File:KangnidoMap.jpg." *Wikimedia Commons*, 15 Jan. 2010, upload.wikimedia.org/wikipedia/commons/7/75/KangnidoMap.jp.

[92]"File:Kunyu Wanguo Quantu by Matteo Ricci Plate 1-3.jpg." *Wikimedia Commons, 23 Oct. 2020,* commons.wikimedia.org/wiki/File:Kunyu_Wanguo_Quantu_by_Matteo_Ricci_Plate_1-3.jpg.

[93]"File:Mercator 1569 world map composite.jpg." *Wikimedia Commons*, 26 Nov. 2016, upload.wikimedia. org/wikipedia/commons/4/4b/Mercator_1569_world_map_composite.jpg.

[94]"File:OrteliusWorldMap1570.jpg." *Wikimedia Commons*, 12 July 2022, upload.wikimedia.org/wikipedia/ commons/e/e2/OrteliusWorldMap1570.jpg.

[95]"File:1594 double hemisphere world map by Petrus Plancius.jpg." *Wikimedia Commons*, 2 Sept. 2022, upload.wikimedia.org/wikipedia/commons/0/0d/1594_double_hemisphere_world_map_by_Petrus_ Plancius.jpg.

[96]Sheng-Wei Wang. *Op. cit.*

[97]Gari Ledyard. "Chapter 10: Cartography in Korea." *Op. cit.*

[98]Ilyon. *Op. cit.*

[99]Ki-baik Lee. Edward W. Wagner and Edward J. Schultz, translators. *Op. cit.*

[100]*Ibid.*

[101]Stella Yingzi Xu. *That Glorious Ancient History of Our Nation: The Contested Re-readings of "Korea" in Early Chinese Historical Records and Their Legacy on the Formation of Korean-ness*, Ann Arbor., MI, USA: University of Michigan Press, 2007, p. 220.

[102]Captivating History. *Op. cit.*

[103]Samcheok City Hall staff. "History." www.samcheok.go.kr, Samcheok City Hall, 2025, samcheok. go.kr/01696/01703/01705.web.

[104]*Ibid.*

[105]Gina L. Barnes. *Op. cit.*

[106]Ki-baik Lee. Edward J. Schultz and Edward W. Wagner, translators. *Op. cit.*

[107]Mark E. Yington. *The Ancient State of Puyŏ in Northeast Asia: Archaeology and Historical Memory.* Cambridge, MA, USA: Harvard University Asia Centre, 2016.

[108]Captivating History. *Op. cit.*

[109]Jurchen is a term used to collectively describe a number of East Asian Tungusic-speaking people; they were descendants of both the nomadic Tungus Malgal peoples and remnants of the defunct Balhae kingdom of Manchuria and northern Korea. They lived in northeastern China, also known as Manchuria, before the eighteenth century.

[110]Their locations can be found at https://web.archive.org/web/20070614134516/http://dicimg.paran. com/100_img/jpg/180/p18050200003.jpg.

[111]*Ibid.*

[112]Seok-Woo Lee and Chang-Hoon Shin. "Implications of the border regime between North Korea and China." in Jin-Hyun Paik, *et al.*, eds. *Asian Approaches to International Law and the Legacy of Colonialism: The Law of the Sea, Territorial Disputes and International Dispute Settlement*, Oxfordshire, England, UK: Routledge, 2013, pp. 109–118.

[113]Sun Jin ji and Sun Hong (孫進己和孫泓). *Nu Zhen Min Zu Shi* (女真民族史) or *Jurchen National History*. Guilin, Guangxi, China: Guangxi Normal University Press (广西师范大学出版社), 1991.

[114]Gertraude Roth Li. "State Building before 1644." In Willard J. Peterson, ed. *The Ch'ing Empire to 1800, Cambridge History of China. Vol. 9*, Cambridge, UK: Cambridge University Press, 2002, pp. 11–14.

[115]*Ibid.*

[116]John J. Stephan. *The Russian Far East: A History*, Stanford, CA, USA: Stanford University Press, 1996, pp. 16–17; Shih-shan Henry Tsai. *The Eunuchs in the Ming Dynasty*, Albany, New York, USA: SUNY Press, 1996, pp. 129–130.

[117]"File:Ravenstein-Tyr-monument-194.png." *Wikimedia Commons*, 27 Mar. 2015, upload.wikimedia.org/ wikipedia/commons/f/f4/Ravenstein-Tyr-monument-194.png.

[118]"File:Map of Ming Chinese empire 1415.jpg." *Wikimedia Commons*, 30 June 2022, upload.wikimedia.org/wikipedia/commons/c/c6/Map_of_Ming_Chinese_empire_1415.jpg.

[119]"File:Ming Empire cca 1580 (en).svg.." *Wikimedia Commons*, 12 May 2023, upload.wikimedia.org/wikipedia/commons/thumb/b/ba/Ming_Empire_cca_1580_(en).svg/2304px-Ming_Empire_cca_1580_(en).svg.png.

[120]"File:History of Korea-1592-1597.svg." *Wikimedia Commons*, 23 Mar. 2012, upload.wikimedia.org/wikipedia/commons/thumb/e/e0/History_of_Korea-1592-1597.svg/1157px-History_of_Korea-1592-1597.svg.png.

[121]Embassy of the People's Republic of China in the Democratic People's Republic of Korea. "朝鲜半岛八道地名的由来和各地人物性格特点 (The Origins of the Place Names in the Korean Peninsula and the Character Traits of the People in These Places)." http://kp.china-embassy.gov.cn/, Embassy of the People's Republic of China in the Democratic People's Republic of Korea, 2025, kp.china-embassy.gov.cn/chn/cxgk/201807/t20180713_1089033.htm.

[122]"File:Eight provinces of Korea-en.svg." *Wikimedia Commons*, 28 Feb. 2022, upload.wikimedia.org/wikipedia/commons/thumb/1/13/Eight_provinces_of_Korea-en.svg/1263px-Eight_provinces_of_Korea-en.svg.png.

[123]Embassy of the People's Republic of China in the Democratic People's Republic of Korea. *Op. cit.*

[124]*Ibid.*

[125]*Ibid.*

[126]*Ibid.*

[127]*Ibid.*

[128]*Ibid.*

[129]*Ibid.*

[130]*Ibid.*

[131]Guy Delisle. Helge Dascher, trans. *Pyongyang: A Journey in North Korea.* Montreal, QC, Canada: Drawn and Quarterly, 2018.

[132]UNESCO World Heritage Convention. "Properties inscribed on the World Heritage List." https://whc.unesco.org/, UNESCO World Heritage Center, 2025, whc.unesco.org/en/statesparties/kr.

[133]Ilyon. *Op. cit.*

[134]Captivating History. *Op. cit.*

[135]Ki-baik Lee. Edward W. Wagner and Edward J. Schultz, translators. *Op. cit.*

[136]*Ibid.*

[137]*Ibid.*

[138]사단법인신라문화진흥원 (The Silla Cultural Promotion Agency). "신라의 역사와 문화 (History and Culture of Silla)." https://web.archive.org/, The Silla Cultural Promotion Agency, 2025, https://web.archive.org/web/20080321070426/http://www.shilla.or.kr/shilla_culture/

[139]Yang Jun (楊軍). *Fu Yu Shi Yan Jiu* (夫餘史研究) or *Buyeo History and Research.* Lanzhou, Gansu Province, China (中国甘肃省兰州市): Lanzhou University Press (兰州大学出版社), 2011.

[140]*Ibid.*

[141]"File:Buyeo.svg." *Wikimedia Commons*, 26 Nov. 2021, upload.wikimedia.org/wikipedia/commons/thumb/8/8b/Buyeo.svg/1365px-Buyeo.svg.png.

Chapter 3

An Ancient World Map Depicts China in 1433, at the Early Stage of the European Age of Discovery

Abstract

In this chapter, I investigate an ancient world map — Kunyu Wanguo Quantu (坤輿萬國全圖/坤輿万国全图) (abbreviated as KWQ) or Complete Geographical Map of All the Kingdoms of the World — to determine the era revealed by the political condition of the China portion on this map (abbreviated as the China-KWQ). Kunyu Wanguo Quantu was published by Matteo Ricci in 1602 in China.

In Chapter 2, I have shown that (1) the Korea portion of this map shows that its northern border was established in 1433 by Sejong of Joseon (朝鮮世宗/朝鲜世宗) with the Jurchen people (女真人); the year 1433 was long before the first Europeans landed on the Korean mainland; and (2) the China-KWQ is a map of 1433 because: (i) my previous book titled *Chinese Global Exploration in the Pre-Columbian Era: Evidence from an Ancient World Map* has shown that the entire KWQ was drawn in 1433; and (ii) the Nurgan Regional Military Commission (奴兒干都指揮使司/奴儿干都指挥使司) depicted on the Jurchen land was last inspected by the Ming envoy in 1433. Although the Commission existed in 1409–1435, the years 1434–1435 are after the map was drawn and the years 1409-1432 are before the Joseon-Jurchen border was established.

In this chapter, after a thorough analysis of the locations and histories of the total 132 geographical items and four annotations depicted on the China-KWQ, the results firmly confirm the map's Chinese origin and reveal China in 1433.

Keywords: China, Jurchen, Korea, Kunyu Wanguo Quantu, Matteo Ricci, Nurgan Regional Military Commission

1. Introduction

For over four hundred years, the ancient world map known as Kunyu Wanguo Quantu — (坤輿萬國全圖/坤輿万国全图),[1] abbreviated here as KWQ — written with Chinese characters and having latitudinal and longitudinal lines, has been generally regarded as a European map. The map was published in 1602 by Matteo Ricci (an Italian Jesuit priest and one of the founding figures of the Jesuit China missions) and was believed to be based on the European maps which Ricci brought with him to China in 1582.[2] In my previous book titled *Chinese Global Exploration in the Pre-Columbian Era: Evidence from an Ancient World Map*[3] and in Chapters 1 and 2 of this book, I have made thorough analyses, and presented concrete evidence to show how I deduced the eras from the geographical and historical information on several major portions of the KWQ. My findings are:

(1) The Chinese explored Cape Breton Island on Canada's Atlantic Coast long before the Europeans.

(2) The information on Europe was obtained by Chinese in the Southern Song Dynasty (南宋; 1127–1279).

(3) The information on both Mo Wa La Ni Jia (墨瓦蠟泥加/墨瓦蜡泥加; referring to the regions of Australia, New Zealand, *Tierra del Fuego*/Land of Fire in South America, and Antarctica) and the Americas was obtained by the Ming (1368–1644) mariners in the 1420s during their sixth voyage to the Western Ocean (which Chinese denomination refers to the Indian Ocean and beyond). These mariners belonged to the Treasure Fleets commanded by Admiral Zheng He (鄭和/郑和; 1371–1433/1435; some sources say 1433, but the author of this book has some reservation on that date) in the early fifteenth century.

(4) The information on Africa was obtained by the Ming mariners in 1433 during their seventh voyage to the Western Ocean; as in (3), these mariners also belonged to the Treasure Fleets commanded by Admiral Zheng He in the early fifteenth century.

(5) The information on the map of Japan on the KWQ reveals Japan's political condition in 897 A.D. which was during the Chinese Tang Dynasty (618–907, with an interregnum between 690 and 705), almost 650 years before the first Europeans visited Japan.

(6) The era of the Korea-KWQ is 1433, based on the Jurchen-Joseon border[4] established in that year and the KWQ was a map drawn in 1433. The era of the China-KWQ is also 1433, based on the establishment of the Jurchen – Joseon border in 1433, the fact that China ruled Jurchen land at that time, the Nurgan Regional Military Commission (奴兒干都指揮使司/奴儿干都指挥使司) depicted on the Jurchen land was last inspected by the Ming envoy in 1433, and the fact that the KWQ was a map drawn in 1433. Although the Commission existed from 1409 to 1435, the era 1434–1435 was after the KWQ had been drawn and the era 1409–1432 was before the Joseon-Jurchen border was established.

The above findings strongly support that these portions of the KWQ used Chinese maps as source maps (they no longer exist), instead of having been copied from other contemporary European maps.

All these findings represent my earnest effort in seeking the true maritime world history. It has produced fruitful results which merit attention. In this chapter, a thorough analysis will be carried out of all the annotations (four in total), and the locations and histories of the 131 geographical items depicted on the China portion of the KWQ, abbreviated herein as China-KWQ.

2. The China-KWQ is of Chinese Origin Based on Comparison with the Three Major European Maps in the Sixteenth Century

Some scholars think that the China-KWQ is a copy or adapted copy of the three major European maps in the sixteenth century shown in Fig. 3.1. However, I will also show that the China-KWQ is a far superior and more ancient map than these three European maps in terms of China's coastal shape, rivers, lakes, and other geographical, historical and political information depicted on the map. Such comparisons are detailed below.

The four maps in Fig. 3.1 are, in this book's terminology: (a) the China-KWQ extracted from the KWQ;[5] (b) the China-Mercator (1569) extracted from the 1569 World Map by Gerardus Mercator;[6] (c) the China-Ortelius (1570) extracted from the 1570 World Map by Abraham Ortelius;[7] and (d) the China-Plancius (1594) extracted from the 1594 World Map by Petrus Plancius.[8]

The following are the two aspects which show that the China-KWQ is of Chinese origin after comparing it with these three major European maps in the sixteenth century:

(a)　　　　　　　　　　(b)

(c)　　　　　　　　　　(d)

Fig. 3.1.   (a) shows the China-KWQ extracted from the KWQ (public domain); (b) shows the China-Mercator (1569) extracted from the 1569 World Map by Gerardus Mercator (public domain); (c) shows the China-Ortelius (1570) extracted from the 1570 World Map by Abraham Ortelius (public domain); and (d) shows the China-Plancius (1594) extracted from the 1594 World Map by Petrus Plancius (public domain).

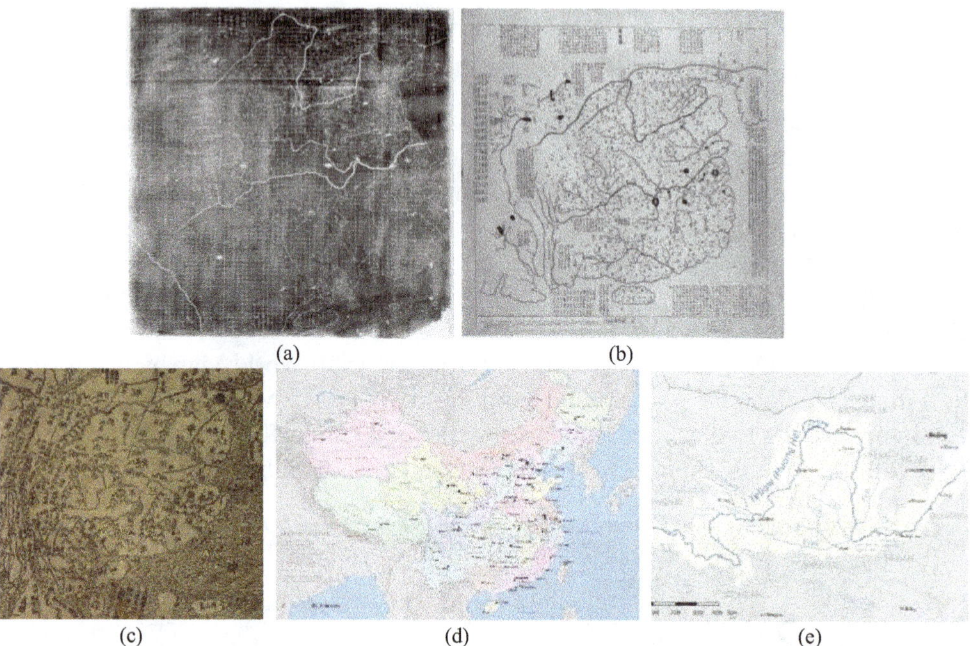

Fig. 3.2.   (a) The Yu Ji Tu (禹跡圖/禹迹图) or Map of the Tracks of Yu Gong (public domain),[13] carved into stone in 1137,[14] located in the Stele Forest of Xian, Shaanxi, China. China's coastline, and river and lake systems are clearly defined and precisely positioned on the map. On the reverse side of the engraving is another map shown in (b) called the 1136 Huayi Tu (華夷圖/华夷图) or Map of China and Barbaric Countries (public domain).[15] (c) is extracted from the KWQ (public domain). (d) is a modern geographic map of China (*Source*: SY, under CC BY-SA 4.0, https://commons.wikimedia.org/wiki/File:Geographic_Map_of_China.png).[16] (e) is a modern map of the Yellow River (*Source*: Shannon1, under CC BY-SA 4.0 International, 3.0 Unported, 2.5 Generic, 2.0 Generic and 1.0 Generic license, https://commons.wikimedia.org/wiki/File:Yellowrivermap.jpg).[17]

2.1 *The China-KWQ is a far superior map in presenting China's geography*

We can start by comparing China's coastline, river and lake systems, and the Great Wall (萬里長城 / 万里长城) depicted on the four maps.

(1) The coastline: the China-KWQ in Fig. 3.1(a) clearly defines and precisely pinpoints China's coastline on the map in comparison with those on a modern map shown in Fig. 3.2(d), whereas the three European maps show major discrepancies: the sharp triangular edge in the coastline on the China-Mercator (1569) and China-Ortelius (1570), though smoothed out somewhat on the China-Plancius (1594), is not observed on the China-KWQ and the modern map.

(2) The rivers: The large northerly rectangular bend of the Yellow River that forms the Yellow River's entire middle section is nicely depicted on the China-KWQ, see Fig. 3.1(a). It is in excellent agreement with that on a modern map in Fig. 3.2(e). However, such a large northerly

rectangular bend does not appear on the China-Mercator (1569) and the China-Ortelius (1570). On the China-Plancius (1594), there is a river (presumably the Yellow River) which forms a huge, closed loop in its entire middle section. This cannot exist.

(3) The lakes: Qinghai Lake (青海湖; 37°N, 100°E) is the largest lake in China. On the China-KWQ, the lake's coordinates are c. 36.8°N (the KWQ latitude is converted to the modern latitude after multiplying it by 0.9856; will be abbreviated as "KWQ latitude → modern latitude" from now on), c. 94°E, showing a latitude accuracy around 0.2° and longitude error around 6° in comparison with those on a modern map. This lake cannot be identified on the three European maps.

Dongting Lake (洞庭湖) is the second largest lake of China, next to Qinghai Lake. The waters of the entire lake system discharge into the Yangtze River at Yueyang (岳陽/岳阳; 29.4°N, 113.1°E). On the China-KWQ, that discharge location (unnamed on the map) is at c. 29.7°N (KWQ latitude → modern latitude), 106.0°E, showing a latitude accuracy of 0.3° and a longitude error around 7°. This lake is not depicted on the China-Mercator (1569) and China-Ortelius (1570). On the China-Plancius (1570) there is an unnamed lake whose waters discharge into an unnamed river (presumably the Yangtze River) at c. 33°N, c. 121°E, showing a latitude error of c. 2.9° and longitude error around 15°. They are far inferior to the coordinates on the China-KWQ.

You may wonder: why is the China-KWQ so much better than the three major sixteenth century European maps?

The simplest and most logical answer is that the ancient Chinese cartographers already knew how to depict China's coastlines, Yellow River, Yangtze River, and lakes, etc. since ancient days as shown in Fig. 3.2(a) and 3.2(b).

In Fig. 3.2(a), the Yu Ji Tu (禹跡圖/禹迹图) or Map of the Tracks of Yu Gong was made by an unknown Chinese mapmaker in the Southern Song Dynasty (南宋; 1127–1279) in honour of Yu (禹) or Yu the Great (大禹; second millennium B.C.), who controlled floods in China.[9] The coastline of China, and the river and lake systems are clearly defined and precisely located on the map. Such an exceptional precision of the representation was facilitated by the use of a rectilinear grid system named Ji Li Hua Fang (計里畫方/计里画方) first used by Chinese maps and much later by European cartographers. This square division divides China into nearly 5000 tiles. Each square represents 100 *li* by 100 *li* (in the Southern Song Dynasty, the *li* was about 416 m; today it is equivalent to 500 m).[10] The map covers a huge distance of about 3000 km in width and meticulously identifies roughly 380 administrative districts, nearly 80 rivers, 70 mountains, and 5 lakes.[11] Dated 1137, the contents of the map suggest a cartography started in the Northern Song Dynasty (北宋; 960–1127), more than 60 years before the completion date.[12]

In Fig. 3.2(b), the 1136 (some sources say 1137 or even as early as 1034) Huayi Tu (華夷圖/华夷图) or Map of China and Barbaric Countries has been explained in detail in Chapters 1 and 2 of this book. It was engraved at the back of the same stele that has the Yu Ji Tu (禹跡圖/禹迹图) or Map of the Tracks of Yu Gong. Huayi Tu has very accurate latitudes and longitudes for cities in the interior part of China, which may indicate the use of spherical trigonometry projection.[18] The map is based on Hainei Huayi Tu (海內華夷圖/海内华夷图) or Map of China and the Barbarian Countries within the Seas produced in 801 A.D. by Jia Dan (賈耽/贾耽).[19]

The 1136 Huayi Tu has the Great Wall of China depicted on the northern edge of the country. The China-KWQ also depicts the Great Wall from c. 38°N near the deserts of Ningxia (寧夏/宁夏) to c. 42.9°N (KWQ latitude → modern latitude) near Kaiping (開平/开平) in today's Liaoning Province (遼寧省/辽宁省). The Great Wall is not depicted on the China-Mercator (1569) and the China-Ortelius (1570); while it does show up on the China-Plancius (1594), the Great Wall starts from c. 39°N at a location in western China, to c. 50°N near the coast of Japan Sea crossing the entire northern Korean Peninsula. It is obvious that Plancius did not know where the eastern end of the Great Wall of China was at the time and depicted it deeply and mistakenly in the Jurchen land.

(4) The Great Wall of China: According to historical records, several walls were built from as early as 1000 B.C.,[20] with several stretches later joined by Qin Shi Huang (秦始皇; 220–206 B.C.), the first emperor of China. Later, many successive dynasties built and maintained multiple stretches of border walls to prevent the raids from the northern steppe invaders. The best-known sections of the wall were built by the Ming Dynasty (1368–1644) in the sixteenth century.

Both maps in Fig. 3.2(a) and 3.2(b) show concrete evidence of Chinese cartographers' knowledge and capability in depicting the geography of China, many hundreds of years before the European explorers reached China. Being able to draw the China-KWQ the way it was simply demonstrates that these ancient Chinese cartographers had not lost their ability in cartography inherited from their ancestors. Hence, there is no basis, and it is illogical to say that the superior China-KWQ is a copy of any of the three inferior European maps or the combination of them in the sixteenth century.

The truth revealed in this subsection is that the Ming cartographers had kept and mastered their ancestors' skills to make the China-KWQ, hence this map is of Chinese origin.

2.2 *The China-KWQ presents the Ming administrative units and regions in 1433*

Chapter 2 of this book and the Introduction of this chapter have provided strong evidence in suggesting that the China-KWQ is a map of 1433. Then, the important goal of this chapter and the Appendix is to verify that all the remaining geographical items (other than the Nurgan Regional Military Commission last inspected in 1433) on the China-KWQ were not established after 1433. Such a verification task involves great effort, but it is completed in this chapter. The analysis and results are detailed in the Appendix at the end of this chapter. A general summary will be offered here. To that end, we must have some knowledge about the early Ming history.

The Ming Hongwu Emperor (明洪武帝; 1328–1398; reigned 1368–1398) conquered China proper, ended the Mongol-led Yuan Dynasty and forced the remnant Yuan Court (known as Northern Yuan) to retreat to the Mongolian Plateau. The Hongwu Emperor established the Ming Dynasty at the beginning of 1368 and initiated the basic framework of the Ming administrative structure.[21]

Figure 3.3 shows the Ming Dynasty administrative units and regions during the late period (1421–1424) of the Yongle era, from the time when the Yongle Emperor moved the Ming capital from Nanjing to Beijing in 1421 to his death in 1424. The Yongle Emperor (永樂帝/永乐帝; 朱棣; 1360–1424; reigned 1402–1424) was the fourth son of the Hongwu Emperor and the third emperor of the Ming Dynasty. During the Yongle era, Ming reached its peak in national strength.

After Yongle Emperor passed away in 1424, his eldest son, Hongxi Emperor (洪熙帝; 1378–1425) was short lived and only reigned a year from 1424 to 1425. Then, Yongle Emperor's

Fig. 3.3.   Ming Dynasty administrative units and regions (*Source*: Jason22 at Chinese Wikipedia, under CC Attribution 2.5 Generic license, https://commons.wikimedia.org/wiki/File:Ming_Dynasty_Administrative_division.jpg)[24] during the late Yongle period (1421–1424).

grandson, Xuande Emperor (宣德帝; 1399–1435), reigned from 1425 to 1435, and Ming entered a relatively peaceful period within the Ming history.

Historical records show that in 1428 (the third year of Xuande) Jiaozhi (交阯; 1407–1428) ceased to be Ming's province,[22] and in 1433 the Nurgan Regional Military Commission (奴兒干都指揮使司/奴儿干都指挥使司 or abbreviated as 奴兒干都司/奴儿干都司) was lastly inspected.[23] Hence in 1433. the whole country should have similar administration units and regions as shown in Fig. 3.3, except with thirteen provinces instead of fourteen provinces. Hence, if the China-KWQ is of the era 1433, it should also show the similar administration units and regions as shown in Fig. 3.3, except Jiaozhi. This is true on the China-KWQ.

The item numbers of these administrative units as listed in the Appendix are given here, following their names:

1 Thirteen provinces (with name in Chinese; and position in the Appendix):

(1)   Shandong (山東/山东; Item 78)
(2)   Shanxi (山西; Item 62)
(3)   Henan (河南; Item 74)
(4)   Shaanxi (陝西; Item 63)

(5)   Sichuan (四川; Item 116)
(6)   Huguang (湖廣/湖广; Item 87)
(7)   Jiangxi (江西; Item 96)
(8)   Zhejiang (浙江; Item 90)
(9)   Guangdong (廣東/广东; Item 102)
(10) Guangxi (廣西/广西; Item 105)
(11) Yunnan (雲南/云南; Item 110)
(12) Guizhou (貴州/贵州; Item 107)
(13) Fujian (福建; Item 92)

As mentioned, the Jiaozhi (交趾) Province established in 1407 by the Yongle Emperor is no longer on the list, because in 1428 the Ming Xuande Emperor abolished the Jiaozhi Chengxuan Provincial Administration Commission (交趾承宣布政使司; the Chinese name 承宣布政使司 can be abbreviated as 布政使司 or 司, lit. "division", or commonly known as 省, lit. "province").[25] Afterwards, the Ming Emperor immediately named Chen Hao (陳暠/陈皓) — the puppet monarch that Lê Lợi (黎太祖; 1384/1385–1433; reigned 1428–1433) supported before he came to the throne — as the "King of Annan" and the Annan Kingdom became independent.[26]

After the Great Ming gave up Jiaozhi, its influence in Southeast Asia was greatly weakened, and many foreign countries did not come to pay tribute anymore. Hence, in 1430 (the fifth year of Xuande), the Xuande Emperor ordered Zheng He (鄭和/郑和; 1371–1433/1435) to go overseas again for his seventh and final voyage to the Western Ocean (Indian Ocean and beyond).[27]

2 Two capitals: Jingshi Shuntianfu (京師順天府/京师顺天府; Jingshi, lit. "capital") in today's Beijing, parts of Hebei Province (河北省), Tianjin City (天津市); and Nanjing Yingtianfu (南京應天府/南京应天府). The Ming founding Emperor — Hongwu Emperor — set his capital at Yingtianfu or Jingshi Yingtianfu in today's Nanjing. In 1421, the Yongle Emperor changed Beijing to Jingshi (京師)[28] and moved the Ming capital from Jingshi Yingtianfu (京師應天府/京师应天府) to Jingshi Shuntianfu (京師順天府/京师顺天府),[29] and Jingshi Yingtianfu (京師應天府/京师应天府) became Nanjing Yingtianfu (南京應天府/南京应天府), the secondary capital.

However, Nanjing remained important and, as John King Fairbank and Merle Goldman point out,[30] its shipyard is known to have constructed two thousand vessels from 1403 to 1419, including treasure ships measuring 112 m/367 ft to 134 m/440 ft in length and 45 m/148 ft to 54 m/177 ft in width (the largest shipwrecks discovered to date are in New Zealand with dimensions c. 120 m × 48 m/394 ft × 157 ft).[31] These vessels were used to service seven different tributary voyages[32] led by Admiral Zheng He to the Western Ocean (Indian Ocean and beyond) to promote the Ming diplomacy and trade with foreign countries.

3 Two regions directly ruled by the central government: North Zhili (北直隸/北直隶) was the region comprising the present Beijing City, Tianjin City, most of Hebei Province, and small parts of Henan Province (河南省) and Shandong Province (山東省/山东省), which was under the jurisdiction of the northern capital, Beijing or Jingshi Shuntianfu; South Zhili (南直隸/南直隶; 1421–1645) was the region equivalent to the two provinces of Jiangsu (江蘇/江苏), Anhui (安徽) and the city of Shanghai (上海), which was under the jurisdiction of the southern capital, Nanjing or Nanjing Yingtianfu.

4 Seven guard battalions in the region to the west of Jiayuguan (Jiayu Pass; 嘉峪關/嘉峪关). It is the present northwest of Gansu (甘肅/甘肃), northern Qinghai (青海), and eastern Xinjiang (新疆). They were also known as "Seven Garrisons West of Jiayu Pass (關西七衛/关西七卫)", "Seven Guards of Northwest (西北七衛/西北七卫)" or "Seven Guards of Mongolia (蒙古七衛/蒙古七卫)" (because the leaders of the seven guards were all Mongolian aristocrats in this system of hereditary soldiery). The Guard Battalion was in the wei (衛/卫)-suo (所) system. The Jiayuguan or Jiayu Pass (嘉峪關/嘉峪关) was the first frontier fortress at the west end of the Ming Dynasty Great Wall (this end can be identified on the China-KWQ).

Although the Seven Guard Battalions were established in the Yongle era, they are not depicted in Fig. 3.3. However, they are depicted on the China-KWQ except for the Anding Wei (安定衛/安定卫).

Here are the names of the seven guard battalions, listed next with their dates of establishment and numbered position in the Appendix:

(1) Aduan Wei (阿端衛/阿端卫; 1375;[33] Item 44).
(2) Quxian Wei (曲先衛/曲先卫; 1371;[34] Item 43): During the Hongxi (洪熙) Period (1424–1425), the Ming envoys were robbed and killed in that region. When the Ming army conquered the tribes in that region, the tribesmen fled. When Xuande Emperor sent officials to appease them, more than 42,000 people returned to Quxian Wei.
(3) Anding Wei (安定衛/安定卫; 1375);[35] it is the only one missing on the China-KWQ.
(4) Handong Wei (罕東衛/罕东卫; 1397;[36] Item 45).
(5) Shazhou Wei (沙州衛/沙州卫; 1404;[37] Item 49).
(6) Chijing Menggu Wei (赤斤蒙古衛/赤斤蒙古卫;[38] 1404; Item 48).
(7) Hami Wei (哈密衛/哈密卫; 1406;[39] Item 53).

Later, Shazhou Wei moved inland, and Handong Zuowei (罕東左衛/罕东左卫; 1465–1487) was established in its old place as the eighth guard battalion. But that was after the middle of the fifteenth century. This is why it is not depicted on the China-KWQ.

5 Four Regional Military Commissions:

(1) The Nurgan Regional Military Commission (奴兒干都指揮使司/奴儿干都指挥使司 or abbreviated as 奴兒干都司/奴儿干都司; 1409–1435; after 1435 the Commission moved inland; Item 10): It was established in 1409 (the seventh year of Yongle) as the highest local military and political unit in the Ming Dynasty in the lower reaches of Heilongjiang (黑龍江/黑龙江; Amur River) to govern areas such as the Amur-Heilong River Basin (the Amur-Heilong is the longest river in Northeast Asia. It flows through China, Mongolia, and Russia) and the Ussuri River Basin (烏蘇里江流域/乌苏里江流域), see Fig. 3.4.

Fig. 3.4. A map of the Amur River drainage basin (*Source*: Kmusser, under CC BY-SA 4.0, https://commons. wikimedia.org/wiki/File:Amurrivermap.png).[40]

(2) The Liaodong Regional Military Commission (遼東都指揮使司/辽东都指挥使司 or abbreviated as 遼東都司/辽东都司; 1375–1644; Item 1): It was established in 1375 (the eighth year of Hongwu) in the southern part of the Northeast region.[41]

In fact, there was also the Daning Regional Military Commission (大寧都指揮使司/大宁都指挥使司), which was a military and political institution established in 1387 (the 20th year of Hongwu) by the Ming Dynasty in the present northern part of Hebei Province and the southeastern part of the Inner Mongolia Autonomous Region (內蒙古自治區/内蒙古自治区). In 1403 (the first year of Yongle), the Daning Regional Military Commission moved its seat inland and the original site became an abandoned city which was gradually occupied by the Northern Yuan Dynasty;[42] hence, it does not appear in Fig. 3.3. But the abandoned Daning city (Item 25) still shows up on the China-KWQ.

(3) The Duogan Xingdu Regional Military Commission (朵甘行都指揮使司/朵甘行都指挥使司): The Duogan Xingdu Regional Military Commission (朵甘行都指揮使司/朵甘行都指挥使司 or abbreviated as 朵甘都司/朵甘都司; 1374–1565; Item 125)[43] was established in 1374 (the seven-year of Hongwu) in the Western Sichuan Province (四川省西部). It was the highest military and political authority of the Ming Dynasty in China, covering Western Sichuan Province, Northwestern Yunnan Province (雲南省西北部/云南省西北部), Eastern Tibet Autonomous Region (西藏自治區東部/西藏自治区东部), and Southwestern Qinghai Province (青海省西南部).[44]

(4) The Wusizang Regional Military Commission (烏思藏都指揮使司/乌思藏都指挥使司 or abbreviated as 烏思藏都司/乌思藏都司; 1374–1446; Item 122). It is the name of the Ming Dynasty's *tusi*-type[45] of military jurisdiction in the central and western parts of the Qinghai-Tibet Plateau (青藏高原).[46]

Among the four Regional Military Commissions, the Nurgan Regional Military Commission is of particular interest and relevant to the dating of the China-KWQ. The Commission was established on the Jurchen land in 1409 and abolished in 1435.[47] But Ming official's last inspection of the Nurgan Command Post occurred in 1433[48] by Eunuch Yishiha (a Jurchen eunuch of the Ming Dynasty of China; fl. 1409–1451). The era of the Korea-KWQ is determined as 1433 in Chapter 2, because the Jurchen-Joseon border was established in that year and the KWQ was drawn in 1433; then the era of the China-KWQ should also be 1433. This is based on the Jurchen-Joseon border established in 1433, the fact that the Nurgan Regional Military Commission (奴兒干都指揮使司/奴儿干都指挥使司) depicted on the Jurchen land and by China as part of China was last inspected by the Ming envoy in 1433, and the fact that the KWQ was drawn in 1433. Although the Commission existed from 1409 to 1435, the era 1434–1435 was after the KWQ being drawn and the era 1409–1432 was before the Joseon-Jurchen border being established, so that both periods can be ruled out.

3. Conclusions

In summary, in the Introduction, I stated that the purpose of this chapter is to analyse the map of China on Kunyu Wanguo Quantu —（坤輿萬國全圖/坤舆万国全图）or Complete Geographical Map of All the Kingdoms of the World — published by Matteo Ricci in 1602 in China, abbreviated here as the China-KWQ, to determine the origin of this map and its era. Then, in Section 2, I have shown that the China-KWQ is of Chinese origin after comparing it with the three major European maps of the sixteenth century, because: (1) the China-KWQ is a far superior map in presenting China's geography than the contemporary European maps, supported by the remarkably accurate map of Chinese cartographers made since the early ninth century; and (2) the China-KWQ presents the accurate Ming administrative units and regions in 1433, which none of the three major European maps in the sixteenth century has done. Hence, there is no merit in holding the viewpoint that the China-KWQ is a direct or adapted copy of the three major European maps of the sixteenth century.

In the Appendix, after analysing the four annotations and examining the locations and histories of the 132 geographical items depicted on the China-KWQ, I firmly conclude that the era of China revealed by this map is 1433. This was the date when these places were last explored before the information was used in 1433 for drawing the source map of the China-KWQ. This date supports the conclusion I have already suggested in Chapter 2 that both the China-KWQ and the Korea-KWQ are maps of the year 1433.

Appendix

The geographical items on the China-KWQ are analysed in detail in this Appendix. On the map, they are all written in Chinese traditional characters. Since many readers use the Chinese simplified characters, the author of this book has decided to use both: all the Chinese characters in the main text and in the Appendix of this chapter are expressed first by the modern Chinese traditional characters followed by the Chinese simplified characters, separated by a slash "/" between them. If both characters are the same, that expression will be presented only once without repetition. Moreover, all the geographical items are numbered in sequence in this Appendix. The Chinese "pinyin" spelling of each item is given right after it is introduced. My comments are given in the last column of the Appendix.

Item #	Name in Chinese	Pinyin	Name in English	History, Geography and My Comments
			Zone I: Northeast China	

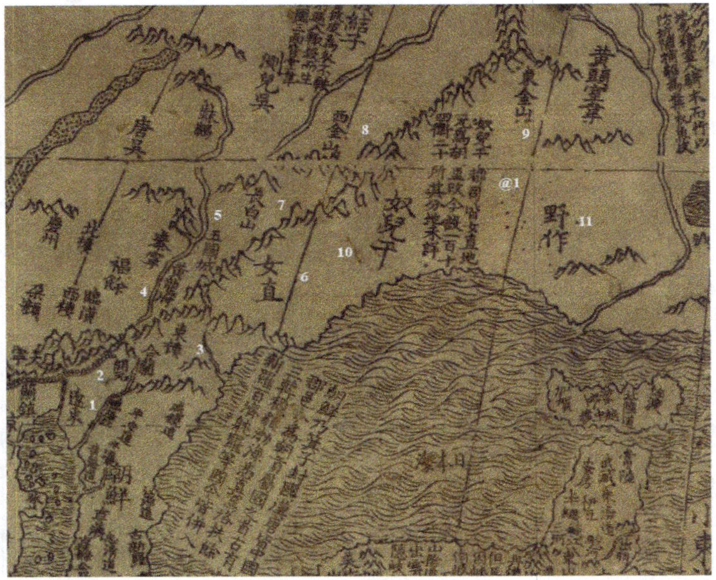

Fig. A3.1.   This figure is extracted from Fig. 3.1(a) (public domain); Items 1–11 and annotation @1 are depicted on this Figure.

Item #	Name in Chinese	Pinyin	Name in English	History, Geography and My Comments
1	遼東	Liao Dong	Liaodong or Liaotung Peninsula	Liaodong or Liaotung Peninsula (遼東半島/辽东半岛) is located between the mouths of the Daliao River (大遼河/大辽河 or Great Liao River, the historical lower section of the 遼河/辽河, lit. "Liao River") in the west and the Yalu River (鴨綠江/鸭绿江) in the east. On the China-KWQ in Fig. A3.1, the unnamed river to the left of the label "遼東" correlates well with the Daliao River. The modern usage of "Liaodong" simply refers to the half of China's Liaoning Province to the left/east bank of the Great Liao/Daliao River.

Item #	Name in Chinese	Pinyin	Name in English	History, Geography and My Comments
				Notice that on the China-KWQ, the Great Wall extends to the North of the Liaodong Peninsula and near Kaiyuan (開元/开元). Rulers of the Ming Dynasty (1368–1644) have ceaselessly maintained and strengthened the Great Wall to prevent the Mongolian invasions. On this map, south of this Wall is the Ming territory. In the main text, I have introduced the Liaodong Regional Military Commission (遼東都指揮使司/辽东都指挥使司 or abbreviated as 遼東都司/辽东都司; 1375–1644). It was established in 1375 (the eighth year of the Ming Hongwu perio) in the southern part of the Northeast region.[49] It was one of the three major administrative agencies in the Northeast during the Ming Dynasty.
2	開元	Kai Yuan	Kaiyuan City	Kaiyuan (開元/开元)[50] was the present Kaiyuan (開原/开原) City in northeastern Liaoning Province. On the China-KWQ, 開元is depicted close to the source of the Yalu River. On a modern map, it is close to the Liao River (遼河/辽河). On the China-KWQ, the city's old name 開元 established in the Yuan Dynasty is depicted. The name was changed to 開原/开原in 1388[51] (but the Kaiyuan Old Town still exists today) in the Ming Dynasty. On the China-KWQ, the Great Wall[52] built by the Ming Dynasty ends close to this city, see Fig. A3.1.
3	東樓	Dong Lou	Donglou	東 means "east" and 樓/楼 "building or mansion". This place name cannot be identified on a modern map. But the name may be derived from the fact that there is a Xilou (西樓/西楼 (Item 24; "西" lit. "west") depicted to the west of it.

| 4 | 黃龍府 | Huang Long Fu | Huanglong Fu | Huanglong Fu (黃龍府/黄龙府; 926–975, 1020–1140; "Fu" is an administrative unit) existed for two periods in Chinese history: from 926 to 975, and from 1020 to 1140.[53] It was established by the Liao Dynasty (遼朝/辽朝; 916–1125) founded by Khitans (契丹), later was destroyed by the Jurchen-led Jin Dynasty in 1125. But in 1140 of the Jin Dynasty (金朝 or 大金, lit. "Great Jin"; 1115–1234), its name was changed to Ji Zhou (濟州/济州), then renamed as Long Zhou (隆州) in 1189.[54] Together with Kaiyuan (開原/开原) and Helan (合蘭/合兰), Huanglong became an administrative division of the Yuan Dynasty (1271-1368).[55] As explained in Chapter 2 of this book, Helan was annexed by Goryeo (高麗/高丽; 918–1392) near the end of the Yuan Dynasty (1271–1368), when Yuan was waning.

After the Kaiyuan government office was moved to Xianping (today's Kaiyuan City), the ancient city of Huanglong Fu slowly turned into ruins. In 1375 (the eighth year of the Hongwu period in the Ming Dynasty), it belonged to the Sanwanwei of Liaodong Regional Military Command (遼東都司三萬衛/辽东都司三万卫; the Wei here means "guard battalion"; and later belonged to Yidong Hewei (亦東河衛/亦东河卫), and finally it was set up as Longan (農安/农安) Station.[56]

At the end of the Ming Dynasty, this area became a nomadic place for Mongolians.[57] |

(Continued)

Item #	Name in Chinese	Pinyin	Name in English	History, Geography and My Comments
5	五國城	Wu Guo Cheng	Wuguo City	The site of Wuguo City (五國城/五国城; 五 lit. "five" and 國/国 lit. "country or kingdom")[58] is also known as the site of "looking at the sky from the bottom of a well (坐井觀天/坐井观天)". This Chinese metaphor describes the situation of a person with a very limited outlook on life. The city was located on the south bank of the Songhua River (松花江) to the northwest of Yilan County (依蘭縣/依兰县) in Harbin City (哈爾濱市/哈尔滨市) of Heilongjiang Province (黑龍江省/黑龙江省).[59] In Fig. A3.1, the location of Wuguo City (五國城/五国城) seems depicted too much to the west than its true geographical location that it is hard to identity the unnamed river flowing at the western side of the city. In Chapter 2, I have discussed Jurchen (女真) which was an ancient nation in Northeast China. Jurchens had tribes in five big cities and formed the famous five kingdoms in history. Jurchens are divided into "Wild" Jurchens or Sheng Jurchens (生女真人) and "Tamed" Jurchens or Shu Jurchens (熟女真人).[60] These names came as follows: in 926, Yelü Abaoji (耶律阿保機/耶律阿保机; 872–926), Taizu of the Liao Dynasty (遼朝/辽朝; 916-1125) destroyed the Kingdom of Bohai (渤海; 698–926) in the north of modern Manchuria (modern province of Heilongjiang). As a result, some Jurchens moved south with the people of Bohai and were incorporated into the Liao nationality; they were called "Shu Jurchens (熟女真人)". Others stayed at their homeland and were called "Sheng Jurchens (生女真人)"; they were not incorporated into the Liao household registration.[61]

				Jin (金; 1115-1234) known as the Great Jin (大金) was an imperial Dynasty of China[62] that drove the Liao to the Western Regions where they became known as the Western Liao (西遼/西辽; 1124–1218). After Jin destroyed the Northern Song Dynasty in 1127, Emperor Huizong (徽宗) and Emperor Qinzong (欽宗/钦宗) were sent under escort to the north. After this Jingkang Incident (靖康事變/靖康事变),[63] they arrived in Wuguo City (五國城/五国城) in 1130 and were imprisoned in the city. In 1135 and 1156, respectively, Huizong and Qinzong died of illness, and they were buried in Wuguo City. Wuguo City was known as the city where the two emperors could only look at the sky from the bottom of a well, to describe the grim future of the two emperors.[64]
6	女直	Nu Zhi	Jurchen	女真 (Nu Zhen) and 女直 (Nu Zhi) are terms used to collectively describe the Jurchen, a number of East Asian Tungusic-speaking people, descended from the Donghu (東胡/东胡; an ancient civilized tribe that lived in the northeast of ancient China, later they were incorporated incorporated into the Xiongnu Empire and branched off into the Xianbei and Wuhuan)[65] people. The Jurchens lived in the northeast of China, later known as Manchuria, before the eighteenth century.[66] The Ming Dynasty categorized the Jurchens into three groups: the Jianzhou Jurchens (建州女真), the Haixi Jurchens (海西女真), and the Wild Jurchens (野人女真 or 東海女真/东海女真). Qing Emperor Taizu (Nurhachi; 努爾哈赤/努尔哈赤; 559–1626)[67] unified the Jurchen tribes, he declared himself Khan in 1616 and founded the Jin (Great Jin) Dynasty. In 1635, Emperor Taizong (Huang Taiji; 皇太極/皇太极; 1592–1643)[68] of Qing Dynasty (1636–1912) changed the name of Jurchen to Manchurian in Shengjing (today's Shenyang). In 1636, Huang Taiji changed the country name from Great Jin (大金) to Great Qing (大淸). In 1644, the Qing Dynasty established by the Manchus ruled the Central Plains and became the second dynasty ruled by ethnic minorities in Chinese history.[69]

(Continued)

Item #	Name in Chinese	Pinyin	Name in English	History, Geography and My Comments
7	長白山	Chang Bai Shan	Changbai Mountains or Changbai Shan; Mountain (山, lit. "Mountain" or "Shan"). The current Chinese name Changbai Shan was first used in the Liao Dynasty (916–1125) of the Khitans[70] and then the Jin Dynasty (1115–1234) of the Jurchens.[71]	The Changbai Mountains (長白山/长白山), also known as Baekdu Mountains). are a major mountain range in Northeast Asia that extends from the Northeast Chinese Provinces of Heilongjiang, Jilin, and Liaoning to the Korean Peninsula and across the present China-North Korea border (41.7° to 42.9°N; 127.7° to ł28.3°E). Most of its peaks exceed 2,000 m/6,600 ft in height. The Changbai Mountain range is regarded as a "sacred mountain range" in the myths and legends of many northern Chinese peoples[72] including the Jurchens, and the Korean people also regard it as the birthplace of their nation.[73] The mountain range was first recorded in *Shan Hai Jin* (山海經/山海经) or *The Classic of Mountains and Seas* under the name Buxian Shan (不咸山). It is also called Dan Dan Daling (單單大嶺/单单大岭) in *Hou Han Shu* (後漢書/后汉书) or *The Book of the Later Han.* In *Xin Tang Shu* (新唐書/新唐书) or *The New Book of Tang*, it was called Taibai Shan (太白山). On the China-KWQ, the name 長白山 (Changbai Mountain Range; 49.3° N, 118.5°E, the KWQ latitudinal data have been multiplied by 0.9856 to compare with modern latitudinal data; KWQ latitude → Modern latitude) is depicted at c. 49.3° N, too far north, and 118.5°E, too far west than its real position (42.0°N 128.0°E; Modern coordinates). Mount Peaktu (白頭山/白头山) is the tallest mountain of the Changbai Mountain Range (also the tallest mountain in North Korea and Northeast China), and the source of the Yalu (鴨綠/鸭绿) and Yumen (圖們/图们) rivers.

8	西金山	Xi Jin Shan	Altai Occidentali, monti[74]	*Altai* is derived from the underlying form "*altañ*", meaning, "gold, golden". 金 is "gold", 山 "mountain", and 西 "west". 西金山means the Gold Mountains in the west or the western part of the Gold Mountains/Altai Mountains The Altai Mountains or Altay Mountains (阿爾泰山脈/阿尔泰山脉), are a mountain range in Central and East Asia, where Russia, China, Mongolia and Kazakhstan converge. On the China-KWQ, 西金山 may mean the western part of the Gold Mountains/Altai Mountains but being mistakenly depicted in Northeast China.
9	東金山	Dong Jin Shan	Altai Orientali, monti[75]	東 means "east". Similar to西金山 listed above, 東金山may mean the eastern part of the Gold Mountains/Altai Mountains but being mistakenly depicted in Northeast China.
10	奴兒干	Nu Er Gan	Nurgan	In Section 6 of Chapter 2, I have introduced Nurgan (奴兒干城 or abbreviated as 奴兒干) which was an ancient Chinese city located at Telin (modern Tyr, Russia), near the mouth of the Amur River (Heilong Jiang; 黑龍江/黑龙江, meaning, "Black Dragon River"), about 100 km from the sea. Since the Tang Dynasty, the city has been under the rule of the Chinese dynasties. Entering the Ming Dynasty, the Jurchens came under the nominal administration of the Nurgan Regional Military Commission (奴兒干都指揮使司/奴儿干都指挥使司; 1409–1435),[76] abbreviated as Nurgan Command Post (奴兒干都司/奴儿干都司). The Command Post was a Chinese seat or administrative unit established in 1409 on the Jurchen land by Emperor Yongle and abandoned in 1435 by Emperor Xuande.

(Continued)

Item #	Name in Chinese	Pinyin	Name in English	History, Geography and My Comments
				The governing office of the Nurgan Command Post was in Nurgan City. In 1413 Yishiha (亦失哈), a Jurchen Ming eunuch, during his third patrol of the Nurgan Command Post, built a temple to worship Goddess Guanyin (觀音/观音). (There are remains of a Yuan era temple unearthed at the site which was dated by modern archaeologists to the 1260s.) During this eunuch's ninth patrol in 1432 of the Nurgan Command Post, he discovered that the temple he built earlier had been destroyed.[77] Then, he rebuilt the temple and installed a new monument to record the event in 1433[78] during his tenth patrol of the outpost. There is no longer any record of Ming Dynasty officials visiting Nurgan after that, and the Nurgan Command Post was also abolished in 1435.[79] The Nurgan Regional Military Commission is of particular interest and relevant to the dating of the China-KWQ. The Commission was established in 1409 and abolished in 1435 on the Jurchen land.[47] But Ming official's last inspection of the Nurgan Command Post occurred in 1433. The era of the Korea-KWQ is determined as 1433 in Chapter 2, because the Jurchen-Joseon border was established in that year and the KWQ was drawn in 1433; then the era of the China-KWQ should also be 1433. This is based on the Jurchen-Joseon border established in 1433, the fact that the Nurgan Regional Military Commission (奴兒干都指揮使司/奴儿干都指挥使司) depicted on the Jurchen land ruled by China was last inspected by the Ming envoy in 1433, and the fact that the KWQ was drawn in 1433. Although the Commission existed from 1409 to 1435, the era 1434–1435 was after the KWQ being drawn and the era 1409–1432 was before the Joseon-Jurchen border being established, so that both periods can be ruled out.

| @1 | 奴兒干都司皆女直地元爲胡里改今設一百十四衛所其分地不詳 | | Nu Er Gan Du Si Jie
Nu Zhi Di Yuan
Wei Hu Li Gai Jin
She Yi Bai Shi Si
Wei Suo Qi Fen Di
Bu Hiang | My translation and comment (between square brackets) is shown as follows:

The Nurgan Regional .Military Commission [奴兒干都指揮使司/奴儿干都指挥使司] is located on the Jurchen land. In the Yuan Dynasty, the region was occupied by the Huligai tribes. Now the Jurchen tribes are divided into 114 Guard Battalions [in the Ming Dynasty, the Jurchen became vassals to the Ming emperors]. But their detailed ownerships of the lands are unknown.

To avoid repetition, please read Section 6 of Chapter 2 or my comments and detailed analysis of the above annotation. |
| 11 | 野作 | Ye Zuo | Yesso[80] | This place cannot be identified on a modern map. |

Zone II: North and Northeast China

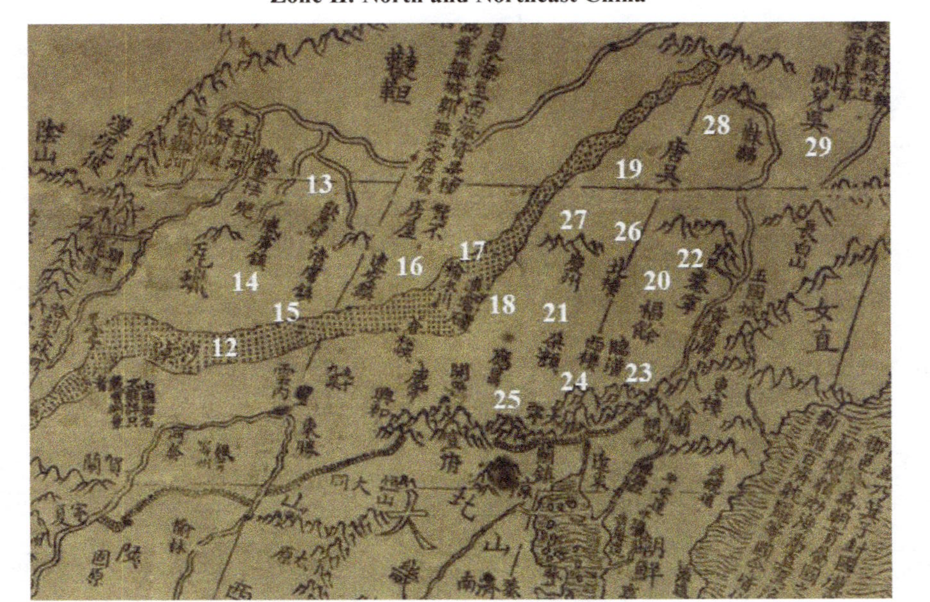

Fig. A3.2.   This figure is extracted from Fig. 3.1(a) (public domain); Items 12–29 are depicted on this Figure.

(*Continued*)

Item #	Name in Chinese	Pinyin	Name in English	History, Geography and My Comments
12	沙漠	Sha Mo	Gobi Desert	The Gobi Desert (戈壁 or 沙漠)[81] is a large desert with exposed bare rock instead of sand, or brushland region, in East Asia. It extends from Northwest China to Northeast China and is the sixth largest desert in the world. The Gobi Desert was only known partially to outsiders. On a modern map, the Gobi stretches from the foot of the Pamirs (帕米爾高原/帕米尔高原; 77°E) to the Greater Khingan Mountains (大興安嶺/大兴安岭; 116–118°E), on the border of Manchuria. The total east-west span is 39°–41°. On the China-KWQ, the Gobi stretches from c. 90.5°E to c. 112.5°E, the total east-west span is c. 22° which is only slightly over 50 percent of the known value. On a modern map, the north-south span of the Gobi is about 20°, whereas on the China-KWQ it is c. 15° (KWQ latitude → modern latitude). Again, the longitudinal error is larger than the latitudinal error. 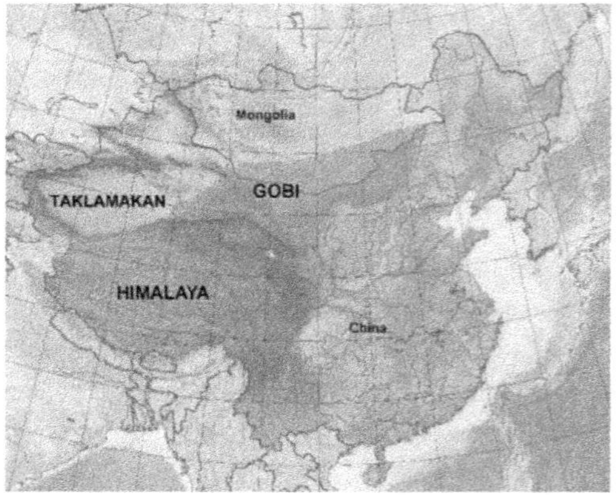

Fig. A3.3. Gobi Desert and Taklamakan Desert (*Source*: TheDrive, under CC BY-SA 4.0, https://commons.wikimedia.org/wiki/File:GobiTaklamakanMap.jpg).[82]

| 13 | 飲馬河 | Yin Ma He | Kherlen River | The old name of Yinma River (飲馬河/饮马河) was Luqu River (臚朐河/胪朐河). It is today's Kherlen River (克魯倫河/克鲁伦河) which flows into Lake Hulun (呼倫湖/呼伦湖) in Inner Mongolia of China. According to *Ming Shi • Cheng Zu Ben Ji* (明史 • 成祖本紀/明史 • 成祖本纪) or *The History of Ming Dynasty: Chengzu*, the name change was made by Ming Yongle Emperor (Chengzu) in 1410.

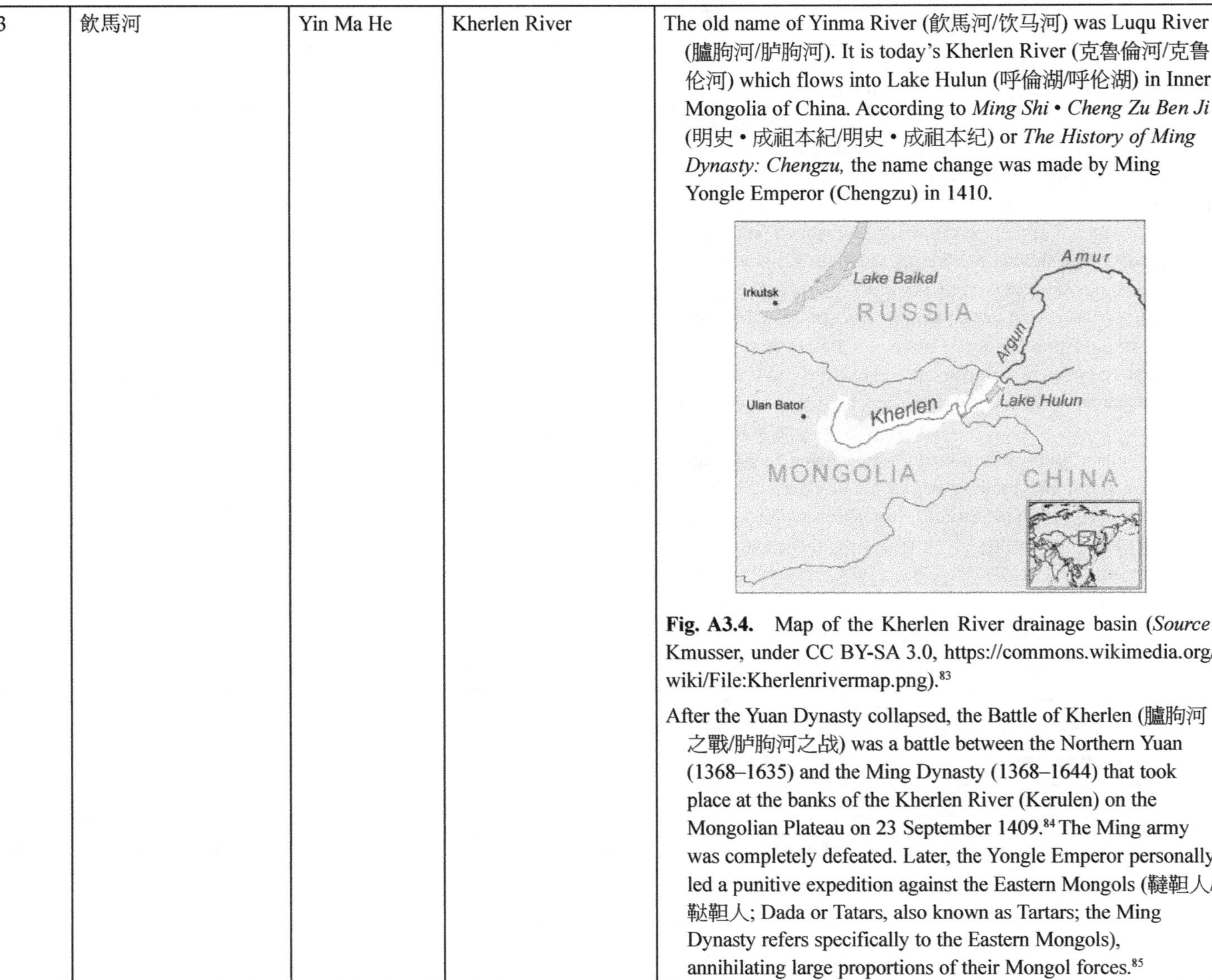

Fig. A3.4.   Map of the Kherlen River drainage basin (*Source*: Kmusser, under CC BY-SA 3.0, https://commons.wikimedia.org/wiki/File:Kherlenrivermap.png).[83]

After the Yuan Dynasty collapsed, the Battle of Kherlen (臚朐河之戰/胪朐河之战) was a battle between the Northern Yuan (1368–1635) and the Ming Dynasty (1368–1644) that took place at the banks of the Kherlen River (Kerulen) on the Mongolian Plateau on 23 September 1409.[84] The Ming army was completely defeated. Later, the Yongle Emperor personally led a punitive expedition against the Eastern Mongols (韃靼人/鞑靼人; Dada or Tatars, also known as Tartars; the Ming Dynasty refers specifically to the Eastern Mongols), annihilating large proportions of their Mongol forces.[85] |

(Continued)

Item #	Name in Chinese	Pinyin	Name in English	History, Geography and My Comments
14	威虜鎮	Wei Lu Zhen	Weilu Town	There was a Weilu Town (威虜鎮/威虏镇) built in the Yuan Dynasty. During the Ming Huangwu Period, the Ming Dynasty established a guard battalion there to protect the country's frontier. However, in 1405 (the third year of Yongle), the guard battalion was abolished.[86]
15	清虜鎮	Qing Lu Zhen	Qinglu Town	This place cannot be located on a modern map.
16	遠安鎮	Yuan An Zhen	Yuanan Town	This place cannot be located on a modern map.
17	榆木川	Yu Mu Chuan	Mongolian: Khailas-ausu, Chinese: 榆木川 (Yumuchuan; in the present-day Duolun County, Inner Mongolia).	The Hailaer River (海拉爾河/海拉尔河) flows into the Hulun Buir Grassland (呼倫貝爾草原/呼伦贝尔草原) through the Dongshan Mountain (東山/东山). During the Ming Dynasty, the river was translated as Yumuchuan (榆木川) in Chinese. The Yumuchuan riverbank has a natural elm forest which is rare in Asia and the area is famous in Chinese history for the following reason: Ming Emperor Yongle (Chengzu) has personally launched five Northern or Mobei (漠北, lit. "extreme cold and barren lands of the Gobi Desert"; the heart of the Mongols) military campaigns against the Northern Yuan force from 1410 to 1424 (1410, 1414, 1422, 1423 and 1424), defeating the Northern Yuan, Eastern Mongols, Oirats, and various other Mongol tribes and maintaining the peace of the Ming border.[87] But he could not eliminate its remnants. In 1424, Ming Chengzu died of illness in his military camp near the Hailar River (榆木川) on the way back to Beijing after winning the fifth personal expedition.[88]
18	連雲磧	Lian Yun Qi	Lianyun Qi	磧 means "desert". But this place cannot be located on a modern map.
19	唐吳	Tang Wu	Tangwu	This place cannot be located on a modern map.

| 20 | 福餘 | Fu Yu | Fuyu | During the Ming Dynasty, in the North, Emperor Yongle founded the Regional Military Commission (*dusi*; 都司) in Manchuria and organised the three Jurchen tribes[89] of Haixi (海西), Jianzhou (建州) and Yeren (野人) into the military guard battalion (*wei-suo*; 衛所) system. Emperor Yongle bestowed nominal titles of military offices on the chieftains of these tribes.[90] The Jurchen chieftains who were bestowed these titles, sent tribute missions every year, along with the chieftains of the Mongolian tribes in Western Manchuria.[91] The latter were organised into three guard units known as the Three Uriankhai Guard Battalions (Wuliangha Sanwei; 兀良哈三衛/兀良哈三卫).[92] They were Taining Wei (泰寧衛/泰宁卫), Duoyan Wei (朵顏衛/朵颜卫) and Fuyu Wei (福餘衛/福余卫).[93] After the fall of the Yuan Dynasty, the Wuliangha Sanwei (兀良哈三衛/兀良哈三卫), Dada/Tatars (韃靼/鞑靼) and Oirats (瓦剌) constituted the three major tribes of Mongolia.[94]

Fuyu Wei was set up in 1389 (the 22nd year of the Hongwu. It got its name because the land was in Fuyu State (the ancient Buyeo as mentioned in Chapter 2), and the grazing land was in the lower reaches of the Nen River (嫩江) in today's Heilongjiang Province (latitude 45.8°N). After the eras of Emperor Xuande (reigned 1426–1435) and Ming Yingzong (reigned 1436–1449), it moved south to areas in the present-day Liaoyang (遼陽/辽阳; latitude 41.3°N), Shenyang (瀋陽/沈阳; latitude 41.8°N), Tieling (鐵嶺/铁岭; latitude 42.2°N), Kaiyuan (開原/开原; latitude 42.5°N) and other cities and counties in Liaoning Province (遼寧/辽宁; latitude 41.8°N).

On the China-KWQ, the latitude of Fuyu at c. 45°N (KWQ latitude → modern latitude) correlates closer with the latitude of the Heilongjiang Province than that of the Liaoning Province, indicating that Fuyu had not moved south yet. Hence, the era of the map is before the end (1435) of the Xuande era of the Ming Yingzong era. After the mid-sixteenth century, Fuyu was under the jurisdiction of the Horqin (科爾沁/科尔沁) tribe of Mongolia. |

(Continued)

Item #	Name in Chinese	Pinyin	Name in English	History, Geography and My Comments
				After the Jiajing period (1522–1566) of the Ming Dynasty, facing the continuous invasion and harassment of the Tatars, both the Duoyan Wei and the Taining Wei were annexed by Tatar Mongolia, and only the Fuyu Wei moved back from the south to the lower reaches of the Nen River.[95]
21	朵顏	Duo Yan	Duoyan	Please see the explanation in Item 20 for Fuyu (福餘/福余). The grazing land of the Duoyan Wei was in the Chuoer River (綽爾河/绰尔河) Basin (46.7°–48.7°N, 120.5°–123.7°E) of the Inner Mongolia Autonomous Region (內蒙古自治區/内蒙古自治区). After the era of Emperor Xuande (reigned 1425–1435), the Duoyan Wei moved south to the Ongniud Banner (翁牛特旗) in Chifeng City (赤峰市; latitude 42.3°N) in the Inner Mongolia Autonomous Region and outside the Great Wall in the north of Hebei Province. On the China-KWQ, the latitude of Duoyan (朵顏/朵颜) is c. 43.9°N (KWQ latitude → modern latitude) which is about halfway between the Chuoer River Basin and Chifeng City, indicating a partial migration from north to south might have taken place during the Xuande period.
22	泰寧	Tai Ning	Taining	Taining (泰寧) was one of the "Three Uriyanghad Guard Battalions (*Wuliangha Sanwei*; 兀良哈三衛/兀良哈三卫) as explained in Item 20. At first, the Taining Wei lived as nomads in the present-day Taoan (洮安; 45.3°N) in Jilin Province (43.9°N). After the reign of Xuande, the Wuliangha Sanwei started to move south. Around the middle of the sixteenth century, all the "Three Uriankhai Guard Battalions" were attached to other Mongolian tribes.[96] On the China-KWQ, the location of 泰寧 (Taining) is depicted at the northeastern position, c. 45.8°N (KWQ latitude → modern latitude), among the three guard battalions, placing it in Heilongjiang, instead of Jilin. This is incorrect. The correct order from north to south should be Fuyu (福餘/福余), Duoyan (朵顏/朵颜) and Taining (泰寧/泰宁).

| 23 | 臨潢 | Lin Huang | Linhuang | Linhuang Fu (臨潢府/临潢府; "Fu" was an administrative unit) was in today's southern suburb of Lindong Town (林東鎮/林东镇), Balin Zuo Banner (巴林左旗), Chifeng City (赤峯市) in Inner Mongolia Autonomous Region.[97] It was called Shangjing (上京) in the Liao Dynasty (916–1125). It was the first capital city built on the homeland in the early days of Khitan (契丹), and the political, economic, military, and cultural centre of the early Liao Kingdom.[98]

In 1120 A.D., the Jin soldiers captured Shangjing, and during the Jin Dynasty, the city was renamed Beijing Linhuang Lu (北京臨潢路/北京临潢路; "Lu" is also an administrative unit). But Linhuang was later gradually abandoned in the Yuan Dynasty.[99]

On the China-KWQ, this obsolete historical site, Linhuang (臨潢/临潢), is still kept on the map. |
| 24 | 西樓 | Xi Lou | Xilou | Xilou (西樓/西楼; 西 means west and 樓 building or mansion) was an ancient place name of Khitan (契丹; a historical nomadic people from Northeast Asia; they inhabited an area corresponding to parts of modern Mongolia, Northeast China and the Russian Far East from the fourth century). The site was in Stone House Village (石房子村), southwest of Baarin/Bairin Left Banner (巴林左旗; the ancient name was Linhuang; 44°N, 119.4°E; Item 23), in today's Inner Mongolia, China. Emperor Taizu of Liao Abaoji's autumn hunting was mostly held here, and he made the place his capital.[100]

In 938, Xilou was officially renamed Shangjing "Upper (Northern, Supreme) Capital (上京臨潢府/上京临潢府)" and it was Linhuang (臨潢/临潢), and the fact is recorded in *Liao Shi* (遼史/辽史) or *The History of Liao* and *Zizhi Tongjian* (資治通鑑/资治通鉴) or *Comprehensive Mirror in Aid of Governance*. |

(*Continued*)

Item #	Name in Chinese	Pinyin	Name in English	History, Geography and My Comments
				On the China-KWQ, Xilou (西樓/西楼) is depicted at 43.9°N, 112.5°E (KWQ latitude → modern latitude) in comparison with Stone House Village (石房子村), the ancient Linhuang, at 44°N, 119.4°E, on a modern map. On the China-KWQ, Xilou is depicted next to Linhuang (臨潢/临潢; Item 23), side by side, see Fig. A3.2. We now know that in fact they were the same place but having different names in different eras.
25	大寧	Da Ning	Daning	The Daning Regional Military Commission (大寧都指揮使司/大宁都指挥使司) was a military and political institution established in 1387 (the 20th year of Hongwu) by the Ming Dynasty in the present northern part of Hebei Province and the southeastern part of the Inner Mongolia Autonomous Region (內蒙古自治區/内蒙古自治区). The following year, it was renamed. In 1403 (the first year of Yongle), the Daning Regional Military Commission was restored, but its seat was changed to Baoding Prefecture (保定府; now Baoding City, Hebei Province). Later, all the guards under its jurisdiction were abolished, and some were moved to the south of the Great Wall.[101] The original site became an abandoned city which was gradually occupied by the Northern Yuan Dynasty; hence, it does not appear in Fig. 3.3. But the abandoned Daning city (Item 25) still shows up on the China-KWQ. The Daning Regional Military Commission was completely abolished in the early Qing Dynasty.[102]
26	北樓	Bei Lou	Beilou	北 means "north". The place may be named because it was located to the north of Xilou (西樓/西楼; Item 24). But this place cannot be identified on a modern map.

27	慶州	Qing Zhou	Qingzhou	Qingzhou (慶州/庆州) was in the present-day Bairin Right Banner (巴林右旗) in Inner Mongolia.[103] The city of Qingzhou was established in order to service the Qing Mausoleum (慶陵/庆陵) in the nearby mountains where Emperors Shengzong (聖宗/圣宗; r. 982–1031), Xingzong (興宗/兴宗; r. 1031–1055), and Daozong (道宗; r. 1055–1101), the sixth, seventh and eighth emperors of the Liao Dynasty (916–1125), are buried.[104]
28	靺鞨	Mo He	Mohe[105]	Mohe (靺鞨) refers to Mohe people, a Tungusic people of ancient Manchuria. As discussed in Chapter 2, Jurchen (女真) is the other term used to collectively describe a number of East Asian Tungusic-speaking people who lived in northeastern China, also known as Manchuria, before the eighteenth century. When the Jurchens first entered Chinese records in 748, they inhabited the forests and river valleys of the land which is now divided between China's Heilongjiang Province and Russia's Primorsky Krai Province. But in earlier records, this area was known as the home of the Sushen (肅慎人/肃慎人; c. 1100 B.C.), the Yilou (挹婁人/挹娄人; around 200 A.D.), the Wuji (勿吉人; c. 500 A.D.), and the Mohe (靺鞨人; c. 700 A.D.).[106] On the China-KWQ, 靺鞨 (Mohe) is depicted to show the region where the Mohe people used to live. The Ming Dynasty set up in the vast Jurchen region more than 380 administrative units under the Jimi (羈縻/羁縻) System,[107] and set up the Nurgan Regional Military Commission's outpost in Telin area (特林; modern Tyr, Russia). The Ming emperors gave Jurchen heads various ministerial positions to strengthen the direct Ming rule over the Northeast.[108]
29	測兒吳	Ce Er Wu	Zeru[109]	This geographical term cannot be identified on a modern map.

(Continued)

Item #	Name in Chinese	Pinyin	Name in English	History, Geography and My Comments
			Zone III: Northeast, North and Northwest China	

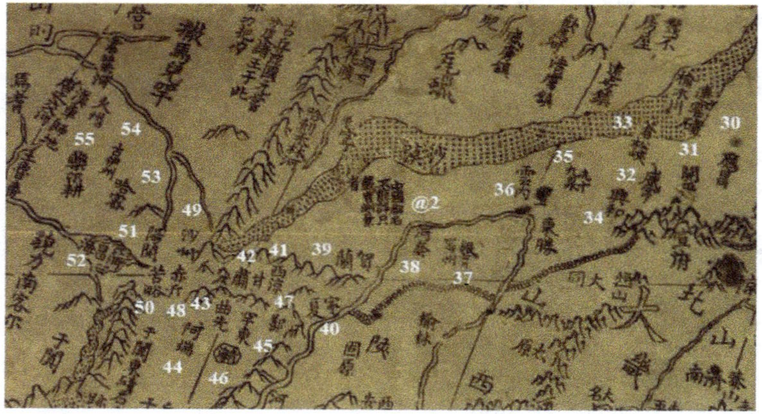

Fig. A3.5.   This figure is extracted from Fig. 3.1(a) (public domain); Items 30–55 and annotation @2 are depicted on this Figure.

Item #	Name in Chinese	Pinyin	Name in English	History, Geography and My Comments
30	應昌	Ying Chang	Yingchang	Yingchang (應昌/应昌) was one of the important cities in the Yuan Dynasty. It was in modern Heshigten Banner (赫什騰旗/赫什騰旗), Inner Mongolia, China.[110] The city of Yingchang was built by the Khongirad Mongols (弘吉剌蒙古人) in 1271,[111] the same year that Kublai (忽必烈 Emperor Shizu; 1215–1294) established the Yuan Dynasty. Yingchang became the capital of the Northern Yuan shortly after the last Yuan emperor Toghon Temür (安權貼睦爾/安欢貼睦爾; 元惠宗 or 元順帝/元顺帝; Emperor Huizong or Emperor Shun of Yuan) lost Dadu (大都) and Shangdu (上都) to the Ming Dynasty in 1368 and 1369, respectively. After the death of Toghon Temür in this city in 1370, the Ming armies managed to capture Yingchang, one of the major cities still in the hands of the Northern Yuan.[112] The Northern Yuan once took back Yingchang in 1374, but the Ming recaptured the city in 1380.

31	開平	Kai Píng	Kaiping	Kaiping (開平/开平) was the second grassland capital established in the Yuan Dynasty; it was built from 1256[113] to 1260, by Kublai Khan (忽必烈汗), the ancestor of the Yuan Dynasty. The city was also known as Xanadu (夏拿都)[114] or Shangdu, lit. "Upper Capital (上都)", and located in Inner Mongolia, northern China. It was first the capital (1263–1273) and then the summer capital (1274–1364)[115] of the Mongol Empire. In 2012, at the 36th session of the World Heritage Committee held in St. Petersburg, Russia, "Yuan Shangdu" was included in the World Heritage List. Xanadu received lasting fame in the western world, because of Venetian explorer Marco Polo's description of it in his celebrated book *Travels* (c. 1300). Marco Polo is widely believed to have visited Shangdu in about 1275.[116] After capturing Kaiping in 1369, the Ming Hongwu Emperor set up Kaiping as a grassland military town, ready to deal with the counterattack of the Northern Yuan Dynasty. Later the Ming Yongle Emperor conquered the grassland five times in response to the rising Tatar and Oirat forces and passed through Kaiping four times.[117] From the middle to the late Yongle era (1402–1424), the weather turned cold during the expeditions, and the Ming army's farmland near Kaiping could not be maintained; the cost of transporting grain and grass was too high. In 1430, the Ming Emperor Xuande ordered the seat of the Kaiping Wei to be moved to Dushibao (獨石堡/独石堡) in the present Hebei Province. The border defence situation was greatly weakened, and the old site of Kaiping City gradually reduced to ruins.

(*Continued*)

Item #	Name in Chinese	Pinyin	Name in English	History, Geography and My Comments
32	威寧	Wei Ning	Weining	On the China-KWQ, Weining (威寧/威宁) in the Ming Dynasty should be a place in the present Inner Mongolia, but it cannot be identified on a modern map.
33	蒼松峽	Cang Song Xia	Cangsong Xia	Here on the China-KWQ, 峽 means "narrow deep valley", not "strait", because it is next to a desert. 蒼 means "green" and 松 "pines". 蒼松峽 (transliterated as Cangsong Xia) means "a narrow deep valley with growing green pines". The place should be in today's Inner Mongolia, but this name cannot be identified on a modern map.
34	興和	Xing He	Xinghe	Xinghe (興和/兴和) should still be in the present Inner Mongolia, but this name cannot be identified on a modern map.
35	九十九泉	Jiu Shi Jiu Quan	99-Springs	九 means "nine", 十 "ten" and 泉 "lake or spring". Jiushijiuquan (九十九泉), lit. "99-Spings". It is in the Huitengxile Grassland (辉腾锡勒草原/辉腾锡勒草原) in the south of Chahar Right Middle Banner (察哈爾右翼中旗/察哈尔右翼中旗), Ulanqab City (烏蘭察布市/乌兰察布市), Inner Mongolia Autonomous Region. The 99- Springs is a lake formed by numerous pits (use "99" to express "many") after the eruption of a volcano. Jiushijiuquan was first recorded in *Wei Shu* (魏书)[118] or *The Book of Wei* compiled by Wei Shou (魏收; 507–572) in the Northern Qi Dynasty (550–577). Since the big tent of the second Mongol Emperor Ögedei Khagan (窩闊台汗/窝阔台汗; c. 1186–1241)[119] was set up in the 99-Springs, it became the heart of his commanding operations.[120]

| 36 | 雲內 | Yun Nei | Yunnei | Yunnei Prefecture (雲內州/云内州) was established in 1055; it was a state from the Liao Dynasty (916–1125) to the Ming Dynasty in China.[121] In 1372 (some other sources say that it was in 1368), Emperor Hongwu abolished the Yunnei Prefecture.[122] But in 1426 (the first year of Xuande), Yunnei Prefecture was changed into a county under the jurisdiction of Feng Zhou/Feng Prefecture (豐州/丰州) when the latter was reestablished.[123] Nevertheless, during the Zhengtung (正统) period (1436–1449), the Yunnei Prefecture (雲內州/云内州) was abolished permanently.
On the China-KWQ, Yunnei (雲內/云内) is depicted side by side with Feng Zhou/Feng Prefecture (豐 [州]), see Fig. A3.5, indicating that the era of the map must be between 1426 and 1449. The conclusion drawn in Chapter 2 that the China-KWQ is a map of 1433 falls nicely in this time range. |
| 37 | 銀宥等州 | Yin You Deng Zhou | Yin, You and other commanderies; commandery was an administrative division in Medieval China. | The Yin Commandery[124] was established in 536 during the Northern and Southern dynasties (南北朝; 420–589). Its administrative office was in the present Shaanxi Province (陝西省/陕西省), and its jurisdiction varied with different Chinese dynasties. The Yin Commandery was abolished in 1106 in the Northern Song Dynasty (北宋; 960–1127).
The You Commandery[125] was established in 738 during the Tang Dynasty (618–907, with an interregnum between 690 and 705). Its administrative office was in today's Inner Mongolia Autonomous Region (內蒙古自治區/内蒙古自治区), and its jurisdiction varied with different Chinese dynasties. The You Commandery was abolished in the Yuan Dynasty (1271–1368).
The China-KWQ still keeps some outdated ancient geographical names on the map. |

(Continued)

Item #	Name in Chinese	Pinyin	Name in English	History, Geography and My Comments
38	河套	He Tao	Hetao	Hetao (河套; river loop)[126] is a C-shaped region in northwestern China consisting of a collection of fertile flood plains stretching from the banks of the northern half of the Ordos Loop (鄂爾多斯環線/鄂尔多斯环线), a large northerly rectangular bend of the Yellow River that forms the river's entire middle section. The name began in the Ming Dynasty. Before the Ming Dynasty, Hetao was called "Henan Land (河南地)". Hetao was frequently used by northern invaders, particularly during the Han and Tang Dynasties, as a staging area for raiding east into the Central Plain (中原) and south into the Loess Plateau (黃土高原) and Guanzhong (關中/关中; in the central part of Shaanxi Province) Region. Hence, this region was of major strategic importance.[127]
@2	中國郡名不能詳只載曾測景者		Zhong Guo Jun Ming Bu Neng Xiang Zhi Zai Zeng Ce Jing Zhe	My translation and comment (between square brackets) are as follows: The names of the Chinese commanderies [established in the region] cannot be detailed; only those which have been surveyed are recorded. As explained in the main text, the Chinese dynasties established many commanderies (the guard battalions established were at the commandery level) in this region as defence frontiers against the Eurasian invaders. Only those which have been surveyed are recorded. Nevertheless, some of the ancient commanderies that were abolished long time ago are kept on the China-KWQ.

| 39 | 賀蘭 | He Lan | The Helan Mountain | The Helan Mountains (賀蘭山/贺兰山), also called Alashan Mountains (阿拉善山), are an isolated desert mountain range forming the border of Inner Mongolia's Alxa League (阿拉善盟; one of twelve prefecture level divisions and three extant leagues of Inner Mongolia) and Ningxia (寧夏/宁夏).
In 1725 (of Qing Dynasty), Ningxia County (寧夏縣/宁夏县) was established. In 1941, it was renamed Helan County賀蘭縣/贺兰县) because it had the same name as Ningxia Province. Helan County is now a county in Ningxia Hui Autonomous Region (寧夏回族自治區/宁夏回族自治区), China. It borders Inner Mongolia to the northwest.
The county is home to the famous Helan Mountains (賀蘭山/贺兰山). It is close to the end of the Great Wall of China. |
| 40 | 寧夏 | Ning Xia | Ningxia | Ningxia (寧夏/宁夏) is now an autonomous region in the northwest of China. Ningxia was the core area of the Western Xia (西夏 or 大夏; 1038–1227) established by the Tangut (黨項/党項) people on the outskirts of the Song Dynasty (960–1279).[128] In 1227, the area came under Mongol domination[129] after Genghis Khan (成吉思汗) conquered Yinchuan (銀川/银川). After the Yuan Dynasty destroyed Western Xia, Ningxia Fu Lu (寧夏府路/宁夏府路) was established,[130] marking the beginning of the use of the name "Ningxia", lit. the "Peaceful Xia".
In 1376 (of the fifth year of Hongwu), Ningxia Wei (Ningxia Guard Battalion; 寧夏衛/宁夏卫) was established,[131] but abolished in 1399 (during the 28th year of Hongwu). It was reestablished in 1403 (during the first year of Yongle).[132] Muslims from Central Asia also began moving into Ningxia from the west. |

(*Continued*)

Item #	Name in Chinese	Pinyin	Name in English	History, Geography and My Comments
41	西涼	Xi Liang	Liang or Western Liang	Lian or Western Liang (涼/凉 or 西涼/西凉), also known as Liangzhou (涼州/凉州), is now Wuwei City (武威市) in today's Gansu Province (甘肅省/甘肃省) of China. During the period of the Sixteen Kingdoms (十六國/十六国; 304–439), five regimes emerged successively along the Silk Road in the Hexi Corridor (河西走廊) including Liang or Western Liang (涼/凉 or 西涼/西凉; 400–421) ruled by the Han people).[133] In 1226, Genghis Khan (c. 1162–1227; the Mongolian Khanate) seized Western Liang.[134] In 1372, Ming Emperor Zhu Yuanzhang (朱元璋; 1328–1398) sent his general Feng Sheng (馮勝/冯胜) to defeat the Yuan army, pacified Gansu, and recovered Liangzhou which was ruled by foreigners for 344 years.[135] The land was finally brought under the rule of the Han nationality, as a Ming territory.[136]
42	甘肅	Gan Su	Gansu	Gansu (甘肅/甘肃) is a province in Northwest China. Its eastern part forms part of one of the cradles of ancient Chinese civilisation. In the Han dynasty, due to the travel of Zhang Qian (張騫/张骞; 164 B.C.–114 B.C.) to the Western Regions (西域),[137] the Hexi Corridor (河西走廊) in Gansu became the main route of the Silk Road (絲綢之路/丝绸之路). It is also famous for temples and Buddhist grottoes such as the Caves of the Thousand Buddhas at Mogao Caves (莫高窟) and the Maijishan Caves (麥積山石窟/麦积山石窟) which contain artistically and historically revealing murals.[138] The State of Qin (秦),[139] known in China as the founding state of the Chinese empire, grew out from the southeastern part of Gansu. When Genghis Khan's army attacked Xixia, he died in Gansu.[140] During the Ming Dynasty, most of the Gansu jurisdiction was inherited from the Yuan Dynasty.

43	曲先	Qu Xian	Quxian	Quxian (曲先) Wei, established in 1371, was one of the Seven Guard Battalions of Quanxi (關西七衛/关西七卫) located to the West of Jiayuguan (嘉峪關/嘉峪关) or Jade Pass in the Ming Dynasty.[141] I have briefly introduced them in Section 2 of the main text. In 1512, forced by Mongolia and others in Qinghai Province (青海省), the local tribes scattered and died, and Quxian Wei (曲先衛/曲先卫) was abolished.[142]
44	阿端	A Duan	Aduan	Aduan (阿端) Wei was one of the Seven Guard Battalions of Quanxi (關西七衛/关西七卫) in the Ming Dynasty. It was located at the present junction of Duoskule Lake (朵斯庫勒湖/朵斯库勒湖) in Qinghai Province (青海省) and Xinjiang Uygur Autonomous Region (新疆維吾爾自治區/新疆维吾尔自治区). In 1375, Aduan Wei (阿端衛/阿端卫) was established but was abolished temporarily. In 1406, it was reestablished. In 1425, during the reign of Hongxi (洪熙; the era of 朱高熾/朱高炽; 1424–1425), when the Quxian (曲先; Item 43) chief tried to kidnap the Ming envoy, they threatened Aduan's commander to go along with them. From Section 2 of the main text, we know that, when the Ming army conquered the tribes, the tribesmen fled. When Xuande Emperor sent officials to appease them, more than 42,000 people returned to Quxian Wei. But Aduan's commander did not dare to return and went to live with the Quxian tribes. In 1432 (the seventh year of Xuande), the son of the Quxian chief sent in tribute and Emperor Xuande was very happy to restore his position. Emperor Xuande also allowed the Aduan chief to stay where he was. Hence, both Quxian and Aduan tribes retained their Guard Battalion status that year.[143] The above history shows that the *Aduan Guard Battalion was restored in 1432, hence, it is logical that this guard battalion is depicted on the China-KWQ which I date to 1433.*

(*Continued*)

Item #	Name in Chinese	Pinyin	Name in English	History, Geography and My Comments
45	罕東	Han Dong	Handong	Handong Wei (罕東衛/罕东卫)[144] was established in 1397 (the thirtieth year of Hongwu), and it was abandoned during the Ming Zhengde (正德) period (he reigned from 1505–1521).[145] In addition, during the Chenghua (成化) period (1465–1487), Ming Emperor Xianzhong (明憲宗/明宪宗) set up Handong Zuowei (罕東左衛/罕东左卫) in Shazhou (沙州; now敦煌 or Dunhuang). The above history tells that Handong Zuowei (罕東左衛/罕东左卫) did not exist in 1433, but Handong Wei (罕東衛/罕东卫) should be on the China-KWQ; this is exactly what we observe on the map.
46	青海	Qing Hai	Qinghai Lake, formerly known as Koko Nur or Kukunor	Qinghai Lake (青海; 37°N, 100°E) is the largest lake in China and an alkaline salt lake. On the China-KWQ, its coordinates are c. 36.8°N (KWQ latitude → Modern latitude), c. 94°E. 青 means "green or greenish blue" and 海 "sea". The Chinese name, using "sea" instead of "lake", is to express the enormous size of the lake.
47	鄯	Shan	Shanzhou	Shan (鄯) was Shanzhou (鄯州).[146] Shanzhou was a place name in China from the Northern Wei Dynasty (北魏; 386–535) to the Song Dynasty (宋; 960–1279).[147] It was established in 526 (the second year of Xiaochang; 孝昌二年) in the Northern Wei Dynasty.[148] In 1104 (the third year of Chongning; 崇寧三年/崇宁三年) in the Northern Song Dynasty, Shanzhou was changed to Xining Prefecture (西寧州/西宁州),[149] it is now Xining (西寧/西宁). The cartographer of the China-KWQ seemed unaware of the name change of Shan Zhou (鄯州) in the Northern Song Dynasty.

48	赤斤	Chi Jin	Chijin	Chijin (赤斤)[150] was established in 1404 (of the second year of Yongle). Eight years later, it was promoted as the Chijing Mengu Wei (赤斤蒙古衛/赤斤蒙古卫). It was in the present Chijin Fort in the northwest of Yumen City (玉門市/玉门市), Gansu. During the Chenghua (成化) period (1465–1487), Turufan (吐魯番/吐鲁番) invaded Chijin Wei and the city became empty.[151] However, that was over 20 years after the China-KWQ was drawn in 1433.
49	沙州	Sha Zhou	Shazhou	In 1227, the Mongolian army destroyed Western Xia or Great Xia (西夏 or 大夏; 1038–1227), conquered Shazhou (沙州) and other places, and the Hexi (河西) area was owned by the Yuan Dynasty (Great Yuan; 1271–1368).[152] In 1277, the Yuan Shazhou was established, and soon became Shazhou Circuit (沙州路). Later it was promoted into a prefecture. In 1404, Shazhou Wei (沙州衛/沙州卫) was established[153] by the Ming Dynasty and the seat of the government was in Dunhuang (敦煌). In the eleventh year of Zhengde (正德十一年; 1516), Dunhuang was occupied by Turpan (吐魯番/吐鲁番).[154] Shazhou Wei was abolished in 1524.
50	苦峪	Ku Yu	Kuyu	Kuyu (苦峪)[155] is an ancient place name. It was in the present southeast of Anxi County (安西縣/安西县), Gansu Province. During the Tang Dynasty, it was the seat of Jinchang County (晉昌縣/晋昌县) in Guangzhou (瓜州). In 1435 of the Ming Dynasty, the commander of the Shazhou Wei (沙州衛/沙州卫) sought help to defend the invasion by Hami (哈密) when he was in trouble. He was ordered to govern Kuyu City.[156]

(*Continued*)

Item #	Name in Chinese	Pinyin	Name in English	History, Geography and My Comments
51	陽関	Yang Guan	Yangguan or Yang Pass	陽 means "sun" or "sunny", but it can also be used to mean "south" (the sunny side of a hill is the southern side), because it lies to the south of the Yumen Pass (玉門關/玉门关) or Jade Pass. Yangguan (陽關/陽関)[157] is a mountain pass that was fortified by Emperor Wu of the Western Han Dynasty (202 B.C.–9 A.D.). It is located approximately 70 km/43 mi southwest of Dunhuang, in the Gansu territory to the west of Shaanxi (陝西) Province in far Northwest China. It was in ancient times in the westernmost administrative centre of China. Yangguan was one of China's two most important western passes, the other being Yumenguan (玉門關/玉门关) or Jade Pass. It was an important landmark on the Silk Road.
52	蒲昌海	Pu Chang Hai	Lop Nur or Lop Nor	蒲昌海 is Lop Nur or Lop Nor (羅布泊/罗布泊).[158] It is a former salt lake, now largely dried up, located in the eastern fringe of the Tarim Basin (塔里木盆地) between the Taklamakan (塔克拉瑪幹/塔克拉玛干) and the Kumtag (庫姆塔格/库姆塔格) deserts in the southeastern portion of the Xinjiang Uygur Autonomous Region. The region was the gateway to the east of the Western Regions and was the main route of east-west traffic at that time. The Kingdom of Loulan (樓蘭/楼兰) since the second century B.C. was an ancient civilisation along the Silk Road, nurtured by the former water resources of the Tarim River (塔里木河) and Lop Nur (羅布泊/罗布泊). Fa Xian (法顯/法显; 337–c. 422) was a Chinese Buddhist monk and translator who travelled by foot from China to India to acquire Buddhist scriptures via the Lop Desert (羅布泊沙漠/罗布泊沙漠) from 395 A.D. to 414 A.D. Later Chinese pilgrims followed his footsteps. Marco Polo during his travels to Asia also passed through the Lop Desert.[159]

53	哈密	Ha Mi	Hami	Hami (哈密; Kumul)[160] is a prefecture-level city in Eastern Xinjiang, China. It is well known for sweet Hami melons. Since the Han Dynasty, Hami has been known for its production of agricultural products and raw resources. The place was the important gate of China to the West. The Ming Dynasty established this region as Kumul Hami in 1404 after the Mongol kingdom Qara Del accepted its supremacy. In 1514, Hami was occupied by Turpan Khan (吐魯番汗/吐鲁番汗).[161] Xuanzang (玄奘; 602–664) was a Chinese Buddhist monk, scholar, traveller, and translator. He is known for the epoch-making contributions to Chinese Buddhism; he visited the oasis town, famous for its melons.[162] Marco Polo reported visiting "Camul" in the early fourteenth century[163] and that was the name it first appeared on European maps during the sixteenth century.
54	古伊州	Gu Yi Zhou	Ancient Yizhou	Yizhou (伊州)[164] was a place name in ancient China. Emperor Wen (隋文帝) of the Sui Dynasty established Yizhou in the fourth year (584 A.D.) of the Kaihuang (開皇/开皇) period. In the Yuan Dynasty, Yizhou was called Hamili (哈密力). In 1406, the Ming Yongle Emperor set up Hami Wei (哈密衛/哈密卫), the westernmost guard battalion in Yizhou.[165]
55	藥尔耕	Yao Er Geng	Yaerkend	This place name cannot be identified on a modern map, but it should be in today's Xinjiang.

(Continued)

Item #	Name in Chinese	Pinyin	Name in English	History, Geography and My Comments
			Zone IV: China: South of Yellow River and North of Yangtze River	

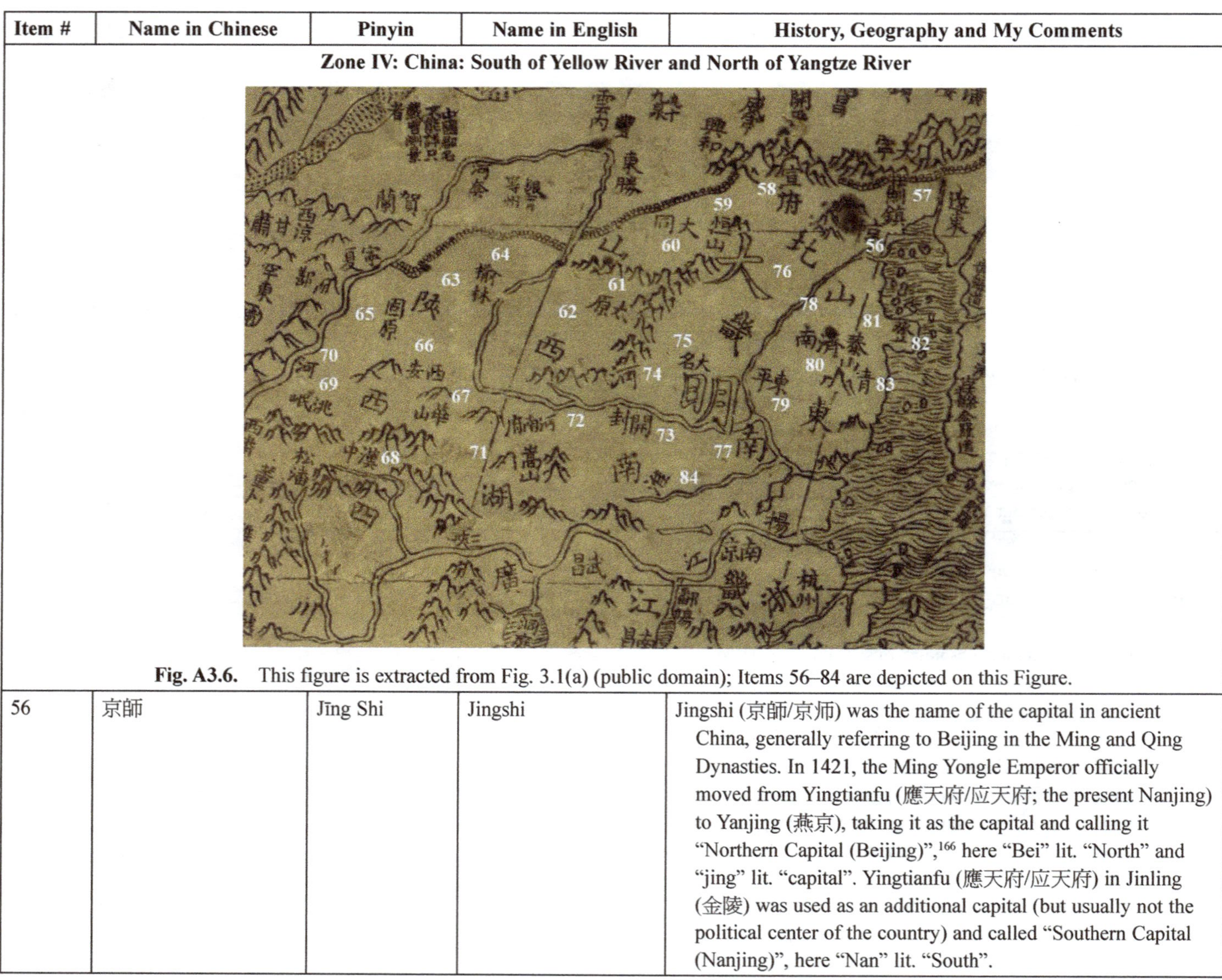

Fig. A3.6. This figure is extracted from Fig. 3.1(a) (public domain); Items 56–84 are depicted on this Figure.

Item #	Name in Chinese	Pinyin	Name in English	History, Geography and My Comments
56	京師	Jīng Shi	Jingshi	Jingshi (京師/京师) was the name of the capital in ancient China, generally referring to Beijing in the Ming and Qing Dynasties. In 1421, the Ming Yongle Emperor officially moved from Yingtianfu (應天府/应天府; the present Nanjing) to Yanjing (燕京), taking it as the capital and calling it "Northern Capital (Beijing)",[166] here "Bei" lit. "North" and "jing" lit. "capital". Yingtianfu (應天府/应天府) in Jinling (金陵) was used as an additional capital (but usually not the political center of the country) and called "Southern Capital (Nanjing)", here "Nan" lit. "South".

| 57 | 薊鎮 | Ji Zhen | Ji Zhen (Town) or Garrison of Ji | Ji Zhen (薊鎮/蓟镇)[167] was one of the nine important garrisons or frontier fortresses in the Ming Dynasty. It took on the important task of defending the capital against the Mongolian invasion and played a role in supporting each other with the Beijing military camp.[168]

The Hongwu Emperor, the founding emperor of the Ming Dynasty, established the first garrison in Gansu (甘肅/甘肃), in 1398. His grandson, Jianwen Emperor, established the Garrison of Ningxia or Ningxia Zhen (寧夏鎮/宁夏镇) in 1402. Then, the Yongle Emperor went to Mobei (漠北; the heartland of the Mongols) five times and set up more garrisons along the Ming border and stationed troops there. Initially, four military towns were established: Liaodong (遼東/辽东), Xuanfu (宣府), Datong (大同), and Yansui (延綏/延绥), then three towns were established in Ningxia (寧夏/宁夏), Gansu (甘肅/甘肃), and Jizhou (薊州/蓟州), and two more garrisons were established in Shanxi (山西) and Guyuan (固原). Altogether they were called Jiu Bian (九邊/九边) or Liu Zhen (九鎮/九镇). Here Jiu is "nine" and Bian lit. "side wall, Great Wall"; Jiu Bian or Jiu Zhen were the nine garrisons in the Chinese northern border along the Great Wall (they guarded the Great Wall in nine sections) which stretched thousands of miles from near the Yalu River in the east to Jiayuguan or the Jade Pass in the west.[169]

The nine garrisons were strengthened through the Hongxi (洪熙; reigned 1424–1425), Xuande (宣德; reigned 1425–1435), Jingtai (景泰; reigned 1449–1457), Tianshun (天順/天顺; reigned 1457–1464) and Chenghua (成化) periods to their final form in the Hongzhi (弘治) period of Xiaozong Emperor (明孝宗; reigned 1487–1505). |

(Continued)

Item #	Name in Chinese	Pinyin	Name in English	History, Geography and My Comments
				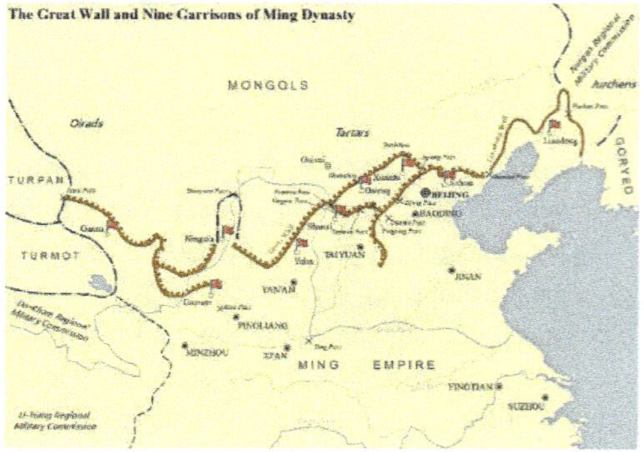 Fig. A3.7. Map of the Great Wall and the Nine Garrisons in the Ming Dynasty (*Source*: SY, under CC BY-SA 4.0, https://commons.wikimedia.org/wiki/File:Ming_Great_Wall.png).[170] The nine garrisons were: 1. Garrison of Liaodong (遼東鎮/辽东镇). 2. Garrison of Ji (薊鎮/蓟镇) 3. Garrison of Xuanfu (宣府鎮/宣府镇) 4. Garrison of Datong (大同鎮/大同镇) 5. Garrison of Shanxi (山西鎮/山西镇), also known as Garrison of Sanguan (三關鎮/三关镇) or Garrison of Piantou (偏頭關/偏头关) or Garrison of Taiyuan (太原鎮/太原镇). 6. Garrison of Yulin (榆林鎮/榆林镇), also known as Garrison of Yanshui (延綏鎮/延绥镇). 7. Garrison of Ningxia (寧夏鎮/宁夏镇) 8. Garrison of Guyuan (固原鎮/固原镇), also known as Garrison of Shaanxi (陝西鎮/陕西镇) 9. Garrison of Gansu (甘肅鎮/甘肃镇) All the places where these garrisons were located are depicted on the China-KWQ.

58	宣府	Xuan Fu	Xuanfu Zhen or Garrison of Xuanfu	Like Ji Zhen (Town) (薊鎮/蓟镇; Item 57), Xuanfu (宣府) also had one of the nine important garrisons in the Ming Dynasty to defend the capital against the Mongolian invasion and played a role in supporting each other with the Beijing military camp. Xuanfu (宣府) was established in 1409 (the seventh year of Yongle).[171] It was in the present Xuanhua District (宣化區/宣化区), Zhangjiakou City (張家口市), Hebei Province.
59	恆山	Heng Shan	Heng Mountain	Heng Mountain (恆山/恒山 or 恆山山脈/恒山山脉)[172] is the general term for a series of peaks in the region to the southeast of Datong City (大同市) in Shanxi Province (山西省) and the southeast of Zhangjiakou City (張家口市/张家口市) in Hebei Province (河北省). These peaks are in the region between the Sanggan River (桑乾河/桑干河) and the Hutuo River (滹沱河). Heng Mountain is known as the northern mountain of the Five Great Mountains of China.
60	大同	Da Tong	Datong	Like Ji Zhen (Town) (薊鎮/蓟镇; Item 57) and Xuanfu (宣府; Item 58), Datong (大同) also had one of the nine important garrisons in the Ming Dynasty to defend the capital against the Mongolian invasion and played a role in supporting each other with the Beijing military camp.
61	太原	Tai Yuan	Taiyuan	Taiyuan (太原) is now the capital and largest city of Shanxi Province, People's Republic of China. Taiyuan is an ancient city with a history of more than 2,500 years and one of the famous historical ancient capitals in China.[173] In the early Ming Dynasty, Taiyuan was established as Taiyuan Prefecture (太原府). Taiyuan (太原) also had one of the nine important garrisons in the Ming Dynasty.

(Continued)

Item #	Name in Chinese	Pinyin	Name in English	History, Geography and My Comments
62	山西	Shan Xi	Shanxi Province; 山 is "mountain" and 西 "west"; 山西 means "West of the Mountains", referring to the province's location west of the Taihang Mountains.[174] Shanxi Province is distinct from Shaanxi Province.	The Mongol Yuan Dynasty administered China into provinces but did not establish Shanxi as a province.[175] In the early Ming Dynasty, the provincial system of the Yuan Dynasty was followed. In 1376 (the ninth year of Hongwu), the province was changed to Shanxi Chengxuan Provincial Administration Commission (山西承宣布政使司 or abbreviated as 布政使司 or 司, meaning, "division" for short, or commonly known as 省, meaning, "province"), with seat at Yangqu (陽曲/阳曲; today's urban area of Taiyuan City).[176] Shanxi only gained its present name and approximate borders during the Ming Dynasty (1368–1644). This took place in 1369 when Ming Emperor Hongwu set up Shanxi and other places as Branch Secretariats (行中書省/行中书省), or simply provinces (行省 or 省).[177]
63	陝西	Shaan Xi	Shaanxi; 陝西 means "Shan's West", referring to Shan plateau (陝塬/陝塬), now known as Zhanbian plateau. Shaanxi Province is distinct from Shanxi Province.	Shaanxi (陝西)[178] is a landlocked province in Northwest China. Shaanxi is considered one of the cradles of Chinese civilization. More than ten feudal dynasties established their capitals in the province during a span of more than 1,000 years, from the Zhou Dynasty (c. 1046 B.C.–256 B.C.) to the Tang Dynasty (618–907, with an interregnum between 690 and 705).[179] The province's principal city and current capital, Xian (西安), is one of the four great ancient capitals of China and is the eastern terminus of the Silk Road, which leads to Europe, the Arabian Peninsula and Africa. Under the Ming Dynasty, Shaanxi was established with the founding of the Shaanxi Chengxuan Provincial Administration Commission (陝西承宣布政使司/陝西承宣布政使司 or abbreviated as 陝西布政使司/陝西布政使司) in 1376 (the ninth year of Hongwu), whose administration also included the modern provinces of 甘肅/甘肃, Ningxia (寧夏/宁夏), and part of Qinghai (青海). In the Ming Dynasty, most visitors from Central Asia and West Asia entered the country via Shaanxi.

| 64 | 榆林 | Yu Lin | Yulin | Yulin (榆林) is a prefecture-level city in the Shaanbei region of Shaanxi Province, China, bordering Inner Mongolia to the north, Shanxi to the east, and Ningxia to the west, see Fig. A3.6.[180]
Yulin Village was built in Hongshan (紅山/红山; today's 雄石峽 or Xiongshixia) in 1408 (another source says as early as 1369), and the name of Yulin first appeared in history. The local soil is especially suitable for planting elms and willows, hence the name.[181] |
| 65 | 固原 | Gu Yuan | Guyuan | Guyuan (固原; 固, meaning, "solidly protected") is a prefecture-level city in the Ningxia Hui Autonomous Region (寧夏回族自治區/宁夏回族自治区) of the People's Republic of China. It occupies the southernmost section of the region, bordering Gansu Province to the east, south, and due west.
The ancient Guyuan City was in the south of Ningxia (寧夏/宁夏; Item 40), which was an important town on the northern road of the eastern section of the ancient Silk Road. In 114 B.C. or the third year of Yuanding (元鼎) in the Western Han Dynasty, Gaoping City (高平城; now Guyuan County) was established. This was when Guyuan City was clearly recorded in history books.
Guyuan in the Ming Dynasty had one of the nine important garrisons set up by the Ming government in the northwest border area, see Fig. A3.7.
In the early Ming Dynasty, the abandoned city of "Guyuan Prefecture" was regarded as Guyuan Li (固原里; 里 was an ancient administrative unit of the township level), which belonged to Kaicheng County (開成縣/开成县), Pingliang Prefecture (平涼府/平凉府), Shaanxi Chengxuan Provincial Administration Commission (陝西布政司/陕西布政司). In 1445 (the tenth year of Zhengtong; 正统十年), the Guyuan Inspection Commission (固原巡檢司/固原巡检司) was added. |

(*Continued*)

Item #	Name in Chinese	Pinyin	Name in English	History, Geography and My Comments
				In 1451 (the second year of Jingtai; 景泰二年), the old city was repaired and garrisoned to protect against Mongolian invasion. In the next year (1452), taking the meaning of "as solidly protected as gold (固)", the newly repaired city was named Guyuan to replace the name of Old Yuan Zhou (古原州 or 故原州; 古 means "ancient" and 故 means "previous") in the ancient times.[182]
66	西安	Xi An	Xian; 西安 means "Western Peace" in Chinese. The old name was Chang'an or Changan; name was adopted in 1369 under the early Ming Dynasty.[183]	Xian or Xi'an (西安) is the capital of Shaanxi Province. Known as Changan throughout much of its history, Xian is one of the Chinese Four Great Ancient Capitals. The city was the starting point of the Silk Road in the Han Dynasty (206 B.C.–220 A.D.), a renowned metropolis in the world in the Tang Dynasty (618–907, with an interregnum between 690 and 705) and home to the Terracotta Army (兵馬俑/兵马俑) and the Mausoleum of the First Qin Emperor – Qin Shi Huang (秦始皇) – both of which are listed as UNESCO World Heritage sites.[184] The city gained the name Xian in 1369 (the second year of Hongwu) in the Ming Dynasty.[185]
67	華山	Hua Shan	Mount Hua or Huasan	Mount Hua (華山/华山) is a mountain located near the city of Huayin (華陰市/华阴市) in Shaanxi Province (陝西), about 120 km/75 mi east of Xian. It is the "Western Mountain (西嶽/西岳)" of the Five Great Mountains (五嶽/五岳) of China and has a long history of religious significance.

				The name of Mount Hua or Huasan first appeared in *Shan Hai Jing* (山海經/山海经) or *The Classic of Mountains and Seas/ Shanhai Jing* (versions of the text may have existed since as early as the fourth century B.C.),[186] and *Yu Gong* (禹貢/禹贡) or *Tribute of Yu* (it is a chapter of *Xia Shu* (夏書/夏书) or the *Book of Xia* section of *Shu Jing* (書經/书经) or *Book of Documents* which is one of the Five Classics of ancient Chinese literature and served as the foundation of Chinese political philosophy for over 2,000 years).[187] As early as the second century B.C., there was a Daoist (道家) temple known as the Shrine of the Western Peak located at the base of Mount Hua. Daoists believed that in the mountain lives the god of the underworld.
68	漢中	Han Zhong	Hanzhong	Hanzhong (漢中/汉中) is a prefecture-level city in the southwest of Shaanxi Province, China, bordering the provinces of Sichuan to the south and Gansu to the west. The founder of the Han Dynasty, Liu Bang (劉邦/刘邦; c. 256 or 246 B.C.–195 B.C.), was once enfeoffed as the king of the Hanzhong region after overthrowing the Qin Dynasty (秦朝). During the Chu-Han contention (楚漢相爭/楚汉相争), Liu Bang (劉邦/刘邦) had the title of the King of Han (漢王/汉王) and later used this title as the name of his imperial dynasty. Hence, Hanzhong was responsible for the naming of the Han Dynasty,[188] and the dynasty was later hailed as the first golden age in imperial Chinese history. From then on, the principal ethnic group in China has been called the Han people. Hanzhong is located at the headwater of the Han River (漢水/汉水), the largest tributary of the Yangtze River (長江/长江). This is also what we see on the China-KWQ, except the Han River is unnamed on this map. In 1370 (the third year of Hongwu) of the Ming Dynasty, Hanzhong became Hanzhong Fu (Prefecture)[189] and this administrative decision was changed in the Qing Dynasty (清朝).

(*Continued*)

Item #	Name in Chinese	Pinyin	Name in English	History, Geography and My Comments
69	洮岷	Tao Min	Taomin	Taomin (洮岷) is a region in Gansu Province. Tao (洮) refers to Tao Zhou (洮州), and Min refers to Min Zhou (岷州). Geographically speaking, they are adjacent to each other. They are in the northwest of China, and the main region where the Yellow River (黄河) and its tributary Tao River (洮河) must pass through. Historically, the Taomin area's strategic position was very important, because it had been a place of long-term competition between the Central Plains dynasties and the northwest ethnic minorities. In the Ming Dynasty, this importance became even more prominent, marking the dividing line between the Han and the Tibetan cultures.
70	河	He	Yellow River or He (Huang He)	The Yellow River or He (河) or Huang He (黄河) is the second-longest river in China,[190] after the Yangtze River (揚子江/扬子江 or Chang Jiang, lit. "長江/长江", lit. "long river"), and the sixth-longest river system in the world.[191] Originating in the Bayan Har Mountains (巴顏喀拉山/巴颜喀拉山) in Qinghai Province (青海省) of Western China, the Yellow River flows through nine provinces, and empties into the Bohai Sea (渤海; the northwestern and innermost extension of the Yellow Sea, to which it connects to the east via the Bohai Strait)[192] near Dongying City (東營市/东营市) in Shandong Province (山東省/山东省).

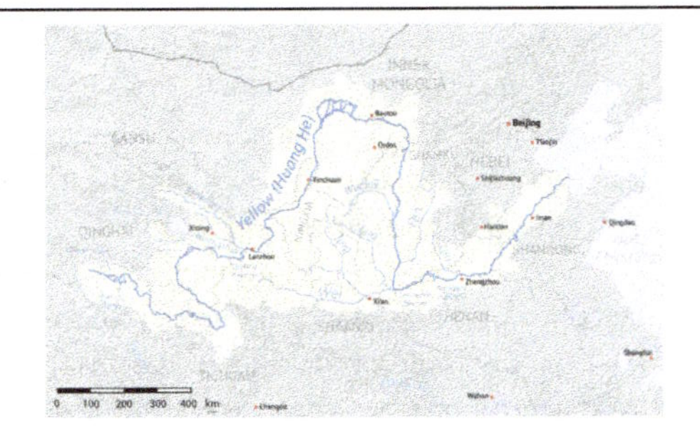

Fig. A3.8.   Map of the Yellow River, whose watershed covers most of northern China (with approximate borders of its basin); it drains to the Yellow Sea (*Source*: Shannon1, under CC BY-SA 4.0 International, 3.0 Unported, 2.5 Generic, 2.0 Generic and 1.0 Generic license, https://commons.wikimedia.org/wiki/File:Yellowrivermap.jpg).[193]

The Yellow River's basin was the cradle of ancient Chinese civilization, and, by extension, the Far Eastern civilizations,[194] and it was the most prosperous region in early Chinese history. There are frequent devastating floods and course changes produced by the continual elevation of the riverbed, sometimes above the level of its surrounding farm fields.

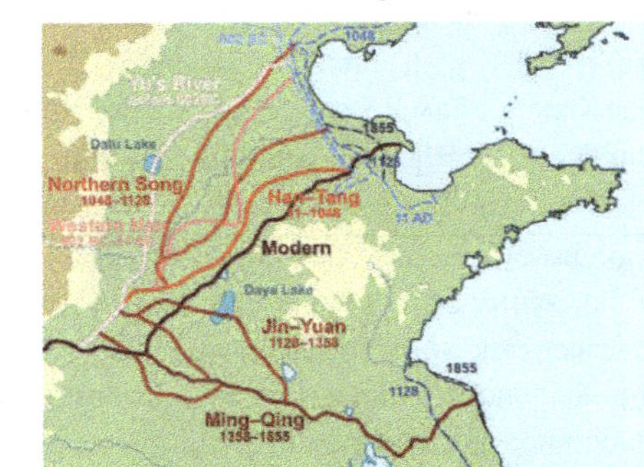

Fig. A3.9.   Historical courses of the Yellow River (*Source*: Stevenliuyi, under CC BY-SA 4.0, https://commons.wikimedia.org/wiki/File:Yellow_River_watercourse_changes_en.png).[195]

(*Continued*)

Item #	Name in Chinese	Pinyin	Name in English	History, Geography and My Comments
				A flood in 1344 returned the Yellow River south of Shandong. The Yuan emperor forced enormous teams to build new embankments for the river under terrible conditions. As a result, the fuelled rebellions led to the founding of the Ming Dynasty.[196] Figure A3.9 shows the historical courses of the Yellow River from ancient days to the Ming-Qing era as results of each flood affecting the river's lower course.
71	嵩山	Song Shan	Mount Song or Songshan	Mount Song (嵩山)[197] is an isolated mountain range in north central China's Henan Province (河南省), along the southern bank of the Yellow River. It is known in literary and folk tradition as the central mountain (中嶽/中岳) of the Five Great Mountains (五嶽/五岳) of China.[198] Since at least as early as the early first millennium, Chinese astronomical mythology had acquired the idea that Mount Song is "the center of Heaven and Earth". It was respected as such by the successive dynasties of the Chinese Empire.[199]
72	河南府	He Nan Fu	Henan Fu or Henan Prefecture	Henan Fu (河南府; Henan Prefecture) was an ancient prefecture referring to the area near Luoyang (洛陽/洛阳).[200] Luoyang has been the capital of many Chinese dynasties, for example, for most of the Tang Dynasty (before 907) it was known as the "Eastern Capital" (東都/东都); during the Later Tang Dynasty (后唐; 923–937), it was the national capital; during the Northern Song Dynasty (北宋; 960–1127) it was known as the "Western Capital" (西京); from 1127 to 1234, the Jurchen (女真) conquerors named it Jinchang Fu (金昌府; a prefecture), also known as "Central Capital" (中京), etc. For these reasons, Henan Prefecture was also colloquially called Luojing (洛京; "Luo Capital"). It was not until the establishment in 1376 (of the ninth year of Hongwu) that the Henan Chengxuan Provincial Administration Commission (河南承宣布政使司 or abbreviated as 河南布政司), then the concept of Henan came into being a province, hence, the Henan Province.

73	開封	Kai Feng	Kaifeng	Kaifeng (開封/开封)[201] is a prefecture-level city in east-central Henan Province, China. It was one of the Eight Ancient Capitals of China, having been the capital eight times in history, while it is best known for having been the Chinese capital during the Northern Song Dynasty. Kaifeng has a history of more than 4,100 years. It gave birth to the far-reaching "Song Culture (宋文化)" which inherited from those of the Han and Tang Dynasties and inspired the later Ming and Qing dynasties. When it was the Eastern Capital of the Song Dynasty, it was the largest city in the world at that time, and it was the place where the famous painting — Qingming Shanghe Tu (清明上河圖/清明上河图)[202] or Along the River During the Qingming Festival—was created.
74	河南	He Nan	Henan; the name means "South of the [Yellow] River", but approximately a quarter of the province actually lies north of the Yellow River.	Henan (河南) is a landlocked province of China, in the central part of the country. The province is an important birthplace of the Chinese nation and Chinese civilization. Henan Province was home to four of the Eight Great Ancient Capitals of China: Luoyang (洛陽/洛阳), Anyang (安陽/安阳), Kaifeng (開封/开封) and Zhengzhou (鄭州/郑州), are all in Henan. As explained in Item 72, the concept of Henan as a province started from the Ming Dynasty when the Henan Provincial Administration Commission (河南布政使司) was established in 1376 (of the ninth year of Hongwu).
75	大名	Da Ming	Daming	Daming (大名) is a county under the jurisdiction of Handan City (邯鄲市/邯郸市) in far southern Hebei Province, China. It was formerly one of the capitals — "Northern Capital" — of the Northern Song Dynasty. The Song Empire had four capitals, with Kaifeng (開封; modern Kaifeng, Henan) serving as the main capital.

(Continued)

Item #	Name in Chinese	Pinyin	Name in English	History, Geography and My Comments
				The city served as an important center for learning during Imperial China. In 782 of the Tang Dynasty, Weizhou (魏州) became the Daming Fu, which was the beginning when "Daiming" was used as a place name. After changing names a few times in later dynasties, it was renamed to Daming Fu (大名府) in 1382 during the Ming Dynasty and stayed unchanged from 1382 until the Republican era.[203]
76	北畿	Bei Ji	Beiji	北 means "north" and 畿 "imperial domain or area near the capital". On the China-KWQ, 北畿 is depicted as an area near 京, i.e., Beijing. Throughout the Ming Dynasty, there were two regions where their Fu and Zhou were directly administered by the Central Government. These two regions were named as the Southern Zhili (南直隸/南直隶) and Northern Zhili (北直隸/北直隶), commonly known as Nanjing and Beijing and their surrounding areas. Nanjing can also specifically refer to Yingtian Prefecture (應天府/应天府), and Beijing can also specifically refer to Shuntian Prefecture (順天府/顺天府); they were commonly known as Nanji (南畿) and Beiji (北畿). Together with the thirteen Chengxuan Provincial Administration Commissions (承宣布政使司), the Ming political or administrative structure can be collectively referred to as "Two Zhilis and Thirteen Chengxuan Provincial Administration Commissions" or abbreviated as "Two Capitals and Thirteen Provinces". This is consistent with the historical record that Beijing was designated as the Ming capital from 1421 to the modern times.
77	南畿	Nan Ji	Nanji	南 means "south" and 畿 "imperial domain or area near the capital". On the China-KWQ, 南畿 is depicted as an area enclosing Nanjing. Item 76 gives detailed explanations for both Beiji (北畿) and Nanji (南畿).

| 78 | 山東 | Shan Dong | Shandong; 山 means "mountain" and 東/东 "east"; Shandong, lit. "east of the Taihang (太行) mountains" | Shandong (山東/山东) is a coastal province of the People's Republic of China. Shandong has played a major role in Chinese history since the beginning of Chinese civilization along the lower reaches of the Yellow River.
The Song Dynasty was forced to cede northern China to the Jurchen (女真) Jin Dynasty (金朝) in 1142 and Shandong was administered by Jin as Shandong East Circuit and Shandong West Circuit – the first use of its current name.
The modern province of Shandong was created by the Ming Dynasty,[204] where it had a more expansive territory, including the agricultural part of Liaoning (遼寧/辽宁). The Ming Dynasty established the Shandong Provincial Administration Commission (which at that time included Liaodong, Beijing, Tianjin, and Hebei). In the early Ming Dynasty, when Shandong recovered from the Yuan Dynasty, Shandong was "mostly uninhabited land". By 1393 (the twenty-sixth year of Hongwu), the cultivated land area became more than double that in the Northern Song Dynasty. After Emperor Yongle moved the capital to Beijing in 1421, Shandong cities along the Beijing-Hangzhou Grand Canal prospered due to the development of water transportation.[205] |
| 79 | 東平 | Dong Ping | Dongping; the name means "eastern peace". The name "東平" first appeared in the Han Dynasty. In the Tang Dynasty, Sucheng County (宿城縣/宿城县) was changed to Dongping County, which was the beginning of the name of Dongping County. It is still used today.[206] | Dongping (東平/东平) is a county in the west of Shandong Province, China.
In 2007 a remarkable series of well-preserved frescoes dating to the Western Han Period (206 B.C.–25 A.D.) was discovered in a tomb. The frescoes show vividly one of the earliest pictorial representations of Confucius (孔子) meeting Laozi (老子), see Fig. A3.10.
During the Yuan Dynasty, after the opening of the Grand Canal (大運河/大运河), Dongping benefited from its location at the intersection of the Grand Canal and Ji River (濟水/济水) and became an important water transport hub. Marco Polo called it "a majestic and magnificent city".[208] After the Ming Dynasty, as the canal was silted up, Dongping's economic status declined. |

(*Continued*)

Item #	Name in Chinese	Pinyin	Name in English	History, Geography and My Comments
				Fig. A3.10. Fresco showing Confucius and Laozi meeting (public domain).[207]
80	濟南	Ji Nan	Jinan; lit. "south of the Ji"; Ji refers to the old Ji River (濟水/济水) that had flowed to the north of the city. The river disappeared in 1852 when the Yellow River changed its course.[209]	Jinan (濟南/济南) is the capital of Shandong Province in Eastern China. The area of present-day Jinan has played an important role in the history of the region from the earliest beginnings of civilization. In the first year of Wu (31 January 1367–23 January 1368) before Zhu Yuanzhang (朱元璋) founded the Ming Dynasty in 1368, he changed Jinan Lu (used in the Yuan Dynasty) to Jinan Fu.[210] And after the Ming Dynasty (1368–1644) created Shandong Province, Jinan became its capital in 1376.

			Jinan was the site of a siege during the Jingnan Campaign (靖難之役/靖难之役; 1399–1402) where the city was defended by loyalists of Emperor Jianwen (建文帝; reigned from 1398 to 1402) against the army of the rebel Zhu Di (朱棣), Prince of Yan (燕王; the future Yongle Emperor; reigned from 1402–1424).[211]
81	泰山	Tai Shan	Mount Tai or Tai San; lit. "exalted mountain".	Mount Tai (泰山) — located north of the city of Taian — is a mountain of historical and cultural significance. It is the highest point in Shandong Province, China.[212] Mount Tai is known as the eastern mountain (東嶽/东岳) of the Sacred Mountains of China. It is associated with sunrise, birth, and renewal, and is often regarded as the foremost of the five (五嶽/五岳). Mount Tai has been a place of worship for at least 3,000 years.[213] Because of its sacred importance and dramatic landscape, it was made a UNESCO World Heritage Site in 1987.[214]
82	登	Deng	Dengzhou or Deng Prefecture	Dengzhou (登州) — an administrative district centering on modern Penglai (蓬莱), Shandong, China — began in the early Tang Dynasty and ended in the Republic of China, which lasted more than 1200 years. In 1376 (the ninth year of Hongwu), the district was promoted as Dengzhou Fu (Dengzhou Prefecture; 登州府).[215]
83	青	Qing	Qingzhou or Qing Prefecture	Qingzhou (青州) is a county-level city in Shandong Province, China. In 1368 (the first year of Hongwu), the Yidu Lu (益都路) in the Yuan Dynasty was changed to Qingzhou Fu (青州府; a prefecture).[216]

(Continued)

Item #	Name in Chinese	Pinyin	Name in English	History, Geography and My Comments
84	淮	Huai	Huai River	The Huai River (淮河) is a major river in China. It is in the region about midway between the Yellow River and the Yangtze River (the two longest rivers in China, see Fig. A3.11). The three rivers form the largest drainage basin in China. All three rivers run from west to east. Due to floods, the Huai River no longer drains directly into the Yellow Sea, because floods have changed the course of the river such that it now primarily discharges into the Yangtze. But on the China-KWQ, the Huai River is depicted as a tributary of the Yellow River, see Fig. A3.6. **Fig. A3.11.** Map of the Huai River (淮河), whose watershed drains much of eastern China between the Yellow River (黃河; Huang He) and Yangtze River (長江/长江; Chang Jiang), eventually flowing into the Yangtze River (*Source*: Shannon, under CC BY-SA 4.0 International, 3.0 Unported, 2.5 Generic, 2.0 Generic and 1.0 Generic license, https://commons.wikimedia.org/wiki/File:Huairivermap.jpg).[217]

Zone V: Central, Southern, Eastern and Western China

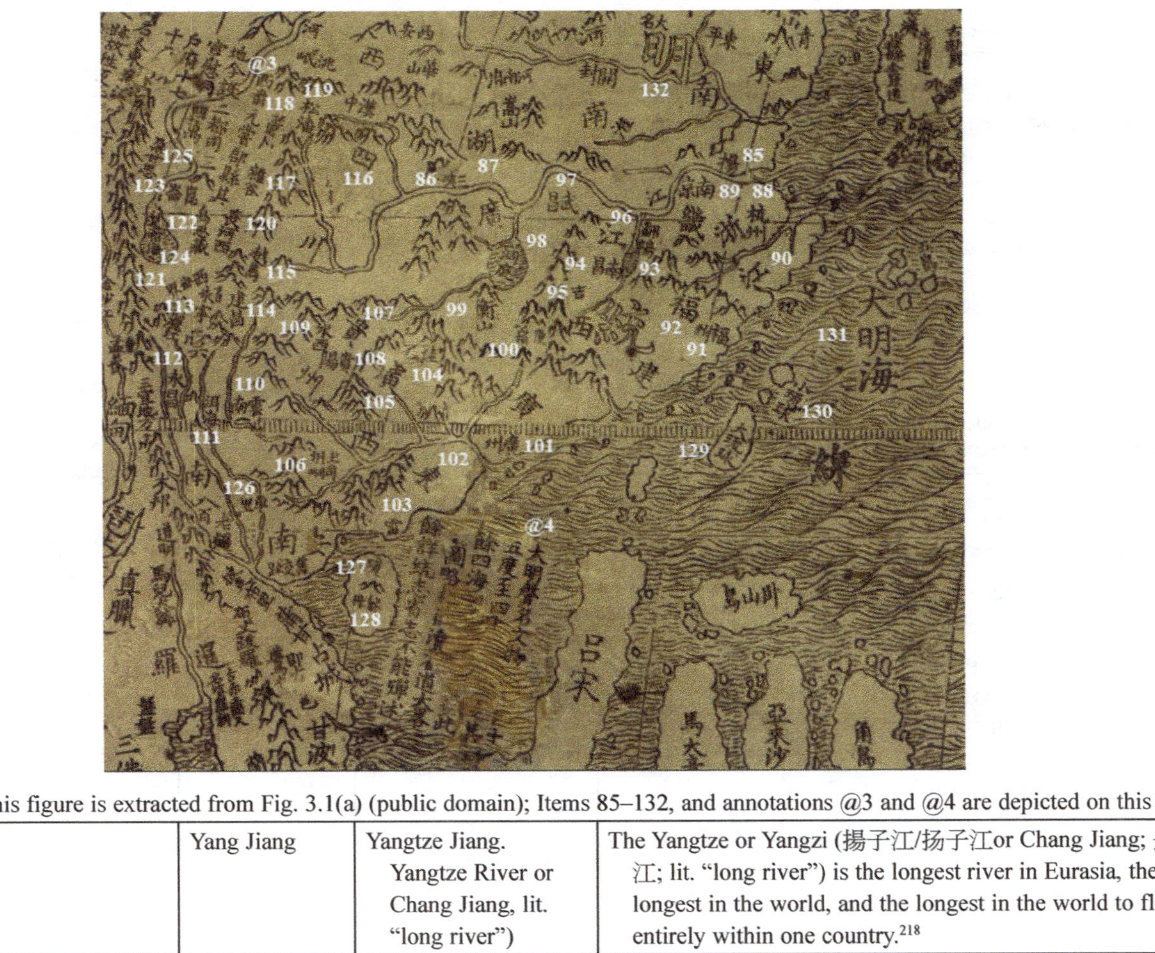

Fig. A3.12.   This figure is extracted from Fig. 3.1(a) (public domain); Items 85–132, and annotations @3 and @4 are depicted on this Figure.

85	揚江	Yang Jiang	Yangtze Jiang. Yangtze River or Chang Jiang, lit. "long river")	The Yangtze or Yangzi (揚子江/扬子江or Chang Jiang; 長江/长江; lit. "long river") is the longest river in Eurasia, the third-longest in the world, and the longest in the world to flow entirely within one country.[218]

(Continued)

Item #	Name in Chinese	Pinyin	Name in English	History, Geography and My Comments
				Fig. A3.13. Map of the Yangtze River (Chang Jiang) drainage basin, with major tributaries (including the Huai River) and cities (*Source*: User:Shannon 1, under CC BY-SA 4.0 International, 3.0 Unported, 2.5 Generic, 2.0 Generic and 1.0 Generic license, https://commons.wikimedia.org/wiki/File:Yangtze_river_map.png).[219] It has its source in the Tanggula Mountains (唐古拉山脈/唐古拉山脉) of the Tibetan Plateau (青藏高原) and flows 6,300 km/3,900 mi[220] in a generally easterly direction to the East China Sea (東海/东海).
86	三峽	San Xia	Sanxia	Sanxia (三峽/三峡) is the Three Gorges (the three adjacent gorges along the middle reaches of the Yangtze River) on the Yangtze River. They are known for their scenery.
87	湖廣	Hu Guang	Huguang	Huguang (湖廣/湖广) was a province of China during the Yuan and Ming Dynasties.[221] It was founded by the Yuan Dynasty in 1274, and its scope of jurisdiction was very large. Historical records show that in the Ming Dynasty, it included initially the modern provinces of Hubei (湖北) and Hunan (湖南), and in the process added areas north of the Yangtze.[222] Huguang was partitioned in 1644 by the newly established Qing Dynasty, becoming the provinces of Hubei and Hunan, which were administered by the Viceroy of Lianghu (兩湖總督/两湖总督). On the China-KWQ, Huguang (湖廣/湖广) is depicted as one province and includes areas north of the Yangtze, indicating that the China-KWQ is a Ming map, not a Qing map.

| 88 | 杭州 | Hang Zhou | Hangzhou | Hangzhou (杭州) grew to prominence as the southern terminus of the Grand Canal and is listed as one of the Seven Ancient Capitals of China. The city is also known as "the House of Silk"[223] with silk fabrics uncovered in the region dating back 4,700 years.
It is believed that Hangzhou was the largest city[224] in the world from 1180 to 1315 and from 1348 to 1358. In Marco Polo's book, he records that the city was "greater than any in the world" and that "the number and wealth of the merchants, and the amount of goods that passed through their hands, was so enormous that no man could form a just estimate thereof."[225]
The Moroccan traveler Ibn Battuta visited Hangzhou in 1345 and was particularly impressed by the beautiful West Lake and a large number of well-crafted and well-painted Chinese wooden ships with colored sails and silk awnings in the canals.[226]
The city remained an important port until the middle of the Ming Dynasty era when its harbor slowly silted up. |
| 89 | 南京 | Nan Jing | Nanjing | Nanjing (南京) has a prominent place in Chinese history and culture. It has served as the capital of various Chinese dynasties, kingdoms and republican governments dating from the third century to 1949 and was the home to one of the world's largest inland ports.
The first emperor of the Ming Dynasty — Zhu Yuanzhang (朱元璋; the Hongwu Emperor) who overthrew the Yuan Dynasty — renamed the city Yingtian (應天/应天), rebuilt it, and made it the dynastic capital in 1368.[227] It is believed that Nanjing was the largest city in the world at the time.[228] |

(*Continued*)

Item #	Name in Chinese	Pinyin	Name in English	History, Geography and My Comments
				In 1421, the Yongle Emperor (永樂帝/永乐帝) relocated the capital to Beijing (北京; 北 means "north" and 京 lit. "capital"). The city began to be called the "southern capital", Nanjing (南京; 南 lit. "south"), in comparison to the capital in the north. Yongle Emperor's successor — the Hongxi Emperor (洪熙帝; reigned from 1424 to 1425) — wished to revert the location of the imperial capital from Nanjing to Beijing. In 1424, he cancelled the maritime expeditions led by Zheng He (鄭和/郑和) permanently, arguably burned down the fleet or left the ships to decompose, and in 1425 appointed Admiral Zheng He as the defender of Nanjing. Then he died suddenly in 1425. His son became the Xuande Emperor (宣德帝) at age 26. The succeeding Xuande Emperor preferred to remain in Beijing, leaving it the primary and *de facto* capital[229] and Nanjing as permanent secondary or reserve capital.[230] However, in the official Ming documents of 1425 to 1441, Nanjing was designated as the capital and Beijing was designated as the temporary capital, owing to the continuing importance of the ancestral injunctions.[231]
90	浙江	Zhe Jiang	Zhejiang; the province's name derives from the Zhe River (浙江; lit. "Zhe Jiang")[232]	Zhejiang (浙江) is an eastern, coastal province of the People's Republic of China. Its capital and largest city is Hangzhou. Zhejiang in South China was conquered by the Mongols in the late thirteenth century who later established the short-lived Yuan Dynasty (1271–1368). Zhejiang became part of the much larger Jiangzhe Province. The Ming Dynasty, which drove out the Mongols in 1368, finally established the present-day province of Zhejiang with its borders having few changes since this establishment.

| 91 | 福州 | Fu Zhou | Fuzhou | Fuzhou (福州) is the capital and one of the largest cities in Fujian Province, China.
From 1405 to 1433, a fleet of the Ming Imperial Navy under Admiral Zheng He sailed from Fuzhou to the Western Ocean (the Indian Ocean and beyond) seven times. Before the last sailing, Zheng erected, between 5 December 1431 and 3 January 1432, a stele dedicated to the Goddess Tian Fei (Matsu) near the seaport.[233]
The Ryukyu Kingdom (琉球王國/琉球王国; 1429–1879) was a kingdom in the Ryukyu Islands, facing Fujian Province (福建省) across the East China Sea (東海/东海). It was ruled as a tributary state of imperial Ming China. In 1392, during the Hongwu Emperor's reign, at the request of the Ryukyuan King, the Ming Chinese sent thirty-six Chinese families from Fujian to manage oceanic dealings in the kingdom. Many Ryukyuan officials were descended from these Chinese immigrants, being born in China or having Chinese grandfathers.[234] The Ryukyu Kingdom established an embassy (Ryūkyū-kan; 琉球館/琉球馆) in Fuzhou (福州). |
| 92 | 福建 | Fu Jian | Fujian | Fujian (福建) is a province on the southeastern coast of China, its capital is Fuzhou.
During the Yuan Dynasty, Fujian was part of the Jiangzhe Province, whose capital was at Hangzhou. After the establishment of the Ming Dynasty, Fujian became a province in 1376, with its capital at Fuzhou.[235] In the early Ming era, Fuzhou Changle (長樂/长乐) was the staging area and supply depot of Zheng He's naval expeditions.[236] |

(Continued)

Item #	Name in Chinese	Pinyin	Name in English	History, Geography and My Comments
93	鄱陽	Po Yang	Poyang Lake	Poyang Lake (鄱陽湖/鄱阳湖) located in Jiujiang City, Jiangxi Province (江西), is the largest freshwater lake in China.[237] The lake is fed by the Gan (贛/赣), Xin (信), and Xiu (修) rivers, etc., which connect to the Yangtze through a channel. On the China-KWQ, see Fig. A3.12, the unnamed river which feeds Poyang Lake (鄱陽湖/鄱阳湖) is the Gan River; the Gan River connects with a river named Yangtze River (揚江/扬江). In 1363, the Battle of Lake Poyang (鄱陽湖之戰/鄱阳湖之战)[238] was a naval conflict which took place between the rebel forces of Zhu Yuanzhang (朱元璋; the founding emperor of the Ming Dynasty) and Chen Youliang (陳友諒/陈友谅) during the Red Turban Rebellion (紅巾起義/红巾起义).[239] This war led to the fall of the Yuan Dynasty. The battle is claimed to be the largest naval battle in history.[240]
94	南昌	Nan Chang	Nanchang	Nanchang (南昌) is the capital of Jiangxi Province, People's Republic of China. The city is bounded on the west by the Jiuling Mountains, (九嶺山/九岭山) and on the east by Poyang Lake (鄱陽湖/鄱阳湖). On the China-KWQ, see Fig. A3.12, there is an unnamed mountain range to the west of Nanchang which correlates well with the Jiuling Mountains, and Poyang Lake is depicted correctly to the east of Nanchang. Nanchang's name changed a few times; until in 1164, it was named Longxing Prefecture (隆興府/隆兴府), which name it retained until 1368. During the Yuan Dynasty, it was the capital of Jiangxi Province, an area that included Guangdong as well. At the end of the Yuan Dynasty (1271–1368), Nanchang became a battleground between Zhu Yuanzhang, the founder of the Ming Dynasty (1368–1644), and the rival local warlord, Chen Youliang. In 1595, Matteo Ricci came to Nanchang while trying to gain entry to Beijing.[241]

95	吉	Ji Zhou	Jizhou	Jizhou (吉州) is a district in Jian City (吉安市) of Jiangxi Province, People's Republic of China. It was a state established in 590 A.D. during the Sui Dynasty (581–618).[242] Jizhou was famous for producing Jizhou ware. Production seems to have begun in the late Tang Dynasty or under the Five Dynasties, and continued until the Yuan Dynasty,[243] but ended suddenly in the fourteenth century.[244]
96	江西	Jiang Xi	Jiangxi	Jiangxi (江西) is a landlocked province in the east of the People's Republic of China. During the Yuan Dynasty, Jiangxi Province was established for the first time. This province also includes the majority of modern Guangdong. Jiangxi acquired (more or less) its modern borders during the Ming Dynasty after Guangdong was separated out. There has been little change to the borders of Jiangxi since.[245] On the China-KWQ, Guangdong and Jiangxi are two separate provinces, showing that it is not a Yuan map.
97	武昌	Wu Chang	Wuchang	Wuchang (武昌) stood on the right (southeastern) bank of the Yangtze River, opposite the mouth of the Han River (漢水/汉水) and faces Hankou (漢口/汉口) and Hanyang (漢陽/汉阳) across the river.[246] The Wuchang Commandery (武昌郡) was set up in 221, when warlord Sun Quan (孫權/孙权) moved the capital of Eastern Wu (東吳/东吴; 222–280) to E County (鄂城; in today's Ezhou City) and renamed it to Wuchang (lit. "prospering from military". This was before the Battle of Red Cliffs (赤壁之戰/赤壁之战).[247] In 223, the Commandery was renamed to Jiangxia Commandery (江夏郡), and the capital of the commandery moved to Xiakou (夏口; in present Wuchang Town). In 280, the Jin Dynasty (266–420) changed Jiangxia Commandery (江夏郡)[248] to Wuchang County (武昌郡), which was the beginning of the official name of Wuchang City. The name of the town switched back and forth between Wuchang and Jiangxia several times in the following centuries.

(*Continued*)

Item #	Name in Chinese	Pinyin	Name in English	History, Geography and My Comments
98	洞庭	Dong Ting	Dongting Lake	Dongting Lake (洞庭湖)[249] is a large, shallow lake in northeastern Hunan Province, China, see Fig. A3.14. It is the second largest lake of China, next to Qinghai Lake (青海湖). It is in the northern part of Hunan Province. It is a flood basin of the Yangtze River, so its volume depends on the season.[250] The present provinces of Hubei and Hunan are named after their location relative to the lake: Hubei means "North of the Lake" and Hunan, "South of the Lake". **Fig. A3.14.** Map showing Lake Dongting and the major rivers flowing into it (*Source*: Kmusser, under CC BY-SA 3.0, https://commons.wikimedia.org/wiki/File:Dongtingriversmap.png).[251] The lake lies in a basin to the south of the Yangtze River and is connected to the Yangtze. The lake is also fed from the south by almost the entire drainage of Hunan Province, with the Xiang River (湘江) flowing in from the south and the Zi (資/资), Yuan (沅), and Li (澧) rivers from the southwest and west. The waters of the entire lake system discharge into the Yangtze at Yueyang (岳陽/岳阳; 29.4°N, 113.1°E). On the China-KWQ, see Fig. A3.12, that discharge location is at 29.7°N (KWQ latitude → modern latitude), 106.0°E, showing a latitude error of 0.3° and a longitude error around 7°.

99	衡山	Heng Shan	Mount Heng	Mount Heng (衡山) is a mountain in southcentral China's Hunan Province known as the Southern Mountain (Chinese: 南岳) of the Five Great Mountains (五嶽/五岳) of China. At the foot of the mountain stands the largest temple in southern China, the Grand Temple of Mount Heng (南嶽大廟/南岳大庙; Nanyue Damiao),[252] showing the coexistence of Buddhism and Taoism; it is where the largest group of ancient buildings in Hunan Province is located.
100	庾嶺	Yu Ning	Dayu Mountains or Dayuling	Dayuling (大庾嶺/大庾岭) or Yuling (庾嶺/庾岭) or Dayu Mountains (大庾嶺/大庾岭) is one of the "Five Mountain Ranges (五嶺/五岭)" in southern China: (1) Yuecheng Mountains (越城嶺/越城岭); (2) Dupang Mountains (都龐嶺/都庞岭); (3) Mengzhu Mountains (萌渚嶺/萌渚岭); (4) Qitian Mountains (騎田嶺/骑田岭); and (5) Dayu Mountains (大庾嶺/大庾岭). Dayuling (大庾嶺/大庾岭) is in the southwest corner of the present Jiangxi Province (江西省), bordering Guangdong Province (廣東省/广东省), hence, it is the main transportation route between Guangdong (廣東/广东) and Jiangxi.
101	廣州	Guang Zhou	"Guangzhou" is the official romanization of the Chinese name 廣州 or 广州 (廣 means "broad" or "expansive".	Guangzhou (廣州/广州) is the capital and largest city of Guangdong Province (廣東省/广东; formerly Romanised as Canton or Kwangtung) in southern China. It has a history of over 2,200 years and was a major terminus of the Maritime Silk Road. **Fig. A3.15.** The course of the Pearl River or Zhujiang (珠江) system through China and Vietnam (*Source*: Kmusser, under CC BY-SA 3.0, https://commons.wikimedia.org/wiki/File:Zhujiangrivermap.png).[253]

(*Continued*)

Item #	Name in Chinese	Pinyin	Name in English	History, Geography and My Comments
				Figure A3.15 shows that Guangzhou is located on the Pearl River or Zhujiang (珠江). The river is an extensive river system in southern China; the Xi (西; lit. "West"), Bei (北; lit. "North"), and Dong (東/东; lit. "East") rivers of Guangdong are all tributaries of the Pearl River. On the China-KWQ, see Fig. A3.12, the unnamed river near 廣州 (Guangzhou) is the Pearl River. The Moroccan traveler Ibn Battuta visited Guangzhou during his fourteenth-century journey around the world;[254] He specifically detailed the process by which the Chinese constructed their large ships in the port's shipyards.[255]
102	廣東	Guang Dong	Guangdong Province	Guangdong (廣東/广东) is a coastal province in South China on the north shore of the South China Sea. The capital of the province is Guangzhou. Guangdong (廣東/广东) and Guangxi (廣西/广西) are called Liang Guang (兩廣/两广 or Liangguang"; Liang, lit. "Two"). In 997 of the Song Dynasty (960–1279), the Two Guangs were formally separated as Guangnan Donglu (廣南東路/广南东路; "East Circuit in Southern Guang") and Guangnan Xīlu 廣南西路/广南西路; "West Circuit in Southern Guang"), which became abbreviated as Guangdong Lu (廣東路/广东路) and Guangxi Lu (廣西路/广西路). "Guangnan East" (廣南東) is the source of the name "Guangdong" (廣東; 广东) and "Guangnan West" (廣南西) is the source of the name "Guangxi" (廣西; 广西).[256] During the Mongol Yuan Dynasty, large parts of current Guangdong belonged to Jiangxi; Item 96).[257] Its present name, "Guangdong Province" was given in 1369 of the Ming Dynasty (1368–1644). Since the sixteenth century, Guangdong has had extensive trade links with the rest of the world.

103	雷	Lei Zhou	Leizhou	Leizhou (雷州) is a county-level city in Guangdong Province, China. In 634 of the Tang Dynasty (618–907, with an interregnum between 690 and 705), the Governor Chen Wenyu (陳文玉/陈文玉) asked to change Donghezhou (東合州) to Leizhou.[258] The name "Leizhou" began then. Leizhou is a famous national historical and cultural city with a long history and flourishing humanities.[259]
104	桂林	Gui Lin	Guilin	Guilin (桂林) is a prefecture-level city in the northeast of China's Guangxi Zhuang Autonomous Region (廣西壯族自治區/广西壮族自治区). Its name means "forest of sweet osmanthus", owing to the large number of fragrant sweet osmanthus trees located in the region. Guilin is one of China's most popular tourist destinations, and is associated with the phrase "Guilin's landscape is the best in the world" (桂林山水甲天下). The name "桂林" was first seen in the *Shan Hai Jing* (山海經/山海经)[260] or *The Classic of Mountains and Seas* in the Warring States Period (481–221 B.C.), in which its *Hainei Nan Jing* (海內南經/海内南经) or *The Classic of Regions within the Seas: The South*, records that "Guilin has eight trees; they are in the East of Fanyu (番隅)".[261] In 997 of the Northern Song Dynasty (960–1127), Guangnan West Circuit (廣南西路/广南西路; see Item 102 for more details), predecessor of modern Guangxi, was established, with Guizhou (貴州/贵州) as the capital. In 1133, Guizhou was renamed Jingjiang Prefecture (靜江路/静江路). The name was changed in 1368 (the first year of Hongwu) to Guilin Prefecture (桂林府).[262]

(*Continued*)

Item #	Name in Chinese	Pinyin	Name in English	History, Geography and My Comments
105	廣西	Guang Xi	Guangxi	Guangxi (廣西/广西) is an autonomous region of the People's Republic of China, located in South China and bordering Vietnam (越南) and the Gulf of Tonkin (北部灣/北部湾). The region was originally inhabited by a collection of tribal groups known to the Chinese as the Baiyue (百越; lit. "Hundred Yue").[263] Guangxi was given provincial level status during the Yuan Dynasty and was one of the thirteen provinces in the Ming Dynasty. Readers can read about Guangdong (廣東/广东; Item 102) to see how Guangxi (廣西/广西) was related to Guangdong (廣東) in the Song Dynasty.
106	土司州峒	Tu Si Zhou Tong	Tusizhou Tong	峒, lit. "cavē"; Tusizhou Tong (土司州峒) refers to the region ruled by the local chieftains. According to the records of *Ming Shilu* (明實錄/明实录) or *Veritable Records of Ming*, chieftain (土司), also known as native official (土官), was an official position in Ming's frontiers. It was first established in the Yuan Dynasty. In the Yuan, Ming, and Qing dynasties, it was a hereditary official position China granted to the leaders of ethnic minority tribes in the northwest and southwest regions. "Ruling native people with local officials (以夷治夷)" was a method of conferring officials and titles on the leaders of various ethnic groups in the border areas for them to rule their own people. During the Yongle period of the Ming Dynasty, chieftains were abolished in some areas of Yunnan (雲南/云南) and Guizhou (貴州/贵州), instead, officials were sent by the central government as chief envoys to implement direct management. However, in the Ming Dynasty, this policy was not implemented thoroughly. In Guanxi, there were still huge numbers of chieftains (土司), according to the records of *Gujin Dushu Jicheng* (古今图书集成/古今圖書集成) or *Complete Collection of Illustrations and Writings from the Earliest to Current Times*.

107	貴州	Gui Zhou	Guizhou	Guizhou (貴州/贵州) is a landlocked province in the southwest region of the People's Republic of China. Guizhou was formally made a province in 1413 (the eleventh year of Yongle) during the Ming Dynasty,[264] with a capital then also called "Guizhou" but now known as Guiyang. Guizhou is a mountainous province, with its higher altitudes in the west and center. It lies at the eastern end of the Yungui Plateau (雲貴高原/云贵高原) which is a highland region primarily spreading over the provinces of Yunnan (雲南/云南) and Guizhou. On the China-KWQ, see Fig. A3.12, 贵州 (Guizhou) is indeed depicted as a mountainous region.
108	貴陽	Gui Yang	Guiyang	Guiyang (貴陽/贵阳) is the largest city and capital of Guizhou Province. It is in the centre of the province. By the time Guizhou became a full province in 1413, the Guizhou Provincial Administration Office was established in Guiyang.[265]
109	水西	Shui Xi	Shuixi	Mu'ege (慕俄格; 300–1698) was a Nasu Yi (納蘇彝族/纳苏彝族) chiefdom in modern Guizhou that existed from 300 to 1698. Since 1279, Mu'ege was conquered by the Yuan Dynasty and became the Chiefdom of Shuixi (水西土司)[266] under the Chinese *tusi* system. Shuixi (水西) was one of the most powerful clans in Southwestern China; the chiefdoms of Bozhou (播州; 876–1600), Sizhou (思州; 582–1413), Shuidong (水東/水东; east of Yachi River or 鴨池河/鸭池河; 975–1630), and Shuixi (水西; west of Yachi River; 1372–1698) are collectively called the "Four Great Native Chiefdoms of Guizhou" (貴州四大土司/贵州四大土司)[267] in Chinese historiography.

(Continued)

Item #	Name in Chinese	Pinyin	Name in English	History, Geography and My Comments
110	雲南	Yun Nan	Yunnan	Yunnan (雲南/云南) is a landlocked province in the southwest of the People's Republic of China. Yunnan is situated in a mountainous area, see Fig. A3.12, with high elevations in the northwest and low elevations in the southeast. Parts of Yunnan formed the Dian Kingdom (滇國/滇国; 278 B.C.–109 B.C.) during the third and second centuries B.C. The Han Dynasty conquered the Dian Kingdom in the late second century B.C.[268] The area was later ruled by the Sino-Tibetan-speaking Kingdom of Nanzhao (南詔國/南诏国; 738–902),[269] followed by the Bai-ruled Dali Kingdom (白族统治的大理國/白族统治的大理国; 937–1094, 1096–1254).[270] After the Mongol conquest of the region in the thirteenth century, Yunnan was conquered and ruled by the Ming Dynasty as a province.
111	洱海	Er Hai	Erhai	Erhai or Er Lake (洱海) is an alpine fault lake in Yunnan Province. Erhai was also known as Yeyuze (葉榆澤/叶榆泽) or Kunming Lake (昆明池) in ancient times. It is the second largest freshwater lake in Yunnan Province and the seventh largest freshwater lake in China.[271]
112	永昌	Yong Chang	Yongchang	Yongchang (永昌) was a city in Yunnan, the present name is Baoshan (保山). Yongchang was established by the Kingdom of Nanzhao (南詔/南诏; 738–902), then inherited by the Bai-ruled Dali Kingdom (白族统治的大理國/白族统治的大理国; 937–1094, 1096–1254). In 1390 (the 23rd year of Hongwu), the Jinchi Regional Military and Civilian Commission (金齒軍民指揮使司/金齿军民指挥使司)[272] was set up with a jurisdiction including Yongchang.

113	永寧	Yong Ning	Chiefdom of Yongning	Chiefdom of Yongning (永寧土司/永宁土司; 1381–1956) was a Mosuo Autonomous Tusi Chiefdom (摩梭土司自治部落) during the Ming and Qing dynasties. The chiefdom was in today's Ninglang Yi Autonomous County (寧蒗彝族自治縣/宁蒗彝族自治县) at the convergence of Yunnan, Sichuan and Tibet. According to legend, the ancestor of the Yongning chieftains was from Tibet. He arrived at Yongning in 24 A.D. Yongning was a part of Nanzhao and later a part of the Dali Kingdom.[273] Mongolians invaded Dali in 1253 and Yongning was then administered by the Yuan Dynasty.[274] Yongning swore allegiance to the Ming Dynasty since 1371.[275] Chieftain Budu Geji (卜都各吉) went to the Ming capital to have an audience with the Hongwu Emperor in 1381; from then on, Yongning joined the Ming *Tusi* System.[276] From 1406, the hereditary chieftains received the official position "Yongning Zhifu" (永寧知府/永宁知府; Zhifu is the name of a local official in ancient China; it is a local official in a prefecture-level administrative region) from the Ming emperor.
114	建昌	Jian Chang	Jianchang	In the ninth century, the Kingdom of Nanzhao (南詔/南诏; 738–902) established Jianchang Fu (a prefecture; 建昌府) and its government was in the present Xichang City (西昌市), Sichuan Province. In 1382 of the early Ming Dynasty, Jianchang Lu was changed into Jianchang Fu, and in 1387 (the 20th year of Hongwu, the Jianchang Ancient City began to be built.[277]

(*Continued*)

Item #	Name in Chinese	Pinyin	Name in English	History, Geography and My Comments
				The *Xichang Xian Zhi* (西昌縣志/西昌县志) or *The Xichang County Chronicle* records that "宁远府城, 西昌县附郭, 即建昌卫。明洪武中建土城, 宣德二年包砌砖石". It is translated by Sheng-Wei Wang as: The Ningyuan City in the Xichang County is the Jianchang Wei. The city began to be built in the middle of the Hongwu period in the Ming Dynasty, and it was finished with covered bricks and stones in the second year of the Xuande period. *The second year of the Xuande period was 1427. This statement shows that Jianchang (建昌) was rebuilt in the Xuande period.*
115	越雋	Yue Jun	It should be "越嶲", pronounced as Yue Juan.	In 1392 (the 25th year of Hongwu), the Yuejuan Wei (越嶲衛/越嶲卫) was established; the seat was in the present Yuexi County (越西縣/越西县) in Sichuan Province (四川省)).[278]
116	四川	Si Chuan	Sichuan Province	Sichuan (四川) is a province in Southwest China occupying most of the Sichuan Basin (四川盆地) and the easternmost part of the Tibetan Plateau (青藏高原)[279] between the Jinsha River (金沙江; lit. "Gold Sand River" on the west, the Daba Mountains (大巴山) in the north and the Yungui Plateau (雲貴高原/云贵高原) to the south. Sichuan is also known to the Chinese as Ba-Shu (巴蜀), because of the ancient states Ba (巴) and Shu (蜀) that once occupied the Sichuan Basin. Shu continued to be used to refer to the Sichuan region all through its history right up to the present day. Currently, both the characters for *Shu* and *Chuan* are commonly used as abbreviations for Sichuan.[280] During the Hongwu era of the Ming Dynasty, the Sichuan Province was set up, and its jurisdiction covered even the present Zunyi (遵義/遵义) in Guizhou Province (貴州省/贵州省), Northeast Yunnan (雲南/云南), and Northwest Guizhou.[281]
117	雜谷	Za Gu	Zagu	In 1407, the Ming Yongle Emperor established an appeasement department (安撫司/安抚司) in Zagu (雜谷/杂谷).[282] The seat of the government is in today's Zagunao (雜穀腦/杂谷脑), Li County (理县), Sichuan.

118	董卜	Dong Pu	Dongpu	In 1415 of the Ming Dynasty, Yongle Emperor established an appeasement department Dongpu Hanhu Xuanwei Commission (董卜韓胡宣慰司/董卜韩胡宣慰司)[283] which was the name of the chieftain organization in the Tibetan area of the Ming Dynasty.[284] The seat of the government was in a county in the northwest of today's Sichuan Province, China. From then on, the chieftain sent in tribute continuously until 1626.[285]
119	松藩	Song Fan	Songfan	Songfan (松藩),[286] also known as Songzhou in Tang Dynasty, is a famous historical city in Sichuan Province. It was a famous frontier town in history, known as the "Gateway of Western Sichuan ("川西門户/川西门户")", and was used as a military base in ancient times. In 1378 (the eleventh year of Hongwu), Songfan Wei (松藩衛/松藩卫) was established in the region.[287]
120	長河西	Chang He Xi	Changhexi	Changhexi (長河西/长河西) is a place name[288] and Changhexi Qianhusuo (長河西千戶所/长河西千户所)[289] was the name of the chieftain organization (*tusi* institution; 土司機構/土司机构) for the Tibetan region in the Ming Dynasty. Changhexi Qianhusuo was established in 1374 (the seventh year of Hongwu) by Hongwu Emperor of the Ming Dynasty, and the government was in Garzê Tibetan Autonomous Prefecture (甘孜藏族自治州; often shortened to Ganzi Prefecture) in today's Sichuan Province. Starting from 1384, the native Changhexi officials started to send in tribute.[290]
121	西番界	Xi Fan Jie	Xifan Territorial Boundary	西 means "west", 番 "foreigners" and 界 "territorial boundary". Xifan Territorial Boundary (西番界) refers to the territorial limit where the Western foreigners should not enter China without permission.

(Continued)

Item #	Name in Chinese	Pinyin	Name in English	History, Geography and My Comments
122	烏思藏	Wu Si Zang	Wusizang	Wusizang (烏思藏/乌思藏) was the administrative district established in the present Tibet area in the Yuan Dynasty (1279–1368) of China. The Ming Dynasty continued the use of Wusizang, and according to the record in *Ming Shi* (明史) or *The History of Ming*, the Ming Hongwu Emperor set up *tusi*-type[291] of military jurisdiction — Wusizang Regional Military Commission (烏思藏都指揮使司/乌思藏都指挥使司) — in the central and western parts of the Qinghai-Tibet Plateau, Northeast India, and parts of Bhutan.[292]
123	崑崙	Kun Lun	Kunlun Mountains	The Kunlun Mountains (崑崙/昆仑)[293] constitute one of the longest mountain chains in Asia, which forms the northern edge of the Tibetan Plateau (青藏高原) south of the Tarim Basin (塔里木盆地).[294]
124	星宿海	Xing Xiu Hai	Stars Sea or Xingxiu Hai	Xingxiu Hai or Stars Sea (星宿海)[295] is a high plateau in the present Qinhai Province (青海省) close to the source of the Yellow River. Xingxiu Sea is in the west of Zaling Lake (扎陵湖) and it is a region where the three sources of the Yellow River merge from the southwest, west, and north, respectively, forming a large swamp and many small lakes. These densely packed lakes are like stars in the sky,[296] so the place is called Xingxiu Hai or Stars Sea.
@3	西番元尝郡县地今设二都司三宣慰司四万户府十七千户所		Xi Fan Yuan Chang Jun Xian Di Jin She Er Du Si San Xuan Wei Si Si Wan Hu Fu Shi Qi Qian Hu Suo	My translation and comments (between square brackets) are as follows: In the land of the Western foreigners, the Yuan Dynasty used to establish commanderies. In the present days [Ming Dynasty], there are two Command Posts, three Xuanwei Divisions [Appeasement Departments], a prefecture of 40,000 households and seventeen divisions each having a thousand households.

				This annotation says that in the Ming Dynasty, under the *tusi* system, the Ming government effectively established an administrative system in the Tibetan region to maintain peace with the neighbouring Aboriginals.
125	朵甘思	Duo Gan Si	Duogansi	Duogansi (朵甘思),[297] also known as Duogan (朵甘), was the name of a region in the Yuan Dynasty, which was equivalent to the eastern part of the Qamdo (昌都) area of the present Tibet Autonomous Region (西藏自治區); it was a part of the Sichuan Ganzi Tibetan Autonomous Prefecture (四川甘孜藏族自治州) and the Aba Tibetan and Qiang Autonomous Prefecture (阿壩藏族羌族自治州/阿坝藏族羌族自治州; an autonomous prefecture of northwestern Sichuan, bordering Gansu to the north and northeast and Qinghai to the northwest). The Ming Dynasty started to set up Duogan Wei (朵甘衛/朵甘卫) locally in 1373 (the sixth year of Hongwu) and promoted it to Duogan Xingdu Regional Military Commission (朵甘行都指揮使司/朵甘行都指挥使司) in 1374 and later Duogan Regional Military Commission (朵甘都指揮使司/朵甘都指挥使司).[298]
126	車里	Che Li	Cheli	Cheli (車里/车里) was a Chiefdom[299] and an administrative unit of the Yuan Dynasty. In the Ming Dynasty, it was set up as a Regional Military and Civilian Commission (an appeasement department; 軍民宣慰使司/军民宣慰使司). The seat of the government was in the present Jinghong (景洪), Yunnan; it was abolished after the Tianqi (天啓/天启) period (1621–1627) in the Ming Dynasty.
127	瓊	Qiong	Hanan Island or Qiongzhou	Qiong (瓊/琼) is an abbreviation for Hainan (海南). It is also known as Qiongzhou (瓊州/琼州; a prefecture). The name means "south of the sea", where the "sea" refers to Qiongzhou Strait (瓊州海峽/琼州海峡) which separates Hainan (海南) from Leizhou Peninsula (雷州半島/雷州半岛). The island is today's southernmost province of the People's Republic of China, consisting of various islands in the South China Sea.

(*Continued*)

Item #	Name in Chinese	Pinyin	Name in English	History, Geography and My Comments
				Hainan Island (海南島/海南岛) was recorded by Chinese mandarin officials in 110 B.C., when the Han Dynasty of China established a military garrison there.[300] After the eleventh century A.D., more and more poor peasants looked for land and moved from other parts of Guangdong/Canton Province to Leizhou Peninsula and Hainan Island.[301]
128	黎母	Li Mu	Limu Mountain	On the China-KWQ, Limu (黎母) refers to the mountain located in the central mountainous terrain of the Hainan Island, which is in Qiongzhong (瓊中/琼中), within the Li and Miao Autonomous County of Hainan Province (海南省黎族苗族自治縣/海南省黎族苗族自治县).
@4	大明聲名文物之盛自十五度至四十二度皆是其餘四海朝貢之國甚多此總圖略載嶽瀆省道大略餘詳統志省志不能殫述		Da Ming Sheng Ming Wen Wu Zhi Sheng Zi Shi Wu Du Zhi Si Shi Er Du Jie Shi Qi Yu Si Hao Zhao Going Zhi Guo Shen Duo Ci Zong Tu Lue Zai Yue Du Sheng Dao Da Lue Yu Xiang Tong Zhi Sheng Zhi Bu Neng Dan Shu	My translation and comments (between square brackets) are as follows: The fame and rich cultural relics of the Great Ming Dynasty are shown [on this map] from the fifteenth degree to the forty-second degree [in latitude]. Many countries from all over the world came to send in tribute. This comprehensive map briefly contains the outline of the mountains, rivers, and provincial roads and structures. The rest of the detailed general records, provincial records cannot be stated here in detail. This annotation clearly states that the China-KWQ coverers the geographical information of the Great Ming territories and its administrative units and regions in a vast area covered by 15°N to 42°N in latitude. It also explicitly states that the China-KWQ can only list some general information, and the more detailed information from the Chinese local general records and provincial records cannot be covered on this map. *This is Mattero Ricci's confession that he has used Chinese records to draw the China-KWQ.*

| 129 | 大琉球 | Da Liu Qiu | Greater Ryukyu or the present Okinawa | In the Ming Dynasty, Taiwan was called "Little Ryukyu".[302] This is because during this period, Okinawa's king from the Kingdom of Chūzan (中山國/中山国; 1314–1429) paid tribute to the Ming Dynasty and the kingdom was called "Ryukyu". The Chinese also distinguished them by calling Okinawa "Greater Ryukyu (大琉球)" and Taiwan "Little Ryukyu (小琉球)".[303]

Historically, the country name of the Kingdom of Ryukyu was given by China: in 1372 (the fifth year of Hongwu), Emperor Zhu Yuanzhang (Hongwu Emperor) of the Ming Dynasty sent his envoy to Ryukyu with an edict in which the place was called "Ryukyu" and formally included Ryukyu as a vassal state of the Ming Dynasty. "Ryukyu" has since become the official name of the place.[304] In 1429, the Kingdom of Chūzan (中山國/中山国; 1314–1429) unified Ryukyu, and the king was canonized as the "King of Ryukyu" by the Ming Dynasty.[305]

Beginning in the fifteenth century, Ryukyu benefited greatly from the Sino-Ryukyu tributary trade and became an important maritime trade hub in Asia. As a result, Chinese civilisation has also been deeply embedded in the cultural genes of Ryukyu.

On the China-KWQ, see Fig. A3.12, the Greater Ryukyu (大琉球), also known as Okinawa (沖繩/冲绳), is depicted to the southwest of the Little Ryukyu (小琉球) or Taiwan (台灣/台湾). This is incorrect; the opposite is true, see Fig. A3.16. This shows that China-KWQ is a map drawn when the Ming knowledge about Taiwan's geographical position was very limited and even incorrect before the Jiajing (嘉靖) period (1521–1567). |

(*Continued*)

Item #	Name in Chinese	Pinyin	Name in English	History, Geography and My Comments
				Fig. A3.16.　Location of Taiwan and the Ryukyu islands (public domain).[306]
130	小琉球	Xiao Liu Qiu	Little Ryukyu or Taiwan	Little Ryukyu (小琉球) was Taiwan as appeared in the historical records of the Ming Dynasty, and it also had a great relationship with Ryukyu. When the Ming Dynasty sent envoys to Ryukyu, the sea navigation was difficult, and they often suffered shipwrecks. Therefore, it was very important and necessary to rely on the navigational position coordinates on the sea, that is, using islands as a guide for direction and distance. It was in this situation that Taiwan appears in the Ming official voyages' records. On the On the China-KWQ, see Fig. A3.12, the labels "大琉球" and "小琉球" should be switched to be correct. Today on a modern map, Little Ryukyu (小琉球) is an outer island in the southwestern part of the Taiwan Island.
131	大明海	Da Ming Hai	East China Sea	大 means "great", 明 "Ming Dynasty" and 海 "sea". 大明海 means the "Sea of the Great Ming". It is today's East China Sea, see Fig. A3.17, which is an arm of the Western Pacific Ocean.

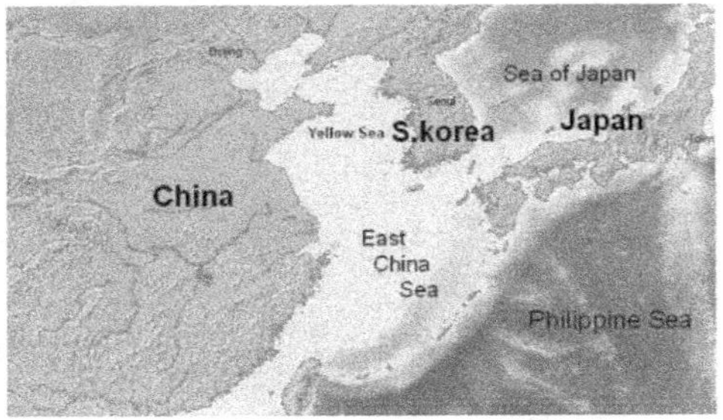

Fig. A3.17. The East China Sea, showing surrounding regions, islands, and seas; *Source*: Map created from DEMIS Mapserver, which are public domain, and then edited and location names added by Map-fanr, under CC BY-SA 3.0, https://commons.wikimedia.org/wiki/File:East_China_Sea.PNG.[307]

132	大明一统	Da Ming Yi Tong	Da Ming or Great Ming	The China-KWQ, see Fig. A3.12, depicts the Great Ming (1368–1644) as "大明一统", meaning "a unified Great Ming".

Endnotes

[1]Library of Congress. "Kun yu wan guo quan tu." www.loc.gov, Library of Congress, 2025, https://loc.gov/item/2010585650

[2]Ronnie Po-Chia Hsia. *Matteo Ricci and the Catholic Mission to China, 1583–1610: A Short History with Documents (Passages: Key Moments in History)*. Indianapolis, IN, USA: Hackett Publishing Company, Inc., 2016.

[3]Sheng-Wei Wang. *Chinese Global Exploration in the Pre-Columbian Era: Evidence from an Ancient World Map*. Singapore: World Scientific Publishing Co., 2023.

[4]Their locations can be found at https://web.archive.org/web/20070614134516/http://dicimg.paran.com/100_img/jpg/180/p18050200003.jpg

[5]"File:Kunyu Wanguo Quantu by Matteo Ricci Plate 1-3.jpg." *Wikimedia Commons: The Free Media Repository*, Wikimedia Foundation, 23 Oct. 2020, commons.wikimedia.org/wiki/File:Kunyu_Wanguo_Quantu_by_Matteo_Ricci_Plate_1-3.jpg

[6]"File:Mercator 1569 world map composite.jpg." *Wikimedia Commons*, 26 Nov. 2016, upload.wikimedia.org/wikipedia/commons/4/4b/Mercator_1569_world_map_composite.jpg

[7]"File:OrteliusWorldMap1570.jpg." *Wikiimedia Commons*, 12 July 2022, upload.wikimedia.org/wikipedia/commons/e/e2/OrteliusWorldMap1570.jpg

[8]"File:1594 double hemisphere world map by Petrus Plancius.jpg." *Wikimedia Commons*, 2 Sept. 2022, upload.wikimedia.org/wikipedia/commons/0/0d/1594_double_hemisphere_world_map_by_Petrus_Plancius.jpg

[9]Library of Congress, Geographicus. "Yu Ji Tu, an example of traditional Chinese cartography." www.sensesatlas.com, Senses Atlas, 2025, https://www.sensesatlas.com/territory/yu-ji-tu-an-exemple-of-chinese-cartography/

[10]*Ibid.*

[11]*Ibid.*

[12]*Ibid.*

[13]"File:Song Dynasty Map.JPG." *Wikimedia Commons*, 27 Mar. 2023, upload.wikimedia.org/wikipedia/commons/4/48/Song_Dynasty_Map.JPG

[14]Library of Congress, Geographicus. *Op. cit.*

[15]"File:華夷圖附注.jpg." *Wikimedia Commons*, 26 Mar. 2023, upload.wikimedia.org/wikipedia/commons/a/ac/華夷圖附注.jpg

[16]"File:Geographic Map of China.png." *Wikimedia Commons*, 5 May 2022, upload.wikimedia.org/wikipedia/commons/6/64/Geographic_Map_of_China.png

[17]"File:Yellowrivermap.jpg." *Wikimedia Commons*, 30 Sept. 2024, upload.wikimedia.org/wikipedia/commons/8/8b/Yellowrivermap.jpg

[18]Charles H. Hapgood. "Chapter 5: The ancient maps of the East and West." *Maps of the Ancient Sea Kings: Evidence of Advanced Civilization in the Ice Age*, Kempton, IL, USA: Adventures Unlimited Press, 1997, pp. 135–147.

[19]Leo Agrow. *History of Cartography*, London, UK: Transaction Publishers, 1963, p. 199.

[20]Julia Lovell. *The Great Wall: China Against the World, 1000 B.–AD 2000*. New York, NY, USA: Grove Press, 2007.

[21]Ali Humayun Akhtar. *1368: China and the Making of the Modern World*. Stanford, CA, USA: Stanford University Press, 2022.

[22]Wu Shilian (吳士連), *et al. Da Yue Shio Ji Quan Shu* (大越史記全書) or *The Complete History of the Great Viet*, Dai Shan Tang (埴山堂), Japan: 1884 (明治17), p. 545; an electronic version can be found at https://ctext.org/wiki.pl?if=gb&res=442611

[23]Gertraude Roth Li. "State Building before 1644." In Willard J. Peterson, ed. *The Ch'ing Empire to 1800, Cambridge History of China. Vol. 9*, Cambridge, UK: Cambridge University Press, 2002, pp. 11–14.

[24]"File:Ming Dynasty Administrative division.jpg." *Wikimedia Commons*, 15 Aug. 2023, upload.wikimedia. org/wikipedia/commons/4/4b/Ming_Dynasty_Administrative_division.jpg

[25]Zhang Tingyu (張廷玉; 1672–1755), *et al. Ming Shi • Juan Qi Shi Wu • Zhi Di Wu Shi Yi • Zhi Guan Si* (明史 • 卷七十五 • 志第五十一 • 職官四) or *History of Ming, Juan (Volume) 75, Record 51, Official 4*; an electronic version can be found at https://ctext.org/wiki.pl?if=gb&res=410835

[26]Kathlene Baldanza. *Ming China and Vietnam: Negotiating Borders in Early Modern Asia*. Cambridge, UK: Cambridge University Press, 2016.

[27]Sheng-Wei Wang. *The Last Journey of the San Bao Eunuch, Admiral Zheng He*. Hong Kong, China: Proverse Publishing, 2019.

[28]Zhang Tingyu (張廷玉; 1672–1755), *et al. Ming Shi • Juan Si Shi • Zhi Di Shi Liu • Di Li Yi* (明史 • 卷四十 • 志第十六 • 地理一) or *History of Ming, Juan (Volume) 40, Record 16, Geography 1*; an electronic version can be found at https://ctext.org/wiki.pl?if=gb&res=410835

[29]*Ibid.*

[30]John King Fairbank and Merle Goldman. *China: A New History*. Cambridge, MA, USA: Harvard University Press, 2006.

[31]Sheng-Wei Wang. "Chapter 2: Chinese explored Australia, New Zealand, Land of Fire and Antarctica long before the Europeans." *Chinese Global Exploration in the Pre-Columbian Era: Evidence from an Ancient World Map, op. cit.*, pp. 31–69.

[32]John King Fairbank and Merle Goldman. *Op. cit.*

[33]Zhang Tingyu (張廷玉; 1672–1755), *et al. Ming Shi • Juan San Bai San Shi • Lie Chuan Di Er Bai Shi Ba • Xi Yu Er Xi Fan Zhu Wei: Xi Ning He Zhou Tao Zhou Min Zhou Den Fan Zu Zhu Wei* (明史 • 卷三百三十 • 列傳第二百十八 • 西域二西番諸衛 • 西寧河州洮州岷州等番族諸衛) or *History of Ming, Juan (Vololume) 330, Biography 218, Western Region 2, Western Tribal Guards Battalions: Xining, Hezhou, Taozhou, Minzhou and so on*; an electronic version can be found at https://ctext.org/wiki.pl?if=gb&res=410835

[34]*Ibid.*

[35]*Ibid.*

[36]*Ibid.*

[37]*Ibid.*

[38]*Ibid.*

[39]Zhang Tingyu (張廷玉; 1672–1755), *et al. Ming Shi • Juan San Bai Er Shi Liu • Lie Chuan Di Er Bai Shi Qi • Xi Yu Yi* (明史 • 卷三百二十九 • 列傳第二百十七 • 西域一) or *History of Ming, Juan (Volume) 329, Biography 217, Western Region 1*; an electronic version can be found at https://ctext.org/wiki.pl?if=gb&res=410835

[40]"File:Amurrivermap.png." *Wikimedia Commons*, 13 Aug. 2024, upload.wikimedia.org/wikipedia/commons/f/fa/Amurrivermap.png

[41]Zhang Tingyu (張廷玉; 1672–1755), *et al. Ming Shi • Juan Si Shi Yi • Zhi Di Shi Qi • Di Li Er* (明史 • 卷四十一 • 志第十七 • 地理二) or *History of Ming, Juan (Volume) 41, Record 17, Geography 2*; an electronic version can be found at https://ctext.org/wiki.pl?if=gb&res=410835

[42]*Ibid.*

[43]Zhang Tingyu (張廷玉; 1672–1755), *et al. Ming Shi • Juan San Bai San Shi Yi • Lie Chuan Di Er Bai Shi Qi • Xi Yu San* (明史 • 卷三百三十一 • 列傳第二百十七 • 西域三) or *History of Ming, Juan (Volume) 331, Biography 217, Western Region 3*; an electronic version can be found at https://ctext.org/wiki.pl?if=gb&res=410835

[44]*Ibid.*

[45]*Ibid*

[46]*Ibid.*

[47]Gertraude Roth Li. *Op. cit.*

[48]John J. Stephan. *The Russian Far East: A History*, Stanford, CA, USA: Stanford University Press, 1996, pp. 16–17; Shih-shan Henry Tsai. *The Eunuchs in the Ming Dynasty*. Albany, New York, USA: SUNY Press, 1996, pp. 129–130.

[49]Zhang Tingyu (張廷玉; 1672–1755), *et al. Ming Shi • Juan Si Shi Yi • Zhi Di Shi Qi • Di Li Er* (明史 • 卷四十 一 • 志第十七 • 地理二) or *History of Ming, Juan (Volume) 41, Record 17, Geography 2*; an electronic version can be found at https://ctext.org/wiki.pl?if=gb&res=410835

[50]Editing Office of Liaoning Place Names Series (遼寧地名叢書編輯室). *Liao Ning Sheng Xian Ming Zhi • Fu • Sheng Ji Ge Shi Di Ming Cheng Jian Zhi* (遼寧省縣名志 • 附 • 省暨各市地名稱簡志) or *County Names of Liaoning Province with Attachment: Brief Annals of Place Names of Provinces and Cities*. Liaoning, China: Editing Office of *Liaoning Place Names Series* (遼寧地名叢書編輯室), December 1982.

[51]*Ibid.*

[52]Waldron, Arthur. *The Great Wall of China: From History to Myth*. Cambridge, UK: Cambridge University Press, 1990.

[53]Zhou Zhenhai and Tu Wei (周振鶴、余蔚). *Zhong Guo Xing Zheng Qu Hua Tong Shi • Liao Jin Juan* (中国行政区划通史 • 辽金卷) or *A General History of China's Administrative Divisions: Liao and Jin Juan (Volumes)*. Shanghai, China: Fudan University Press (复旦大学出版社), 2012.

[54]*Ibid.*

[55]Song Lian and Wang Hui (宋濂; 1310–1381). *Yuan Shi* (元史) or *History of Yuan*; an electronic version of Juan (Volume) 59 can be found at https://ctext.org/wiki.pl?if=gb&chapter=632486

[56]People's Government Office of Nong'an County (农安县人民政府办公室). "History (历史沿革)." http://www.nongan.gov.cn/, People's Government Office of Nong'an County (农安县人民政府办公室), 2025, http://www.nongan.gov.cn/zjna/lsyg/

[57]*Ibid.*

[58]Qianpian Chinese Dictionary (千篇汉语词典). "Wu Guo Cheng (五国城)." https://cidian.qianp.com/, QIANP.com, 2025, https://cidian.qianp.com/ci/五国城

[59]*Ibid.*

[60]Tuo Tuo (脱脱; 1314–1355). *Liao Shi* (遼史) or *History of Liao*; an electronic version of can be found at https://ctext.org/wiki.pl?if=gb&res=845139

[61]*Ibid.*

[62]Tuo Tuo (脱脱), *et al.*, eds. *Jin Shi* (金史) or *History of Jin*; an electronic version of can be found at https://ctext.org/wiki.pl?if=gb&res=270779

[63]Zhang Jiyin (张吉英). *Bian Liang Bei Ge (Jingkang Zhi Bian)* (汴梁悲歌（靖康之变）) or *The Elegy of Kaifeng (Jingkang Incident)*. Shenyang, Liaoning, China: Liaoning People's Publishing House (辽宁人民出版社), 2021.

[64]*Ibid.*

[65]Northeastern Ethnic and Folklore Museum (东北民族民俗博物馆). "古代民族源流与建置 (The Origin and Construction of Ancient Ethnicities)." https://dbmzms.nenu.edu.cn/, Northeastern Ethnic and Folklore Museum (东北民族民俗博物馆), 2025, dbmzms.nenu.edu.cn/info/1008/1066.htm

[66]Sun Jin ji and Sun Hong (孫進己、孫泓). *Nu Zhen Min Zu Shi* (女真民族史) or *Jurchen National History*. Guilin, Guangxi, China: Guangxi Normal University Press (广西师范大学出版社), 2013.

[67]Liu En Ming (刘恩铭). *Nu Er Ha Chi* (努爾哈赤) or *Nurhachi Biography*. Beijing, China: Huaxia Publishing House (华夏出版社), 2013.

[68]Yan Guang Liang. *Huang Tai Ji* (皇太极) or *Emperor Huang Taiji*. Shengyang, Liaoning, China: Shenyang Press, 1991.

[69]Writing team of *A Brief History of the Manchus* (满族简史编写组). *Man Zu Jian Shi* (满族简史) or *A Brief History of the Manchus*. Beijing, China: The Ethnic Publishing House (民族出版社), 2009.

[70]The *Qi Dan Guo Zhi* (契丹國志) or *The Records of Khitan Empire* states that "長白山在冷山東南千餘里……禽獸皆白。" (English translation by Sheng-Wei Wang: "Changbai Mountain is a thousand miles to the southeast of Cold Mountain…All birds and animals there Look white.")

[71]In the "Canonical Historical Records of the Jurchen Jin Dynasty (女真金正史)" of *Jin Shi • Juan Di San Shi Wu* (金史 • 卷第三十五) or *The History of Jin, Juan (Volume) 35*, it states that "長白山在興王之地，禮合尊崇，議封爵，建廟宇 (English translation by Sheng-Wei Wang: Changbai Mountain is in the old Jurchen land, highly Respectful and suitable for building temples" and "厥惟長白，載我金德，仰止其高，實惟我舊邦之鎮 (English translation by Sheng-Wei Wang: Only the Changbai Mountain can carry the Jin Dynasty's spirit; it is so high and it is a unique part of our old land)"; an electronic version can be found at https://ctext.org/wiki.pl?if=gb&chapter=179424.

[72]Yu Jiyuan (于濟源). *Chang Bai Shan Min Jian Gu Shi* (長白山民間故事) or *Changbai Mountains Folktales*. Taipei, Taiwan, China: Cheng Ming Culture Co, Ltd. (昌明文化有限公司), 2018.

[73]*Ibid.*

[74]Huang Shijian (黄时鉴) and Gong Yingyan (龚缨晏). *Li Ma Dou Shi Jie Di Tu Yan Jiu* (利玛窦世界地图研究) or *Research on Matteo Ricci's World Map*. Shanghai, China: Shanghai Chinese Classics Publishing House (上海古籍出版社), 2004, p. 190.

[75]Huang Shijian (黄时鉴) and Gong Yingyan (龚缨晏). *Op. cit.*, p. 194.

[76]Gertraude Roth Li. "State Building before 1644." In Willard J. Peterson, ed. *The Ch'ing Empire to 1800, Cambridge History of China, Vol. 9*. Cambridge, UK: Cambridge University Press, 2002, pp. 11–14.

[77]Wang Zi Chang (王子长). "What Chinese histories are found in overseas historical materials but not in Chinese historical records? (有哪些中国历史发现于境外史料而未见于中国史籍？)." https://www.zhihu.com/, zhihu, 2025, zhihu.com/question/265508227/answer/299350508

[78]*Ibid.*

[79]*Ibid.*

[80]Huang Shijian (黄时鉴) and Gong Yingyan (龚缨晏). Op. cit., p. 201.

[81]Sung-chiao Chao. "The sandy deserts and the gobi of China." https://link.springer.com/, Springer, https://link.springer.com/chapter/10.1007/978-94-009-6080-0_5

[82]"File:GobiTaklamakanMap.jpg." *Wikimedia Commons*, 15 Sept. 2020, upload.wikimedia.org/wikipedia/commons/0/0c/GobiTaklamakanMap.jpg

[83]"File:Kherlenrivermap.png." *Wikimedia Commons*, 6 June 2021, upload.wikimedia.org/wikipedia/commons/2/2b/Kherlenrivermap.png

[84]Shih-Shan Henry Tsai. *Perpetual Happiness: The Ming Emperor Yongle*, Seattle, Washington, USA: University of Washington Press, 2011, p.167.

[85]Hok-lam Chan. "The Chien-wen, Yung-lo, Hung-hsi, and Hsüan-te reigns, 1399–1435." In Frederick W. Mote and Denis *Twitchett, eds. The Cambridge History of China, Juan (Volume) 7: The Ming Dynasty, 1368–1644, Part 1*. Cambridge, UK: Cambridge University Press. *1988.*

[86]IICC. "Ruins of Weilu City (威虏城遗址)." www.silkroads.org.cn, Silk roads World Heritage, 2025, http://www.silkroads.org.cn/portal.php?mod=view&aid=25411

[87]Shang Chuan. *Yongle Emperor.* Guilin, Guangxi, China: Guangxi Normal University Press, 2010.

[88]*Ibid.*

[89]Mao Ruizheng (茅瑞徵). *Dong Yi Kao Lue* (東夷考略) or A *Brief Study of Dongyi*. China: Tang Feng Lou (唐風樓), 1601; an electronic version can be found at https://ctext.org/wiki.pl?if=en&chapter=4521948#%E5%A5%B3%E7%9B%B4

[90]David M. Robinson. *Ming China and its allies: Imperial rule in Eurasia*. Cambridge, UK: Cambridge University Press, 2020. DOI: https://doi.org/10.1017/9781108774253

[91]G. Nakajima. "The Structure and Transformation of the Ming Tribute Trade System." In: Manuel Perez Garcia and Lucio De Sousa, eds. *Global History and New Polycentric Approaches: Europe, Asia and the Americas in a World Network System (Palgrave Studies in Comparative Global History)*. London, UK: Palgrave Macmillan, 2018. https://doi.org/10.1007/978-981-10-4053-5_7

[92]Chifeng University (赤峰学院). "Research on Several Issues in the History of Chahar during the Ming and Qing Dynasties (明清时期察哈尔历史若干问题研究)." https://web.archive.org/, Inner Mongolia Social Science Planning Website (内蒙古社科规划网), 2025, http://nmgskghw.nmgnews.com.cn/system/2013/11/05/011201321.shtml

[93]*Ibid.*

[94]*Ibid.*

[95]*Ibid.*

[96]*Ibid.*

[97]Staff Writer. "The protection of the Liao Shangjing ruins in Inner Mongolia will draw on the model of the Daming Palace ruins (内蒙古辽上京遗址保护将借鉴大明宫遗址模式)." http://artsbuy.com, 中艺网, http://artsbuy.com/InfoDetail.aspx?Id=18409

[98]Tuo Tuo (脫脫). *Liao Shi • Juan San Shi Qi • Zhi Di Qi • Di Li Zhi Yi • Shang Jing Dao* (遼史 • 卷三十七 • 志第七 • 地理志一 •上京道) or *History of Liao, Juan (Volume) 37, Record 7, Geography 1: Shangjing Dao*; an electronic version can be found at https://ctext.org/wiki.pl?if=gb&res=845139

[99]*Ibid.*

[100]CiDianWang (词典网). "Xilou (西楼)." www.cidianwang.com, CiDianWang.com, 2025, cidianwang.com/lishi/diming/1/28071za.htm

[101]Zhang Tingyu (張廷玉; 1672–1755), *et al. Ming Shi • Juan Si Shi Yi • Zhi Di Shi Qi • Di Li Er* (明史 • 卷四十一 • 志第十七 • 地理二) or *History of Ming, Juan (Volume) 41, Record 17, Geography 2*; an electronic version can be found at https://ctext.org/wiki.pl?if=gb&res=410835; Jin Kar Er (今咯儿). "Five minutes to understand the 'Daning Dusi' military and political institutions established by the Ming Dynasty in the western part of the Northeast region (五分钟了解(大宁都司)明朝在东北地区西部设立的军政机构)." www.163.com, NetEase (網易), 2025, 163.com/dy/article/HGQO6FHC0553GONE.html

[102]*Ibid.*

[103]Gao Wende and Cai Zhichun (高文德、蔡志纯). *Meng Gu Shi Xi* (蒙古世系) or *Mongolian Lineage*. Beijing, China: China Social Sciences Press (中国社会科学出版社), 1979.

[104]Babel Stone. "Qingzhou White Pagoda and Qingling Mausoleum." www.babelstone.co.uk, Diary of a Rambling Antiquarian, 2025, https://www.babelstone.co.uk/BabelDiary/2017/09/qingzhou-white-pagoda.html

[105]Huang Shijian (黄时鉴) and Gong Yingyan (龚缨晏) Op. cit., p. 205.

[106]*Mark C. Elliott. The Manchu Way: The Eight Banners and Ethnic Identity in Late Imperial China (illustrated, reprint ed.),* Stanford University Press, 2001, pp. 47–48.

[107]Qian Pian Chinese Dictionaries (千篇汉语词典). "羁縻 (Jimi)." https://cidian.qianp.com/, QIANP.com, 2025, cidian.qianp.com/ci/羁縻

[108]National Ethnic Affairs Commission of the People's Republic of China (中华人民共和国国家民族事务委员会). "Manchu (满族)." www.neac.gov.cn, National Ethnic Affairs Commission of the People's Republic of China, 2025, https://www.neac.gov.cn/seac/ztzl/manz/lsyg.shtml

[109]Huang Shijian (黄时鉴) and Gong Yingyan (龚缨晏). Op. cit., p. 204.

[110]Emil Bretschneider. *Mediæval Researches from Eastern Asiatic Sources: Fragments towards the knowledge of the geography and history of central and western Asia from the 13th to the 17th century. Volume 1,* Tampa, FL, USA: Adamant Media Corporation, 2002, p.162.

[111]Alfred Schinz. *The Magic Square: Cities in Ancient China,* Fellbach, Baden-Württemberg, Germany: Edition Axel Menges, 1996, p. 286.

[112]Luc Kwanten. *Imperial Nomads: A History of Central Asia, 500–1500,* Philadelphia, PA, USA: University of Pennsylvania Press, 1979, p. 243.

[113]Song Lian (宋濂; 1310–1381), *et al. Yuan Shi • Juan Si • Shi Zu Yi* (元史 • 卷四 • 世祖一) or *History of Yuan, Juan (Volume) 4, Shizu 1*; an electronic version can be found at https://ctext.org/wiki.pl?if=gb&res=434890&remap=gb

[114]Ministry of Culture and Tourism of the People's Republic of China. "Ruins of the Yuan Shangdu (元上都遗址)." www.travelchina.org.cn, Ministry of Culture and Tourism of the People's Republic of China, 2025, http://www.ncha.gov.cn/art/2021/7/23/art_2539_170155.html

[115]Masuya Tomoko. "Seasonal capitals with permanent buildings in the Mongol empire." In David Durand-Guédy, ed. *Turko-Mongol Rulers, Cities and City Life,* Leiden, Netherlands: Brill, 2013, p. 239.

[116]Marco Polo. Ronald Latham, trans. *The Travels of Marco Polo.* London, UK: Folio Society, 1958; Marco Polo. Nigel Cliff, ed., trans., and Intro. Coralie Bickford-Smith, lllust. *The Travels.* London, UK: Penguin Classics, 2015.

[117]Shih-shan Henry Tsai. *Perpetual Happiness: The Ming Emperor Yongle. Op. cit.*

[118]An electronic version can be found at https://so.gushiwen.cn/guwen/book_46653FD803893E4F701E75CB779B6F3A.aspx

[119]Song Lian (宋濂; 1310–1381), *et al. Yuan Shi • Juan Er • Taizong* (元史 • 卷二 • 太宗) or *History of Yuan, Juan (Volume) 2: Taizong*; an electronic version can be found at https://ctext.org/wiki.pl?if=gb&chapter=612126

[120]*Ibid.*

[121]Sun Li (孙莉). "Yunnei State City (云内州城)." www.zgbk.com, Encyclopedia of China Publishing House, 2025, https://www.zgbk.com/ecph/words?SiteID=1&ID=506550&Type=bkztb&SubID=991

[122]*Ibid.*

[123]*Ibid.*

[124]Zhou Zhenhe (周振鶴), ed. *Zhong Guo Xing Zheng Qu Hua Tong Shi* (中國行政區劃通史) or *General History of China's Administrative Divisions, Second Edition.* Shanghai, China: Fudan University Press (復旦大學出版社), 2017.

[125]*Ibid.*

[126]Zhou Song (周松). *Ming Chu He Tao Zhou Bian Bian Zheng Yan Jiu* (明初河套周邊邊政研究) or *Research on Frontier Affairs around Hetao in the Early Ming Dynasty.* Gansu People's Publishing House (甘肃人民出版社), 2008.

¹²⁷*Ibid.*

¹²⁸Chen Bangzhan (陈邦瞻; 1567–1623)., ed. *Song Shi Ji Shi Ben Mo* (宋史纪事本末) or *Chronicle of Song History*; an electronic version can be found at https://ctext.org/wiki.pl?if=gb&res=74688&remap=gb

¹²⁹The People's Government of the Ningxia Hui Autonomous Region. "History of Ningxia Hui Autonomous Region (宁夏回族自治区的历史)." http:// www.nx.gov.cn /, The People's Government of the Ningxia Hui Autonomous Region, 2025, https://www.nx.gov.cn/ssjn/nxgk/lsyg/202304/t20230406_4021695.html

¹³⁰*Ibid.*

¹³¹*Ibid.*

¹³²Chen Menglei (陳夢雷; 1650–1741), ed. Complied by Fang Yu (方輿). *Gu Jin Tu Shu Ji Cheng • Zhi Fang Dian • Di 0576 Juan* (古今圖書集成 • 職方典 • 第0576卷) or *Collection of Ancient and Modern Books: Zhi Fang Dian, Juan (Volume) 0576.* Shanghai, China: Chung Hwa Book Co. (中華書局), 1934; an electronic version can be found at https://ctext.org/library.pl?if=gb&res=81155

¹³³Deng Huijun (邓慧君). "The History of Xiliang and Huangniangniangtai (西凉与皇娘娘台的历史往事)." https://gs.ifeng.com/, 2025, https://gs.ifeng.com/c/8G8bcb9aH7S

¹³⁴Frank McLynn. *Genghis Khan: His Conquests, His Empire, His Legacy.* Boston, MA, USA: Da Capo Press, 2016.

¹³⁵Zhang Tingyu (張廷玉; 1672–1755), *et al. Ming Shi• Ben Ji Di Er •Tai Zu Er* (明史 • 本纪第二 • 太祖二) or *History of Ming; Chronicle 2, Taizu 2;* an electronic version can be found at https://ctext.org/wiki.pl?if=gb&res=410835

¹³⁶*Ibid.*

¹³⁷Sima Qian (司馬遷; 145 B.C. –the first century B.C.). *Shi Ji • Juan Yi Bai Er Shi San • Da Wan Lie Chuan Di Liu Shi San* (史記 • 卷一百二十三 • 大宛列傳第六十三) or *Records of the Grand Historian, Juan (Volume) 123, Biography of Da Wan 63;* an electronic version can be found at https://ctext.org/shiji/zh

¹³⁸Dr. Alok Shrotriya and Zhou Xue-ying. "Artistic treasures of Maiji Mountain caves." www.asianart.com, Asia Art, 2025, asianart.com/articles/alok/index.html

¹³⁹Sima Qian. Burton Watson, trans. *Records of the Grand Historian: Qin Dynasty.* New York, NY, USA: Columbia University Press, 1999.

¹⁴⁰Frank McLynn. *Op. cit.*

¹⁴¹Zhang Tingyu (張廷玉; 1672–1755), *et al. Ming Shi • Juan San Bai San Shi • Lie Chuan Di Er Bai Shi Ba • Xi Yu Er Xi Fan Zhu Wei: Xi Ning He Zhou Tao Zhou Min Zhou Den Fan Zu Zhu Wei* (明史 • 卷三百三十 • 列傳第二百十八 • 西域二西番諸衛 • 西寧河州洮州岷州等番族諸衛) or *History of Ming, Juan (Volume) 330, Biography 218, Western Region 2, Western Guard Battalions: Xining, Hezhou, Taozhou, Minzhou and so on;* an electronic version can be found at https://ctext.org/wiki.pl?if=gb&res=410835

¹⁴²*Ibid.*

¹⁴³*Ibid.*

¹⁴⁴*Ibid.*

¹⁴⁵*Ibid.*

¹⁴⁶Zhou Zhenghe (周振鶴), *et al.,* eds. *Zhong Guo Xing Zheng Qu Hua Tong Shi* (中國行政區劃通史) or *General History of the Administrative Divisions in China. Op. cit.*

¹⁴⁷*Ibid.*

¹⁴⁸*Ibid.*

¹⁴⁹*Ibid.*

¹⁵⁰Zhang Tingyu (張廷玉; 1672–1755), *et al. Ming Shi • Juan San Bai San Shi • Lie Chuan Di Er Bai Shi Ba • Xi Yu Er Xi Fan Zhu Wei: Xi Ning He Zhou Tao Zhou Min Zhou Den Fan Zu Zhu Wei* (明史 • 卷三百三十

• 列傳第二百十八 • 西域二西番諸衛 • 西寧河州洮州岷州等番族諸衛) or *History of Ming, Juan (Volume) 330, Biography 218, Western Region 2, Western Guard Battalions: Xining, Hezhou, Taozhou, Minzhou and so on*; an electronic version can be found at https://ctext.org/wiki.pl?if=gb&res=410835

[151]*Ibid.*

[152]*Ibid.*

[153]Zhang Tingyu (張廷玉; 1672–1755), *et al. Ming Shi • Juan San Bai Er Shi Liu • Lie Chuan Di Er Bai Shi Qi • Xi Yu Yi* (明史 • 卷三百二十九 • 列傳第二百十七 • 西域一) or *History of Ming, Juan (Volume) 329, Biography 217, Western Region 1*; an electronic version can be found at https://ctext.org/wiki.pl?if=gb&res=410835

[154]*Ibid.*

[155]*Ibid.*

[156]Gao Wende (高文德), *et al.*, ed. *Zhong Guo Shao Shu Min Zu Shi Da Ci Dian* (中国少数民族史大辞典) or *Dictionary of the History of Chinese Minorities*, Jilin, China: Jilin Education Press (吉林教育出版社),1995, p. 1330.

[157]Zhang Tingyu (張廷玉; 1672–1755), *et al. Ming Shi • Juan San Bai Er Shi Liu • Lie Chuan Di Er Bai Shi Qi • Xi Yu Er* (明史 • 卷三百二十九 • 列傳第二百十七 • 西域二) or *History of Ming, Juan (Volume) 329, Biography 217, Western Region 2*; an electronic version can be found at https://ctext.org/wiki.pl?if=gb&res=410835; Han Dian (汉典). "Yangguan (阳关)." https://www.zdic.net/, Han Dian, 2025, https://www.zdic.net/hans/%E9%98%B3%E5%85%B3

[158]Sima Qian (司馬遷; 145 B.C–the first century B.C.). *Shi Ji • Juan Yi Bai Er Shi San • Ds Wan Lie Chuan Di Liu Shi San* (史記 • 卷一百二十三 • 大宛列傳第六十三) or *Records of the Grand Historian, Juan (Volume) 123, Biography of Da Wan 63*; an electronic version can be found at https://ctext.org/shiji/zh

[159]Marco Polo. Ronald Latham, trans. *Op. cit.*

[160]E. Bretschneider, M.D. "Notices of the medieval geography and history of central and west Asia, drawn from Chinese and Mongol writings and compared with the observation of western authors in the Middle Ages." *Journal of the North-China Branch of the Royal Asiatic Society*, vol. 10, 1876, p. 184; an electronic version can be found at https://books.google.com.hk/books?id=C4hJAAAAYAAJ&pg=PA184&redir_esc=y#v=onepage&q&f=false

[161]*Ibid.*

[162]Sally Hovey Wriggins. "Xuanzang on the Silk Road." www.lifelonglearningcollaborative.org/, lifelonglearningcollaborative, 2025, https://www.lifelonglearningcollaborative.org/silkroads/articles/xuanzang-on-the-silk-road.html

[163]Marco Polo. Ronald Latham, trans. *Op. cit.*

[164]City Population. "China: Xīnjiāng: Prefectures, Cities, Districts and Counties." http://www.citypopulation.de/, City Population, 2025, http://www.citypopulation.de/en/china/xinjiang/admin/

[165]Zhang Tingyu (張廷玉; 1672–1755), *et al. Ming Shi • Juan San Bai Er Shi Liu • Lie Chuan Di Er Bai Shi Qi • Xi Yu Yi* (明史 • 卷三百二十九 • 列傳第二百十七 • 西域一) or *History of Ming, Juan (Volume) 329, Biography 217, Western Region 1*; an electronic version can be found at https://ctext.org/wiki.pl?if=gb&res=410835

[166]Shih-shan Henry Tsai. *Perpetual Happiness: The Ming Emperor Yongle. Op. cit.*

[167]Lu Jie and Meng Zhaoyong (鲁杰、孟昭永). "A brief examination of the Chief Military Officers of Ji Town in the Ming Dynasty (明蓟镇总兵官考略)." *Wen Wu Chun Qiu* (文物春秋) or *Annals of Cultural Relics*, Juan (Volume) 4, 2008; an electronic version can be found at http://www.thegreatwall.com.cn/phpbbs/index.php?id=114021&threadid=114020

[168]*Ibid.*

[169]Julia Lovell. *Op. cit.*

[170]"File:Ming Great Wall.png." *Wikimedia Commons*, 2 Oct. 2020, upload.wikimedia.org/wikipedia/commons/8/82/Ming_Great_Wall.png

[171]Cheng Daosheng (程道生). *Jiu Bian Tu Kao • Xuan Fu* (九邊圖考 • 宣府) or *Examine the Maps of the Nine Frontier Fortresses: Xuanfu*. Shanghai, China: The Commercial Press (商務印書館), 1919.

[172]Vincent Goossaert, "Hengshan." In Fabrizio Pregadio, ed. *The Encyclopedia of Taoism*, London, UK: Routledge, 2008, pp. 481–482.

[173]Taiyuan Local Chronicles Compilation Committee (太原市地方志编纂委员会). *Taiu Yuan Shi Zhi* (太原市志) or *Taiyuan City Chronicles*. Taiyuan, Shanxi Province, China: Shanxi Ancient Books Publishing House (山西古籍出版社), 1999.

[174]Endymion Wilkinson. *Chinese History: A New Manual. Harvard–Yenching Institute Monograph Series. Vol. 84*, Cambridge, MA, USA: Harvard University Asia Center, 2012, p. 234.

[175]Ke Shaowu (柯劭忞)." Xin Yuan Shi (新元史) or A New History of Yuan. Shanghai Ancient Books Publishing House (上海古籍出版社), 2018.

[176]Zhang Tingyu, (張廷玉; 1672–1755), *et al. Ming Shi • Juan Si Shi Yi • Zhi Di Shi Qi • Di Li Er* (明史 • 卷四十一 • 志第十七 • 地理二) or *History of Ming, Juan (Volume) 41*, Record 17, Geography 2; an electronic version can be found at https://ctext.org/wiki.pl?if=gb&res=410835

[177]Xia Bo Ming, ed. *The history of Chinese Administrative Division*: Ming Dynasty. Shanghai, China: Fudan University Press (復旦大學出版社), 2010.

[178]Shaanxi Local Chronicles Office (陕西省地方志办公室). "Provincial Overview (省情概况)." https://baike.baidu.com/, Shaanxi Local Chronicles Office (陕西省地方志办公室), 2025, https://baike.baidu.com/reference/19657132/1287iR68WB9uI3Q5AjSt1VMFqDlXfmkUujoRC9b1T4vk1h_jNp5A2LOhn6ccCc JxQmDpHkZzQ33s3CKZEXBCjxOJeqeBiBR6OeW8REszpXiHOB2XkfdLRw

[179]*Ibid.*

[180]Yulin Municipal People's Government (榆林市人民政府). "Historical Evolution (历史沿革)." www.yl.gov.cn, Yulin Municipal People's Government (榆林市人民政府), 2025, https://baike.baidu.com/reference/2677749/e6a6GcLoUIfA-N-nTRCQnQjraRPQnrwLs2KiQV0kgqloD3-BQ6ogVXX4A0ezR7 YlchYb9YCBRRtlkeBD_KJfxjoo83Cq93S_BYBgkCU

[181]Mo Xiusui (墨修岁). "Why do Yuci, Yulin and Yuzhong all have the character 'yu' in their names (为什么榆次、榆林、榆中都带一个榆字)?." www.zhihu.com, Zhi Hu, 2025, zhihu.com/question/617078588

[182]Awei (阿威). "Guyuan, Ningxia: An in-depth interpretation of the origin of the name 'Guyuan' will subvert your cognition (宁夏固原: 深度解读 '固原'名字的由来，颠覆你的认知)." www.163.com, NetEase (網易), 2025, https://www.163.com/dy/article/HDSDPLJI0542HDTE.html

[183]Mark Cartwright. "Chang'an." www.worldhistory.org, World History Foundation and World History Publishing, 2025, https://www.worldhistory.org/Chang%27an/

[184]UNESCO World Heritage Centre "Mausoleum of the First Qin Emperor." https://whc.unesco.org/, UNESCO, 2025, whc.unesco.org/pg.cfm?cid=31&id_site=441

[185]Jean-Baptiste Du Halde. *Description Géographique, Historique, Chronologique, Politique, et Physique de l'Empire de la Chine et de la Tartarie Chinoise [A Geographical, Historical, Chronological, Political, and Physical Description of the Empire of China and Chinese Tartary] (in French)*. The Hague, Netherlands: H. Scheurleer, 1976.

[186]Anonymous. Anne Birrell, ed. and trans. *The Classic of Mountains and Seas*. London, U. K.: Penguin Classics, 1999.

[187]An electronic version of *Yu Gong* (禹貢) or *Tribute of Yu* can be found at https://ctext.org/shang-shu/tribute-of-yu/zh

[188]Hu Axiang (胡阿祥). "On the Source of 'Han', the Name of Liu Bang's Dynasty (刘邦汉国号考原)." *Journal of Historical Science* (史学月刊), vol. 6, 2001, pp. 57–62.

[189]Hanzhong Local Chronicles Editorial Committee (汉中地方志编辑委员会). *Han Zhong Di Fang Zhi* (汉中地方志) or *Hanzhong Local Chronicles, Juan (Volume) 1*. Xian, Shaanxi, China: Sanqin Press (三秦出版社), 2022.

[190]Jun Hou, *et al.* "Attribution identification of terrestrial ecosystem evolution in the Yellow River Basin." *Open Geosciences*, vol. 14, no. 1, 2022, pp. 615-628. https://doi.org/10.1515/geo-2022-0385

[191]Huh *et al.*, 2004. "Yellow River (Huang He) Delta, China, Asia." http://www.geol.lsu.edu/, The World Delta Database, 2025, http://www.geol.lsu.edu/WDD/ASIAN/Huanghe/huange_he.htm

[192]*Ibid.*

[193]"File:Yellowrivermap.jpg." *Wikimedia Commons*, 23 Oct. 2021, upload.wikimedia.org/wikipedia/commons/8/8b/Yellowrivermap.jpg

[194]Archibald John Little. *The Far East*, Oxford, UK: Clarendon Press, 1905, p. 53.

[195]"File:Yellow River watercourse changes en.png." *Wikimedia Commons*, 22 Oct. 2022, upload.wikimedia.org/wikipedia/commons/3/3d/Yellow_River_watercourse_changes_en.png

[196]Bamber Gascoigne and Christina Gascoigne. *The Dynasties of China*. New York City, NY, USA: Perseus Books Group, 2003.

[197]Vincent Goossaert. "Songshan." In Fabrizio Pregadio, ed. *The Encyclopedia of Taoism. Vol. II*. London, UK: Routledge, 2008.

[198]*Ibid.*

[199]ICOMOS. "Historic Monuments of Dengfeng in 'The Centre of Heaven and Earth'." https://whc.unesco.org/, UNESCO World Heritage Centre, 2025, whc.unesco.org/en/list/1305

[200]Sima Qian (司馬遷; 145 B.C.–the first century B.C.). *Shi Ji • Juan Yi Bai Er Shi Jiu • Lie Chuan Di Liu Shi Jiu* (史記 • 卷一百二十九 • 列傳第六十九) or *Records of the Grand Historian, Juan (Volume) 129, Biography 69;* an electronic version can be found at https://ctext.org/shiji/zh

[201]Zhang Tingyu (張廷玉; 1672–1755), *et al. Ming Shi • Juan Si Shi Er • Zhi Di Shi Ba • Di Li San • Kai Feng Fu* (明史 • 卷四十二 • 志第十八• 地理三：開封府) or *History of Ming, Juan (Volume) 42, Record 18, Geography 3: Kaifeng Fu* ; an electronic version can be found at https://ctext.org/wiki.pl?if=gb&res=410835

[202]"File:Bianjing city gate.JPG." *Wikimedia Commons*, 18 Nov. 2023, upload.wikimedia.org/wikipedia/commons/d/d2/Bianjing_city_gate.JPG

[203]Daming County Chronicle Compilation Committee (大名县县志编纂委员会), ed. *Da Ming Xian Zhi* (大名县志) or *Daming County Chronicle*. Beijing, China: Xinhua Publishing House (新华出版社); an electronic version can be found at https://ctext.org/library.pl?if=gb&res=92314&remap=gb

[204]*Ibid.*

[205]Xu Fang (徐方), ed. "The History of Shandong (山东历史)." https://baike.baidu.com/, CNR.CN (中国广播网), 2025, baike.baidu.com/item/山东省/209822

[206]Dongping News (东平新闻). "Our county has been recognized as a Chinese place-name cultural heritage and a thousand-year-old county')." https://web.archive.org/, Dongping TV Broadcast Station, 2025, web.archive.org/web/20171210231942/http://www.dongping.gov.cn/contents/1191/158979.html

[207]"File:Confucius and Laozi, fresco from a Western Han tomb of Dongping County, Shandong province, China.jpg." *Wikimedia Commons*, 31 Oct. 2020, en.wikipedia.org/wiki/Dongping_County#/media/

File:Confucius_and_Laozi,_fresco_from_a_Western_Han_tomb_of_Dongping_County,_Shandong_province,_China.jpg

[208]Hu Meng and Li Jing (胡蒙李静). "Walking the Grand Canal: Marco Polo once praised Dongping as a 'majestic and beautiful metropolis' (行走大运河：马可·波罗曾赞东平'雄伟美丽的大都')." https://web.archive.org/, Qilu.com (齐鲁网), 2025, https://web.archive.org/web/20171210175828/http://news.163.com/14/0923/15/A6RC6DOV00014SEH.html

[209]Anthony J. Barbieri-Low and Robin D. S. Yates. *Law, State, and Society in Early Imperial China (2 Juan (Vol)s): A Study with Critical Edition and Translation of the Legal Texts from Zhangjiashan Tomb No. 247 (Sinica Leidensia) Bilingual Edition*. Leiden, Netherlands: Brill, 2015.

[210]Zhang Tingyu (張廷玉; 1672–1755), *et al. Ming Shi • Juan Si Shi Yi • Zhi Di Shi Qi • Di Li Er* (明史 • 卷四十一 • 志第十七 • 地理二) or *History of Ming, Juan (Volume) 41, Record 17, Geography 2*; an electronic version can be found at https://ctext.org/wiki.pl?if=gb&res=410835

[211]Zhang Tingyu (張廷玉; 1672–1755), *et al. Ming Shi • Juan Yi Bai Si Shi Er • Lie Chuan Di San Shi* (明史 • 卷一百四十二 • 列傳第三十) or *History of Ming, Juan (Volume) 142, Biography 30*; an electronic version can be found at https://ctext.org/wiki.pl?if=gb&res=410835

[212]Professor Li Cunxiu (李存修教授). "Why is Mount Tai called 'The Best Mountain in the World' (泰山爲何被稱爲「天下第一山」)" https://chiculture.org.hk/, Academy of Chinese Studies (中國文化研究院), 2025, https://chiculture.org.hk/tc/china-five-thousand-years/3814

[213]*Ibid.*

[214]*Ibid.*

[215]Zhang Tingyu (張廷玉; 1672–1755), *et al. Ming Shi • Juan Si Shi Yi • Zhi Di Shi Qi • Di Li Er* (明史 • 卷四十一 • 志第十七 • 地理二) or *History of Ming, Juan (Volume) 41, Record 17, Geography 2;* an electronic version can be found at https://ctext.org/wiki.pl?if=gb&res=410835

[216]*Ibid.*

[217]"File:Huairivermap.jpg." *Wikimedia Commons*, 27 July 2022, upload.wikimedia.org/wikipedia/commons/3/32/Huairivermap.jpg

[218]Simon Winchester. *The River at the Center of the World: A Journey Up the Yangtze, and Back in Chinese Time*. London, UK: Picador, 2004.

[219]"File:Yangtze river map.png." *Wikimedia Commons*, 24 Oct. 2022, upload.wikimedia.org/wikipedia/commons/f/fc/Yangtze_river_map.png

[220]The World Atlas Editorial Team. "Yangtze River." www.worldatlas.com/, WorldAtlas, 2025, https://www.worldatlas.com/rivers/yangtze-river.html

[221]Anonymous author. "Where is the modern location of historical 'Huguang' that fills Sichuan? (历史中的'湖广填四川'中的'湖广'是指的现在哪里?)." https://web.archive.org/, rcwin.com, 2025, web.archive.org/web/20110707023209/http://www.cqhakka.cn/ymwh/hgtsc/200705/109.html

[222]Anonymous author. *Op. cit.*; Zhang Tingyu (張廷玉; 1672–1755), *et al. Ming Shi • Juan Si Shi Si • Zhi Di Er Shii • Di Li Wu* (明史 • 卷四十四 • 志第二十 • 地理五) or *History of Ming, Juan (Volume) 44, Record 20, Geography 5*; an electronic version can be found at https://ctext.org/wiki.pl?if=gb&res=410835

[223]Janet L. Abu-Lughod. *Before European Hegemony: The World System A.D. 1250–1350*, Oxford, UK: Oxford University Press, 1991, p. 137.

[224]Matt Rosenberg. "Largest Cities Throughout History: Determining population prior to census-taking was no easy task." www.thoughtco.com, Thoughtco, 2025, https://www.thoughtco.com/largest-cities-throughout-history-4068071; Janet L. Abu-Lughod. *Op. cit.*

[225]Marco Polo. Ronald Latham, trans. *Op. cit*; Marco Polo. Nigel Cliff, ed., trans., and Intro. *Op. cit*.

[226]Tim Mackintosh-Smith. *The Travels of Ibn Battutah*. London, UK: Macmillan Collector's Library, 2016; Michael Elliott. "Summer Journey 2011: The Enduring Message of Hangzhou." https://content.time.com/time/specials/packages/article/0,28804,2084273_2084272_2084481,00.html

[227]Nanjing Yearbook (南京年鉴), 2012. "Nanjing Historical Evolution (南京历史沿革)." https://web.archive.org/, www.nanjing.gov.cn, 2025, https://web.archive.org/web/20130609184340/http://www.nanjing.gov.cn/njgk/csgk/csgk3/

[228]*Ibid.*

[229]Hok-lam Chan. "The Chien-wen, Yung-lo, Hung-hsi, and Hsüan-te reigns, 1399–1435." *The Cambridge History of China, Juan (Volume) 7: The Ming Dynasty, 1368–1644, Part 1*, Cambridge, UK: Cambridge University Press, 1998, pp. 282–283.

[230]Edward L. Dreyer. *Zheng He: China and the Oceans in the Early Ming Dynasty, 1405–1433*, New York, NY, USA: Pearson Longman, 2007, pp. 140–141.

[231]Edward L. Dreyer. *Op, cit.*, p. 168.

[232]Paul A. Cohen. "Chiang Kai-Shek." *The History*, New York, NY, USA: Columbia University Press, 2007, p. 71.

[233]Sheng-Wei Wang. "Chapter 3: Luo Maodeng's *Tianfeng*/Yun Chong was Mecca in Saudi Arabia." *The Last Journey of the San Bao Eunuch, Admiral Zheng He*, op. cit., p. 92.

[234]Shih-shan Henry Sai. *The eunuchs in the Ming Dynasty*, Albany, NY, USA: SUNY Press, 1996, p. 145; an electronic version can be found at https://books.google.com.hk/books?id=Ka6jNJcX_ygC&pg=PA145&redir_esc=y#v=onepage&q&f=false

[235]Fujian Local Records Editorial Committee (福建省地方志编纂委员会), ed. *Fu Jian Sheng Zhi Zong Gai Shu* (福建省志总概述) or *An Overview of Fujian Province*. Fujian, China: China Local Records Publishing (方志出版社), 2002.

[236]Sheng-Wei Wang. "Chapter 3: Luo Maodeng's *Tianfeng*/Yun Chong was Mecca in Saudi Arabia." *Op. cit.*

[237]Wang Shumin and Dòu Hongshen (王蘇民、竇鴻身), eds. *Zhong Guo Hu Bo Zhi* (中國湖泊志) or *China Lake Chronicle*. Beijing, China: China Science Publishing & Media, Ltd. (科学出版社), 1998.

[238]Zhang Xiuping (张秀平), *et al.* "Battle of Lake Poyang (鄱阳湖之战)." https://web.archive.org/, http://military.china.com/, 2025, web.archive.org/web/20140106040700/http://military.china.com/zh_cn/history2/06/11027560/20050407/12224277.html

[239]*Ibid.*

[240]*Ibid.*

[241]Dr. Mary Laven. *Mission to China: Matteo Ricci and the Jesuit Encounter with the East*. London, UK: Faber and Faber, Ltd., 2011, p. 103.

[242]Zhou Zhenhai and Tu Wei (周振鹤、余蔚). *Zhong Guo Xing Zheng Qu Hua Tong Shi* (中国行政区划通史) or *A General History of China's Administrative Divisions. Op. cit.*

[243]Suzanne G. Valenstein. *A Handbook of Chinese Ceramics*. New York, NY, USA: Metropolitan Museum of Art, 1988.

[244]British Museum. "Jizhou tea bowl." https://artsandculture.google.com/, Google Arts & Culture, 2025, artsandculture.google.com/asset/jizhou-tea-bowl/jwGxEOdtn58mkw

[245]Gu Ao Kuang Sheng (古傲狂生). "Which were the 13 provinces (13省是哪13省)？" https://baike.baidu.com/, HefeiWanbao, 2025, baike.baidu.com/reference/19438118/b235llYgbssAlW2yyf26S7eep8TwlUvlos6jG38IkNxH7h83F1zZjPqG09YMP5nk8IY9VGGVkyxf135looTQNhQIs0ZSlEcn-J2j-4NZ2lEXIHcDh-kEVa5ZfITw

[246]Google Maps. "Wuchang District." www.google.com, Google Maps, 2025, https://www.google.com/maps/place/Wuchang+District,+Wuhan,+Hubei,+China/@30.5640498,114.345795,12z/data=!3m1!4b1!4m6!3m5!1s0x342eaf978840440d:0x2109edeb11bec27e!8m2!3d30.5538599!4d114.31599!16zL20vMDEzcTlu?entry=ttu

[247]Anonymous. Wilt L. Idema and Stephen H. West, translators. *Records of the Three Kingdoms in Plain Language*. Indianapolis, IN, USA: Hackett Publishing Company, Inc., 2016.

[248]Fang Xuanling (房玄齡; 579–648), *et al. Jin Shu • Juan Shi Wu • Zhi Di Wu • Di Li Xia* (晉書 • 卷十五 • 志第五 • 地理下) or *Book of Jin, Juan (Volume) 15, Record 5, Geography 2*; an electronic version can be found at https://ctext.org/wiki.pl?if=gb&res=788577

[249]Wei Ming. *Famous Lakes in China* (in English and Chinese). Huangshan City, Anhui: Huangshan Publishing House, 2014.

[250]Ibid.

[251]"File:Dongtingriversmap.png." *Wikimedia Commons*, 2 June 2021, upload.wikimedia.org/wikipedia/commons/5/57/Dongtingriversmap.png

[252]People's Government of Hunan Province. "Grand Temple of Mount Heng (南岳大庙)." https://baike.baidu.com/, www.Hunan.gov.cn, 2025, https://baike.baidu.com/reference/2074603/395bn9_iqq-Ks8EPk2boi1pVUwG9JGCrb4D8Abg3FBAY_TCt9bWtqvSACzDFqNP2iOeInBbYovwo6Gv1doX6XJHroWXVCAnSbhZYS1i__WsEMbFTOaILyZLbJwl_9koUtoESfzp4QQoeFcg

[253]"File:Zhujiangrivermap.png." *Wikimedia Commons*, 25 June 2022, upload.wikimedia.org/wikipedia/commons/3/3b/Zhujiangrivermap.png

[254]Ross E. Dunn. *The Adventures of Ibn Battuta*. Berkeley, CA, USA: University of California Press, 1986.

[255]*Ibid.*

[256]Tuo Tuo and A Lu Tu (脫脫、阿魯圖). *Song Shi • Juan Jiu Shi Yi • Di Li Zhi Liu* (宋史 • 卷九十一 • 地理志六) or *History of Song, Juan (Volume) 91, Record 43, Geography 6*; an electronic version can be found at https://ctext.org/wiki.pl?if=gb&res=975976

[257]Zhang Tingyu (張廷玉; 1672–1755), *et al. Ming Shi • Juan Si Shi Wu • Zhi Di Er Shi Yi • Di Li Liu* (明史 • 卷四十五 • 志第二十一 • 地理六) or *History of Ming, Juan (Volume) 45, Record 21, Geography 6*; an electronic version can be found at https://ctext.org/wiki.pl?if=gb&res=410835

[258]Guo Shengbo (郭声波), ed. *Zhong Guo Xing Zheng Qu Hua Tong Shi: Tang Dynasty* (中国行政区划通史: 唐代卷) or *General History of China's Administrative Divisions: Tang Dynasty*. Shanghai, China: Fudan University Press (复旦大学出版社), 2012.

[259]Wen Er (文二), ed. "Introduction to the history and culture of Leizhou, a famous historical and cultural city (历史文化名城雷州历史和文化介绍)." https://baike.baidu.com/, www.861sw.com (中国历史网), 2025, https://baike.baidu.com/reference/7098496/f4e1D60690e6oFa9ti6zu6RwDubYuexppOMCnEquRyUodFDnYVBoyQVtgbdHe_Zho4Q-B4N3jHXdPJgrZeXONY0

[260]An electronic version can be found at https://ctext.org/shan-hai-jing/zh

[261]Zhang Tingyu (張廷玉; 1672–1755), *et al. Ming Shi • Juan Si Shi Wu • Zhi Di Er Shi Yi • Di Li Liu* (明史 • 卷四十五 • 志第二十一 • 地理六) or *History of Ming, Juan (Volume) 45, Record 21, Geography 6;* an electronic version can be found at https://ctext.org/wiki.pl?if=gb&res=410835

[262]*Ibid.*

[263]William Meacham. "Defining the Hundred Yue." *Bulletin of the Indo-Pacific Prehistory Association*, Vol. 15, 1996, pp. 93–100. DOI:10.7152/bippa.v15i0.11537.

[264]Guizhou Provincial People's Government. "Historical Evolution (历史沿革)." https://web.archive.org/, http://www.guizhou.gov.cn/dcgz/gzgk/lsyg/, 2025, https://web.archive.org/web/20190226173107/http://www.guizhou.gov.cn/dcgz/gzgk/lsyg/

[265]Endymion Wilkinson. *Chinese History: A New Manual*, Harvard-Yenching Institute Monograph Series 84, Cambridge, MA, USA: Harvard-Yenching Institute, Harvard University Asia Centre, 2012, p. 233.

[266]Yan Binzhen (颜丙震). *Ming Hou Qi Qian Shu Pi Lin Di Qu Tu Si Fen Zheng Yan Jiu* (明后期黔蜀毗邻地区土司纷争研究) *Research on Tusi Disputes in the Adjacent Areas of Guizhou and Shu in the Late Ming Dynasty*. Beijing, China: People's Daily Publishing House (人民日报出版社), 2018.

[267]*Ibid.*

[268]Sima Qian (司馬遷; 145 B.C. –the first century B.C.). *Shi Ji • Juan Yi Bai Yi Shi Liu • Xi Nan Yi Lie Chuan* (史记 • 卷一百一十六 • 西南夷列传) or *Historical Records, Juan (Volume) 116, Biography of Southwest Yi*; an electronic version can be found at https://ctext.org/shiji/xi-nan-yi-lie-zhuan/zhs

[269]Liao Deguang (廖德广). *Nan Zhao Guo Shi Tan Jiu* (南诏国史探究) or *Research on the History of Nanzhao*. Kunming, China: Yunnan Nationalities Publishing House (云南民族出版社), 2006.

[270]Duan Yuming (段玉明). *Da Li Guo Shi* (大理国史) or *History of the Dali Kingdom*. Kunming, China: Yunnan Nationalities Publishing House (云南民族出版社), 2003.

[271]China Culture Editor. "Erhai Lake." http://en.chinaculture.org/, ChinaCulture.org, 2025, http://en.chinaculture.org/library/2003-09/24/content_34359.htm

[272]Song Lian (宋濂; 1310–1381), *et al. Yuan Shi • Juan Liu Shi Yi* (元史 • 卷六十一) or *History of Yuan, Juan (Volume) 61*; an electronic version can be found at https://ctext.org/wiki.pl?if=gb&res=434890&remap=gb

[273]Chuan-kang Shih. *Quest for Harmony: The Moso Traditions of Sexual Union and Family Life*, Stanford, CA, USA: Stanford University Press, 2009, p. 52.

[274]Chuan-kang Shih. *Op. cit.,* p. 53.

[275]Joseph F. Rock. "Chapter V: Yung-ning territory: Its history and geography." *The ancient Na-khi Kingdom of southwest China*. Cambridge, MA, USA: Harvard University Press, 1947, pp. 355–434.

[276]Chuan-kang Shih. *Op. cit.,* pp. 52–53

[277]Zhou Zhenhe (周振鹤). *Zhong Guo Xing Zheng Qu Hua Tong Shi • Ming Dai Juan* (中国行政区划通史 • 明代卷) or *General History of Chinese Administrative Divisions: Ming Dynasty*. Shanghai, China: Fudan University Press (复旦大学出版社), 2007.

[278]Guoxuedashi. "Yuejuan Wei (越巂卫)." https://m2.guoxuedashi.net/, Collection of Classical Books by Masters of Chinese Studies (国学大师 古典图书集成), 2025, https://m2.guoxuedashi.net/diming/84255aqiu/

[279]Archibald Little. *The Far East, 1905.* Cambridge, UK: Cambridge University Press, 2010; an electronic version can be found at https://books.google.com.hk/books?id=kxNxhSy1BZUC&pg=PA63&redir_esc=y#v=onepage&q&f=false

[280]Sichuan Provincial People's Government. "Historical Geographical Characteristics." https://web.archive.org/, Sichuan Yearbook Press, 2025, web.archive.org/web/20120819023648/http://english.sc.gov.cn/SichuaninPerspective/BriefingaboutSichuan/200906/t20090624_770675.shtml

[281]*Ibid.*

[282]Zhang Tingyu (張廷玉; 1672–1755), *et al. Ming Shi • Juan San Bai San Shi Yi • Lie Chuan Di Er Bai Shi Qi • Xi Yu San* (明史 • 卷三百三十一 • 列傳第二百十七 • 西域三) or *History of Ming, Juan (Volume) 331, Biography 217, Western Region 3*; an electronic version can be found at https://ctext.org/wiki.pl?if=gb&res=410835

[283]*Ibid.*

[284]*Ibid.*

[285]*Ibid.*

[286]Shi Weile (史为乐), *et al. Zhong Guo Li Shi Di Ming Da Ci Dian* (中国历史地名大辞典) or *Dictionary of Historical Place Names in China*. China Social Sciences Publishing House (中国社会科学出版社), 2005.

[287]Chen Menglei (陳夢雷; 1650–1741). *Qin Ding Gu Jin Tu Shu Ji Cheng • Fang Yu Hui Bian • Zhi Fang Dian • Song Pan Wei Bu* (欽定古今圖書集成•方輿彙編•職方典•松潘衛部) or *Collection of Ancient and Modern Books: Zhi Fang Dian: Songpan Wei, Compiled by Fang Yu* ; an electronic version can be found at https://ctext.org/library.pl?if=gb&res=81155

[288]Gao Wende (高文德), *et al. Zhong Guo Shao Shu Min Zu Da Ci Dian* (中国少数民族史大辞典) or *Dictionary of Chinese Minority Histories. Op. cit.*

[289]*Ibid.*

[290]*Ibid.*

[291]Zhang Tingyu (張廷玉; 1672–1755), *et al. Ming Shi • Juan San Bai San Shi Yi • Lie Chuan Di Er Bai Shi Qi • Xi Yu San* (明史•卷三百三十一•列傳第二百十七•西域三) or *History of Ming, Juan (Volume) 331, Biography 217, Western Region 3*; an electronic version can be found at https://ctext.org/wiki.pl?if=gb&res=410835

[292]*Ibid.*

[293]National Geographic. *National Geographic Atlas of China*. Washington, D.C., USA: National Geographic, 2007.

[294]*Ibid.*

[295]LinkSea. Randy, ed. "Xingxiu Hai Sea: The Mysterious and Beautiful Source of the Yellow River (星宿海：黄河源头神秘而美丽所在)." www.scieau.com, Scieau Examiner, 2025, https://www.scieau.com/articles/2017013047

[296]*Ibid.*

[297]Zhang Tingyu (張廷玉; 1672–1755), *et al. Ming Shi • Juan San Bai San Shi Yi • Lie Chuan Di Er Bai Shi Qi • Xi Yu San* (明史•卷三百三十一•列傳第二百十七•西域三) or *History of Ming, Juan (Volume) 331, Biography 217, Western Region 3*; an electronic version can be found at https://ctext.org/wiki.pl?if=gb&res=410835

[298]*Ibid.*

[299]Bin Yang. "Chapter 4: Rule Based on Native Customs." *Between Winds and Clouds: The Making of Yunnan (Second Century BCE to Twentieth Century CE)*. Columbia University Press, 2008.

[300]Hainan Provincial Committee of the Chinese Communist Party (中共海南省委机关). "Han Dynasty: From 'being part of the territory' to 'being abandoned' (汉代：从'初入版图'到'弃置珠崖')." https://hndaily.cn/#/news, Hainan Ribao (海南日报), 2025, https://baike.baidu.com/reference/533000/76c6znxES-GCExRwRy0exc2S5cGbvRqsb9oAsXi0y2b-6jLiyk4TWLXpQsBgLVZjYSt_KR2Rc5sP74q9KnBmJyh4OwLIWX0Dfld4L8aVLDznm1U; Office of Local History of Hainan and Compilation Committee of Hainan Historical Chronicles (海南省地方史志办公室、海南省地方志编纂委员会). *Hainan Sheng Zhi* (海南省志) or *Hainan Province Chronicles, Juan (Volume) 11, Part 1*. Hainan, China: Nan Hai Publishing Co. (南海出版公司), 1993.

[301]*Ibid.*

[302]Zi Fan (子樊). "Greater Ryukyu or Little Ryukyu? The historical relationship between Okinawa and Taiwan (大琉球？小琉球？ 沖繩與台灣的歷史淵源)." www.epochtimes.com, Epoch Times Taiwan, 2025, https://www.epochtimes.com.tw/n186118/大琉球-小琉球--沖繩與台灣的歷史淵源.html

[303]*Ibid.*

[304]*Ibid.*
[305]*Ibid.*
[306]"File:Location of the Ryukyu Islands.JPG." *Wikimedia Commons*, 11 Apr. 2022, upload.wikimedia.org/wikipedia/commons/c/c6/Location_of_the_Ryukyu_Islands.JPG
[307]"File:East China Sea.PNG." *Wikimedia Commons*, 31 July 2022, upload.wikimedia.org/wikipedia/commons/3/35/East_China_Sea.PNG

Chapter 4

An Ancient World Map Depicts Chinese Exploration of Southeast Asia in 1432–1433, Long Before the First Europeans Began to Conquer that Region in 1511

Abstract

The map Kunyu Wanguo Quantu （坤輿萬國全圖/坤舆万国全图） or Complete Geographical Map of All the Kingdoms of the World was published in China by Italian missionary Matteo Ricci in 1602. It has been generally regarded as a European map.

However, in this chapter, I will show that the part of Kunyu Wanguo Quantu showing Mainland Southeast Asia and the islands surrounding it (excluding islands in Oceania), abbreviated as Southeast Asia-KWQ, is of Chinese origin, not a copy or adapted copy of the major sixteenth-century European maps. This finding is obtained after analysing the geographical information and histories of the 60 geographical items and eleven related annotations on the map.

Moreover, historical records show that the Ming Treasure Fleets visited Southeast Asia, both when outbound in 1432 and when inbound in 1433, during their seventh (and last) voyage to the Western Ocean. Hence, the Ming mariners most likely obtained their latest geographical data of the region in the period 1432–1433; hence, the era revealed by the Southeast Asia-KWQ should also be in the period 1432–1433. This era falls within the period 1428–1470 which is the era deduced directly from the analysis of the Southeast Asia-KWQ. The period 1432–1433 is long before the arrival of the European explorers and colonizers in this region.

Keywords: China, Kunyu Wanguo Quantu, Matteo Ricci, Southeast Asia, Treasure Fleet

1. Introduction

For over four hundred years, the ancient world map — Kunyu Wanguo Quantu （坤輿萬國全圖/坤舆万国全图），[1] abbreviated here as KWQ — written with Chinese characters and having latitudinal and longitudinal lines, has been generally regarded as a European map. The map was published in 1602 by Matteo Ricci (an Italian Jesuit priest and one of the founding figures of the Jesuit

China missions) and was believed to be based on the European maps which Ricci brought with him to China in 1582.[2] In my previous book titled *Chinese Global Exploration in the Pre-Columbian Era: Evidence from an Ancient World Map*,[3] I have made thorough analyses to show how the eras of several major portions of the KWQ can be deduced from their geographical and historical information. My findings were: (1) the information on Europe was obtained by Chinese in the Southern Song Dynasty (南宋; 1127–1279); (2) the information on both Mo Wa La Ni Jia (墨瓦蠟泥加/墨瓦蜡泥加; referring to the regions of Australia, New Zealand, *Tierra del Fuego* /Land of Fire in South America, and Antarctica) and the Americas was obtained by the Ming (1368–1644) mariners in the 1420s during their sixth voyage to the Western Ocean; (3) the information on Africa was obtained by the Ming mariners in 1433 during their seventh voyage to the Western Ocean; these mariners belonged to the Treasure Fleets commanded by Admiral Zheng He (鄭和/郑和; 1371–1433/1435); (4) the information on the map of Japan reveals Japan in 897 A.D. of the Chinese Tang Dynasty (618–907, with an interregnum between 690 and 705), almost 650 years before the first Europeans visited Japan; (5) the information on Korea reveals the era 1433 when the northern border of Joseon with Jurchen was established by Sejong the Great (朝鮮世宗/朝鲜世宗) after he successfully expanded in 1433 his country's northernmost boundary to the river of Tuman through the subjugation of the Jurchens; and (6) the era of the China portion of the KWQ is also 1433; it was the same year when the Nurgan Command Post (1409-1435) China established on the Jurchen land was lastly inspected by a Jurchen eunuch of the Ming Dynasty before it was abolished in 1435. My findings point out that the afore-mentioned portions of the KWQ most likely used Chinese maps as source maps (they no longer exist), instead of being copied from other contemporary European maps.

In this chapter, I shall use the same methodology for other regions: a thorough analysis of all the annotations, and the locations and histories of all the geographical items depicted on the mainland of Southeast Asia (it is in fact an extension of the Asian continent and includes the present-day countries of Cambodia, Laos, Myanmar, Thailand, Vietnam, and Peninsular Malaysia), as well as its surrounding islands on the KWQ. Chinese sources referred to the entire region as Nanyang, "南洋", lit. "Southern Ocean", whereas the Europeans referred to the mainland of Southeast Asia as Indochina due to its location between China and the Indian subcontinent and its cultural influence by both India and China. Here the mainland of Southeast Asia and its surrounding islands on the KWQ is abbreviated as the Southeast Asia-KWQ. The information on the Southeast Asia-KWQ will be compared with those on four major sixteenth-century European maps: (1) the "Southeast Asia-Mercator (1569)" extracted from the 1569 World Map by Gerardus Mercator; (2) the "Southeast Asia-Ortelius (1570; #1)" extracted from the 1570 World Map by Abraham Ortelius;[4] (3) the "Southeast Asia-Plancius (1594)" extracted from the 1594 World Map by Petrus Plancius; and (4) the "Southeast Asia-Ortelius (1570; #2)" extracted from the 1570 Ortelius Map of Asia (first edition)-*Asiae Nova Descriptio*.

In the Appendix of this chapter, I analyse the geographical information and histories of all the items depicted on the Southeast Asia-KWQ in Fig. 4.1(a) and compare them with those on the four major sixteenth-century European maps in Figs. 4.1(b)–4.1(e) whenever possible. The results of these comparisons are used, as explained in the following sections, as the basis for determining the origin and era revealed by the Southeast Asia-KWQ.

Fig. 4.1. (a) shows the Southeast Asia-KWQ extracted from the world map — Kunyu Wanguo Quantu（坤輿萬國全圖/坤舆万国全图） or Complete Geographical Map of All the Kingdoms of the World — published by Matteo Ricci in 1602 in China (public domain);[5] (b) shows the Southeast Asia-Mercator (1569) extracted from the 1569 World Map by Gerardus Mercator (public domain);[6] (c) shows the Southeast Asia-Ortelius (1570; #1) extracted from the 1570 World Map by Abraham Ortelius (public domain);[7] (d) shows the Southeast Asia-Plancius (1594) extracted from the 1594 World Map by Petrus Plancius;[8] and (e) shows the Southeast Asia-Ortelius (1570; #2) extracted from the 1570 Ortelius Map of Asia (first edition)-*Asiae Nova Descriptio* (public domain).[9]

In this Chapter, I include two maps by Abraham Ortelius, the Southeast Asia-Ortelius (1570; #1) extracted separately from the 1570 Ortelius World Map and the Southeast Asia-Ortelius (1570; #2) extracted from the 1570 Ortelius Map of Asia (first edition)-*Asiae Nova Descriptio*. Because on the former map, Ortelius skipped many geographical names perhaps because they are less significant on the scale of a world map; the latter is a map of Asia, hence, these places are relatively more significant.

2. The Southeast Asia-KWQ is of Chinese Origin, not a Copy or Adapted Copy of the Four Major Sixteenth-Century European Maps

The above section title is based on the following findings made in the Appendix to be summarised here. For brevity, all the endnotes will be denoted in the Appendix to avoid repetitions here.

2.1 *Among the 60 geographical items depicted on the Southeast Asia-KWQ, 23 of them (c. 38%) are not depicted on any of the four European maps*

Since 23 (about 38%) of the geographical items depicted on the Southeast Asia-KWQ are absent on any of the four European maps, it is not reasonable to say that the Southeast Asia-KWQ is a copy or adapted copy of any or the combined four European maps. Among the group of 23, seven of them are of Chinese origin, another seven of them are native names, nine of them cannot be identified with their origins (cannot identify them on modern maps); and none of them is of European origin. This reveals that the Southeast Asia-KWQ has its own source — a Chinese source — since Matteo Ricci made this map in China and had access to Chinese maps and records that he could not have in Europe.

Moreover, there is also a fundamental reason in supporting the Chinese origin of the Southeast Asia-KWQ, because China and the Southeast Asia have had close contacts since Antiquity; hence, Chinese cartographers had maps and records on Southeast Asia. For example, from 111 B.C. to 938 A.D. northern Vietnam (越南) was under Chinese rule. Vietnam was governed by a series of Chinese dynasties including the Han (漢/汉; 202 B.C.–220 A.D.), Eastern Wu (東吳/东吴; 222 A.D.–280 A.D.), Cao Wei (曹魏; 220 A.D.–266 A.D), Jin (晉/晋; 266 A.D.–420 A.D.), Liu Song (劉宋/刘宋; 420 A.D.–479 A.D.), Southern Qi (南齊/南齐; 479 A.D.–502 A.D.), Liang (梁; 502 A.D.–557 A.D.), Sui (隋; 581–618), Tang (唐; 618–907, with an interregnum between 690 and 705), and Southern Han (南漢/南汉; 917–971). However, European influence on Southeast Asia only started to enter in the sixteenth century, with the arrival of the Portuguese in Malacca, Maluku, and the Philippines. As described in the Appendix, many of the Southeast Asian countries had tributary relationships with ancient China. If the Southeast Asia-KWQ were truly a copy or adapted copy of the four European maps, how come there is such a high percentage of geographical items not sharing with the four European maps, and none of these geographical items has a European origin? Table 4.1 shows these 23 items.

Table 4.1. Here are the 23 geographical items on the Southeast Asia-KWQ, which are not depicted on any of the four major sixteenth-century European maps; they are separated in three categories, and the item positions in the Appendix are also given.

Category	Origin (Number)	Geographical Item (name; Item position in the appendix)
1	China (6)	1. Lao Country (寮國/寮国 or 老撾/老挝; Item 2); 2. Babai (八百; Item 3); 3. Mubang (木邦; Item 4); 4. Da Ni (大泥; Item 19); 5. An unrecognised Chinese geographical item (左 "耳", 右 "兮"; Item 46); 6. Southern Sea (南海; Item 55)
2	Local (8)	1. Prachinburi (布丈跋羅/布丈跋罗 or 巴真; Item 9); 2. Pagan (蒲甘; Item 11); 3. Pan Pan (盤盤/盘盘; Item 15); 4. Samboja Kingdom (三佛齊王國/三佛齐王国; old name was Srivijaya 室利佛逝; Item 16); 5. Johor, also spelled as Johore (藥兒/药儿 or 柔佛; Item 21); 6. Matan or Mactan (馬大音/马大音; Item 25); 7. Pekalongan (把挈路曷; Item 43); 8. Surabaia or Surabaya (雅罷牙/雅罢牙; Item 44)
3	Unknown origin, and cannot be identified on a modern map (9)	1. Daoming (道明; Item 5); 2. Lanbaidao (蘭白道/兰白道; Item 14); 3. Bingdong (丙東/丙东; Item 17); 4. *Corno, isola del* (角島/角岛; Item 28); 5. Palawan (巴那馬/巴那马; Item 30); 6. Papuas (巴布亞私/巴布亚私; Item 38); 7. Wumitule (兀彌突勒/兀弥突勒; Item 57); 8. Tunian (大泥俺; Item 58); 9. Bazhen (把珎; Item 60)

2.2 Examples of the geographical items on the Europeans maps, decades before they were reached by the Europeans

Table 4.1 gives the complete list of the remaining 37 geographical items depicted on any of the European maps and Table 4.2 gives examples showing Chinese transliterated geographical names, instead of the European transliterated ones, further revealing the Chinese origin of this map. Among the 37 geographical items, Timor (地木島/地木岛; Item 47) deserves special attention. It was one of the places incorporated into the trading networks of the ancient Javanese, Chinese and Indians as early as the fourteenth century as an exporter. The Portuguese settlement came as late as the end of the sixteenth century and the Dutch in the mid-seventeenth century. Yet, the Southeast Asia-Mercator (1569), the Southeast Asia-Ortelius (1570; #2) and the Southeast Asia-Plancius (1594) already depict "Timor" before the Europeans' arrival. One wonders how Mercator, Ortelius and Plancius knew the existence of Timor before it was "discovered" by the Europeans.

Similarly, Banten (板淡; Item 45; 6.4°S, 106.1°E) used to be a port town near the western end of Java in Indonesia. The first known European to land on the island of Java was Dutch explorer Cornelis de Houtman (1565–1599). He arrived at Banten in June 1596 and returned to Holland in 1597. He died in 1599 during his second trip to the East. Yet, on the Southeast Asia-Ortelius (1570; #2), there is "Banta" on the western end of "IAVA MAIOR". You wonder again: how did Ortelius know "Banta (板淡)" in 1570 decades before Dutch explorer Cornelis de Houtman first arrived at Banten in June 1596 and returned to Holland in 1597?

Table 4.2.   Here are the origins of the remaining 37 geographical items which are depicted on the Southeast Asia-KWQ and also on at least one of the four European maps; they are separated into five categories, and the item positions in the Appendix are also given.

Category	Origin (Number)	Geographical Item (name; Item position in the Appendix)
1	China (3)	1. Annan/the old name was Jiaozhi (安南/舊交阯; Item 1); 2. Siam (暹羅/暹罗; Item 8); 3. Zhan Cheng or the city of the Cham (占城; Item 12)
2	Local (17)	1. Mottama (馬兒大莽; Item 6); 2. Chenla (真臘/真腊; Item 7); 3. Bicipuri (披支布里; Item 10); 4. Kambuja (甘波牙; Item 13); 5. Pahang (彭亨; Item 18); 6. Martaban (馬大邦/马大邦; Item 20); 7. Malacca (滿剌加/古哥羅富沙; Item 22); 8. Luzon (呂宋/吕宋; Item 23); 9. *Maynilà* (瑪泥兒訝/玛泥儿讶; Item 24); 10. Maluku Islands (馬路古地方/马路古地方; Item 36); 11. Biba or Bima (皮馬/皮马; Item 41); 12. Banten (板淡; Item 45); 13. Timor (地木島/地木岛; Item 47); 14. Samar (娑麻剌; Item 49); 15. Lamuri or Lambri (覽比/览比; Item 50); 16. Sumatra (蘇門答臘/苏门答腊; Item 56); 17. Achen (大真; Item 59)
3	Europe (7)	1. Mindanao (茗答鬧/茗答闹; Item 29); 2. Felipinas (非利皮那; Item 31); 3. Island of Borneo (波爾匿何/波尔匿何; Item 32); 4. Celebes (色力皮; Item 34); 5. Celebes (食力百私; Item 35); 6. *Jave la Grande* (大爪哇, 即訶陵曰闍婆/大爪哇, 即诃陵曰阇婆; Item 42); 7. *Jave mine* (小爪哇; Item 54) **Notie:** Celebes (食力百私; Item 35) is a repetition of Celebes (色力皮; Item 34); *Jave la Grande* (大爪哇, 即訶陵曰闍婆/大爪哇, 即诃陵曰阇婆; Item 42) and *Jave mine* (小爪哇; Item 54) are names taken from Marco Polo's travel book: *Jave la Grande* is the present-day Java Island (it is one of the islands in Indonesia); Jave mine is the present-day Sumatra, one of the Sunda Islands of western Indonesia. Hence, only four more names remain: the name "Mingdanao" first appeared in European writing in 1521; the name Felipinas *was* given by a Spanish explorer in 1542; the name "Borneo" appeared after a Portuguese contacted the native people of Boneo; and Celebes might be considered a Portuguese rendering of the native name "Sulawesi". These European-derived names in the sixteenth century may be added by Matteo Ricci in 1602 to the source map of the Southeast Asia-KWQ to update the map, because Section 2 will show that *the Southeast Asia-KWQ is in fact a map based on the political landscape of this region in 1433.*
4	Unknown origin; cannot be identified on a modern map (9)	1. Gosas (卧山島/卧山岛; Item 26); 2. Arrecifes (亞來沙/亚来沙; Item 27); 3. Chinabalo (止男巴洛; Item 33); 4. Cainam (蓋南/盖南; Item 37); 5. *Zolot, isola* (梭羅島/梭罗岛; Item 39); 6. Palabar or Baiabar (巴亞巴/巴亚巴; Item 40); 7. Petan (伯旦; Item 48); 8. Basman (把西蠻/把西蛮; Item 51); 9. Feriech (弗爾色/弗尔色; Item 53)
5	Arabic (1)	1. Barus (番蘇爾/番苏尔; Item 52)

Table 4.3. This table shows two examples of the Southeast Asia-KWQ using Chinese transliterated geographical names, instead of the European transliterated ones, further revealing the Chinese origin of this map.

	Geographical Names	Item Number	Origin	Chinese Transliteration	European Transliteration
1	Pahang	18	Khmer word for "tin"	彭亨 (Peng Heng), 彭坑 (Peng Keng), or 彭杭 (Peng Hang)	*Pam, Pan, Paam, Paon, Phaan, Phang, Paham, Pahan, Pahaun, Phaung, Phahangh*
2	Luzon	23	Tagalog word *lusong*	吕宋 (Lu Song)	Palahan Island, Paloban, Luçonia or Luconia

2.3 *The Southeast Asia-KWQ has eleven annotations providing lots of information of countries, cities, places, regions, islands, etc., in terms of their names, city walls, histories, grains, products, people's living styles, birds, animals, reptiles, climates, etc., whereas the four European maps have no such annotation*

The detailed information reveals that the Chinese cartographer(s) who drew the source map(s) of the Southeast Asia-KWQ had good knowledge and data about these geographical items obtained by the Chinese explorers who had gone to these places.

In summary, based on the above analysis and evidence it is reasonable to conclude that the Southeast Asia-KWQ is of Chinese origin, not a copy or adapted copy of the four major sixteenth-century European maps.

3. Determining the Political Era Revealed by the Southeast Asia-KWQ

There are five steps for determining the political era revealed by the Southeast Asia-KWQ in this section. Again, for brevity, most endnotes in this section are denoted in the Appendix to avoid repetitions:

3.1 *The political era revealed by the Southeast Asia-KWQ is not before 1428*

From 1407 to 1427 (from the fifth year of Yongle to the second year of Xuande), Vietnam was ruled by the Chinese Ming Dynasty as the province of Jiaozhi (交趾). Historical records show that from 1428 (the third year of Xuande), Jiaozhi (交趾) ceased to be Ming's province, because in 1428 the Ming Xuande Emperor abolished the Jiaozhi Chengxuan Provincial Administration Commission (交趾承宣布政使司 or abbreviated as交趾布政使司 or 交趾司; 司, lit. "division", or commonly known as 省, meaning "province"). Afterwards, the Ming Emperor immediately named Chen Hao (陳暠/陈皓) — the puppet monarch whom Lê Lợi (黎太祖; 1384/1385–1433; reigned 1428–1433) supported before he came to the throne — as the "King of Annan (安南王)", and the Annan Kingdom (安南王國/安南王国) became independent.

Since the Southeast Asia-KWQ depicts Annan (安南; Item 1), and specifically denotes it as the old Jiaozhi (舊交趾/旧交趾), the political era revealed by this map is not before 1428.

3.2 *The political era revealed by the Southeast Asia-KWQ was in a period 1428–1496/1520*

The Kingdom of Sumatra was called the "Samudra Kingdom" (須文達那國/须文达那国) in the Yuan Dynasty. It was located at the mouth of the Parsei River in today's Sumatra Island, where there is still a small village called Sumandra (須文達那/须文达那). The Arabic transliteration of Sumandra is "Sumatra", because Maghrebi traveler, explorer and scholar Ibn Battutah (1304–1368/1369) visited this country and called it Sumatra. The king of Samudra Kingdom sent envoys and tribute to the Yuan Dynasty. The kingdom was renamed the Kingdom of Sumatra (蘇門答臘國/苏门答腊国; Item 56) during the Hongwu period in the Ming Dynasty. In 1435 (the tenth year of Xuande), the Xuande Emperor granted the young king's son to succeed to the throne.

Later, the Kingdom was destroyed by Aceh (啞齊/哑齐 or 亞齊/亚齐; officially the Kingdom of Aceh Darussalam; c. 1496/1520–1903). The exact origins of the Aceh Sultanate or the Kingdom of Aceh (啞齊/哑齐) Darussalam are still disputed; Portuguese sources claim that the Sultanate started from 1520, not c. 1496, and Sumatra became the name of the entire island. Since the Kingdom of Sumatra (蘇門答臘國/苏门答腊国) is depicted on the Southeast Asia-KWQ, but not the Kingdom of Aceh Darussalam, and the name of Sumatra does not cover the entire island, this political condition of the Southeast Asia-KWQ reveals an era before the founding the Kingdom of Aceh in either 1496 or 1520.

Combining 3.1 and 3.2 gives the political era of the Southeast Asia-KWQ to a period from 1428 to no later than 1496/1520.

3.3 *The political era revealed by the Southeast Asia-KWQ can be further narrowed to a period 1428–1470*

Srivijaya (三佛齊/三佛齐 or old name 室利佛逝; c. 671–1470) was a Buddhist thalassocratic empire based on Sumatra Island (蘇門答臘島/苏门答腊岛; in the present-day Indonesia). In 1397 (the thirtieth year of Hongwu), the Majapahit Kingdom (滿者伯夷王國/满者伯夷王国) in East Java (東爪哇/东爪哇) destroyed the Old Srivijaya Dynasty, and the prince of Srivijaya fled to the Malay Peninsula to establish the Malacca Dynasty. At that time, more than 1,000 Chinese lived in the Old Harbor (舊港/旧港; lit. "Old Port"; the then capital of Srivijaya, also known as Palembang 巨港). They supported Liang Daoming (梁道明), a local Chinese from Guangdong Province (廣東省/广东省) as king and established a New Srivijaya Dynasty, which was destroyed later by the Majapahit Empire (1293–1527) in 1470. But Srivijaya (三佛齊/三佛齐) is depicted on the

Southeast Asia-KWQ (but erroneously on the Malay Peninsula, instead of the Sumantra Island), so the map reveals a political era from 1428 to 1470.

Combining the facts in 3.1, 3.2 and 3.3, we know that the Southeast Asia-KWQ is a map revealing the political era of 1428–1470.

3.4 *The era revealed by the Southeast Asia-KWQ can be finally narrowed to the period 1432–1433*

The Southeast Asia-KWQ cannot be drawn without actual exploration of the geographical locations depicted on it. The most likely explorers after 1428 were the mariners from the Ming Treasure Fleets led by Admiral Zheng He during their seventh (and last) voyage.

I have investigated *Qian Wen Ji • Xia Xi Yang* （前聞記 •下西洋/前闻记 •下西洋）[10] or *A Record of History Once Heard: Down to the Western Ocean* written by the Ming historian Zhu Yunming (祝允明; 1461–1527) and the 1597 epic work by Luo Maodeng (羅懋登/罗懋登) titled *San Bao Tai Jian Xi Yang Gi* （三寶太監西洋記/三宝太监西洋记）[11] or *An Account of the Western World Voyage of the San Bao Eunuch*. Both dealt with this seventh voyage and present consistent results for a major portion of the Ming Treasure Fleet's navigational routes and timelines. These have been analysed in my book titled *The Last Journey of the San Bao Eunuch, Admiral Zheng He.*[12] Since both the outbound (leaving home) and inbound (returning home) voyages of the Ming Treasure Fleets to the Western Ocean en route Southeast Asian countries and their surrounding islands, the years 1432 and 1433 are the years of Zheng He mariners' final exploration of the Southeast Asia for them to collect geographical data to update their older map. Soon after 1433, the *Haijin* (海禁) or sea ban (maritime ban) started and lasted until the middle of the sixteenth century. Their updated map later was used by Mattero Ricci as the source map for drawing the Southeast Asia-KWQ. Hence, the era revealed by the Southeast Asia-KWQ should be in the period 1432–1433. The Treasure Fleet arrived at Taicang on 7 July 1433[13] (21 June of the eighth year of Xuande), and on 22 July 1433,[14] they arrived in the capital Beijing. On 27 July, the Xuande Emperor bestowed ceremonial robes and paper money to the fleet's personnel.[15]

The period 1432–1433 falls within the period of 1428–1470 deduced directly from analysing the Southeast Asia-KWQ. This period is several decades before the Bartolomeu Dias and Vasco da Gama sailed around the southern tip of Africa (the Cape of Good Hope) to find a direct sea route to Asia in 1488 and 1498, respectively. All the eras of the geographical items analysed and depicted on the Southeast Asia-KWQ are also consistent with the period 1432–1433, except those added by Matteo Ricci.

4. Conclusions

The map Kunyu Wanguo Quantu (坤輿萬國全圖/坤舆万国全图) (abbreviated as KWQ) or Complete Geographical Map of All the Kingdoms of the World was published in China by Italian missionary Matteo Ricci in 1602. It has been generally regarded as a European map.

However, in Section 2 of this chapter, I have shown that the part of this map containing Mainland Southeast Asia and the islands surrounding it (excluding islands in Oceania), which is abbreviated as the Southeast Asia-KWQ, is of Chinese origin, not a copy or adapted copy of the major sixteenth-century European maps. This finding is obtained by a detailed analysis of the geographical information and histories of the 60 geographical items and eleven related annotations on the map, as listed in the Appendix of this chapter.

Among the 60 geographical items depicted on the Southeast Asia-KWQ, 23 of them (c. 38%) are not depicted on any of the four major European maps discussed in this chapter, revealing that the source map of the Southeast Asia-KWQ has a non-European origin. The most likely origin was Chinese, because among the 23 geographical items, six of them are of Chinese origin, eight of them are native names, and nine of them cannot be identified with their origins (cannot identify them on modern maps), whereas none of them is of European origin. Hence the Southeast Asia-KWQ cannot be a direct or adapted copy of the major sixteenth-century European maps. The main reasons are three-fold: (1) Matteo Ricci was in China, and he had opportunities to see the Chinese maps drawn before the European Age of Discovery, which he could not have accessed while he was in Europe; (2) China was then the greatest sea power in the world and the Ming Treasure Fleets led by Admiral Zheng He sailed seven times to the Western Ocean en route Southeast Asia each time; Zheng He's mariners took the opportunities to explore the region; and (3) China and Southeast Asia have had close contacts since Antiquity, hence Chinese cartographers knew Southeast Asia well to update their maps of the region.

Next in Section 3 of this chapter, I have deduced the political landscape of the Southeast Asia-KWQ after analysing, in the Appendix, the geographical information and the histories of all the items on this portion of the KWQ. From this political condition, I was able to deduce step by step its era to be in the period 1428–1470 with the historical information on Annan and Samboja Kingdom (old name was Srivijaya). However, historical records show that after 1428 the Ming Treasure Fleets visited Southeast Asia in the 1430s during their seventh (and last) voyage to the Western Ocean, both on their way out in 1432 and on the way home in 1433. Hence, the geographical data depicted on the Southeast Asia-KWQ could only be obtained during the period 1432–1433. The era 1432–1433 falls in the period 1428–1470, and it was long before the arrival of the European explorers and colonisers in this region.

All the eras of the geographical items analysed and depicted on the Southeast Asia-KWQ are also consistent with the period 1432–1433.

The main contribution Matteo Ricci made to the Southeast Asia-KWQ was the introduction of four sixteenth-century European-derived place names: (1) Mindanao (茗答鬧/茗答闹; Item 29); (2) Felipinas (非利皮那; Item 31); (3) Island of Borneo (波尔匿何; Item 32); and (4) Celebes (色力皮; Item 34) which refers to the same place as Celebes (食力百私; Item 35). However, the Southeast Asia-KWQ is not a map of 1602, because the political condition of the map does not support that date.

The other two European-derived names — *Jave la Grande* (大爪哇/即訶陵曰闍婆; Item 42) and *Jave mine* (小爪哇; Item 54) — were taken from Marco Polo's travel book; the former corresponds to today's Java Island and the latter corresponds to today's Sumatra Island of Western Indonesia.

Appendix

The geographical items on the Southeast Asia-KWQ are analysed in detail in this appendix. On the map, they are almost all written in Chinese traditional characters. Since many readers use the Chinese simplified characters, this book uses both: all the Chinese characters in the main text and in the appendix of this chapter are expressed first by the modern Chinese traditional characters, then followed by the Chinese simplified characters, separated by a slash "/" between them. If both characters are the same, that expression will be presented only once without repetition. Moreover, all the geographical items are numbered in sequence in this Appendix. The annotations are numbered as well with an "@" sign in front of the number. The "•" in front of an item number denotes that this item does not appear on any of the four sixteenth-century European maps. The Chinese "pinyin" version of each item is given right after it is introduced. My comments are given in the last column of the Appendix.

Item #	Name in Chinese	Pinyin	Etymology	History, Geography and My Comments
The Mainland Southeast Asia and surrounding islands				

Fig. A4.1.   This figure is extracted from Fig. 4.1(a) (public domain); Items 1-60 and annotations @1-11 are depicted on this Figure.

| 1 | 安南/舊交阯 | An Nan (Jiu Jiao Zhi) | Annan (the old name was Jiaozhi) | Annan (安南) was an imperial protectorate and the southernmost administrative division of the Tang Dynasty and Wu Zhou Dynasty of China from 618–907.[16] It was in the present-day Vietnam.

From 1407 to 1427 (from the fifth year of Yongle to the second year of Xuande), Vietnam was ruled by the Chinese Ming Dynasty as the province of Jiaozhi.[17] Then, historical records show that from 1428 (the third year of Xuande) Jiaozhi (交阯) ceased to be Ming's province,[18] because in 1428 the Ming Xuande Emperor abolished the Jiaozhi Chengxuan Provincial Administration Commission (交阯承宣布政使司 or abbreviated as 交阯布政使司 or 交阯司; 司, lit. "division", or commonly known as 省, meaning "province").[19] Afterwards, the Ming Emperor immediately named Chen Hao (陳暠/陈皓) — the puppet monarch whom Lê Lợi (黎太祖/黎利; 1384/1385–1433; reigned 1428–1433) supported before he came to the throne — as the "King of Annan (安南王)" and the Annan Kingdom (安南王國/安南王国) became independent.[20]

Since the Southeast Asia-KWQ depicts Annan (安南), and states its old name as Jiaozhi (交阯), the political era revealed by this map cannot be before 1428.

After the Great Ming gave up Jiaozhi, its influence in Southeast Asia was greatly weakened, and many foreign countries did not come to pay tribute anymore. Hence, in 1430 (the fifth year of Xuande), the Xuande Emperor ordered Zheng He (鄭和/郑和; 1371–1433/1435) to go overseas again for his seventh (and last voyage) to the Western Ocean (Indian Ocean and beyond) to revive China's tributary system and promote Ming's maritime diplomacy.[21]

On the Southeast Asia-Mercator (1569), Annan is depicted as "Gachuchina". On the Southeast Asia-Plancius (1594), there is "Ainao" which sounds close to Annan, but its location is wrong. On the Southeast Asia-Ortelius (1570' #1), Annan is not depicted. But on the Southeast Asia-Ortelius (1570; #2), Annan is depicted as "Gavcin China". |

(Continued)

Item #	Name in Chinese	Pinyin	Etymology	History, Geography and My Comments
				The Portuguese conquered Malacca in 1511; they were also the first Europeans to set foot on Vietnamese soil because Fernão Peres de Andrade (d. 1552) was the first Portuguese to visit the Vietnamese region in 1516.[22]
●2	老撾	Lao Wo	Lao Country; the northern part of the present-day Laos. However, the name "Laos" started after the French united the three Lao kingdoms in French Indochina in 1893;[23] the country's name is *Muang Lao* or *Pathet Lao*, lit. "Lao Country (寮國/寮国 or 老撾/老挝)", because the dominant ethnic group in the country are the Lao people. According to *Lao Wo Ji Nian* (老撾紀年/老挝纪年) or *Chronicles of Laos*, they immigrated from the southern China in ancient days.	The Lao Country (寮國/寮国 or 老撾/老挝; the northern part of the present-day Laos) is the only landlocked country in Southeast Asia. It borders China to the north. The present-day Laos traces its historic and cultural identity to Lan Xang or Lancang (瀾滄/澜沧).[24] In 1353, the whole of Lao Country was unified and the Kingdom of Lancang (瀾滄王國/澜沧王国; 1353–1707) was established. It existed as one of the largest kingdoms in Southeast Asia.[25] In ancient Chinese books, the country was once called 老撾 and some other names. In 1416 (the fourteenth year of Yongle), the king of Lancang died at the age of sixty; he was succeeded by his son Lan Kham Daeng. According to *Da Yue Shi Lue* (大越史略) or *The Abridged Chronicles of Viet*, in 1421 during the reign of Lan Kham Daeng, the Lam Sơn Uprising (藍山起義/蓝山起义; 1418–1427) took place in Jiaozhi (交趾) under Lê Lợi (黎利) against the Ming Dynasty. Lê Lợi sought Lan Xang's assistance. An army of 30,000 with 100 elephant cavalry was dispatched, but instead Lancang sided with the Chinese.[26]
●3	八百	Ba Bai	Babai; in the Ming Dynasty, in the northern part of Thailand (in Chiang Mai), there was a country that the Chinese called the Kingdom of Eight Hundred (Babai, lit. "八百"[27]; 1292–1775). In the Yuan Dynasty, the Chinese called it the Kingdom of Eight Hundred	On the Southeast Asia-KWQ, to the east of Babai is the Liao Country (寮國/寮国 or 老撾/老挝; Item 2) and to its west is Mubang (木邦; the present-day northeastern Myanmar; Item 4). The Ming Dynasty followed the Yuan Dynasty and canonised a Babai Xuanwei Commission or Babai Pacification Commission (八百宣慰司; an appeasement department under the *tusi* system). In 1391 (the 24th year of Hongwu), it was changed into Babai-dadian Military-cum-Civilian Pacification Commission (八百大甸軍民宣慰使司/八百大甸军民宣慰使司).[28]

			Wives (八百媳婦國/八百媳妇国"), named after the hearsay that "a minister has eight hundred wives" or the leader of each of the 800 towns or villages in Ching Mai was a woman and these eight hundred female chiefs should be regarded as the wives of the King of Chiang Mai.	In the Ming Dynasty, the northern part of the present-day Thailand was under the *tusi* (土司) system of the Yunnan Province.[29] However, in 1432 (the seventh year of Xuande), when Babai sought military help from the Ming Court to resist Siam invading its land, the Ming Court only issued an edict to stop surrendering without sending any troops.[30] This shows that after the Xuande Emperor gave up Jiaozhi, the Ming influence in Southeast Asia was greatly weakened and could not rule its peripherals as effectively as in the Yongle period.
•4	木邦	Mu Bang	Mubang; it was in today's Myanmar. The first European (English merchant) Ralph Fitch (c. 1550–1611) went to Burma/Myanmar in 1586–1587 and Malacca (in today's Malaysia). He returned to London in 1591.	According to recorded history, the earliest inhabitants in today's Myanmar were the Pyu[31] (驃人/骠人; a Tibeto-Burman-speaking people) who established the Pyu city-states and adopted Theravada Buddhism. The Pyu, who lived near the Qinghai Lake in today's Qinghai and Gansu,[32] entered the Irrawaddy (伊洛瓦底) valley from present-day Yunnan, c. second century B.C. The first-ever unification of the Irrawaddy valley and its periphery was made by the Pagan Kingdom (蒲甘王國/蒲甘王国; 849–1297). After the First Mongol invasion of Myanmar in 1287, several small kingdoms, including the Hanthawaddy Kingdom (1287–1552), came to dominate the region. In 1404 (the second year of Yongle) of the Ming Dynasty, the Mubang Military-cum-Civilian Pacification Commission (木邦軍民宣慰使司/木邦军民宣慰使司) was established in Mubang (木邦).[33] The territory ruled by this commission was within today's Myanmar. At the beginning of 1575, Mubang was captured by Myanmar. But later, the Ming army recovered it. On the Southeast Asia-Mercator (1569), there depicts "Berma"; on the Southeast Asia-Plancius (1594), there depicts "Verma"; on the Southeast Asia-Ortelius (1570; #1), there depicts "Brema"; and on the Southeast Asia-Ortelius (1570; #2), there is "Verma". While we know that the

(*Continued*)

Item #	Name in Chinese	Pinyin	Etymology	History, Geography and My Comments
				official English name Burma was an old name of Myanmar, the Mubang Military-cum-Civilian Pacification Commission (木邦軍民宣慰使司/木邦军民宣慰使司) was not depicted explicitly on the above four European maps.
●5	道明	Dao Ming	Daoming	This place cannot be identified on a modern map.
6	馬兒大莽	Ma Er Da Mang	Modern-day Mottama; old name was Martaban. "Mottama" derives from the Mon language term "Mumaw", which means "rocky spur".[34]	Mottama (馬兒大莽/马儿大莽; formerly Martaban) is a town in Myanmar. It is located on the west bank of the Thanlwin River (also known as Salween River 薩爾溫江/萨尔温江 in Myanmar and Nu Jiang 怒江 or Nu River in China). It should be opposite Mawlamyine (also spelled Mawlamyaing), but Fig. A4.2 is not detailed enough to also show Mottama. 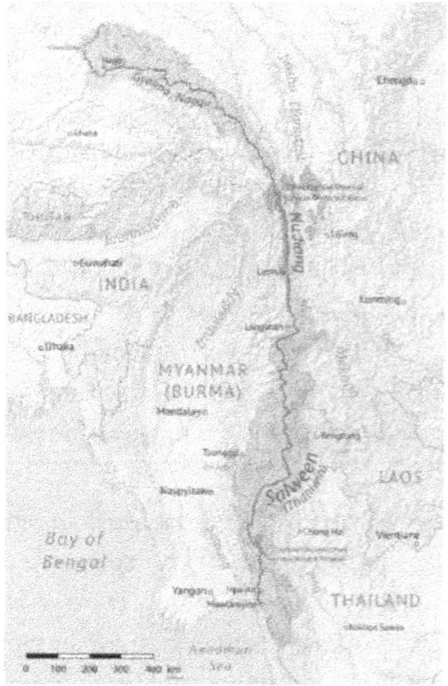

Fig. A4.2. Map of the Salween River basin (*Source*: Shannon1, under CC BY-SA 4.0 International, 3.0 Unported, 2.5 Generic, 2.0 Generic and 1.0 Generic license, https://commons.wikimedia.org/wiki/File:Salween_river_basin_map.png);[35] Mawlamyine is near the mouth of the Salween River.

				Mottama was the capital of the Martaban Kingdom, later known as Hanthawaddy Kingdom (漢達瓦底王國/汉达瓦底王国; 1287–1551/52) from 1287 to 1364.[36] It was an important trading port from the fourteenth century to the early sixteenth century.[37] However, on the Southeast Asia-KWQ, Mottama (馬兒大莽/马儿大莽; formerly Martaban) is mistakenly depicted as landlocked. On the Southeast Asia-Mercator (1569), there depicts "Martaban" on the west bank of the Thanlwin River (also known as Salween River 薩爾溫江/萨尔温江 in Myanmar and Nu Jiang 怒江 or Nu River in China).
7	真臘	Zhen La	Chenla; it is based on the pronunciation of the Cambodian name កម្ពុជាធិបតេយ្យ; later *Cambodia* is an anglicisation of the French *Cambodge*, which in turn is the French transliteration of the Khmer. The term *Cambodia* was already in use in Europe as early as 1524, because in the wok written by Antonio Pigafetta,[38] he cites it as *Camogia*.	Chenla (真臘/真腊; 550–1431) was in the south of the Kingdom of Champa (占城; a collection of independent Cham polities; 137–1832) and can be reached from there in three days and nights with favourable winds by ship in the Ming Dynasty.[39] The Chinese name "真臘" appears on the *Mao Kun Map* (茅坤圖/茅坤图 or 鄭和航海圖/郑和航海图, meaning "Zheng He's Navigation Map"). The kingdom was within the territory of today's Cambodia (柬埔寨). The Kingdom went through three periods: (1) the early Chenla (550–802); it started as the successor polity of the Kingdom of Funan (扶南; 68 A.D.–550 A.D.); this period of Cambodian history is known by historians as the Pre-Angkor period;[40] (2) the Khmer Empire (高棉帝國/高棉帝国 or 吳哥王朝/吴哥王朝; 802–1431; this period of Cambodian history is known as the Angkor period); and (3) the late Chenla (from 1423 to 1593, Cambodia continued its decline; in the nineteenth century, it became a co-subject of Siam and Vietnam).[41]

(*Continued*)

Item #	Name in Chinese	Pinyin	Etymology	History, Geography and My Comments
				In the Angkor period and from the ninth to the twelfth centuries, the Angkor Dynasty was at its peak in Indochina, and it declined after the fourteenth century. During the fourteenth and fifteenth centuries Khmer was invaded by Siam (the former name of Thailand). In 1403 (the first year of Yongle), the Chenla king sent in tribute to become a tributary country to the Ming Dynasty.[42] In 1408 (the sixth year of Yongle) and 1412 (the tenth year of Yongle), the Yongle Emperor sent Zheng He as an envoy to visit the Kingdom of Chenla.[43] In 1431 (the sixth year of Xuande), Siam (暹羅/暹罗) captured Angkor, and the capital was badly damaged. However, in 1434, the Chenla King restored the country and moved its capital to Phnom Penh (金邊/金边).[44] After the Ming Wanli period (1573–1620), the entire country was already called "Cambodia (柬埔寨)".[45] On the Southeast Asia-KWQ, Chenla (真臘), see Fig. A4.1, is depicted at an incorrect location, facing the Bay of Bengal (孟加拉灣/孟加拉湾). It should face the Gulf of Thailand (泰國灣/泰国湾)/Gulf of Siam (暹羅灣/暹罗湾). On the Southeast Asia-Mercator (1569), there is "Camboya" near the geographical location of today's Cambodia. On the Southeast Asia-Ortelius (1570; #2), there is "Camboia".

| 8 | 暹羅 | Xian Luo | Siam; Siam was China's ancient name for the current Southeast Asian country Thailand: สยาม (Sayam) in Thai, and Siam in English. At that time, the Thai people did not call themselves "Siamese", they called themselves "Thai", meaning "free man" in the Thai language. | Siam (暹羅/暹罗) was the former name of Thailand and referred to the historical region of Central Thailand (located at the centre of the Indochinese Peninsula, see Fig. A4.3), usually including Southern Thailand.
In 1377, the Ming Hongwu Emperor canonized King Ayutthaya as "King of Siam (暹羅國王/暹罗国王)", so the name "Siam" was officially established.[46] Between 1370 and 1643 A.D., the country's envoys visited and traded in China 102 times, and Chinese envoys from the Ming Dynasty also made return visits nineteen times.[47]
European contact began in 1511 with a Portuguese diplomatic mission to Ayutthaya (阿育他亞王國/阿育他亚王国 or 大城王國/大城王国; 1351–1767), which became a regional power by the end of the fifteenth century.[48] In the seventeenth century, the French, Dutch, and English followed.
On the Southeast Asia-Mercator (1569), there is "Sian" near approximately the geographical location of the present-day Thailand. On the Southeast Asia-Ortelius (1570; #2), there is "Siam". |

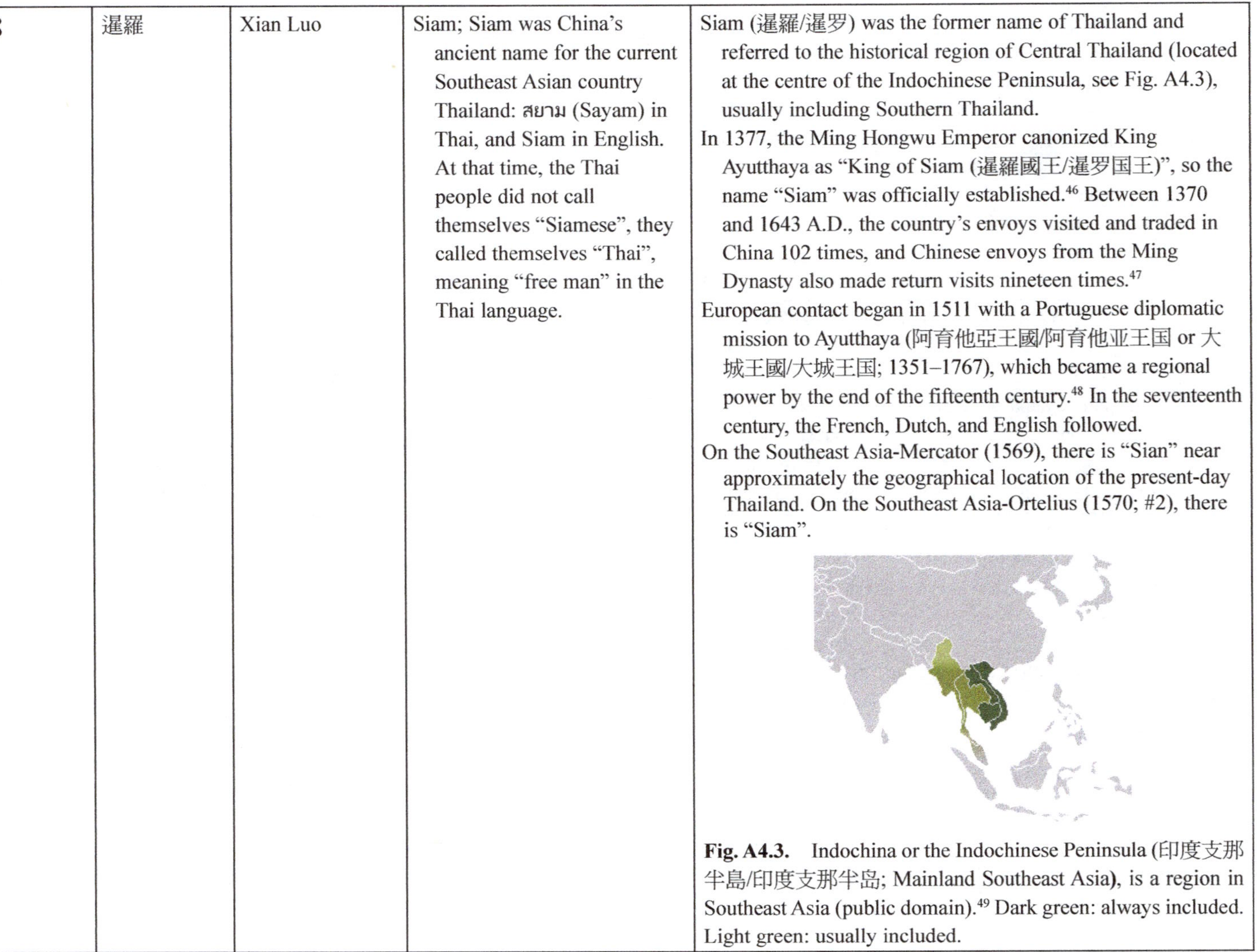

Fig. A4.3. Indochina or the Indochinese Peninsula (印度支那半島/印度支那半岛; Mainland Southeast Asia), is a region in Southeast Asia (public domain).[49] Dark green: always included. Light green: usually included.

(*Continued*)

Item #	Name in Chinese	Pinyin	Etymology	History, Geography and My Comments
@1	古赤土國又名波羅刹	Gu Chi Tu Guo You Ming Bo Luo Sha	Gu Chi Tu Guo You Ming Bo Luo Sha	My translation and comment (between square brackets) are as follows: The ancient Chitu Kingdom [Kingdom of Red Earth; 赤土國/赤土国] is also known as Po Luo Sha [波羅刹/波罗刹]. According to Chinese record,[50] Siam was the Kingdom of Red Earth (Chitu Guo; 赤土國/赤土国) in the Sui (581–618) and Tang (618–907, with an interregnum between 690 and 705) dynasties, because its soil had a reddish colour. The kingdom was in the southwest of Champa and could be reached by ship from Champa in ten days and nights with favourable winds in the Ming Dynasty.[51]
● 9	布丈跋羅	Bu Zhang Ba Luo	Prachinburi; "buri" means "city" in Sanskrit.[52]	It may refer to the area around Prachinburi (巴真) in the present-day Thailand. Portugal was the first European nation to contact the Ayutthaya Kingdom in Siam, or today's Thailand, in 1511.
10	披支布里	Pi Zhi Bu Li	Bicipuri	Bicipuri (披支布里) refers to an ancient city in today's Thailand. On the Southeast Asia-Ortelius (1570; #2), there is "Bicipuri".
● 11	蒲甘	Pu Gan	Bagan or Pagan	Pagan (蒲甘; today's Bagan), an ancient city, is a UNESCO World Heritage Site in the Mandalay Region of Myanmar.[53] The urban region has many pagodas and Buddhist temples built in various historical periods in Myanmar.[54] From the ninth to the thirteenth centuries, the city was the capital of the Pagan Kingdom (蒲甘王國/蒲甘王国 849–1297), the first kingdom that unified the regions that would later become Myanmar (緬甸/缅甸). In the twelfth and thirteenth centuries, Pagan and the Khmer Empire were the two main empires in Mainland Southeast Asia.[55]

				In 1283, Yuan soldiers attacked Bagan from the Yunnan region, and the city of Bagan was conquered. Bagan survived into the fifteenth century as a human settlement,[56] and as a pilgrimage destination throughout the imperial period.
12	占城	Zhan Cheng	Champa; there have been debates about the origin of "Champa", but the official designation of Champa in Chinese historical texts was Zhan Cheng (占城), meaning "the city of the Cham", not "city of the Champa".[57]	The Kingdom of Champa (占城; 137–1832) was a collection of independent Cham polities that extended across the coast of what is contemporary present-day central and southern Vietnam.[58] The old name of Champa was Zhan-po (占婆); the name change took place in the Tang Dynasty. Champa was an important link in the spice trade, which stretched from the Persian Gulf to South China. It served as a supplier of aloe in the Arab maritime routes around the Mainland of Southeast Asia. Champa has always been a friendly vassal in southwest China. During Zheng He's voyages to the Western Ocean, Champa was always the first stop including in the seventh (and last) voyage. While the Treasure Fleet was waiting for the prevailing winds, an inscription entitled Tian Fei Ling Ying Zhi Ji （天妃靈應之記/天妃灵应之记） or A Record of Tianfei Showing Her Presence and Power, was erected at Changle (長樂/长乐) anchorage in Fujian Province (福建省) in early 1431.[59] The inscription also places Champa as the first stop on their voyage. The inscription reads as follows: 。．．。皇上嘉其忠誠，命和等統率官校、旗軍數萬人，乘巨舶百餘艘，賒幣往趙之，所以宣德化而柔遠人也。自永樂三年奉使西洋，迨今七次，所歷番國，由佔城國、爪哇國、三佛齊國、暹羅國，直逾南豚、錫蘭山國、古里國、柯枝國，抵於西域忽魯謨斯國、阿丹國、木骨都束國，大小凡三十餘國，涉滄溟十萬餘裡。

(*Continued*)

Item #	Name in Chinese	Pinyin	Etymology	History, Geography and My Comments
				The translation by J.J.L. Duyvendak is listed as follows (my comment is between square brackets):[60] …From the time when we, Zheng He and his companions at the beginning of the Yongle Period, received the imperial commission as envoys to the barbarians, up until now seven voyages have taken place and each time we have commanded several tens of thousands of government soldiers and more than a hundred ocean-going vessels. Starting from Taicang [this is incorrect; the fleet started from Nanjing] and taking the seas, we have by way of the countries of Champa, Siam, Java, Cochin and Calicut reached Hormuz and other countries of the western regions, more than thirty countries in all. However, from the beginning of the fifteenth century, Champa was constantly invaded by the Kingdom of Annan in the north.[61] The king of Champa sent many envoys to the Ming Dynasty for help. In 1402, Annan captured two prefectures in Champa. In 1407, the Yongle Emperor of the Ming Dynasty sent troops to attack Annan together with the Champa Kingdom and regained the Champa land that had been occupied by Annan not long before. But in 1471, Annan conquered Champa, occupied its capital, and expanded its territory to Quy Nhon (歸仁/归仁). In 1832, Champa completely perished. In 1516 Portuguese traders landed in Da Nang, Champa, naming it Cochinchina (modern Vietnam).[62] On the Southeast Asia-Mercator (1569), Champa is depicted as "Campa", and on the Southeast Asia-Plancius (1594), it is depicted as "Canpa". On the Southeast Asia-Ortelius (1570; #2), there is "CAMPAA".

| @2 | 占城即古林邑產烏木 | Zhang Cheng Ji Gu Lin Yi Chan Wu Mu | My translation and comments (between square brackets) are as follows:

Champa was preceded in the region by the ancient Kingdom of Linyi [林邑; 192–629; it was a kingdom located in central Vietnam and was one of the earliest recorded Champa kingdoms; in the Ming Dynasty, Linyi was part of Annan].[63] The place produces ebony [a kind of tree that produces valuable black wood].[64] |
| 13 | 甘波牙 | Gan Bo Ya | Kambuja; when the Europeans arrived, "Kambuja" became "Cambodia" as documented in 1524.[65] | This is a repetition of 真臘 in Item 7. Kambuja (甘波牙 or 柬埔寨) was the name of the Kingdom of Chenla (真臘王國/真腊王国; Khmer) during the Khmer Empire (吳哥王朝/吴哥王朝or高棉帝國/高棉帝国; 802–1431).
In 802 A.D., Jayavarman II (a Khmer prince) declared himself king, uniting the warring Khmer princes of Chenla under the name "Kambuja", lit. "甘波牙". This marked the beginning of the Khmer Empire, which flourished for over 600 years.
Kambuja (today's Cambodia) was a country located in the southern portion of the Indochinese Peninsula in Southeast Asia, bordered by Thailand to the northwest, Laos to the north, Vietnam to the east, and the Gulf of Thailand (泰國灣/泰国湾) to the southwest. However, on the Southeast Asia-KWQ, 真臘 (the old name before Kambuja) is depicted as facing the Bay of Bengal (孟加拉灣/孟加拉湾). That is incorrect. It should be placed at the same location as Kambuja (甘波牙). Both Chinese characters denote the same country in different dynasties or eras. |

(*Continued*)

Item #	Name in Chinese	Pinyin	Etymology	History, Geography and My Comments
				On the Southeast Asia-Mercator (1569), the Kingdom of Kambuja (甘波牙) is depicted as "Camboya" near the geographical location of today's Cambodia; and on the Southeast Asia-Ortelius (1570; #1), it is depicted as "Caboia", also near the geographical location of today's Cambodia. On the Southeast Asia-Ortelius (1570; #2), there is "Camboia".
● 14	蘭白道	Lan Bai Dao	Lanbaidao	The place cannot be identified on a modern map.
● 15	盤盤	Pan Pan	Kingdom of Panpan or Pan Pan on the east coast of the Malay Peninsula.	Pan Pan (盤盤/盘盘) is a lost small Hindu Kingdom believed to have existed around the third to seventh century.[66] It is believed to have been located on the east coast of the Malay Peninsula. During the period of 424 to 453, Pan Pan sent its first mission to the Chinese Liu Song Dynasty (劉宋/刘宋; 420–479).[67] Pan Pan also sent tribute to the Chinese Liang Dynasty (梁朝; 502–557), the Chen Dynasty (陳朝/陈朝; 557–589) and the Tang Dynasty (唐朝; 618 to 907, with an interregnum between 690 and 705). The Southeast Asia-KWQ depicts this very old and outdated historical information on it.
● 16	三佛齊	San Fo Qi	Srivijaya (室利佛逝) in Sanskrit was transliterated into "Samboja Kingdom (三佛齊王國/三佛齐王国)" by the historical records in the Song Dynasty.	Srivijaya or Sanfoqi (三佛齊/三佛齐, 佛逝、舊港/旧港or old name 室利佛逝; c. 671–1470) was a Buddhist thalassocratic empire based on Sumatra Island (蘇門答臘島/苏门答腊岛; in today's Indonesia), which influenced much of Southeast Asia, see Fig. A4.4. It originated from Palembang (巨港) in southeastern Sumatra (蘇門答臘島/苏门答腊岛).

In 1397 (the thirtieth year of Hongwu), the Majapahit Kingdom (滿者伯夷王國/满者伯夷王国) in East Java (東爪哇/东爪哇) destroyed the Old Srivijaya Dynasty, and the prince of Srivijaya fled to the Malay Peninsula to establish the Malacca Dynasty.[68] At that time, more than 1,000 Chinese lived in the Old Harbour (舊港/旧港; lit. "Old Port"; the then capital of Srivijaya; also known as Palembang 巨港). They supported Liang Daoming (梁道明), a local Chinese from Guangdong Province (廣東省/广东省), as king and established a New Srivijaya Dynasty, which was destroyed later by the Majapahit Empire (1293–1527) in 1470.[69]

In 1407 (the fifth year of Yongle), when Admiral Zheng He (鄭和/郑和) went on his second voyage to the Western Ocean, he returned from Malacca and passed through the old port of Palembang where his fleet was attacked by the pirate Chen Zuyi (陳祖義/陈祖义). The overseas Chinese Shi Jinqing (施進卿/施进卿; born year unknown–1423) in Srivijaya asked Zheng He for help. Zheng He's military defeated Chen Zuyi and took him back to the capital in China to be executed. In the same year, Shi Jinqing sent his son-in-law to Beijing to pay tribute, and the Yongle Emperor ordered Shi Jinqing to be the Envoy of the Palembang Pacification Commission (舊港宣慰使/旧港宣慰使) to rule the place.[70] Peace was finally restored to the Strait of Malacca and the whole region acknowledged Ming supremacy in return for diplomatic recognition, military protection, and trading rights. The Ming emperor also took the opportunity to announce to overseas nations that the Ming Dynasty had replaced the Mongolian Yuan as the ruler of China, and to promote the prosperity of the new dynasty.

(Continued)

Item #	Name in Chinese	Pinyin	Etymology	History, Geography and My Comments
				 Fig. A4.4. The maximum extent of Srivijaya around the eighth to the eleventh century with a series of Srivijayan expeditions and conquests (*Source*: Gunawan Kartapranatar, under CC BY-SA 3.0, https://commons.wikimedia.org/wiki/File:Srivijaya_Empire.svg).[71] On the Southeast Asia-KWQ, 三佛齊 (Samboja Kingdom or Srivijaya) is depicted on the Malay Peninsula (馬來半島/马来半岛), instead of on Sumatra Island (蘇門答臘島/苏门答腊岛), it seems to indicate that Srivijaya (三佛齊/三佛齐) was the dynasty established after 1397 by the prince of Srivijaya (三佛齊/三佛齐), who fled to the Malay Peninsula. But this depiction is incorrect.
@3	即古干陀利今爲舊港宣慰司	Ji Gu Gan Tuo Li Jin Wei Jiu Gang Xuan Wei Si		My translation and comments (between square brackets) are as follows: The place [Srivijaya; 三佛齊/三佛齐 or old name 室利佛逝] was the ancient Gantouli [干陀利]; now it is where the Old Port [舊港] Xuanweisi [Pacification Commission or Proclamation and Consolation Department] is established.

			Palembang is called Kū-káng (舊港/旧港) in Hokkien, meaning "Old Port", 港, lit. "harbor or port"; 巨 is also read "kū" in certain dominant dialects of Hokkien, and was thus borrowed to use in place of 舊 (old). Hence, in modern Chinese, Palembang is also written as 巨港, lit. "giant port". The establishment of the Old Port Pacification Commission was one of the few imperial courts in Chinese history that directly conferred titles on Han Muslims, and it is also one of the few overseas "enclaves" in Chinese history.[72]
@4	舊港地扼諸蕃之會商舶合湊富饒其民沿海架筏蓋屋而居覆以椰葉移則起椿而行其土沃倍於他壤有尼白樹酒比椰酒更佳傍國如古城大泥等皆有之	Jiu Gang Di E Zhu Fan Zhi Hui Shang Bo He Cou Fu Rao Qi Min Yan Hai Jia Fa Gai Wu Er Ju Fu Yi Ye Ye Yi Ze Qi Chun Er Xing Qi Tu Wo Bei Yu Ta Rang You Ni Bai Shu Jiu Bi Ye Jiu Geng Jia Bang Guo Ru Gu Cheng Da Ni Deng Jie You Zhi	My translation and comments (between square brackets) are as follows: The Old Port [舊港/旧港] is an important gathering place for the different foreign tribes, merchants, and ships, and it is an affluent city. People set up rafts along the coast and build houses by covering their rafts with coconut leaves. They move around by pulling out the wooden poles [attached to their rafts]. Their soil is twice as fertile as other soils, and there is Nibaishu wine [made from the Nibai trees], which is better than coconut wine. Other countries like the ancient city state Dani [大泥], etc., all produce this kind of wine. In the seventh century, the Old Port (舊港/旧港) was the birthplace of the Srivijaya Kingdom; from 1407 to 1440 of the Ming Dynasty, it was also where the Old Port Pacification Commission (one of the Ming administrative agencies) was located. The Old Port was an important supply station for Zheng He's voyages to the Western Ocean. As many Arab businessmen and immigrants came to the Indonesian archipelago, Islam began to spread. Shi Jinqing (施進卿/施进卿), the leader of the Chinese armed force in the Old Port, was a Muslim. Under the rule of Yongle Emperor, the national power of the Ming Dynasty reached its height.

(Continued)

Item #	Name in Chinese	Pinyin	Etymology	History, Geography and My Comments
● 17	丙東	Bing Dong	Bingdong	The place cannot be identified on a modern map.
18	彭亨	Peng Heng	Pahang;[73] the Khmer word for tin is Pahang; the sites of the tin mines were on the Malay Peninsula within the sphere of influence of Khmer civilization; there is a hypothesis that the place was named after the Khmer term for the mineral.[74] Arabs and Europeans, on the other hand, transliterated Pahang to *Pam, Pan, Paam, Paon, Phaan, Phang, Paham, Pahan, Pahaun, Phaung, Phahangh*,[75] somewhat different from the Chinese transliterations of "彭坑", "彭杭", or "彭亨".[76] The Southeast Asia-KWQ uses Chinese transliterated "彭亨", revealing that the map is of Chinese origin.	Pahang (彭亨) officially Pahang Darul Makmur, lit. "The Abode of Tranquillity") is a sultanate and a federal state of Malaysia. The book *Xingcha Shenglan* （星槎勝覽/星槎胜览） or *The Overall Survey of the Star Raft* records it as "彭坑". In *Zheng He Hang Hai Tu* （鄭和航海圖/郑和航海图） or *Zheng He's Navigation Map/Zheng He Map*, it is written as "彭杭", see Fig. A4.5; and in *Dong Xi Yang Kao* （東西洋考/东西洋考） or *East and West Foreign Studies*, it is written as "彭亨". In 1378 (the eleventh year of Hongwu), the Pahang King Maharaja Darao sent envoys to the Ming court to pay tribute.[77] In 1412 (the tenth year of Yongle), Zheng He went to Pahang as an envoy. In 1414 (the twelfth year of Yongle), the king of Pahang also sent envoys to the Ming court to pay tribute`. Pahang was Srivijaya's vassal state. In 1528, Pahang joined forces with the Sultan of Malacca (*Malay*: Melaka), Alauddin Riayat Shah II, who established himself in Johor to expel the Portuguese from the Malay Peninsula. Two attempts were made in 1547 at Muar and in 1551 at Portuguese Malacca.[78] On the Southeast Asia-Mercator and the Southeast Asia-Ortelius (1570; #2), there is *Pam*.

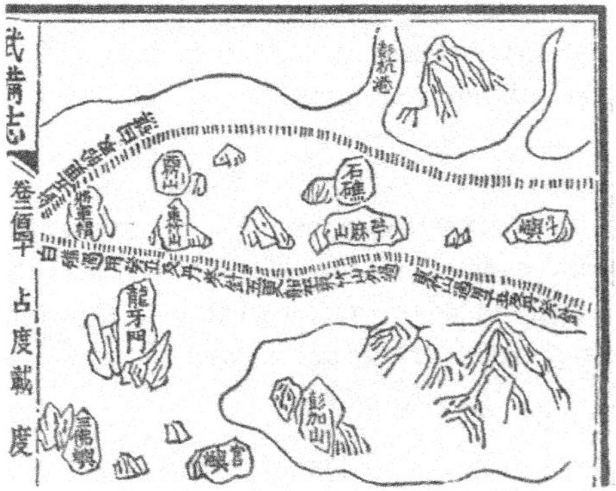

Fig. A4.5.　The seventeenth-century Mao Kun map (茅坤圖/茅坤图) based on the early fifteenth-century navigation maps of Zheng He is showing Pahang Harbour (彭杭港) and the Pahang River estuary (public domain).[79]

• 19	大泥	Da Ni	Da Ni; its old Chinese name was Boni (浡泥).	According to *Ming Shi* (明史) or *History of Ming*, in 1405 (the third year of Yongle), the king of Boni (浡泥) sent envoys to China to pay tribute, and the Yongle Emperor of the Ming Dynasty sent an official there to confer him the title of king.[80] The kingdom ruled fourteen states and was located to the west of the Old Port (舊港/旧港); it could be reached in forty days from Champa (占城; Item 12) by ship in the Ming Dynasty. At first, the kingdom was a vassal state to Java (爪哇), then to Siam (暹羅/暹罗). Later it was renamed to Da Ni (大泥)[81] from Boni. Boni was a medieval state on the Island of Borneo, sometimes considered to be the predecessor of modern Brunei (汶萊; the name was derived from *Baru nah*, lit. "that's it!" or "there") situated on the northern coast of the Island of Borneo (婆羅洲/婆罗洲). The name "Borneo" developed after European contact with the Brunei Kingdom in the sixteenth century; it may originally derive from the Sanskrit word *váruṇa*, meaning either "water" or "Varuna", the Hindu god of rain.[82] Many Chinese people lived on that island.

(*Continued*)

Item #	Name in Chinese	Pinyin	Etymology	History, Geography and My Comments
				However, on the Southeast Asia-KWQ, Da Ni (大泥) is erroneously depicted on the Malay Peninsula.
@5	大泥出極大之鳥名為厄蟇有翅不能飛其足如馬行最速馬不能及羽可為盔纓膽亦厚大可為杯孛露國尤多	Da Ni Chu Ji Da Zhi Niao Ming Wei E Ma You Chi Bu Neng Fei Qi Zu Ru Ma Xing Zui Su Ma Bu Neng Ji Yu Ke Wei Kui Ying Dan Yi Hou Da Ke Wei Bei Bei Lu Guo You Duo	My translation and comments (between square brackets) are as follows: Da Ni has a very large bird, called ema [鶴駝 or 厄蟇 (rhea or cassowary)]; "emu (厄蟇)" is a species of flightless bird endemic to Australia, where it is the tallest native bird; "rhea" is also known as South American ostrich; and "cassowary" is a genus of large flightless birds native to the Australo-Papuan region]. It has wings but cannot fly. Its feet are like a horse, and when it moves at the fastest speed, even the horses cannot catch up with it. Its feathers can be used as helmet tassels, and its gallbladders are also thick and large, which can be used as cups. There are many birds of this kind in the country of Beilu [孛露; Peru in South America]. The Suri, whose scientific name is Rhea pennata,[83] is the largest bird in South America, also known as the ostrich, or Andean ostrich. The bird is non-flying and can reach a height of 5 feet and weigh up to 66 pounds; its beak and head are grey, with its back and neck darker and the ends of its feathers white.[84] The Suri only inhabit South America and thrive in different environments including adverse territories such as Peruvian punas.

				This annotation reveals that the cartographer of the Southeast Asia-KWQ had knowledge about South America. In my previous book,[85] I have shown that Zheng He's mariners explored the Americas during their sixth voyage in the 1420s to the Western Ocean: the date was more than 100 years before the Spanish explorer Francisco Pizarro (1478–1541) founded the first Spanish settlement — San Miguel de Piura in Peru — in July 1532.[86]
20	馬大邦	Ma Da Bang	Martaban	馬大邦/马大邦 (Martaban) is a repetition of 馬兒大莽 in Item 6. However, this Martaban (馬大邦/马大邦) is depicted correctly as a port city. On the Southeast Asia-Mercator (1569), there is "Martaban" on the west bank of the Thanlwin River (also known as Salween River 薩爾溫江/萨尔温江 in Myanmar and Nu Jiang 怒江 or Nu River in China).
● 21	藥兒	Yao Er	Johor; the area was first known to produce gems near the Johor River; Arabic traders referred to it as "Jauhar",[87] a word borrowed from the Persian "gauhar", which also means "precious stone" or "jewel".[88] For easier pronunciation, the local people call it Johor.[89]	Johor, also spelled as Johore (藥兒/药儿 or 柔佛)[90] is a state of Malaysia in the south of the Malay Peninsula. During the Yongle period, when Zheng He travelled across the Western Ocean, there was no place called Johor (柔佛). However, it was said that Zheng He might have passed by the Dongxi Zhu Mountain (東西竺山/东西竺山). The mountain is in today's Johor. Hence, it is suspected that Zheng He passed by Johor.[91] Johor was part of the Malaccan Sultanate before the Portuguese conquered Malacca's capital in 1511.[92] The Johor Sultanate (1528–1855) was founded later by Ala'udin Ri'ayat Shah II.[93]

(*Continued*)

Item #	Name in Chinese	Pinyin	Etymology	History, Geography and My Comments
22	滿剌加/古哥羅富沙	Man La Jia/Gu Ge Luo Fu Sha	Malacca/the ancient Geluofusha; the name "Malacca" originated from a popular legend surrounding the founding of the sultanate; Malacca was a fishing village before the arrival of the first Sultan.	Malacca (滿剌加/满剌加) is a state in Malaysia located in the southern region of the Malay Peninsula, next to the Strait of Malacca. Its capital is Malacca City which has been listed as a UNESCO World Heritage Site since 7 July 2008.[94] The Malacca Sultanate (1400–1511) was a Malay sultanate based in the modern-day state of Malacca, Malaysia. In *Ming Shi* (明史) or *History of Ming*, the 滿剌加/满剌加 (pronounced as "Manlajia") it records was the modern-day "Malacca"[95]: 滿剌加，在占城南。順風八日至龍牙門，又西行二　日即至。或云即古頓遜，唐哥羅富沙。 My translation and comments (between square brackets) are as follows: Manlajia [滿剌加/满剌加] is in the south of Champa [占城]. It takes eight days from there [Champa] in the downwind [by ship] to get to Longyamen [龍牙門/龙牙门, lit. "Dragon Tooth Gate"], and then go west for two more days to reach the place which was named Dunsun [頓遜/顿逊] in ancient days, or Geluofusha [哥羅富沙/哥罗富沙] in the Tang Dynasty. Longyamen (lit. "Dragon Tooth Gate") is a place which sounds so handsome. What it really means is a "gate" sandwiched by Singapore. In contemporary nautical knowledge, the "gate" is the waterway sandwiched between the two sides. However, there are many small islands in this area, and it is said that the "Dragon Tooth" was already destroyed, hence, difficult to determine its location now.

In 1403 (the first year of Yongle), the first official Chinese trade envoy led by Admiral Yin Qing (尹慶/尹庆) arrived in Malacca.[96] In 1405 (the third year of Yongle), Malacca sent envoys to pay tribute,[97] and the Ming Dynasty officially designated Malacca as a country. In 1408 (the sixth year of Yongle), Zheng He visited Malacca and accepted its tributes.[98]

Staring from the early fifteenth century, Malacca officially submitted to Ming China as a protectorate. This relationship enabled Malacca to develop into a major trade settlement on the trade route between China and India, Middle East, Africa, and Europe.[99]

According to *Zheng He Hang Hai Tu* （鄭和航海圖/郑和航海图） or *Zheng He's Navigation Charts/Zheng He Map*, several temporary trading posts were established by Zheng He's navy during their visit to the Western countries including one in Malacca (now Malacca, Malaysia), Sumatra (near Lokshomawe in northern Sumatra, Indonesia), Calicut, and Hormuz. These trading posts were called官廠/官厂 (Official Operational Headquarters): each headquarters was built like a city with double gates (called *Weng Cheng;* 瓮城 or Urn fort)[100] and contained warehouses. These *Weng Chengs* provided operational headquarters to support Zheng He's operations during his fleets' voyages to the Western Ocean.

In April 1511, Alfonso de Albuquerque (c. 1453–1515; a Portuguese general, admiral, and statesman)[101] led an expedition from Goa to Malacca. His force conquered the city on 24 August 1511. After seizing the city Alfonso de Albuquerque spared the Indian, Chinese, and Burmese inhabitants but had the Muslim inhabitants massacred or sold into slavery.[102]

Malacca is depicted on the four European maps discussed in this chapter.

(*Continued*)

Item #	Name in Chinese	Pinyin	Etymology	History, Geography and My Comments
@6	滿剌加迪常有飛龍繞樹龍身不過四五尺人常射之	Man La Jia Di Chang You Fei Long Rao Shu Long Shen Bu Guo Si Wu Chi Ren Chang She Zhi		My translation is as follows: In Malacca, there are often flying dragons circling around the trees. The body of the dragon is no more than four or five feet long, and people often shoot at it. These "flying dragons" refer to snakes. Flying snakes like Chrysopelea paradisi, the paradise tree snake, normally live in the trees of South and Southeast Asia. There, they move along tree branches and, sometimes, get to the ground or another tree.[103]
23	吕宋	Lu Song	Luzon Island Group; the name "Luzon" is thought to derive from a Tagalog word *lusong*, referring to a particular kind of large wooden mortar used in dehusking rice.[104]	In ancient times, the largest and most populous island in the Luzon Island Group of the Philippines was the Luzon Island (吕宋; 16°N, 121°E). On the island, there was a small country called Luzon. From the Song and Yuan dynasties, Chinese merchant ships have often come here for trade. Both the book *Dong Xi Yang Kao* (東西洋考/东西洋考)[105] or *On the Countries in the Eastern and Western Ocean,* and *Ming Shi* (明史) or *History of Ming* have special records on Luzon.[106] From 1372（the fifth year of Hongwu）to 1410 (the eighth year of Yongle), Luzon sent envoys to Ming China twice. In 1405 (the third year of Yongle), Admiral Zheng He led Treasure Fleet to Luzon and the Yongle Emperor appointed an overseas Chinese as the Governor of Luzon;[107] his rule lasted for 20 years (1405–1424).

The Portuguese were the first European explorers who recorded it in their charts as *Luçonia* or *Luçon*. American diplomat Edmund Roberts (1784–1836) visited Luzon in the early nineteenth century and wrote that Luzon was "discovered" in 1521.[108] A permanent Spanish settlement was finally established in 1565. From 1571 to 1898, Luzon Island was occupied by Spain, and it was called Little Luzon (呂宋/呂宋); the Great Luzon referred to the entire Philippines ruled by Spain.

On the Southeast Asia-KWQ, Luzon (呂宋) is depicted at c. 14.8°N (KWQ latitude → modern latitude), 110.5°E, in comparison with Luzon Island (呂宋; 16°N, 121°E) on a modern map.

On the Southeast Asia-Mercator (1569), the name "Palahan Island" (c. 12.5°N, c. 135°E) is depicted roughly where the Luzon Island Group should be. However, Palahan is only a very small islet located at (c. 12.6°N., c. 125'E)[109] in the Luzon Island Group on a modern map, not representing the entire Luzon Island Group.

On the Southeast Asia-Ortelius (1570; #1), the name "Paloban" at c. 12.5°N, c. 129.5°E is depicted. On the Southeast Asia-Plancius (1594), there is an island depicted as Luconia Insula (c. 17°N, c. 129.5°E); just like in old Latin, Italian, and Portuguese maps, the island is often called "Luçonia" or "Luconia".[110] On the Southeast Asia-Ortelius (1570; #2), "Paloban" is also depicted at c. 12°N, c. 137°E.

(Continued)

Item #	Name in Chinese	Pinyin	Etymology	History, Geography and My Comments
24	瑪泥兒訝	Ma Ni Er Ya	*Maynilà* is the Filipino name for the city; it comes from the phrase *may-nilà*, meaning "where indigo is found".[111]	Manila (瑪泥兒訝/玛泥儿讶) is the modern-day City of Manila in the Philippines. By 1258, a Tagalog-fortified polity called Maynila existed on the site of today's Manila. In 1571, after the defeat of the polity's last indigenous Rajah Sulayman in the Battle of Bangkusay,[112] Spanish conquistador Miguel López de Legazpi (1502–1572) began constructing the walled fortification on the ruins of an older settlement from whose name the Spanish-and-English name Manila derives. On the Southeast Asia-Plancius (1594), Manila is depicted as "manilha".
• 25	馬大音	Ma Da Yin	Matan;[113] according to ancient record, the original name of the island was Opong. As the islanders turned to piracy, it was then referred to as Mangatang ("to wait in ambush"), which evolved into Matan, then Mactan (part of the present-day Cebu province).	Matan or Mactan (馬大音/马大音) was in Visayas (one of the three principal geographical divisions of the Philippines, along with Luzon and Mindanao; it is in the central part of the archipelago). At the Battle of Mactan on 27 April 1521, ruler Lapu-Lapu of Mactan and his soldiers defeated and killed Ferdinand Magellan who attempted to circumnavigate the world,[114] and delayed Spanish occupation of the islands by over 40 years.
26	卧山島	Wo Shan Dao	Gosas[115]	On the Southeast Asia-KWQ, the island is off the northeastern coast of today's Luzon Island, Philippines, but the island cannot be identified on a modern map. On the Southeast Asia-Ortelius (1570; #2), there is *Gosas*.
27	亞來沙	Ya Lai Sha	Arrecifes[116]	The island cannot be identified on a modern map. On the Southeast Asia-Plancius (1594), 亞來沙 is depicted as "Arecifes", not "Arrecifes". On the Southeast Asia-Ortelius (1570; #2), there is "Arizifes".
• 28	角島	Jiao Dao	*Corno, isola del.*[117]	On the Southeast Asia-KWQ, *Corno, isola del* cannot be identified on a modern map.

29	茗答鬧	Ming Da Nao	Mindanao; the name is a Spanish corruption of the name of the Maguindanao people in southwestern Mindanao during the Spanish colonial period.[118]	Mindanao (茗答鬧/茗答闹 or 棉蘭老/棉兰老) is the second-largest island in the Philippines, after Luzon Island.[119] In 1521 Venetian scholar and explorer Antonio Pigafetta (c. 1491–c. 1531) described reaching "Maingdano" during his sailing with Magellan on the first circumnavigation of the globe.[120] In 1543, Ruy López de Villalobos (c. 1500–1546) was the first Spaniard to reach Mindanao.[121] "Mindanao" is depicted on the four European maps. On the Southeast Asia-Plancius (1594), it is depicted as Mindana".
• 30	巴那馬	Ba Na Ma	Palawan[122]	Palawan (巴那馬) is an archipelagic province of the Philippines, see Fig. A4.6. The Spanish arrived in the late fifteenth century.
31	非利皮那	Fei Li Pi Na	Felipinas	The Philippines is divided into three major island groups: Luzon, the Visayas, and Mindanao. Luzon and Mindanao are both named after the largest island in their respective groups, while the Visayas (also referred to as the Visayan Islands) is an archipelago.

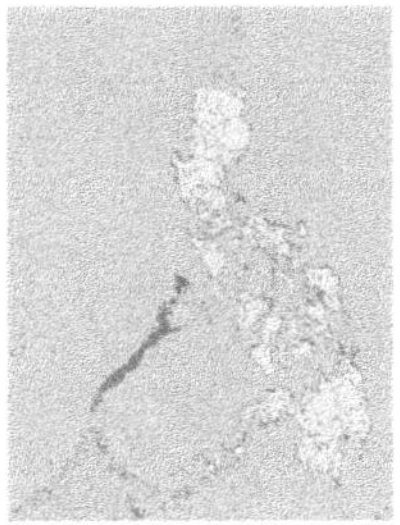

Fig. A4.6 Location of the Palawan archipelago in the Philippines (*Source*: Milenioscuro, under CC BY-SA 4.0, https://commons.wikimedia.org/wiki/File:Palawan_in_Philippines.svg)[123]

(*Continued*)

Item #	Name in Chinese	Pinyin	Etymology	History, Geography and My Comments
				During his 1542 expedition, Spanish explorer Ruy López de Villalobos (c. 1500–1546)[124] named the islands of Leyte (the largest of the six provinces of Eastern Visayas) and Samar (薩馬島/萨马岛; located in the eastern Visayas) as "Felipinas" after King Philip II of Castile (then Prince of Asturias). This name was later extended to the entire archipelago.[125] On the Southeast Asia-KWQ, see Fig. A4.1, the relative positions of Filipinas (非利皮那; Item 31) and Mindanao (茗答鬧/茗答闹 or 棉蘭老/棉兰老; Item 29) are not accurately represented as those on two modern maps shown in Figs. A4.7(a) and A4.7(b). 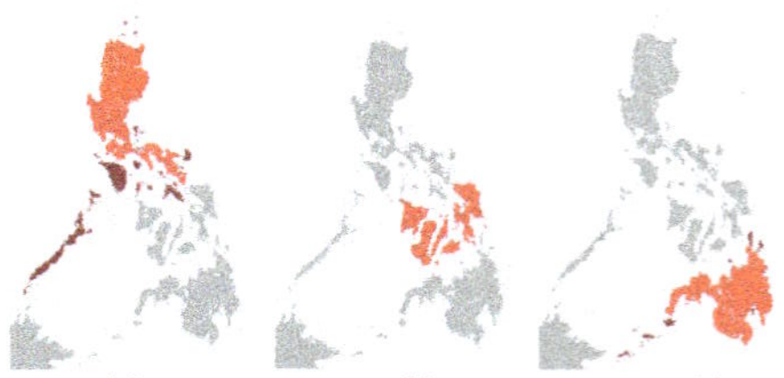 (a) (b) (c) **Fig. A4.7.** (a) Luzon Island (mainland Luzon, Luzon proper) in bright red; associated islands in the Luzon group of islands (Luzon Island group) are in maroon (*Source*: JL 09, under CC BY-SA 3.0, https://commons.wikimedia.org/wiki/File:Luzon_Island_Red.png).[126] (b) The Visaya Islands or Visayas in red (*Source*: JL 09, under CC BY-SA 3.0, https://commons.wikimedia.org/wiki/File:Visayas_Red.png);[127] the rightmost island in (b) is Samar Island and the island to its immediate left is Leyte Island. (c) Mindanao Island (mainland Mindanao, Mindanao proper) in bright red; associated islands in the Mindanao group of islands (Mindanao Island group) are in maroon (*Source*: JL 09, under CC BY-SA 3.0, https://commons.wikimedia.org/wiki/File:Mindanao_Red.png).[128]

| 32 | 波尔匿何 | Bo Er Ni He | Island of Borneo; also known colloquially as Kalimantan Island (加里曼丹島/加里曼丹岛) in the old days. The Native people of Borneo referred to their island as *Pulu K'lemantang*. After Portuguese explorer Jorge de Menezes (c. 1498–1537) made contact in 1526 with the natives,[129] that name became the modern-day Indonesian Borneo.[130] 波尔匿何 is the transliteration of "Borneo". | On the Southeast Asia-Plancius (1594), Felipinas is depicted as "Philippina", and the relative positions of Philippina and Mindanao are better presented then those on the Southeast Asia-KWQ. However, none of the other three European maps depicts Felipinas or Philippina.

Colloquially in Indonesia, the whole island of Borneo (波爾匿何島/波尔匿何島 or 婆羅洲/婆罗洲) is also called "Kalimantan", see Fig. A4.8, which constitutes only 73% of the island's area of the Island of Borneo and is the present-day Indonesian portion of the Island of Borneo. The Island of Borneo is the third largest island in the world. The non-Indonesian parts of Borneo are Brunei and East Malaysia. |

Fig. A4.8. Location of present-day Kalimantan (the Indonesian portion) on Borneo Island (*Source*: Sadalmelik, under CC BY-SA 3.0, https://commons.wikimedia.org/wiki/File:Sulawesi_Locator_Topography.png);[131] at present the island is politically divided among three countries: Malaysia (馬來西亞/马来西亚) and Brunei (汶萊/文莱) in the north, and Indonesia (印度尼西亞/印度尼西亚) to the south.

On the Southeast Asia-KWQ, the ancient Kingdom of Boni (浡泥國/浡泥国) occupied the entire Island of Borneo, but it is also erroneously depicted on the Malay Peninsula as Da Ni (大泥; Item 19), see Fig. A4.1.

(Continued)

Item #	Name in Chinese	Pinyin	Etymology	History, Geography and My Comments
				The earliest written record of navigation between China and Kalimantan appears in *Liang Shu* (梁書/梁书) or *Book of Liang*, in 520 A.D. (the first year of Emperor Wu of Liang Dynasty). In ancient Chinese books, it was then called Boni (淳泥), Poli, or Poluo, and later evolved to become Brunei (婆羅乃/婆罗乃) which is today the common name of Brunei (汶萊/文莱; situated on the northern coast of the island of Borneo). In 977, Chinese records began to use the term *Boni* (淳泥) to refer to Borneo (波爾匿何/波尔匿何). In 1225, Boni was also mentioned by the Chinese official Zhao Rugua (趙汝适/赵汝适).[132] By the end of the fourteenth century, the Kingdom of Boni on the Island of Borneo (婆羅洲/婆罗洲) became a vassal state of Majapahit (滿者伯夷/满者伯夷; 1293–1527; in today's Indonesia).[133] The ancient Kingdom of Boni (淳泥國/淳泥国) was the hub of maritime transportation between the East and the West at that time due to its important geographical location. The country showed its allegiance to the Ming Dynasty of China.[134] In 1405 (the third year of Yongle), Boni's king sent envoys to China; the Yongle Emperor also sent an official to Boni to confer him the title of king. The king was so pleased that he took his wife and relatives to China for a visit.[135] In 1408 (the sixth year of Yongle), they went to the capital to visit the Yongle Emperor,[136] but died there the same year. Before his death, he showed his passion to Ming, and he was buried in Nanjing.[137] Later the Kingdom of Boni was renamed to Da Ni (大泥; Item 19) from Boni. So, Boni was the old name of Da Ni (大泥; Item 19),[138] and Boni was situated on the entire Island of Borneo, see Fig. A4.1. We now know that the Southeast Asia-KWQ should not depict Da Ni (大泥; Item 19) on the Malay Peninsula.

			The Southeast Asia-Mercator (1569) depicts "Burne"; the Southeast Asia-Ortelius (1570; #1) depicts "Burneo"; the Southeast Asia-Ortelius (1570; #2) depicts *Borneo*; and the Southeast Asia-Plancius (1594) also depicts *Borneo*.
@7	即浡泥國地炎熱多風雨木柵為城其田鑄鐵為筒穿之於身有藥樹煎膏塗身兵刃不傷無筆札以刀刺貝多葉行之俗好事佛	Ji Bo Ni Guo Di Yan Re Duo Feng Yu Mu Zha Wei Cheng Qi Tian Zhu Tie Wei Tong Chuan Zhi Yu Shen You Yao Shu Jian Gao Tu Shen Bing Ren Bu Shang Wu Bi Zha Yi Dao Ci Bei Duo Ye Xing Zhi Su Hao Shi Fo	My translation and comments (between square brackets) are as follows: This is to say that the place [Island of Borneo] is the land of the Kingdom of Boni [浡泥國/浡泥国]. Its weather is hot, windy, and rainy, and its city is walled with wooden fences. People there cast iron in the field and make it into a tube to wear around their bodies. There are medicine trees, and when the decoction made from the tree is applied to the human body, even weapons cannot cause injuries. Although there is no pen, people carve cowry shells with a knife many times to write things on them. They have the custom of worshipping Buddha. This annotation clearly states that the kingdom on the Island of Borneo (波爾匿何/波尔匿何; Item 32) was the ancient Kingdom of Boni (浡泥國/浡泥国) or Da Ni (大泥; Item 19) in the Chinese records.
33	止男巴洛	Zhi Nan Ba Luo	Chinabalo[139]	The island cannot be identified on a modern map. On the Southeast Asia-Ortelius (1570; #2), there is "Chinabalo".
34	色力皮	Se Li Pi	Celebes; the name might be considered a Portuguese rendering of the native name "Sulawesi".	Celebes is an island in Indonesia, also known as Sulawesi Island (蘇拉威西島/苏拉威西岛). The Southeast Asia-Mercator (1569) depicts "Celebes" in the southeast of the Island of Borneo. On the Southeast Asia-Ortelius (1570; #2), there is *Celebres*.
35	食力百私	Shi Li Bai Si	Celebes; transliterated from Sulawesi (a native name)[140]	Celebes is an island in Indonesia, also known as Sulawesi. This is a repetition of 色力皮 in Item 34.

(*Continued*)

Item #	Name in Chinese	Pinyin	Etymology	History, Geography and My Comments
				The island's colonial history began when Portuguese sailors in search of gold set foot on Sulawesi in 1511 and named the island "Celebes" transliterated from the native name "Sulawesi".[141] The landmass of Sulawesi includes four peninsulas. On the Southeast Asia-KWQ, there are 色力皮 (Item 34) and 食力百私 (Item 35), both are transliterated as Celebes and each perhaps represents a different peninsula of the Sulawesi Island; the other two peninsulas are not depicted. **Fig. A4.9.** Sulawesi Island (*Source*: Sadalmelik, under CC BY-SA 4.0 International, 3.0 Unported, 2.5 Generic, 2.0 Generic and 1.0 Generic license, https://commons.wikimedia.org/wiki/File:Sulawesi_Locator_Topography.png).[142]
36	馬路古地方	Ma Lu Gu Di Fang	Maluku Islands or Moluccas; the etymology of the word *Maluku* is unclear and has been a matter of debate for many experts.[143] The first recorded word that can be identified with *Maluku* comes from an Old Javanese eulogy of 1365.[144]	The Maluku Islands or the Moluccas are an archipelago in the eastern part of Indonesia. The islands are located east of Sulawesi, west of New Guinea, and north and east of Timor. Here on the southeast Asia-KWQ, the Maluku region (馬路古地方/马路古地方) is denoted instead of Maluku Islands (馬路古群島/马路古群岛), specifically, perhaps because the archipelago covers a huge area, as shown in Fig. A4.10. The Southeast Asia-Ortelius (1570; #1) depicts *Molucce insule* and the Southeast Asia-Ortelius (1570; #2) depicts MOLVCOS.

				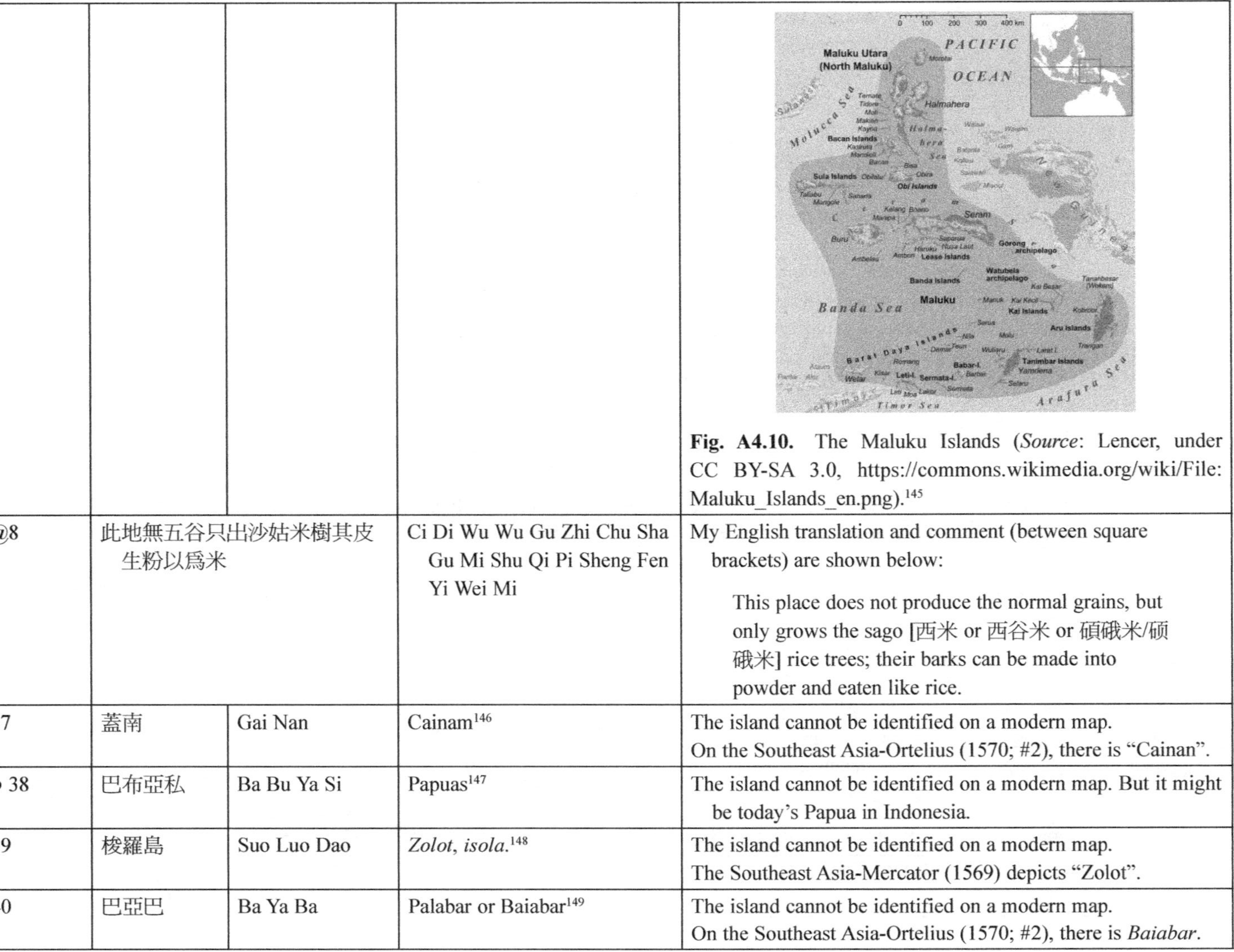Fig. A4.10. The Maluku Islands (*Source*: Lencer, under CC BY-SA 3.0, https://commons.wikimedia.org/wiki/File: Maluku_Islands_en.png).[145]
@8	此地無五谷只出沙姑米樹其皮生粉以爲米	Ci Di Wu Wu Gu Zhi Chu Sha Gu Mi Shu Qi Pi Sheng Fen Yi Wei Mi		My English translation and comment (between square brackets) are shown below: This place does not produce the normal grains, but only grows the sago [西米 or 西谷米 or 碩硪米/硕硪米] rice trees; their barks can be made into powder and eaten like rice.
37	蓋南	Gai Nan	Cainam[146]	The island cannot be identified on a modern map. On the Southeast Asia-Ortelius (1570; #2), there is "Cainan".
● 38	巴布亞私	Ba Bu Ya Si	Papuas[147]	The island cannot be identified on a modern map. But it might be today's Papua in Indonesia.
39	梭羅島	Suo Luo Dao	*Zolot, isola.*[148]	The island cannot be identified on a modern map. The Southeast Asia-Mercator (1569) depicts "Zolot".
40	巴亞巴	Ba Ya Ba	Palabar or Baiabar[149]	The island cannot be identified on a modern map. On the Southeast Asia-Ortelius (1570; #2), there is *Baiabar*.

(Continued)

Item #	Name in Chinese	Pinyin	Etymology	History, Geography and My Comments
@9	此處海島甚多船甚難行其地出檀香丁香金銀香安息香蘇木胡椒片腦	Ci Chu Hai Dao Shen Duo Chuan Shen Nan Xing Qi Di Chu Tan Xiang Ding Xiang Jin Ying Xiang An Xi Su Mu Hu Jiao Pian Nao	My translation and comments (between brackets) are as follows: There are many islands here, and it is very difficult for boats to sail in this area. The land produces sandalwood [檀香], clove [丁香], kemenyan [金銀香/金银香], benzoin [安息香], sappan wood [蘇木/苏木], black pepper [胡椒], and borneol [冰片 or 片腦/片脑].	
41	皮馬	Pi Ma	Biba or Bima;[150] an Old Javanese eulogy to a Javanese king of the Majapahit Empire — the fourteenth-century Nagarakretagama[151] — already mentioned Bima to be on Sumbawa.	Bima (皮馬/皮马) is a city on the eastern coast of the Sumbawa Island[152] in central Indonesia's province West Nusa Tenggara. It is the largest city on that island. On the Southeast Asia-Ortelius (1570; #2), there is *Biba*.
42	大爪哇/即訶陵曰闍婆	Da Zhao Wa/ Ji He Ling Yue She Po	*Jave la Grande*; other names are He Ling and She Po. The name *Jave la Grande* (大爪哇) was from Marco Polo's travel book.	According to Ma Huan's book (*Yingya Shenglan*, published in 1451), which is about the countries visited by the Chinese over the course of the Ming treasure voyages led by Zheng He in the early fifteenth century, the Chinese called this island Java "爪哇", which is pronounced as "Zhao Wa", and the island was called She Po (闍婆) or He Ling (訶陵/诃陵) in the past.[153] In the Chinese records, there has been no use of "大爪哇 (*Jave la Grande*)" or "小爪哇 (*Jave mine*)", but only "爪哇 (Zhao Wa)". The mention of "大爪哇 (*Jave la Grande*)" and "小爪哇 (*Jave mine*)" started from Marco Polo's travel book.[154] However, they are not quite the same as those depicted on the Southeast Asia-KWQ.

				Today's Java Island (7.5°S, 110°E) in Indonesia lies between Sumatra to the west and Bali Island (巴厘岛/巴厘岛; a province of Indonesia and the westernmost of the Lesser Sunda Islands) to the east, with Borneo lying to the north and Christmas Island (聖誕島/圣诞岛) to the south. 大爪哇 (*Jave la Grande*) at 9.4°S (KWQ latitude → modern latitude), 97.5°E on the Southeast Asia-KWQ fits the geographical location of today's Java Island. In 1378 (the eleventh year of Hongwu), She Po (闍婆) sent in tribute to the Ming Court.[155] The Southeast Asia-Mercator (1569) depicts "Iaua maior"; the Southeast Asia-Ortelius (1570; #1) depicts "Iaua maior"; and the Southeast Asia-Plancius (1594) depicts "Iaua". They correspond to today's Java Island. On the Southeast Asia-Ortelius (1570; #2), there is "IAVA MAIOR".
@ 10	爪哇元兵曾到擒其王其地通商船極多其富饒金銀珠寶琿壖瑪瑙犀角象牙木香等俱有	Zhao Wa Yuan Bing Zeng Dao Qin Qi Wang Qi Di Tong Shang Chuan Ji Duo Qi Fu Rao Jin Yin Zhu Bao Hun Qu Ma Nao Xi Jiao Xiang Ya Mu Xiang Deng Ju You		My translation is as follows: The Yuan soldiers have been to Java and captured the king. The place is a trading hub and many merchant ships come to this area. The area is rich in gold, silver, jewelleries, splendid jade, agate, rhinoceros' horn, ivory, costus root, etc. The Yuan Dynasty under Kublai Khan (忽必烈汗) attempted in 1293 to invade Java, with 20,000[156] to 30,000 soldiers.[157] This was intended as a punitive expedition against Kertanegara (the last and most important ruler of the Singhasari Kingdom of Java; reigned from 1268 to 1292) who had refused to pay tribute to the Yuan and maimed one of their envoys. In the end, the invasion ended with Yuan failure and victory for the new state, Majapahit.
• 43	把拏路曷	Ba Na Lu He	Pekalongan	The place may be today's Pekalongan (also transliterated as 北加浪岸) which is a harbour city in Java, Indonesia.

(*Continued*)

Item #	Name in Chinese	Pinyin	Etymology	History, Geography and My Comments
				According to a Chinese twelfth-century geographical treatise — *Ling Wai Dai Da* (嶺外代答/岭外代答) or *Representative Answers from the Region beyond the Mountains* — written in the Southern Song Dynasty by Zhou Qufei (周去非), a Chinese merchant ship set out from Guangzhou in November, sailed day and night, could arrive at Pukalong (莆加龍/莆加龙; the Chinese name for Pekalongan) about a month later.[158]
● 44	雅罷牙	Ya Ba Ya	Surabaia or Surabaya;[159] "Surabaya" comes from the Javanese "sura ing baya", meaning "bravely facing danger"; it connects with a legendary battle between a "shark" and a "crocodile" to establish which was the strongest animal.[160]	Surabaya (雅罷牙/雅罢牙; also known as 泗水 in Chinese), the second largest city in Indonesia, is the capital of East Java. It was officially established in 1293; the date was taken from the glory of Raden Wijaya, the first king of Majapahit who fought against China's troops from the Yuan Dynasty (1271–1368). Ma Huan visited Java during Zheng He's fourth expedition in 1413. He documented the visit of Zheng He's treasure ships in his book *Yingya Shenglan* published in 1451 this way (translation by J.V.G. Mills): "After travelling south for more than 20 *li*, the ship reached *Sulumayi*, whose foreign name is Surabaya. At the estuary, the outflowing water is fresh".[161] Surabaya is located on the edge of the northern coast of Java Island and deals with the Madura Strait (馬都拉海峽/马都拉海峡) and Java Sea, see Fig. A4.11. It is one of the earliest port cities in Southeast Asia. But on the Southeast Asia-KWQ, it is incorrectly depicted at the northwest coast of Java Island.

				The first known European to land on the island of Java was Dutch explorer Cornelis de Houtman (1565–1599).[162] He arrived at Banten, not Surabaya, in June 1596, during a voyage commissioned by the Dutch East India Company (Vereenigde Oostindische Compagnie or VOC) to lead an expedition to find a new route to the lucrative spice trade in the East Indies.[163] He returned to Holland in 1597. Fig. A4.11. Surabaya location in Java, Indonesia (*Source*: NordNordWest, under CC BY-SA 3.0, https://commons.wikimedia.org/wiki/File:Indonesia_Java_location_map.svg).[164]
45	板淡	Ban Dan	Banten, sometimes called Bantam; the name has several possible native origins.	Banten (板淡; 6.0°S, 106.2°E) is a port town near the western end of Java, Indonesia. The coordinates of Banten (板淡) on the Southeast-Asia-KWQ are c. 7.4°S (KWQ latitude → modern latitude), c. 95°E. The site where the city of Banten remains is known as Old Banten which was part of the Kingdom of Tarumanagara (塔魯馬納加拉/塔鲁马纳加拉; c. fifth century–669) during the fifth century.[165] As mentioned above, the first known European to land on the island of Java was Dutch explorer Cornelis de Houtman (1565–1599).[166] He arrived at Banten in June 1596 and returned to Holland in 1597. He died in 1599 during his second trip to the East. On the Southeast Asia-Ortelius (1570; #2), there is *Banta* on the western end of "IAVA MAIOR". You wonder: how did Ortelius know "Banta (板淡)" in 1570 long before Dutch explorer Cornelis de Houtman first arrived at Banten in June 1596 and returned to Holland in 1597?

(*Continued*)

Item #	Name in Chinese	Pinyin	Etymology	History, Geography and My Comments
● 46	左 "耳" 右 "兮"			The Chinese character cannot be identified in the Chinese dictionary.
47	地木島	Di Mu Dao	Timor: the name is a variant of *timur*, Malay for "east", because it lies at the eastern end of the Lesser Sunda Islands (小異他群島/小異他群岛), an archipelago in Indonesian archipelago.	Timor (地木島/地木岛) is an island at the southern end of Maritime Southeast Asia, in the north of the Timor Sea. The earliest historical record about Timor Island is the thirteenth-century Chinese *Zhu Fan Zhi* (諸蕃志/诸蕃志)[167] or *A Description of Barbarian Nations, Records of Foreign People* written by Zhao Rugua (趙汝适/赵汝适) in the Song Dynasty, where it is called *Ti-wu* and is noted for its sandalwood. Timor was in the trading networks of the ancient Javanese, Chinese and Indians in the fourteenth century as an exporter. Yet, the Portuguese settlement came as late as the end of the sixteenth century, and the Dutch, based in Kupang, in the mid-seventh century. Both the Southeast Asia-Mercator (1569) and the Southeast Asia-Plancius (1594) depict "Timor". On the Southeast Asia-Ortelius (1570; #2), there is also *Timor*. One wonders: how did Mercator and Ortelius know the existence of Timor in 1569 and 1570, long before the Portuguese settlement came as late as the end of the sixteenth century?
48	伯旦	Bo Dan	Petan[168]	The island cannot be identified on a modern map. Both the Southeast Asia-Mercator (1569) and the Southeast Asia-Ortelius (1570; #1) depict "Petan".

49	娑麻剌	Suo Ma La	Samar;[169] the name is of native origin.	Samar (娑麻剌 or 薩馬/萨马; 12°N, 125°E) is the third largest and seventh-most populous island in the Philippines. The island is shown in Fig. A4.7(b). Samar was the first island of the Philippines sighted by the Spanish expedition in 1521, led by Ferdinand Magellan (c. 1480–1521).[170] The island was transcribed as *Zamal* in the diary of Antonio Pigafetta (c. 1491–c. 1531), but Magellan did not land there.[171] The Spaniards later called the island Felipinas (Item 31), while the Portuguese called it Lequios. But on the Southeast Asia-KWQ, Samar (娑麻剌) is depicted as a place at c. 35°S (KWQ latitude → modern latitude), c. 118.5°E on the "Small Java Island (*Jave mine*; 小爪哇)" drawn in the Southern Hemisphere. We know that all the 7,641 islands in the Philippines are geographically positioned in the Northern Hemisphere of the Earth. Hence, it is incorrect to depict Samar on a "Small Java Island (小爪哇) in the Southern Hemisphere". The existence of a "Small Java Island (小爪哇) in the Southern Hemisphere" and inside the Gulf of Carpentaria is a huge mistake. We will see more of such mistakes for geographical items depicted on this "Small Java Island (小爪哇) in the Southern Hemisphere". Since there is no record of Chinese visits to any "Small Java Island (小爪哇)", the drawing of such an island in the Southern Hemisphere must be caused by Matteo Ricci's misinterpretation of Marco Polo's *Jave la Grande* (大爪哇) and *Jave mine* (小爪哇) written in his travel book.[172] Another incorrect analysis of Marco Polo's travel itinerary shows that the former may be the present-day Australia and the latter New Guinea.[173]

(*Continued*)

Item #	Name in Chinese	Pinyin	Etymology	History, Geography and My Comments
				The Southeast Asia-Mercator (1569) depicts "Samara" on the "Iaua minor" Island. The Southeast Asia-Ortelius (1570; #1) depicts the "Iaua minor" Island but does not fill in other geographical items on that island. The Southeast Asia-Plancius (1594) does not depict such a "Iaua minor" Island in the Gulf of Carpentaria. The Southeast Asia-Ortelius (1570; #1) does not depict many geographical items south of the equator, and "Iaua minor" Island is omitted.
50	覽比	Lan Bi	Lambri or Lamuri; Lambri was mentioned by the thirteenth–fourteenth century European travellers Marco Polo (c. 1254–1324) and Odoric of Pordenone (c. 1280–1331).[174] The name is of native origin and was known to the Arabs from the ninth century onward. It has various forms by Indians, Chinese, Javanese, Europeans, etc.	Lamuri or Lambri (覽比/览比; 5.6°N, 95.5°E) was a kingdom in northern Sumatra, Indonesia, from the Srivijaya period until the early sixteenth century. In Fig. A4.4, we can see the location of Lambri at the tip of the Sumatra Island, one of the Sunda Islands of western Indonesia, see Fig. A4.12. (a) (b) **Fig. A4.12.** (a) shows the Greater Sunda Islands (*Source*: Kikos, under CC BY-SA 3.0, https://commons.wikimedia.org/wiki/File:LocationGreaterSundaIslands.svg);[175] and (b) shows the Lesser Sunda Islands (*Source*: NordNordWest, under CC BY-SA 3.0, https://commons.wikimedia.org/wiki/File:LocationLesserSundaIslands.svg).[176] Chinese historical records indicate that the ancient Lambri was used as a staging post for traders waiting out the winter monsoon for favourable winds to take them westwards to Sri Lanka, India and the Arab world.

				Among the Chinese records, *Ming Shi* (明史) or *History of Ming* calls the kingdom *Nan-bo-li* (Nan Boli; 南渤利).[177] In 1412 (the tenth year of Yongle), the king of *Nan-bo-li* sent his envoy to accompany Sumatra's envoy to China to send in his tribute.[178] In the 1430s, when Admiral Zheng He's fleets sailed to the Western Ocean, Zheng He granted gifts to all the countries his fleets visited during their seventh voyage. *Nan-bo-li* was also included.[179] Hence, there is no reason for Chinese cartographers not to draw Lambri on the Sumatra Island. Again, it was Matteo Ricci who must have been influenced by Marco Polo (c. 1254–1324) to misinterpret Marco Polo's *Jave mine* (小爪哇) in his travel book, to draw this fanciful island — *Jave mine* — in the Gulf of Carpentaria, and to move the geographical location of Lambri from northern Sumatra in the Northern Hemisphere to the Southern Hemisphere. Lambri was mentioned by early European travelers Marco Polo and Odoric of Pordenone.[180] Marco Polo noted that the people were "idolators" when he passed through in the late thirteenth century.[181] The Southeast Asia-Mercator (1569) mistakenly depicts "Lamburi" on the "Iaua minor" Island.
51	把西蠻	Ba Xi Man	Basman[182]	The place cannot be identified on a modern map. The Southeast Asia-Mercator (1569) mistakenly depicts "Basman" on the "Iaua minor" Island.
52	番蘇尓	Fan Su Er	Barus, also known as Fansur, Pansur and possibly Barusai.[183] Arab sources used the name Fansur or Pansur.[184]	Barus or Fansur or Pansur was a well-known port town or kingdom on the western coast of Sumatra. It was also a regional trade center from around the seventh century or earlier until the seventeenth century.[185]

(*Continued*)

Item #	Name in Chinese	Pinyin	Etymology	History, Geography and My Comments
				The Greek geographer Claudius Ptolemy (c. 100– c. 170 A.D.) in his work *Geography* recorded the name Barusai or Barousai (Ancient Greek: Βαροῦσαι) as a group of five islands, which some scholars believe to refer to islands facing the Western Sumatran coast at Barus.[186] In the fourth-century Chinese records, the name "*Po-lu*" is found, suggested to mean Barus or the northern part of Sumatra. Around 900, a Muslim Persian explorer Ahmad ibn Rustah called Fansur "a well-known country in the Indies", and wrote about its jurisdiction[187] The Southeast Asia-KWQ depicts Fansur (番蘇尔) mistakenly on the Jave mine (小爪哇, Item 54) in the Southern Hemisphere. The Southeast Asia-Mercator (1569) also mistakenly depicts "Fansur" on the "Iaua minor" Island.
53	弗尔色	Fu Er Se	Feriech[188]	The place cannot be identified on a modern map. The Southeast Asia-Mercator (1569) mistakenly depicts "Feriech" on the "Iaua minor" Island.
54	小爪哇	Xiao Zhao Wa	*Jave mine;* lit. "Small Java".	On this fanciful island, the *Jave mine*, lit. "Small Java" (小爪哇) in the deep Southern Hemisphere, Matteo Ricci depicted five geographical items listed from Item 48 to 52 in this Appendix, which should in fact be in the Northern Hemisphere. Specifically, *Ming Shi* (明史) or *History of Ming* clearly states,[189] 南渤利，在蘇門答剌之西。順風三日夜可至。 My translation and comments (between square brackets) are as follows: *Nan-bo-li* [南渤利 or 覽比; Lambri or Lamuri] is to the west of Sumatra [蘇門答剌/苏门答剌] and can be reached from there [Sumatra] in three days and nights with favourable winds [by ship]. This statement clearly shows that Lambri or Lamuri was in the Northern Hemisphere just like Sumatra, and not too far

				away from it. Since Admiral Zheng He's fleets have visited this kingdom, and the above statement matches closely what we know on a modern map, there is no way to support that Lambri or Lamuri was in the Gulf of Carpentaria (see Fig. A4.13) as Matteo Ricci drew it. Both the Southeast Asia-Mercator (1569) and the Southeast Asia-Ortelius (1570; #1) depict "Iaua minor" Island, but the Southeast Asia-Plancius (1594) no longer depicts this fanciful island. 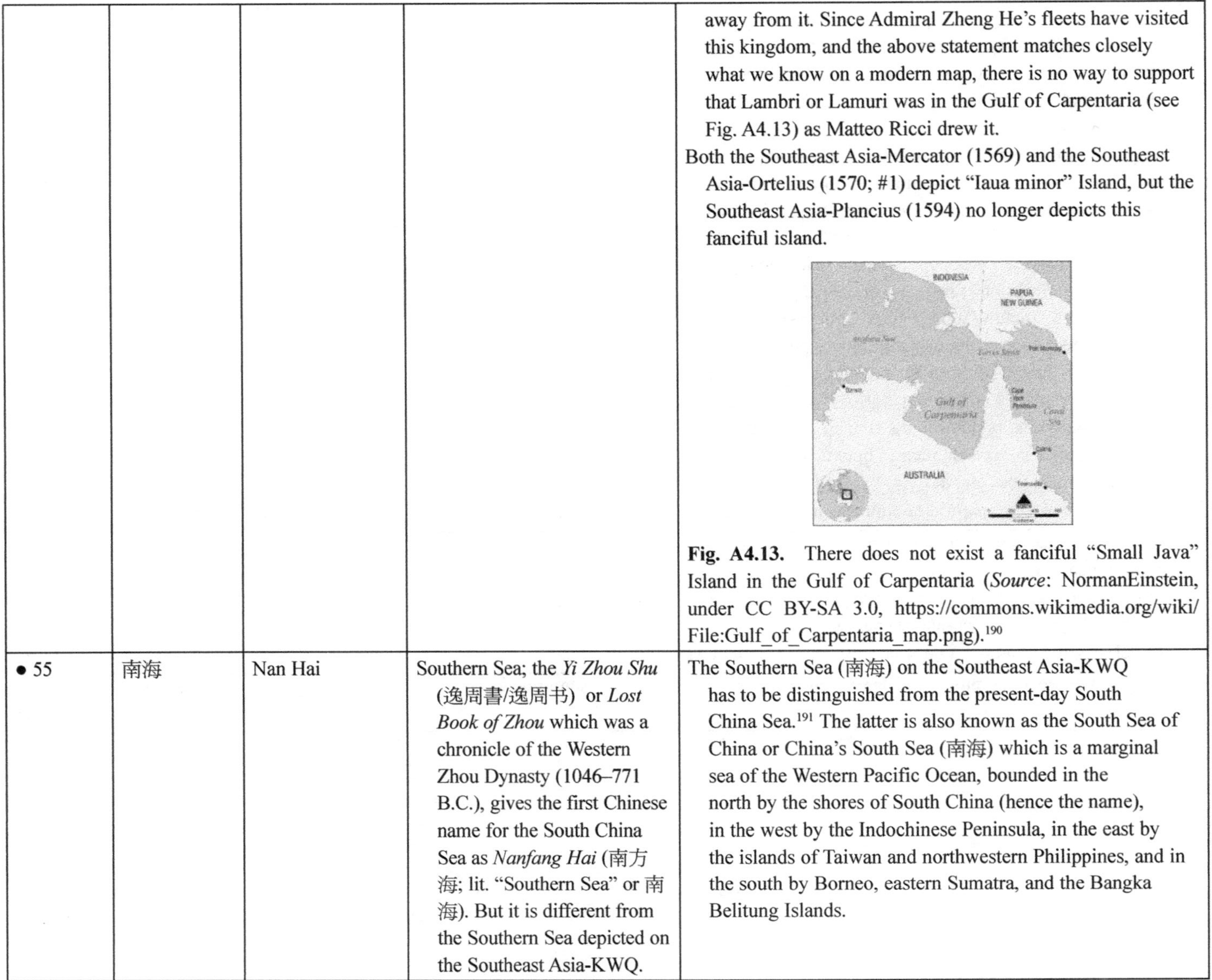 **Fig. A4.13.** There does not exist a fanciful "Small Java" Island in the Gulf of Carpentaria (*Source*: NormanEinstein, under CC BY-SA 3.0, https://commons.wikimedia.org/wiki/File:Gulf_of_Carpentaria_map.png).[190]
• 55	南海	Nan Hai	Southern Sea; the *Yi Zhou Shu* (逸周書/逸周书) or *Lost Book of Zhou* which was a chronicle of the Western Zhou Dynasty (1046–771 B.C.), gives the first Chinese name for the South China Sea as *Nanfang Hai* (南方海; lit. "Southern Sea" or 南海). But it is different from the Southern Sea depicted on the Southeast Asia-KWQ.	The Southern Sea (南海) on the Southeast Asia-KWQ has to be distinguished from the present-day South China Sea.[191] The latter is also known as the South Sea of China or China's South Sea (南海) which is a marginal sea of the Western Pacific Ocean, bounded in the north by the shores of South China (hence the name), in the west by the Indochinese Peninsula, in the east by the islands of Taiwan and northwestern Philippines, and in the south by Borneo, eastern Sumatra, and the Bangka Belitung Islands.

(*Continued*)

Item #	Name in Chinese	Pinyin	Etymology	History, Geography and My Comments
				But on the Southeast Asia-KWQ, the Southern Sea (南海) is depicted at where the present-day Gulf of Carpentaria is. Clearly, the Southern Sea (南海) on this map is not today's South China Sea, but an indication of the ocean far to the south of China and deep into the Southern Hemisphere.
56	蘇門荅臘	Su Men Da La	Kingdom of Sumatra	The Kingdom of Sumatra was called "Sumandra Kingdom" (須文達那國/须文达那国) in the Yuan Dynasty. It was located at the mouth of the Parsei River in today's Sumatra Island, where there is still a small village called Sumandra (須文達那/须文达那). The Arabic transliteration of Sumandra is "Sumatra", because Maghrebi traveler, explorer and scholar Ibn Battutah (1304–1368/1369)[192] visited this country and called it Sumatra. The king sent envoys and tribute to the Yuan Dynasty. The kingdom was renamed the Kingdom of Sumatra (蘇門荅臘國/苏门答腊国) during the Hongwu period in the Ming Dynasty,[193] but its territory did not cover the entire island of Sumatra yet. On the Southeast Asia-KWQ, it is depicted at the southern half of Sumatra Island and just south of the equator. During both the Hongwu and Yongle periods, the King of Sumatra sent envoys and paid tribute to China. Admiral Zheng He visited the Kingdom of Sumatra several times during the Yongle period.[194]

In 1415 (the thirteenth year of Yongle), during Admiral Zheng He's fourth voyage (1413–1415) when he visited Sumatra, he captured the kingdom's former rebellious prince and sent him to Beijing for punishment. The young King was grateful. In 1435 (the tenth year of Xuande), the Xuande Emperor granted the young king's son to succeed to the throne. Later, the Kingdom was destroyed by Aceh (啞齊/哑齐[195] or 亞齊/亚齐; officially the Kingdom of Aceh Darussalam; 1496–1903; the exact origins of the Aceh Sultanate are still disputed; Portuguese sources state that the Sultanate started from 1520[196]), and Sumatra became the name of the entire island.

As mentioned earlier, on the Southeast Asia-KWQ, the Kingdom of Sumatra (蘇門答臘國/苏门答腊国) is depicted south of the Equator, not covering the entire island, and 大真 (Achen; the predecessor of Aceh; see Item 59) is depicted at the northern region of Sumatra Island, before the Kingdom of Aceh Darussalam destroyed the Kingdom of Sumatra down to the south. Hence, the Southeast Asia-KWQ reveals a political condition before 1496 or 1520, not in 1602.

"Sumatra" is depicted on the four European maps; but on the Southeast Asia-Ortelius (1570; #2), it is written as "Samotra". Also, the four maps all depict Sumatra for the entire Sumatra Island. This is because all these maps are sixteenth century maps drawn after the founding of the Kingdom of Aceh Darussalam in 1496 or 1520, and at that time the name "Sumatra" already became the name of the entire island.

(Continued)

Item #	Name in Chinese	Pinyin	Etymology	History, Geography and My Comments
@ 11	此島古名大波巴那周圍共四千里有七王君之土產金子象牙香品甚多	Ci Dao Gu Ming Da Bo Ba Na Zhou Wei Gong Si Qian Li You Qi Wang Jun Zhi Tu Chan Jin Zi Xiang Ya Xiang Pin Shen Duo	Ci Dao Gu Ming Da Bo Ba Na Zhou Wei Gong Si Qian Li You Qi Wang Jun Zhi Tu Chan Jin Zi Xiang Ya Xiang Pin Shen Duo	My translation and comments (between square brackets) are as follows: The ancient name of this island was Dabobana [大波巴那]. Its perimeter totals about four thousand *li* [one *li* is about 0.5 km], and the island is ruled by seven kings. The local products are gold and ivory, and it is rich in incense. In ancient times, Sumatra's Sanskrit names were *Suwarnadwīpa* ("Island of Gold") and *Suwarnabhūmi* ("Land of Gold"), because of the gold deposits in the island's highlands.[197] It was called Jinzhou (金洲, lit. "State of Gold") in Chinese literature[198] because the mountainous areas of Sumatra have been rich in gold since ancient times. In the sixteenth century, its reputation as the "Island of Gold" attracted many Portuguese explorers to Sumatra in search of gold.
• 57	兀弥突勒	Wu Mi Tu Le	Wumitule[199]	Wumitule (兀彌突勒/兀弥突勒) was in today's Southeastern Sumatra Island. The place cannot be identified on a modern map.
• 58	大泥俺	Da Ni An	Tunian[200]	Tunian (大泥俺) was in the northern region of Sumatra Island. The place cannot be identified on a modern map.
59	大真	Da Zhen	Achen[201]	On the Southeast Asia-KWQ, 大真 (Achen) is depicted where the Kingdom of Aceh (啞齊/哑齐[202] or 亞齊/亚齐; officially the Kingdom of Aceh Darussalam; c. 1496–1903) was in today's northwestern Sumatra, Indonesia, see Fig. A4.14.

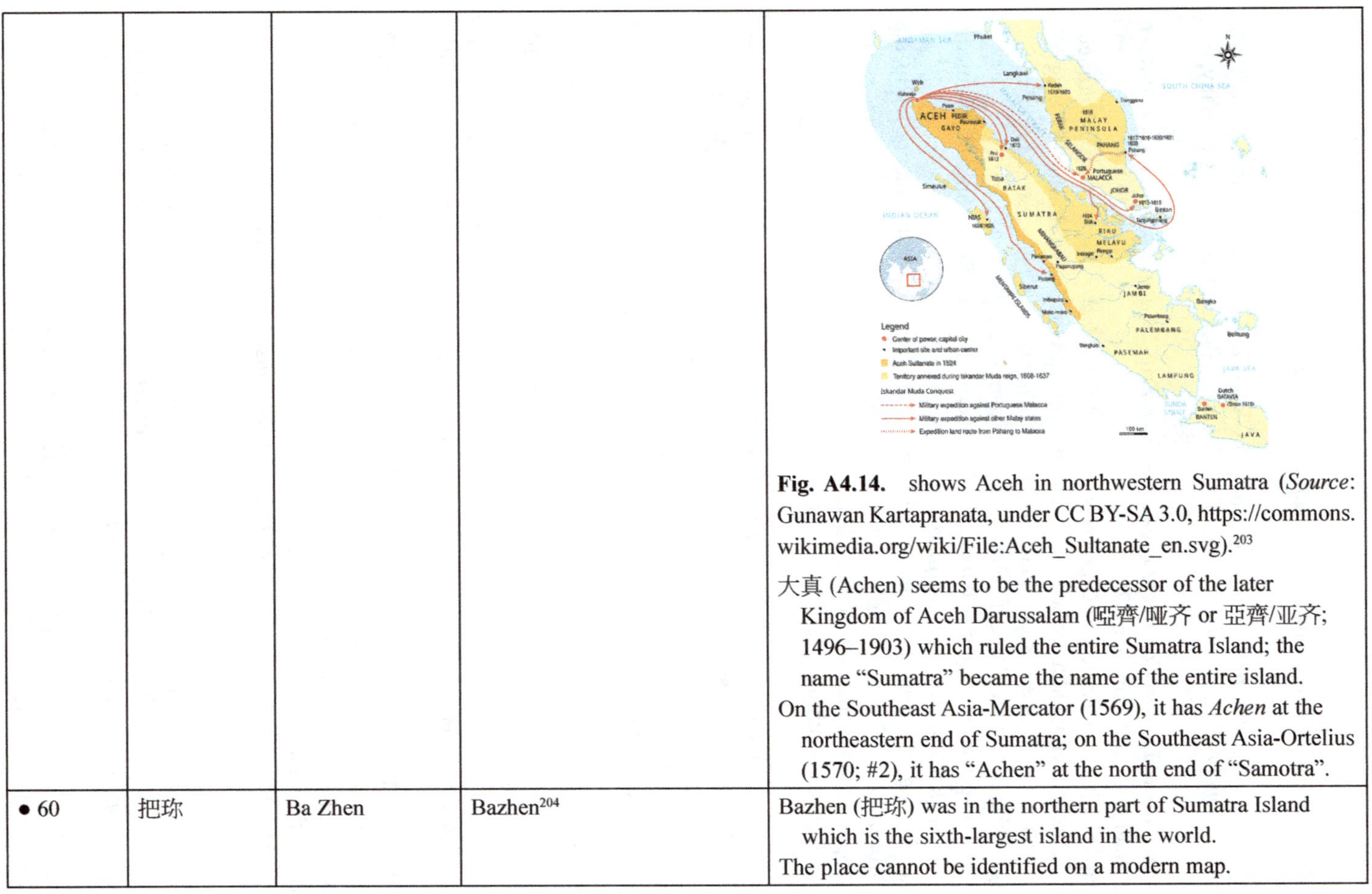

Fig. A4.14.   shows Aceh in northwestern Sumatra (*Source*: Gunawan Kartapranata, under CC BY-SA 3.0, https://commons.wikimedia.org/wiki/File:Aceh_Sultanate_en.svg).[203]

大真 (Achen) seems to be the predecessor of the later Kingdom of Aceh Darussalam (啞齊/哑齐 or 亞齊/亚齐; 1496–1903) which ruled the entire Sumatra Island; the name "Sumatra" became the name of the entire island.

On the Southeast Asia-Mercator (1569), it has *Achen* at the northeastern end of Sumatra; on the Southeast Asia-Ortelius (1570; #2), it has "Achen" at the north end of "Samotra".

● 60	把珠	Ba Zhen	Bazhen[204]	Bazhen (把珠) was in the northern part of Sumatra Island which is the sixth-largest island in the world. The place cannot be identified on a modern map.

Endnotes

[1]Library of Congress. "Kun yu wan guo quan tu." www.loc.gov, Library of Congress, 2025, loc.gov/item/2010585650

[2]Ronnie Po-Chia Hsia. *Matteo Ricci and the Catholic Mission to China, 1583–1610: A Short History with Documents (Passages: Key Moments in History)*. Indianapolis, IN, USA: Hackett Publishing Company, Inc., 2016.

[3]Sheng-Wei Wang. *Chinese Global Exploration in the Pre-Columbian Era: Evidence from an Ancient World Map*. Singapore: World Scientific Publishing Co., 2023.

[4]"File:OrteliusWorldMap1570.jpg." *Wikimedia Commons: The Free Media Repository,* Wikimedia Foundation, 12 July 2022, upload.wikimedia.org/wikipedia/commons/e/e2/OrteliusWorldMap1570.jpg

[5]"File:Kunyu Wanguo Quantu by Matteo Ricci Plate 1-3.jpg." *Wikimedia Commons,* 23 Oct. 2020, commons.wikimedia.org/wiki/File:Kunyu_Wanguo_Quantu_by_Matteo_Ricci_Plate_1-3.jpg

[6]"File:Mercator 1569 world map composite.jpg." *Wikimedia Commons,* 26 Nov. 2016, upload.wikimedia.org/wikipedia/commons/4/4b/Mercator_1569_world_map_composite.jpg

[7]"File:OrteliusWorldMap1570.jpg." *Wikimedia Commons,* 12 July 2022, upload.wikimedia.org/wikipedia/commons/e/e2/OrteliusWorldMap1570.jpg

[8]"File:1594 double hemisphere world map by Petrus Plancius.jpg." *Wikimedia Commons,* 2 Sept. 2022, upload.wikimedia.org/wikipedia/commons/0/0d/1594_double_hemisphere_world_map_by_Petrus_Plancius.jpg

[9]"File:1570 Ortelius Map of Asia (first edition) - Geographicus - AsiaeNovaDescriptio-ortelius.jpg." *Wikimedia Commons,* 22 Mar. 2011, upload.wikimedia.org/wikipedia/commons/a/a3/1570_Ortelius_Map_of_Asia_(first_edition)_-_Geographicus_-_AsiaeNovaDescriptio-ortelius.jpg

[10]Zhu Yunming (祝允明; 1461–1527). *Qian Wen Ji • Xia Xi Yang* (前闻记 •下西洋) or *A Record of History Once Heard: Down to the Western Ocean*; an electronic version can be found at https://ctext.org/wiki.pl?if=gb&chapter=203549

[11]Luo Maodeng (罗懋登; lived around 1596 A.D.). *San Bao Tai Jian Xi Yang Gi* (三宝太监西洋记) or *An Account of the Western World Voyage of the San Bao Eunuch;* an electronic file can be found at https://www.diancang.xyz/wenxueyishu/sanbaotaijianxiyangji/

[12]Sheng-Wei Wang. "Chapter 3: Luo Maodeng's *Tianfang*/Yun Chong was Mecca in Saudi Arabia." *The Last Journey of the San bao Eunuch, Admiral Zheng He,* Hong Kong, China: Proverse Hong Kong, 2019, pp. 91–138.

[13]Zhu Yunming (祝允明; 1461–1527). *Op. cit.*; Ma Huan. J.V.G. Mills, trans. *Ying-yai Sheng-lan: The Overall Survey of the Ocean's Shores [1433]*, Cambridge, UK: Cambridge University Press, 1970, pp. 17–18; Dreyer, Edward L. Dreyer. *Zheng He: China and the Oceans in the Early Ming Dynasty, 1405–1433*, New York, NY, USA: Pearson Longman, 2007, p. 161.

[14]Zhu Yunming (祝允明; 1461–1527). *Op. cit.*; Ma Huan. J.V.G. Mills, trans. *Op. cit.*; Edward L. Dreyer *Op. cit.*, pp. 162–163.

[15]*Ibid.*

[16]Zhang Tingyu (張廷玉; 1672–1755), *et al. Ming Shi • Juan San Bai Er Shi Yi • Lie Chuan Di Er Bai Jiu • Wai Guo Er* (明史 • 卷三百二十一 • 列傳第二百九 • 外國二) or *History of Ming, Juan (Volume) 321, Biography 209, Foreign Country 2*; an electronic version can be found at https://ctext.org/wiki.pl?if=gb&res=410835

[17]*Ibid;* Guo Zhenduo and Zhang Xiaomei (郭振鐸、張笑梅). *Yue Nan Tong Shi* (越南通史) or *General History of Vietnam*. Beijing, China: China Renmin University Press (中國人民大學出版社), 2001.

[18]Zhang Tingyu (張廷玉; 1672–1755), *et al. Ming Shi • Juan Qi Shi Wu • Zhi Di Wu Shi Yi • Zhi Guan Si* (明史・卷七十五・志第五十一・職官四) or *History of Ming, Juan (Volume) 75, Record 51, Official 4*; an electronic version can be found at https://ctext.org/wiki.pl?if=gb&res=410835; Wu Shilian (吳士連), *et al. Da Yue Shio Ji Quan Shu* (大越史記全書) or *The Complete History of the Great Viet*, Dai Shan Tang (埴山堂), Japan: 1884 (明治17), p. 545; an electronic version can be found at https://ctext.org/wiki.pl?if=gb&res=442611

[19]Zhang Tingyu (張廷玉; 1672–1755), *et al. Ming Shi • Juan Qi Shi Wu • Zhi Di Wu Shi Yi • Zhi Guan Si* (明史・卷七十五・志第五十一・職官四) or *History of Ming, Juan (Volume) 75, Record 51, Official 4*; an electronic version can be found at https://ctext.org/wiki.pl?if=gb&res=410835

[20]Kathlene Baldanza. *Ming China and Vietnam: Negotiating Borders in Early Modern Asia.* Cambridge, UK: Cambridge University Press, 2016.

[21]Sheng-Wei Wang. *The Last Journey of the San Bao Eunuch, Admiral Zheng He. Op. cit.*

[22]Bill Hayton. *A Brief History of Vietnam: Colonialism, War and Renewal: The Story of a Nation Transformed.* North Clarendon, VT, USA: Tuttle Publishing, 2022.

[23]Grant Evans. *A Short History of Laos: The land in between.* Crows Nest, NSW, Australia: Allen & Unwin, 2014.

[24]Peter and Sanda Simms. *The Kingdoms of Laos: Six Hundred Years of History.* Oxfordshire, England, UK: Routledge, 1999.

[25]*Ibid.*

[26]Peter and Sanda Simms. *Op. cit.*, pp. 47–48 and pp. 51–52.

[27]Zhang Tingyu (張廷玉; 1672–1755), *et al. Ming Shi • Juan San Bai Yi Shi Wu • Lie Chuan Di Er Bai San • Yunnan Tusi San* (明史・卷三百一十五・列傳第二百三・雲南土司三) or *History of Ming, Juan (Volume) 135, Biography 203, Yunan Tusi 3*; an electronic version can be found at https://ctext.org/wiki.pl?if=gb&res=410835

[28]Volker Grabowsky. "The Northern Tai Polity of Lan Na (Babai-Dadian) Between the Late 13th to Mid-16th Centuries: Internal Dynamics and Relations with Her Neighbours." *Asia Research Institute Working Paper Series*, no. 17, 2004, pp. 1–76. https://ari.nus.edu.sg/wp-content/uploads/2018/10/wps04_017.pdf

[29]Zhang Tingyu (張廷玉; 1672–1755), *et al. Ming Shi • Juan San Bai Yi Shi Wu • Lie Chuan Di Er Bai San • Yunnan Tusi San* (明史・卷三百一十五・列傳第二百三・雲南土司三) or *History of Ming, Juan (Volume) 315, Biography 203, Yunan Tusi 3*; an electronic version can be found at https://ctext.org/wiki.pl?if=gb&res=410835

[30]*Ibid.*

[31]Elizabeth H. Moore. *Early Landscapes of Myanmar.* Bangkok, Thailand: River Books, 2007.

[32]*Ibid.*

[33]Lu Ren (陸韌). *Yuan Ming Shi Qi De Xi Nan Bian Jiang Yu Bian Jiang Jun Zheng Guan Kong* (元明時期的西南邊疆與邊疆軍政管控) or *The Southwest Frontier and the Frontier Military and Political Control in the Yuan and Ming Dynasties*, Beijing, China: Social Sciences Academic Press (社會科學文獻出版社), 2015, p. 192.

[34]Than Un. "Observations on the Translation and Annotation of the Royal Orders of Burma." *Crossroads: An Interdisciplinary Journal of Southeast Asian Studies*, vol. 4, no. 1, 1988, pp. 91–99.

[35]"File:Salween river basin map.png." *Wikimedia Commons*, 15 Nov. 2020, upload.wikimedia.org/wikipedia/commons/3/3c/Salween_river_basin_map.png

[36]George Cœdès. Walter F. Vella, ed. Susan Brown Cowing, trans. *The Indianized States of Southeast Asia.* Hawaii, USA: University of Hawaii Press, 1968; Maung Htin Aung. *A History of Burma.* Cambridge, UK: Cambridge University Press, 1967.

[37]Thant Myint-U. *The River of Lost Footsteps: Histories of Burma*. New York, NY, USA: Farrar, Straus and Giroux, 2006.

[38]"Relazione del primo viaggio intorno al mondo." *Wikisource*, 2 Apr. 2020, https://it.wikisource.org/wiki/Relazione_del_primo_viaggio_intorno_al_mondo

[39]Zhang Tingyu (張廷玉; 1672–1755), *et al. Ming Shi • Juan San Bai Er Shi Si • Lie Chuan Di Er Bai Shi Er • Wai Guo Wu* (明史 • 卷三百二十四 • 列傳第二百十二 • 外國五) or *History of Ming, Juan (Volume) 324, Biography 212, Foreign Country 5*; an electronic version can be found at https://ctext.org/wiki.pl?if=gb&res=410835

[40]Ian Glover. *Southeast Asia: From Prehistory to History*. Oxfordshire, UK: Psychology Press, 2004.

[41]Zhou Zhongjian (周中堅). "A brief history of Cambodia (Part 2) [J]." *Southeast Asia (*東南亞縱橫*)*, vol.1, 1984, pp. 37–42.

[42]Zhang Tingyu (張廷玉; 1672–1755), *et al. Ming Shi • Juan San Bai Er Shi Si • Lie Chuan Di Er Bai Shi Er • Wai Guo Wu* (明史 • 卷三百二十四 • 列傳第二百十二 • 外國五) or *History of Ming, Juan (Volume) 324, Biography 212, Foreign Country 5*; an electronic version can be found at https://ctext.org/wiki.pl?if=gb&res=410835

[43]*Ibid.*

[44]*Ibid.*

[45]*Ibid.*

[46]*Ibid.*

[47]*Ibid.*

[48]Tricky Vandenberg. "The Portuguese in Ayutthaya." www.ayutthaya-history.com, Ayutthaya Historical Research, 2025, https://www.ayutthaya-history.com/Settlements_Portuguese.html

[49]"File:Indochina Peninsula.png." *Wikimedia Commons*, 8 Sept. 2020, upload.wikimedia.org/wikipedia/commons/5/58/Indochina_Peninsula.png

[50]Zhang Tingyu (張廷玉; 1672–1755), *et al. Ming Shi • Juan San Bai Er Shi Si • Lie Chuan Di Er Bai Shi Er • Wai Guo Wu* (明史 • 卷三百二十四 • 列傳第二百十二 • 外國五) or *History of Ming, Juan (Volume) 324, Biography 212, Foreign Country 5*; an electronic version can be found at https://ctext.org/wiki.pl?if=gb&res=410835

[51]*Ibid.*

[52]Anna Giacalone Ramat and Paolo Ramat, eds. *The Indo-European Languages*. Thames, Oxfordshire, UK: Routledge, 2015.

[53]UNESCO World Heritage Centre. "Bagan." https://whc.unesco.org/, UNESCO, 2025, whc.unesco.org/en/list/1588

[54]*Ibid.*

[55]Victor B. Lieberman. *Strange Parallels: Southeast Asia in Global Context, c. 800–1830, volume 1, Integration on the Mainland*. Cambridge, UK: Cambridge University Press, 2003.

[56]Michael Aung-Thwin. *Pagan: The Origins of Modern Burma*. Honolulu, Hawaii, USA: University of Hawaii Press, 1985.

[57]Michael Vickery. "Champa Revised." In Bruce Lockhart and Kỳ Phương Trần, eds. *The Cham of Vietnam: History, Society and Art*, Hawaii, USA: University of Hawaii Press, 2011, pp. 363–420.

[58]Zhang Tingyu (張廷玉; 1672–1755), *et al. Ming Shi • Juan San Bai Er Shi Si • Lie Chuan Di Er Bai Shi Er • Wai Guo Wu* (明史 • 卷三百二十四 • 列傳第二百十二 • 外國五) or *History of Ming, Juan (Volume) 324, Biography 212, Foreign Country 5*; an electronic version can be found at https://ctext.org/wiki.pl?if=gb&res=410835

[59]Sheng-Wei Wang. "Luo Maodeng's *Tienfang*/Yun Chong was Mecca in Saudi Arabia." *The Last Journey of the San Bao Eunuch, Admiral Zheng He, op. cit.*, p. 92.

[60]J.J.L. Duyvendak. "The True Dates of the Chinese Maritime Expeditions in the Early Fifteenth Century." *T'oung Pao, Second Series*, vol. 34, livr. 5, 1939, pp. 341–413. https://www.jstor.org/stable/4527170

[61]Zhang Tingyu (張廷玉; 1672–1755), *et al. Ming Shi • Juan San Bai Er Shi Si • Lie Chuan Di Er Bai Shi Er • Wai Guo Wu* (明史 • 卷三百二十四 • 列傳第二百十二 • 外國五) or *History of Ming, Juan (Volume) 324, Biography 212, Foreign Country 5*; an electronic version can be found at https://ctext.org/wiki.pl?if=gb&res=410835

[62]Tana Li. *Nguyễn Cochinchina: southern Vietnam in the seventeenth and eighteenth centuries*, Ithaca, NY, USA: Southeast Asia Program Publications, 1998, p. 72.

[63]Michael Vickery. "A short history of Champa." In Andrew Hardy, *et al.*, eds. *Champa and the Archaeology of Mỹ Sơn (Vietnam)*, Singapore: NUS Press, 2009, pp. 45–61.

[64]R. S. P. Beekes and Lucien van Beek. *Etymological Dictionary of Greek*, Leiden, Netherlands: Brill, 2009, p. 368.

[65]Antonio Pigafetta (1524). "Relazione del primo viaggio intorno al mondo." *Wikisource*, 2 Apr. 2020, it.wikisource.org/wiki/Relazione_del_primo_viaggio_intorno_al_mondo

[66]Zhang Zhaoyuan (張昭; 894–972) and Jia Wei (賈緯; birth year unknown–952). *Jiu Tang Shu • Juan Yi Bai Jiu Shi Qi • Lie Chuan Di Yi Bai Si Shi Qi • Pan Pan Guo* (舊唐書 • 卷一百九十七 • 列傳第一百四十七 • 盤盤國) or *Old Tang Book, Juan (Volume) 197*, Biography 147 • *Panpan Kingdom*; an electronic version can be found at https://ctext.org/wiki.pl?if=gb&res=456206

[67]George Cœdès. Walter F. Vella, ed. Susan Brown Cowing, trans. *Op. cit.*

[68]Zhang Tingyu (張廷玉; 1672–1755), *et al. Ming Shi • Juan San Bai Er Shi Si • Lie Chuan Di Er Bai Shi Er • Wai Guo Wu* (明史 • 卷三百二十四 • 列傳第二百十二 • 外國五) or *History of Ming, Juan (Volume) 324, Biography 212, Foreign Country 5*; an electronic version can be found at https://ctext.org/wiki.pl?if=gb&res=410835

[69]*Ibid.*

[70]*Ibid.*

[71]"File:Srivijaya Empire.svg." *Wikimedia Commons*, 22 Nov. 2022, upload.wikimedia.org/wikipedia/commons/thumb/4/45/Srivijaya_Empire.svg/2014px-Srivijaya_Empire.svg.png

[72]Pei Lin (霈霖). Thomas, ed. "Why were China's 'overseas enclaves' in Southeast Asia abandoned after only existing for 33 years (中国在东南亚的'海外飞地,' 为何仅存在33年就丢掉了)?" https://baike.baidu.com/, Baidu, 2025, baike.baidu.com/tashuo/browse/content?id=2792313e07198afb6910e871&fromModule=usercenter-userpage_tab-list&fromModule=lemma_middle-info

[73]Huang Shijian (黄时鉴) and Gong Yingyan (龚缨晏). *Li Ma Dou Shi Jie Di Tu Yan Jiu* (利玛窦世界地图研究) or *Research on Matteo Ricci's World Map*, Shanghai, China: Shanghai Chinese Classics Publishing House (上海古籍出版社), 2004, p. 203.

[74]William Linehan. *History of Pahang*. Kuala Lumpur, Malaysia: Malaysian Branch of the Royal Asiatic Society, 1973.

[75]William Linehan. *Op, cit.*, p. 2.

[76]Zhang Tingyu (張廷玉; 1672–1755), *et al. Ming Shi • Juan San Bai Er Shi Si • Lie Chuan Di Er Bai Shi Er • Wai Guo Wu* (明史 • 卷三百二十四 • 列傳第二百十二 • 外國五) or *History of Ming, Juan (Volume) 324, Biography 212, Foreign Country 5*; an electronic version can be found at https://ctext.org/wiki.pl?if=gb&res=410835

[77]*Ibid.*

[78]Ahmad Sarji Abdul Hamid. *The Encyclopedia of Malaysia, vol. 16 - The Rulers of Malaysia*. Singapore: Editions Didier Millet, 2011.

[79]"File:Mao Kun map - Pahang, Pulau Tioman.jpg." *Wikimedia Commons*, 26 Nov. 2016, upload.wikimedia.org/wikipedia/commons/3/3e/Mao_Kun_map_-_Pahang%2C_Pulau_Tioman.jpg

[80]Zhang Tingyu (張廷玉; 1672–1755), *et al. Ming Shi • Juan San Bai Er Shi Wu • Lie Chuan Di Er Bai Shi San • Wai Guo Liu* (明史 • 卷三百二十五 • 列傳第二百十三 • 外國六) or *History of Ming, Juan (Volume) 325, Biography 213, Foreign Country 6*; an electronic version can be found at https://ctext.org/wiki.pl?if=gb&res=410835

[81]*Ibid.*

[82]Luthfia Ayu Azanella and Bayu Galih, eds. "Hari Nusantara, Kenali Nama Lawas 5 Pulau Besar di Indonesia." https://nasional.kompas.com/, Kompas.com, 2025, nasional.kompas.com/read/2018/12/13/18543791/hari-nusantara-kenali-nama-lawas-5-pulau-besar-di-indonesia

[83]Google News. "Meet America's largest bird: the Suri." https://peru.info/en-us/, PromPeru, 2025, peru.info/en-us/tourism/news/3/16/meet-america-s-largest-bird--the-suri

[84]*Ibid.*

[85]Sheng-Wei Wang. *Chinese Global Exploration in the Pre-Columbian Era: Evidence from an Ancient World Map. Op. cit*

[86]John Paul Zronik. *Francisco Pizarro: Journeys Through Peru and South America*. New York, NY, USA: Crabtree, 2005.

[87]Johor State Investment Centre. "Johor History." https://web.archive.org/, Johor State Investment Centre, 2025, web.archive.org/web/20110831122622/http://www.jsic.com.my/linkpage02/his.php

[88]Jonathan Rigg. *A dictionary of the Sunda language of Java*. Open Library: Legare Street Press, 2022.

[89]National Library of Malaysia. "Origin of Place Names – Johor." https://web.archive.org/, National Library of Malaysia, 2025, web.archive.org/web/20080209105902/http://sejarahmalaysia.pnm.my/portalBI/list.php?ttl_id=8§ion=sm03

[90]Zhang Tingyu (張廷玉; 1672–1755), *et al. Ming Shi • Juan San Bai Er Shi Wu • Lie Chuan Di Er Bai Shi San • Wai Guo Liu* (明史 • 卷三百二十五 • 列傳第二百十三 • 外國六) or *History of Ming, Juan (Volume) 325, Biography 213, Foreign Country 6*; an electronic version can be found at https://ctext.org/wiki.pl?if=gb&res=410835

[91]*Ibid.*

[92]Elaine Sanceau. *Indies Adventure; the Amazing Career of Alfonso De Albuquerque, Captain-general and Governor of India (1509-1515)*. Open Library: Hassell Street Press, 2021.

[93]Muzaffar Husain, *et al. Concise History of Islam (unabridged ed.)*, Old Delhi, India: Vij Books India Pvt Ltd., 2011, p. 310.

[94]Rohayah Che Amat. "Historic Cities of The Straits of Malacca UNESCO World Heritage Site: Threats and Challenges." *Journal of World Heritage Studies*, Special Issue, 2019, pp. 9–15.

[95]Zhang Tingyu (張廷玉; 1672–1755), *et al. Ming Shi • Juan San Bai Er Shi Wu • Lie Chuan Di Er Bai Shi San • Wai Guo Liu* (明史 • 卷三百二十五 • 列傳第二百十三 • 外國六) or *History of Ming, Juan (Volume) 325, Biography 213, Foreign Country 6*; an electronic version can be found at https://ctext.org/wiki.pl?if=gb&res=410835

[96]*Ibid.*

[97]*Ibid.*

[98]*Ibid.*

[99]Geoff Wade, trans. "*Southeast Asia in the Ming Shi-lu*: an open access resource." https://web.archive.org/, Asia Research Institute and the Singapore E-Press, National University of Singapore, 2025, https://web.archive.org/web/20160304112601/http://www.epress.nus.edu.sg/msl/entry/1730?hl=Malacca

[100]Sheng-Wei Wang. *Chinese Global Exploration in the Pre-Columbian Era: Evidence from an Ancient World Map. Op. cit.*; *Weng Cheng* is an enclosure built on the outside (or inside; or both) of the city gate with defensive systems to protect towns and cities in China in pre-modern times; it is part of the walls of ancient Chinese cities; its shape is either semi-circular or square. The semi-circular one is called *Weng Cheng* 瓮城 or Weng City, or Weng Cheng; the square one is called *Fang Cheng*方城 or Square City, Fang City or *Fang Cheng*; Sheng-Wei Wang. "Chapter eight: Zheng He's generals force their way through the Fengdu gate, but later decide to return home." *The Last Journey of the San Bao Eunuch, Admiral Zheng He, op. cit.*, pp. 259–260.

[101]Elaine Sanceau. *Indies Adventure; the Amazing Career of Alfonso De Albuquerque, Captain-general and Governor of India (1509-1515).* Open Library: Hassell Street Press, 2021.

[102]Daniel R. Headrick. *Power Over Peoples: Technology, Environments, and Western Imperialism, 1400 to the present.* Princeton, NJ, USA: Princeton University Press, 2010.

[103]Emily Conover. "Here's how flying snakes stay aloft." www.sciencenews.org, Society for Science, 2025, sciencenews.org/article/how-flying-snakes-stay-aloft

[104]Keat Gin Ooi. *Southeast Asia: A Historical Encyclopedia from Angkor Wat to East Timor.* London, UK: Bloomsbury Academic, 2004.

[105]Zhang Xie (張燮; 1574–1640). *Dong Xi Yang Kao • Chuan Wu • Dong Yang Lie Guo Kao* (東西洋考 • 卷五 • 東洋列國考) or *On the Countries in the Eastern and Western Ocean, Juan (Volume) 5: Eastern Countries;* an electronic version can be found at https://ctext.org/wiki.pl?if=gb&res=87006

[106]Zhang Tingyu (張廷玉; 1672–1755), *et al. Ming Shi • Juan San Bai Er Shi San • Lie Chuan Di Er Bai Shi Yi • Wai Guo Si* (明史 • 卷三百二十 三 • 列傳第二百十一 • 外國四) or *History of Ming, Juan (Volume) 323, Biography 211, Foreign Country 4*; an electronic version can be found at https://ctext.org/wiki. pl?if=gb&res=410835

[107]Khai Leong Ho, ed. *Connecting and Distancing: Southeast Asia and China.* Singapore: Institute of Southeast Asian Studies, 2009; Stanley Karnow. In Our Image: America's Empire in the Philippines. New York, NY, USA: Random House LLC., 2010.

[108]Edmund Roberts. *Embassy to the Eastern Courts of Cochin-China, Siam, and Muscat*, New York, NY, USA: Harper & Brothers, 1837, p. 59.

[109]National Geospatial-Intelligence Agency. *Pub. 162: Sailing Directions (Enroute) Philippine Islands.* Bethesda, Maryland, USA: The United States Government, 2004, p. 70.

[110]Jan Huygen Van Linschoten (1596). "Exacta & Accurata Delineatio cum Orarum Maritimarum tum etjam locorum terrestrium quae in Regionibus China, Cauchinchina, Camboja sive Champa, Syao, Malacca, Arracan & Pegu." Barry Lawrence Ruderman Antique Maps Inc. Archived from the original on September 1, 2019. Retrieved June 16, 2021; Matthias Quad and Johann Bussemachaer (1598). "Asia Partiu Orbis Maxima MDXCVIII." Barry Lawrence Ruderman Antique Maps Inc. Archived from the original on September 1, 2019. Retrieved June 16, 2021.

[111]Joseph Baumgartner. "Manila – Maynilad or Maynila?" *Philippine Quarterly of Culture and Society*, vol. 3, no. 1, 1975, pp. 52–54. https://www.jstor.org/stable/29791188

[112]Laura Lee Junker. "Integrating History and Archaeology in the Study of Contact Period Philippine Chiefdoms." *International Journal of Historical Archaeology*, vol. 2, no. 4, 1998.

[113]Huang Shijian (黃时鉴) and Gong Yingyan (龚缨晏). *Op. cit.*, p. 197.

[114]Carmen N. Pedrosa. "The untold stories of Lapu-Lapu and Zheng He." www.philstar.com/, Philstar Global, 2025, https://www.philstar.com/opinion/2017/04/30/1689754/untold-stories-lapu-lapu-and-zheng-he

[115]Huang Shijian (黃时鉴) and Gong Yingyan (龚缨晏). *Op. cit.*, p. 198.

[116]Huang Shijian (黃时鉴) and Gong Yingyan (龚缨晏). *Op. cit.*, p. 194.

[117]Huang Shijian (黃时鉴) and Gong Yingyan (龚缨晏). *Op. cit.*, p. 193.

[118]Gwyn Campbell. *Bondage and the Environment in the Indian Ocean World*, Cham, Switzerland: Springer, 2018, p. 84.

[119]UN System-wide Earthwatch. "Island Directory." https://archive.ph/, UN System-wide Earthwatch Web Site, 2025, archive.ph/TBEHy#selection-15.0-15.34

[120]Antonio Pigafetta. *Magellan's Voyage: A Narrative Account of the First Circumnavigation*. Mineola, NY, USA: Dover Publications, 1994.

[121]The kahimyang Project. "Today in Filipino history, November 1, 1542, Ruy Lopez de Villalobos started his expedition to the Philippines." https://kahimyang.com/, The kahimyang Project, 2025, kahimyang.com/kauswagan/articles/721/today-in-philippine-history-november-1-1542-ruy-lopez-de-villalobos-started-his-expedition-to-the-philippines

[122]Huang Shijian (黄时鉴) and Gong Yingyan (龚缨晏). *Op. cit.*, p. 187.

[123]"File:Palawan in Philippines.svg." *Wikimedia Commons*, 9 Sept. 2023, upload.wikimedia.org/wikipedia/commons/thumb/a/a8/Palawan_in_Philippines.svg/1869px-Palawan_in_Philippines.svg.png

[124]The kahimyang Project. *Op. cit.*

[125]William Henry Scott. *Barangay: Sixteenth-century Philippine Culture and Society*. Quezon City, Philippines: Ateneo de Manila University Press, 1994.

[126]"File:Luzon Island Red.png." *Wikimedia Commons*, 6 May 2023, upload.wikimedia.org/wikipedia/commons/f/f8/Luzon_Island_Red.png

[127]"File:Visayas Red.png." *Wikimedia Commons*, 11 Nov. 2022, upload.wikimedia.org/wikipedia/commons/3/35/Visayas_Red.png

[128]"File:Mindanao Red.png." *Wikimedia Commons*, 6 May 2023, upload.wikimedia.org/wikipedia/commons/c/ce/Mindanao_Red.png

[129]Leigh Wright. "Brunei: An Historical Relic." *Journal of the Hong Kong Branch of the Royal Asiatic Society*, vol. 17, 1977, pp. 12-29. https://www.jstor.org/stable/23889576

[130]Christoph Friedrich von Ammon and Leonhard Bertholdt. *Kritisches Journal der neuesten theologischen Literatur, Volume 6*. Charleston, SC, USA: Nabu Press, 1817.

[131]"File:Kalimantan Locator.svg." *Wikimedia Commons*, 24 Nov. 2022, upload.wikimedia.org/wikipedia/commons/thumb/0/08/Kalimantan_Locator.svg/2560px-Kalimantan_Locator.svg.png

[132]Siti Norkhalbi Haji Wahsalfelah. *Traditional Woven Textiles: Tradition and Identity Construction in the 'New State' of Brunei Darussalam (PDF)*. Bandar Seri Begawan, Brunei Muara, Brunei Darussalam: Universiti Brunei Darussalam, 2005; an electronic file can be found at https://archive.org/details/TraditionalWovenTextilesTraditionAndIdentityConstructionBruneiDarussalam

[133]Ooi Keat Gin and Hoang Anh Tuan. *Early Modern Southeast Asia, 1350–1800*. Oxfordshire, England, UK: Routledge, 2015.

[134]Mohammad Al-Mahdi Tan Kho. *Malaysia-Philippines Territorial Dispute: The Sabah Case*. Taipei, Taiwan, China: National Chengchi University, 2014.

[135]Zhang Tingyu (張廷玉; 1672–1755), *et al. Ming Shi • Juan San Bai Er Shi Wu • Lie Chuan Di Er Bai Shi San • Wai Guo Liu* (明史 • 卷三百二十五 • 列傳第二百十三 • 外國六) or *History of Ming, Juan (Volume) 325, Biography 213, Foreign Country 6*; an electronic version can be found at https://ctext.org/wiki.pl?if=gb&res=410835

[136]*Ibid.*

[137]*Ibid.*

[138]*Ibid.*

[139]Huang Shijian (黄时鉴) and Gong Yingyan (龚缨晏). *Op. cit.*, p. 187.

[140]John Everett-Heath. *The concise dictionary of world place-names (Fourth ed.)*, Oxford, UK: Oxford University Press, 2018, p. 1131.

[141]Staff writer. "Sulawesi History Facts and Timeline." www.world-guides.com, World Guides, 2025, world-guides.com/asia/indonesia/sunda-islands/sulawesi/sulawesi_history.html

[142]"File:Sulawesi Locator Topography.png." *Wikimedia Commons*, 22 June 2021, upload.wikimedia.org/wikipedia/commons/9/96/Sulawesi_Locator_Topography.png

[143]Leonard Andaya. *The world of Maluku*, Honolulu, Hawaii, USA: University of Hawaii Press, 1993, p. 47.

[144]Theodoor Gautier Thomas Pigeaud. *Java in the 14th Century: A Study in Cultural History, Volume III: Translations (3rd revised ed.)*. The Hague, the Netherlands: Martinus Nijhoff, c. 1960.

[145]"File:Maluku Islands en.png." *Wikimedia Commons*, 23 Sept. 2020, upload.wikimedia.org/wikipedia/commons/f/f5/Maluku_Islands_en.png

[146]Huang Shijian (黃时鉴) and Gong Yingyan (龚缨晏). *Op. cit.*, p. 204.

[147]Huang Shijian (黃时鉴) and Gong Yingyan (龚缨晏). *Op. cit.*, p. 187.

[148]Huang Shijian (黃时鉴) and Gong Yingyan (龚缨晏). *Op. cit.*, p. 201.

[149]Huang Shijian (黃时鉴) and Gong Yingyan (龚缨晏). *Op. cit.*, p. 187.

[150]Huang Shijian (黃时鉴) and Gong Yingyan (龚缨晏). *Op. cit.*, p. 189.

[151]George Cœdès. Walter F. Vella, ed. Susan Brown Cowing, trans. *Op. cit.*

[152]Hans Hägerdal. *Held's History of Sumbawa: An Annotated Translation*. Amsterdam, the Netherlands: Amsterdam University Press, 2017.

[153]Zhang Tingyu (張廷玉; 1672–1755), *et al. Ming Shi • Juan San Bai Er Shi Si • Lie Chuan Di Er Bai Shi Er • Wai Guo Wu* (明史 • 卷三百二十四 • 列傳第二百十二 • 外國五) or *History of Ming, Juan (Volume) 324, Biography 212, Foreign Country 5*; an electronic version can be found at https://ctext.org/wiki.pl?if=gb&res=410835

[154]Marco Polo. Ronald Latham, trans. *The Travels of Marco Polo*. London, UK: Folio Society, 1958; Marco Polo. Nigel Cliff, ed., trans., and Intro. Coralie Bickford-Smith, lllust. *The Travels*. London, UK: Penguin Classics, 2015.

[155]Zhang Tingyu (張廷玉; 1672–1755), *et al. Ming Shi • Juan Er • Ben Ji Di Eri • Taii Zhu Er* (明史 • 卷二 • 本紀第二 • 太祖二) or *History of Ming, Juan (Volume) 2, Main History 2, Taizhu Er*; an electronic version can be found at https://ctext.org/wiki.pl?if=gb&res=410835

[156]Jack Weatherford. *Genghis Khan and the Making of the Modern World*, New York, NY, USA: Random House, 2004, p. 239.

[157]David W. Bade. *Of Palm Wine, Women and War: The Mongolian Naval Expedition to Java in the 13th Century*. Singapore: Institute of Southeast Asian Studies, 2013.

[158]Zhou Qufei (周去非; 1135–1189). *Ling Wai Dai Da • She Po* (嶺外代答 • 闍婆) or *Representative Answers from the Region beyond the Mountains*; an electronic version can be found at https://ctext.org/wiki.pl?if=gb&res=711480

[159]Huang Shijian (黃时鉴) and Gong Yingyan (龚缨晏). *Op. cit.*, p. 203.

[160]Akhyari Hananto. "Asal Nama 'Surabaya', Ternyata bukan Hiu dan Buaya." www.goodnewsfromindonesia.id, Google News, 2025, www.goodnewsfromindonesia.id/2016/02/26/asal-nama-surabaya-bukan-hiu-dan-buaya

[161]Ma Huan. J.V.G. Mills, trans. *Op. cit.*, p. 90.

[162]Raymond John Howgego, ed. "Houtman, Cornelis." *Encyclopedia of Exploration to 1800*, Sydny, Australia: Hordern House, 2003, pp. 520–521.

[163]*Ibid.*

[164]"File:Indonesia Java location map.svg." *Wikimedia Commons*, 14 Jan. 2023, upload.wikimedia.org/wikipedia/commons/thumb/9/95/Indonesia_Java_location_map.svg/1382px-Indonesia_Java_location_map.svg.png

[165]George Cœdès. Walter F. Vella, ed. Susan Brown Cowing, trans. *Op. cit.*

[166]Raymond John Howgego, ed. *Op. cit.*

[167]Zhao Rugua (赵汝适; 1170–1231). *Zhu Fan Zhi* (諸蕃志) or *A Description of Barbarian Nations, Records of Foreign People*; an electronic file can be found at https://ctext.org/wiki.pl?if=gb&res=520299

[168]Huang Shijian (黄时鉴) and Gong Yingyan (龚缨晏). *Op. cit.*, p. 193.

[169]Huang Shijian (黄时鉴) and Gong Yingyan (龚缨晏). *Op. cit.*, p. 201.

[170]Charles McKew Parr. *So Noble a Captain: The Life and Times of Ferdinand Magellan*, New York, NY, USA: Thomas Y. Crowell, 1953, p. 431.

[171]*Ibid.*

[172]Marco Polo. Ronald Latham, trans. *Op. cit*; Marco Polo. Nigel Cliff, ed., trans., and Intro. Coralie Bickford-Smith, Illust. *Op. cit.*

[173]Xiao Lang (小狼). "Where are the 'Big Java' and 'Small Java' in Marco Polo's *Travel* ((马可·波罗游记) 中的 '大爪哇' 和 '小爪哇' 在哪)？" https://zhuanlan.zhihu.com/, Zhihu, 2025, zhuanlan.zhihu.com/p/346320283

[174]John Norman Miksic and Goh Geok Yian. *Ancient Southeast Asia.* Oxfordshire, England, UK: Routledge, 2016; https://books.google.com.hk/books?id=R8tCDQAAQBAJ&pg=PT877&redir_esc=y#v=onepage&q&f=false

[175]"File:LocationGreaterSundaIslands.svg." *Wikimedia Commons*, 26 Aug. 2023, upload.wikimedia.org/wikipedia/commons/thumb/0/0c/LocationGreaterSundaIslands.svg/2560px-LocationGreaterSundaIslands.svg.png

[176]"File:LocationLesserSundaIslands.svg." *Wikimedia Commons*, 9 Mar. 2023, upload.wikimedia.org/wikipedia/commons/thumb/c/cd/LocationLesserSundaIslands.svg/2560px-LocationLesserSundaIslands.svg.png

[177]Zhang Tingyu (張廷玉; 1672–1755), *et al. Ming Shi • Juan San Bai Er Shi Wu • Lie Chuan Di Er Bai Shi San • Wai Guo Liu* (明史 • 卷三百二十五 • 列傳第二百十三 • 外國六) or *History of Ming, Juan (Volume) 325, Biography 213, Foreign Country 6*; an electronic version can be found at https://ctext.org/wiki.pl?if=gb&res=410835

[178]*Ibid.*

[179]*Ibid.*

[180]John Norman Miksic and Goh Geok Yian. *Ancient Southeast Asia.* Oxfordshire, UK: Routledge, 2016.

[181]Marco Polo. Henry Yule and Henri Cordier, translators. *The Travels of Marco Polo: The Complete Yule-Cordier Edition*, Mineola, NY, USA: Dover Publications, 1993, p. 299.

[182]Huang Shijian (黄时鉴) and Gong Yingyan (龚缨晏). *Op. cit.*, p. 192.

[183]Jane Drakard. "An Indian Ocean Port: Sources for the Earlier History of Barus." *Archipel, Villes d'Insulinde (II)*, vol. 37, 1989, pp. 53–82. https://www.persee.fr/doc/arch_0044-8613_1989_num_37_1_2562

[184]*Ibid.*

[185]*Ibid.*

[186]G.E. Gerini. *Researches on Ptolemy's Geography of Eastern Asia (further India and Indo-Malay archipelago), Asiatic Society Monographs*, London, UK: Royal Asiatic Society, 1909, pp. 428–430.

[187]L.F. Brakel. "Hamza Pansuri: Notes on: Yoga Practies, Lahir dan Zahir, the 'Taxallos', Punning a Difficult Passage in the Kitāb al-Muntahī, Hamza's likely Place of Birth, and Hamza's Imagery." *Journal of the Malaysian Branch of the Royal Asiatic Society*, vol. 52, no. 1 (235), 1979, pp. 73–98.

[188]Huang Shijian (黄时鉴) and Gong Yingyan (龚缨晏). *Op. cit.*, p. 189.

[189]Zhang Tingyu (張廷玉; 1672–1755), *et al. Ming Shi • Juan San Bai Er Shi Wu • Lie Chuan Di Er Bai Shi San • Wai Guo Liu* (明史 • 卷三百二十 五 • 列傳第二百十三 • 外國六) or *History of Ming, Juan (Volume) 325, Biography 213, Foreign Country 6*; an electronic version can be found at https://ctext.org/wiki.pl?if=gb&res=410835

[190]"File:Gulf of Carpentaria map.png." *Wikimedia Commons*, 31 July 2022, upload.wikimedia.org/wikipedia/commons/e/e0/Gulf_of_Carpentaria_map.png

[191]Jianming Shen. "China's Sovereignty over the South China Sea Islands: A Historical Perspective." *Chinese Journal of International Law*, vol. 1, no. 1, 2002, Pages 94–157, https://doi.org/10.1093/oxfordjournals.cjilaw.a000432

[192]Tim Mackintosh-Smith. *The Travels of Ibn Battutah*. London, UK: Macmillan Collector's Library, 2016.

[193]Zhang Tingyu (張廷玉; 1672–1755), *et al. Ming Shi • Juan San Bai Er Shi Wu • Lie Chuan Di Er Bai Shi San • Wai Guo Liu* (明史 • 卷三百二十 五 • 列傳第二百十三 • 外國六) or *History of Ming, Juan (Volume) 325, Biography 213, Foreign Country 6*; an electronic version can be found at https://ctext.org/wiki.pl?if=gb&res=410835

[194]*Ibid.*

[195]*Ibid.*

[196]J.M. Barwise and N.J. White. *A Traveller's History of Southeast Asia*. New York, NY, USA: Interlink Books, 2002.

[197]Jane Drakard. *A Kingdom of Words: Language and Power in Sumatra*. Oxford, UK: Oxford University Press, 1999.

[198]Yi Jing (义净; 635–713). *Da Tang Xi Yu Qiu Fa Gao Seng Chuan • Zhen Gu Chuan* (大唐西域求法高僧传 • 贞固传) or *The Biography of the Dharma-Seeking Monks in the Western Regions of the Tang Dynasty: Zhen Gu*; an electronic version can be found at https://www.diancang.xyz/foxuebaodian/11602/

[199]Huang Shijian (黄时鉴) and Gong Yingyan (龚缨晏). *Op. cit.*, p. 185.

[200]Huang Shijian (黄时鉴) and Gong Yingyan (龚缨晏). *Op. cit.*, p. 185.

[201]*Ibid.*

[202]Zhang Tingyu (張廷玉; 1672–1755), *et al. Ming Shi • Juan San Bai Er Shi Wu • Lie Chuan Di Er Bai Shi San • Wai Guo Liu* (明史 • 卷三百二十 五 • 列傳第二百十三 • 外國六) or *History of Ming, Juan (Volume) 325, Biography 213, Foreign Country 6*; an electronic version can be found at https://ctext.org/wiki.pl?if=gb&res=410835

[203]"File:Aceh Sultanate en.svg." *Wikimedia Common*, 5 July 2023, upload.wikimedia.org/wikipedia/commons/thumb/2/2f/Aceh_Sultanate_en.svg/1945px-Aceh_Sultanate_en.svg.png

[204]Huang Shijian (黄时鉴) and Gong Yingyan (龚缨晏). *Op. cit.*, p. 192.

Chapter 5

An Ancient World Map Depicts Chinese Exploration of the Indian Subcontinent and Its Eastern Periphery in 1432–1433, Long Before Vasco da Gama Landed in India in 1498

Abstract

The map Kunyu Wanguo Quantu (坤輿萬國全圖/坤輿万国全图) (abbreviated as KWQ) or Complete Geographical Map of All the Kingdoms of the World was published in China by Italian missionary Matteo Ricci in 1602. The map has been regarded as having European origin.

However, after analysing the geographical information and histories of all the 44 geographical items and seven related annotations on the part of KWQ showing the Indian subcontinent and its eastern periphery (abbreviated as ISC&EP-KWQ), I have discovered that the map contains distinct Chinese elements to support its Chinese origin.

Moreover, the political era revealed by the ISC&EP-KWQ was initially deduced by me to be in the period 1394–1493, because the Jaunpur Sultanate depicted on the map existed from 1394 to 1493. But after considering other historical records, the era can be narrowed to the period 1432–1433 in the Ming Dynasty. The reasons are: (1) the Ming Treasure Fleets led by Admiral Zheng He (鄭和/郑和) explored the Indian subcontinent and its eastern periphery during all their seven voyages (1405–1433), in particular, during their seventh (and last) voyage in 1432 on their outgoing journey and in 1433 on their homeward journey; they must have obtained the latest geographical data of the Indian subcontinent and its eastern periphery regions to update their older maps; (2) soon after these mariners returned to China in 1433, the Ming Court stopped future treasure voyages due to financial constraints, political opposition, and a shift in China's foreign policy, until the resulting *Haijin* (海禁) or sea ban was lifted in the mid-sixteenth century; and (3) the political era deduced directly from the ISC&EP-KWQ is in the period 1394–1493, which is no later than 1493, hence, any voyages after the sea ban would make no contribution to the making of the ISC&EP-KWQ. Combining these three constraints leads to the conclusion that the latest geographical data on the map could only be obtained by the Ming mariners in 1432–1433, hence, the era revealed by the map is in the period 1432–1433. This era falls within the period 1394–1493 and precedes the arrival of the first Europeans in India in 1498.

Keywords: Kunyu Wanguo Quantu, Matteo Ricci, Indian subcontinent, Ming treasure voyages, Southeast Asia, Zheng He

1. Introduction

For over four hundred years, the ancient world map — Kunyu Wanguo Quantu (坤輿萬國全圖/坤輿万国全图),[1] abbreviated here as KWQ — drawn with Chinese characters and having latitudinal and longitudinal lines, has been generally regarded as a European map. The map was published in 1602 by Matteo Ricci (an Italian Jesuit priest and one of the founding figures of the Jesuit China missions) and was believed to be based on the European maps which Ricci brought with him to China in 1582.[2]

In my previous book titled *Chinese Global Exploration in the Pre-Columbian Era: Evidence from an Ancient World Map*,[3] and Chapters 1–4 of this book, I have made thorough analyses to show how the origin and the eras of many major portions of the KWQ can be deduced from their geographical and historical information, and the annotations on the map.

Most importantly, my findings have shown that the afore-mentioned portions of the KWQ must have used mainly Chinese maps as source maps (they no longer exist), instead of the contemporary European maps.

In this chapter, I shall continue my effort for the region of the Indian subcontinent and its eastern periphery, using the same approach: (1) a thorough analysis of the seven annotations, and of the locations and histories of all the 44 geographical items depicted on the ISC&EP-KWQ; and (2) compare the information obtained with that on the major sixteenth century European maps to find the true origin and era of the Indian subcontinent and its eastern periphery depicted on the KWQ.

The Indian subcontinent is a physiographical region in Southern Asia. Geopolitically, it spans the present major landmasses from the countries of Bangladesh, Bhutan, India, Maldives, Nepal, Pakistan, and Sri Lanka. It is roughly located between c. 8–37°N in latitude and c. 61–97°E in longitude. Its eastern periphery, in this chapter, refers to the long and somewhat narrow region on the KWQ from Pegu (莲牛; Item 36 in the Appendix) in the south, to Khotan (于闐/于阗; Item 38) in the north, which lies between China and the Indian subcontinent.

The geographical information on the ISC&EP-KWQ will be compared with that on the following four major European maps in the sixteenth century: (1) the ISC&EP-Mercator (1569), extracted from the 1569 World Map by Gerardus Mercator; (2) the ISC&EP-Ortelius (1570; #1), extracted from the 1570 World Map by Abraham Ortelius; (3) the ISC&EP-Plancius (1594), extracted from the 1594 World Map by Petrus Plancius; and (4) the ISC&EP-Ortelius (1570; #2), extracted from the 1570 Ortelius Map of Asia (first edition)-*Asiae Nova Descriptio*.

In the Appendix, I have analysed the geographical information and histories of all the 44 geographical items, and the original annotations depicted on the ISC&EP-KWQ shown in Fig. 5.1(a) and compared them with those on the above-mentioned four major European maps shown in Figs. 5.1(b)–5.1(e) whenever possible. The results of these comparisons are used as the basis for determining the origin and era of the ISC&EP-KWQ.

Fig. 5.1.   The five maps shown in (a)–(e) are: (1) the ISC&EP-KWQ extracted from the KWQ (public domain);[4] (b) the ISC&EP-Mercator (1569) extracted from the 1569 World Map by Gerardus Mercator (public domain);[5] (c) the ISC&EP-Ortelius (1570; #1) extracted from the 1570 World Map by Abraham Ortelius (public domain);[6] (d) the ISC&EP-Plancius (1594) extracted from the 1594 World Map by Petrus Plancius (public domain);[7] and (e) the ISC&EP-Ortelius (1570; #2) extracted from the 1570 Ortelius Map of Asia (first edition)-*Asiae Nova Descriptio* (public domain).[8] In all these figures, the mouth of the Indus River (身毒河; Item 26), the Diu/Dio (丢; Item 24) Island and the approximate Gujarat (吳茶蠟/吳茶蜡; Item 25) region are highlighted by a red curve; the Ganges/Ganga River (恆河/恒河) is pointed out by a red arrow. The Item number inside the brackets is given in the Appendix of this chapter.

2. The ISC&EP-KWQ is of Chinese Origin

Figure 5.1 shows the five maps to be compared for finding the origin of the ISC&EP-KWQ. There are at least the following three main aspects which show that the ISC&EP-KWQ is of Chinese origin.

2.1 *The general geographical shapes on the maps*

The general geographical shapes on the four European maps in Figs. 5.1(b)–5.1(e) appear to correlate closer with one another than with that on the ISC&EP-KWQ in Fig. 5.1(a). Especially on the ISC&EP-KWQ, there is a pronounced peninsula depicted on the west coast of India consistent with that on a modern map, but no peninsula is present on the four European maps. Notice that on the ISC&EP-Ortelius (1570; #2), the pronounced structure on the western coast of India is not a peninsula, but the Diu Island, as shown in Fig. 5.2.

This is the first evidence to show that the ISC&EP-KWQ looks unlike a copy or adapted copy of the four major European maps in the sixteenth century. This aspect will be further discussed in the latter part of this section.

2.2 *The river system*

The Ganges/Ganga River

The second major difference is the total absence of the Ganges (in Chinese: 恆河/恒河; in India: Ganga, in Bangladesh: Padma) on the ISC&EP-KWQ. But the river is depicted on the four European maps as pointed out by the red arrows in Figs. 5.1(b)–5.1(e) and on modern maps in Figs. 5.3(a) and 5.3(b).

The Ganges is the longest river in India and the most sacred river to Hindus.[12] It is worshipped as the goddess *Ganga* in Hinduism.[13] It is also the place where Buddhism arose, and many remains of Buddhist holy sites can be seen there. The river originates in the western Himalayas (喜馬拉雅山/喜马拉雅山) and flows across northern India (印度) into Bangladesh (孟加拉), where it

Fig. 5.2.   Diu Island on the ISC&EP-Ortelius (1570; #2), extracted from the 1570 Ortelius Map of Asia (first edition)-*Asiae Nova Descriptio* (public domain).[9]

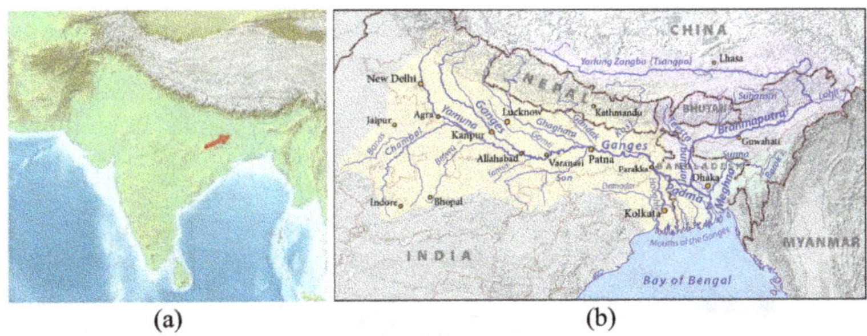

(a)   (b)

Fig. 5.3.   (a) shows the river system in South Asia (*Source*: DEMIS Mapserver, पाटलिपुत्र (talk), under CC BY-SA 3.0, https://commons.wikimedia.org/wiki/File:South_Asia_non_political,_with_rivers.jpg)[10] with the red arrow pointing to the Ganges. (b) is a map of the drainage basins of the Ganges (yellow), the Brahmaputra (布拉馬普特拉河/布拉马普特拉河 or 雅魯藏布江; violet), and the Meghna (梅克納河; green) (*Source*: Pfly, under CC BY-SA 3.0, https://commons.wikimedia.org/wiki/File:Ganges-Brahmaputra-Meghna_basins.jpg).[11]

empties into the Bay of Bengal (孟加拉灣/孟加拉湾) as shown on modern maps in Figs. 5.3(a) and 5.3(b).

In the seventh century, Chinese Buddhist monk Xuanzang (玄奘) began his seventeen-year overland journey to India at age 27 to fetch the Buddhist scriptures.[14] After returning to China, he wrote a famous travelogue, in which he described the Ganges with blue waters, which people believe carry "waters of blessedness", and in which a dip leads to expiation of sins.[15] But Xuanzang was not a surveyor, and might not leave behind any map of the Ganges. The ISC&EP-KWQ seems to tell us that the ancient Chinese had not surveyed the Ganges when the ISC&EP-KWQ was drawn.

For the Europeans, the first traveler to mention the Ganges was the Greek envoy Megasthenes (c. 350–290 B.C.) in his work *Indica* which describes Mauryan India (322 B.C.–184 B.C.) as follows: "India, … possesses many rivers…, which, having their sources in the mountains which stretch along the northern frontier, traverse the level country, and not a few of these, after uniting with each other, fall into the river called the Ganges. Now this river, …, flows from north to south, and empties its waters into the ocean…".[16]

Figure 5.4 is a European map produced around 300 A.D. It is the most detailed map we have of India after Ptolemy's *Geographia*[17] before the Middle Ages. This map shows only about 2% of the total Peutinger Table (Latin for "The Peutinger Map"; also referred to as Peutinger's *Tabula* or *Tabula Peutingeriana*; it is an illustrated ancient Roman road map showing the layout of the road network of the Roman Empire).[18] The map shows that the Ganges flows eastward. The island at the bottom is Ceylon (the present-day Sri Lanka).

The State of India (1505–1961; also referred as the Portuguese State of India or simply Portuguese India) was founded six years after the discovery in the late fifteenth century (1497–1499) of a sea route to the Indian subcontinent by Vasco da Gama (1469–1524); it was part of the Kingdom of Portugal and a state of the Portuguese Empire.

If the ISC&EP-KWQ were a copy or adapted copy of these four European maps, how is it possible for Matteo Ricci, the claimed cartographer of this map in the early seventeenth century, to omit the Ganges entirely from the map? Why would he have done that?

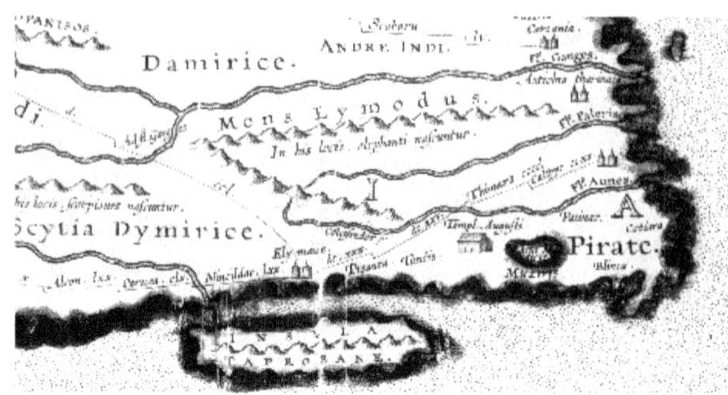

Fig. 5.4.   Ancient map of India (public domain);[19] the Ganges is the topmost river in this figure.

The Indus River

The comparisons of the Indus River are no less startling. To begin with, we need to refresh our knowledge of this river.

The Indus is a transboundary river of Asia and a trans-Himalayan (喜馬拉雅山/喜马拉雅山) river of South and Central Asia. The 3,120 km/1,939 mi[20] river rises in mountain springs northeast of Mount Kailash (岡仁波齊峰/冈仁波齐峰) in Western Tibet. It flows northwest through the present-day disputed region of Kashmir (克什米爾/克什米尔), bends sharply to the left after the Nanga Parbat Massif (南迦帕爾巴特峰/南迦帕尔巴特峰), and flows south-by-southwest through Pakistan (巴基斯坦), before emptying into the Arabian Sea (阿拉伯海; a part of the northern Indian Ocean) near the port city of Karachi (卡拉奇; c. 24.9°N, c. 67.0°E; the capital city of today's Pakistani province of Sindh).[21]

This river was known to the ancient Indians in Sanskrit as *Sindhu*, transliterated by Chinese as "身毒", hence "身毒河" (河, lit. "river"), and the Persians as *Hindu*.

The Chinese name "身毒" or "天竺" appeared in the pre-Qin (先秦) period and through the periods of the Sui (隋; 581–618) and Tang (唐; 618–907, with an interregnum between 690 and 705) dynasties; it was first recorded in *Shi Ji* (史记) or *Records of the Grand Historian*, referring to the ancient country in the Indus River Basin.[22]

In Fig. 5.1(a), the Indus River (身毒河) is the only river depicted in the Indian subcontinent on the ISC&EP-KWQ. On this map and near the river's mouth, the Indus River is erroneously separated into two branches by an island named as 吳茶蠟/吴茶蜡 (Gujarat; Item 25), see Fig. 5.1(a). On a modern map, Gujarat is not an island. It is a state along the western coast of India with its coastline of about 1,600 km/994 mi (the longest in the country), and most of it lies on the Kathiawar Peninsula, see Fig. 5.5(a). However, on the ISC&EP-KWQ, the Kathiawar Peninsula is wrongfully denoted as Diu (丢; Item 24). On a modern map, Diu is an island off the southern tip of the Kathiawar Peninsula, see Fig. 5.5(b).

On the ISC&EP-KWQ, Gujarat (吳茶蠟/吴茶蜡) is not only erroneously depicted as an island, see Fig. 5.6(a), but it is also misplaced toward the estuary of the Indus River. The Indus River empties into the Arabian Sea near the port city of Karachi (c. 24.9°N, c. 67.0°E) on a modern map. The river mouth on the ISC&EP-KWQ is wide and can be estimated from Fig. 5.6(a) as c. 24.1°N (KWQ latitude → modern latitude), c. 71.8°–73.3°E (on a modern map, the width is close

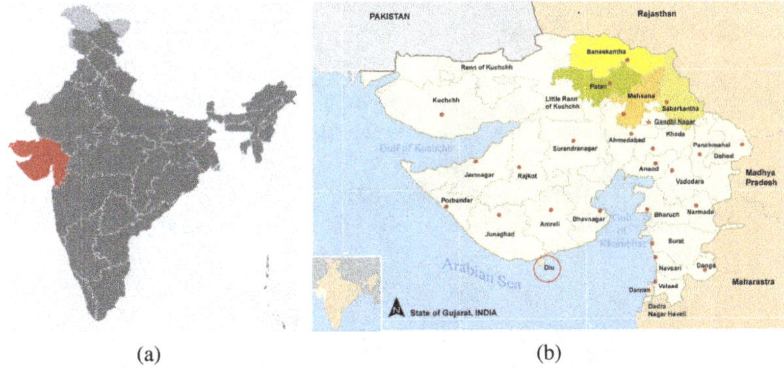

(a)   (b)

Fig. 5.5. (a) Gives the location of Gujarat in India (*Source*: फ़लिपुरो (Filpro), under CC BY-SA 4.0, https://commons. wikimedia.org/wiki/File:IN-GJ.svg);[23] most of it lies on the Kathiawar Peninsula in the middle portion of the red-coloured region. (b) gives the location of the Diu Island (enclosed in a red circle; *Source*: Self Made - w:User:Miljoshi, under CC BY-SA 3.0, https://commons.wikimedia.org/wiki/File:Map_GujDist_North.png)[24] which is off the southern coast of Gujarat's Kathiawar Peninsula in India; the island is separated from the mainland by a tidal creek.

to c. 67.1–68.3°E). The latitude is in reasonably good agreement with the latitude of Karachi, but the longitude is more eastward. Notice that the error in longitude on the ISC&EP-KWQ is often larger than the error in the latitude when compared with a modern map.

However, Gujarat is not an island, and it should not be depicted as an island blocking the estuary of the Indus River. Most of modern-day Gujarat lies on the Kathiawar Peninsula (22°N, 71°E) in India, see Figs. 5.5(a) and 5.5(b). Also, on the ISC&EP-KWQ, the Diu Island located at c. 22.0°N (KWQ latitude → modern latitude), c. 74.0°E, is erroneously depicted as a huge peninsula (the Kathiawar Peninsula). On a modern map, it is a small island located at 20.7°N, 71.0°E, see Fig. 5.5(b); the distance between Diu Harbor in India, and the Port of Karachi (c. 24.9°N, c. 67.0°E) in Pakistan, is 412 nautical miles or 763 km/474 mi.

The ISC&EP-Mercator (1569) in Fig. 5.6(b) depicts the Indus River erroneously as flowing through the Gujarat region, and the river separates into three branches before emptying into the Arabian Sea. Near their estuaries, there are several islands, including the Diu Island at c. 20.5°N, c. 88.0°E, which is too far eastward from the Diu Island on a modern map at 20.7°N 71.0°E. The Diu Island is off the southern coast of Gujarat's Kathiawar Peninsula in India, separated from the mainland by a tidal creek. We can see that the ISC&EP-Mercator (1569) depicts the latitude of the Diu Island accurately, but depicts the longitude poorly, being too far to the east by c. 17°. Nevertheless, there is a big river lying to the west of the erroneously placed "Indus River" on the ISC&EP-Mercator (1569); its mouth is at c. 23°N, c. 84.8°E. This river may be the actual Indus River, but its name is hardly readable on the electronic version of the ISC&EP-Mercator (1569).

The ISC&EP-Ortelius (1570; #1) and the ISC&EP-Ortelius (1570; #2) in Figs. 5.6(c) and 5.6(e), respectively, also depict the "Indus River" erroneously as flowing through the Guzarate/ Gvzarate (Gujarat) region. The qualitative features of these two maps are similar. For brevity, I will only give detailed discussion of the ISC&EP-Ortelius (1570; #1) as follows: the mouth of the Indus River is located at c. 22.5°N, c. 83.0°E, which is distant from the port city of Karachi (c. 24.9°N,

Fig. 5.6. The maps (a)–(d) show the Indus River, the Gujarat region, and the Diu Island, highlighted by a red ellipse curve on the ISC&EP-KWQ, the ISC&EP-Mercator (1569), the ISC&EP-Ortelius (1570; #1), the ISC&EP-Plancius (1594), and the ISC&EP-Ortelius (1570; #2), respectively. They are extracted from Figs. 5.1(a)–5.1(e) (public domain). On the ISC&EP-KWQ, the Indus River is depicted as 身毒河; on the four European maps it is depicted as *Indus flu*. On the ISC&EP-KWQ, Gujarat is denoted as 吴茶蠟; on the ISC&EP-Ortelius (1570; #1), the Gujarat region is denoted as "Guzarate", and on the ISC&EP-Ortelius (1570; #2) as "Gvzarate".

c. 67.0°E) in longitude. Again, there is an unnamed river to the west of this erroneous "Indus River", and this unnamed river may be the correct Indus River, but it is only partially drawn. However, on the ISC&EP-Ortelius (1570; #2), that unnamed river on the ISC&EP-Ortelius (1570; #1), is named as *Ilment fl*.

The Diu Island on the ISC&EP-Ortelius (1570; #1) is depicted at c. 21.5°N, c. 81.8°E, to compare with its coordinates at 20.7°N 71.0°E on a modern map. The longitudinal error is large, although the ISC&EP-Ortelius (1570; #1) correctly depicts Diu as an island. Besides, on this map, most of Guzarate/Gvzarate (Gujarat) does not lie on a peninsula (the Kathiawar peninsula).

The ISC&EP-Plancius (1594) in Fig. 5.6(d) has similar erroneous depiction of the Indus River flowing through the Gujarat (although this name is not denoted on the map) region. There is also an unnamed river to the west of this erroneous "Indus River", which may be the correct Indus River.

The Diu Island is denoted as the "Dio Island" and the mouth of the Indus River is located at 21.5°N, 84.0°E, which is again too far to the city of Karachi (c. 24.9°N, c. 67.0°E).

The above discussion and Figs. 5.6(b)–5.6(e) show that the four European maps share significant similarities in wrongly depicting the path of the Indus River and its mouth location, but they are correct in depicting Diu as an island, and the Gujarat as a region instead of an island. However, the ISC&EP-KWQ mistakes Diu (丢; Item 24) as a peninsula and erroneously places it near the mouth of the Indus River; in addition, the map also mistakes the Gujarat (吳茶蠟/吳茶蜡; Item 25) region as an island in the estuary of the Indus River. Overall, the ISC&EP-KWQ depicts smaller longitudinal errors than those on the four European maps. Figure 5.7 shows the Indus River and the Gujarat region on a modern map.

The above comparisons show that the four European maps correlate closely among themselves, they can be regarded as adapted copies among themselves, whereas the ISC&EP-KWQ

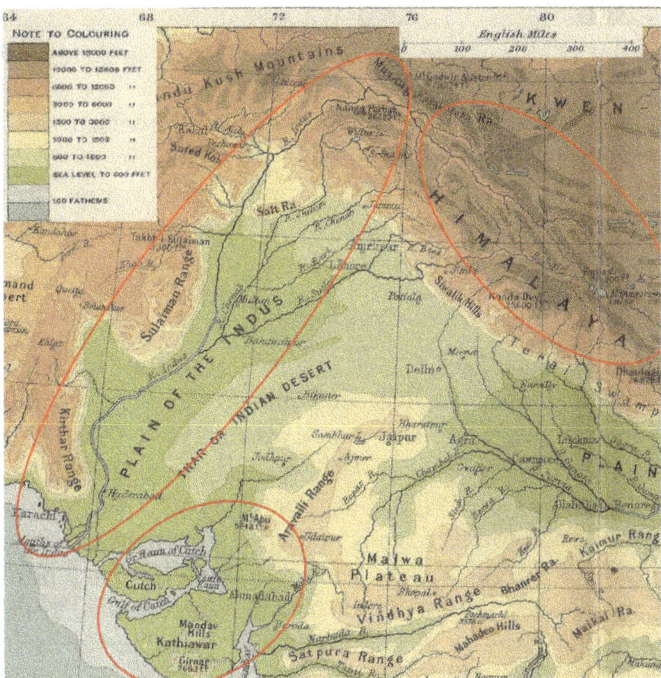

Fig. 5.7. The two red ellipses highlight the course and the major tributaries of the Indus River and other major rivers in a modern geographical map of Upper South Asia (public domain);[25] the circle at the bottom highlights roughly the Gujarat region, most of which lies on the Kathiawar Peninsula.

shows distinct geographical differences from the four European maps, and it is the only map showing the correct path of the Indus River. Hence, it cannot be a direct or adapted copy of the four European maps.

Moreover, the Indus River is expressed as 身毒河 on the ISC&EP-KWQ; 身毒 is the Chinese transliteration of *Sindhu* in Sanskrit and *Hindu* in Persian, meaning "the border river".[26] From the Persian Achaemenid Empire, the name passed to the Greeks as *Indós*,[27] then adopted by the Romans as *Indus*. Later, the name India derived from *Indus*.[28] During British colonial rule (approximately 1757–1947), the British referred to the Indian subcontinent as "India". This term was derived from the river Indus, which also marked the western boundary of British India.

The Chinese name "身毒" or "天竺" appeared in the pre-Qin (先秦; before 221 B.C.) period and through the periods of the Sui (隋; 581–618) and Tang (唐; 618–907, with an interregnum between 690 and 705) dynasties. The name was first recorded in *Shi Ji* (史記/史记) or *Records of the Grand Historian,* published around 90 B.C. Its depiction on the ISC&EP-KWQ indicates the usage of the ancient Chinese transliterated *Sindhu* in Sanskrit and *Hindu* in Persia, rather than "Indus" adopted by the Romans, in representing the Indus River. This adds one more layer of evidence to show the Chinese rather than the European origin of the map.

2.3 *Whence the mystical Lake Chiamay?*

Now we come to the point of discussing the four major rivers in Asia to the east of the Indian subcontinent.

We have just obtained multiple layers of evidence to show that the ISC&EP-KWQ is of Chinese origin. Now, a stunning feature shows up on the five maps we have been discussing: the ISC&EP-KWQ and the four sixteenth-century European maps in Figs. 5.1(a)–5.1(e) coherently show that the great Asian rivers to the east of the Indian Subcontinent originate from a single mystical lake. I shall now discuss this stunning common feature.

Briefly, we see the following:

(1) On the ISC&EP-KWQ shown in Figs. 5.1(a), four rivers originate from 加湖 (transliterated from "Lake Chiamay"; Item 39): one river has the Chinese name 金沙江 (Jinsha River; Item 40), which merges with an unnamed second river in Myanmar (緬甸/缅甸; Item 43) before emptying into the Bay of Bengal (榜葛剌海 or 孟加拉灣/孟加拉湾; Item 2); the third river is also unnamed and passed through Dagula (大古剌; Item 42) before flowing into the Bay of Bengal. Even though the name of the fourth river is not written explicitly, it should refer to Irrawaddy River (the Chinese name is 安義河/安乂河), because its location and flowing path match where the Irrawaddy River is. Besides, the annotation next to it (see detailed discussion in the Appendix under Item @5) states that "安義河受三十水产金沙", which can be translated by Sheng-Wei Wang as: "The Anyi River (安義河/安乂河) receives water from 30 tributaries and produces gold sand". This description matches what we know today about the Irrawaddy River.

(2) On the ISC&EP-Mercator (1569) shown in Fig. 5.1(b), six Asian rivers originate from Lake Chiamay, of which three eastern rivers merge and flows into the present-day South China Sea

(南中國海/南中国海); two other rivers flow into the Andaman Sea (安達曼海/安达曼海; a marginal sea of the northeastern Indian Ocean which is separated from the Bay of Bengal to its west by the Andaman Islands and the Nicobar Islands); the sixth river flows into the Bay of Bengal.

(3) On the ISC&EP-Ortelius (1570; #1) shown in Fig. 5.1(c), only two unnamed Asian rivers originate from an unnamed lake; one river flows into the Bay of Bengal and the other flows into the South China Sea.

(4) On the ISC&EP-Plancius (1594) shown in Fig. 5.1(d), four unnamed Asian rivers originate from an unnamed lake; one of these flows into the Bay of Bengal, two others flow into the present-day Andaman Sea and the fourth flows into the South China Sea.

(5) On the ISC&EP-Ortelius (1570; #2) shown in Fig. 5.1(e), six Asian rivers originate from an unnamed lake: one flows to the north, one unnamed river flows northwest into the inland region north of Comatai, one more flows westward into Bengala, the fourth and the fifth unnamed rivers merge right before flow into the present-day Andaman Sea, and the sixth river named Menan fl., flows into the South China Sea.

After comparing the five maps with the modern map, we know that the depictions in these five ancient maps all deviate from reality to various degrees.

First, we now know that there are four major Asian rivers to the east of the periphery of the Indian subcontinent, but they all have separate sources and do not originate from the same mystical Lake Chiamay. We can offer explanations, one by one from east to west, as follows:

(1) The Mekong River (湄公河) and its upper half called Lancang River (瀾滄江/澜沧江), see Fig. 5.8(b): The Lancang River is the longest river flowing from north to south in China. Originating from the Tanggula Mountain Range (唐古拉山脈/唐古拉山脉) in Qinghai Province, China, it flows through the Valley of the Hengduan Mountains (橫斷縱谷/横断纵谷) and Yunnan-Guizhou Plateau (雲貴高原/云贵高原).[29] The river runs south until it leaves China at the Nanla Bayout (臘河口/腊河口) in Yunnan Province (雲南省/云南省) and from there changes its name from the Lancang River to the Mekong River. The river then flows through Myanmar, Laos, Thailand, Cambodia, and southern Vietnam and finally empties into the South China Sea in the Western Pacific Ocean, see Fig. 5.8(b), at the Mekong Delta in Vietnam (越南) to the south of Ho Chi Minh City (胡志明市). This river is the longest in Southeast Asia. All the source maps (see endnotes 4–8) used for extraction to draw Figs. 5.1(a)–5.1(e), depict the Mekong River correctly, and the river does not originate from Lake Chiamay. The Mekong River is out of the scope of this chapter, hence, it is not included explicitly in Figs. 5.1(a)–5.1(e).

(2) The Jinsha River (金沙江; Item 40), see Fig. 5.8(a): This is the Chinese name for the upper stretches of the Yangtze River from Yushu (玉樹/玉树) in Qinghai Province (青海省) to Yibin (宜賓/宜宾) in Sichuan Province (四川省). From Yibin, where the Min River (岷江) merges into it, it was named the Yangtze River/Yangzijiang (揚子江/扬子江) or Changjiang (長江) and flows into the East China Sea near Shanghai (上海). Hence, none of the five ancient maps

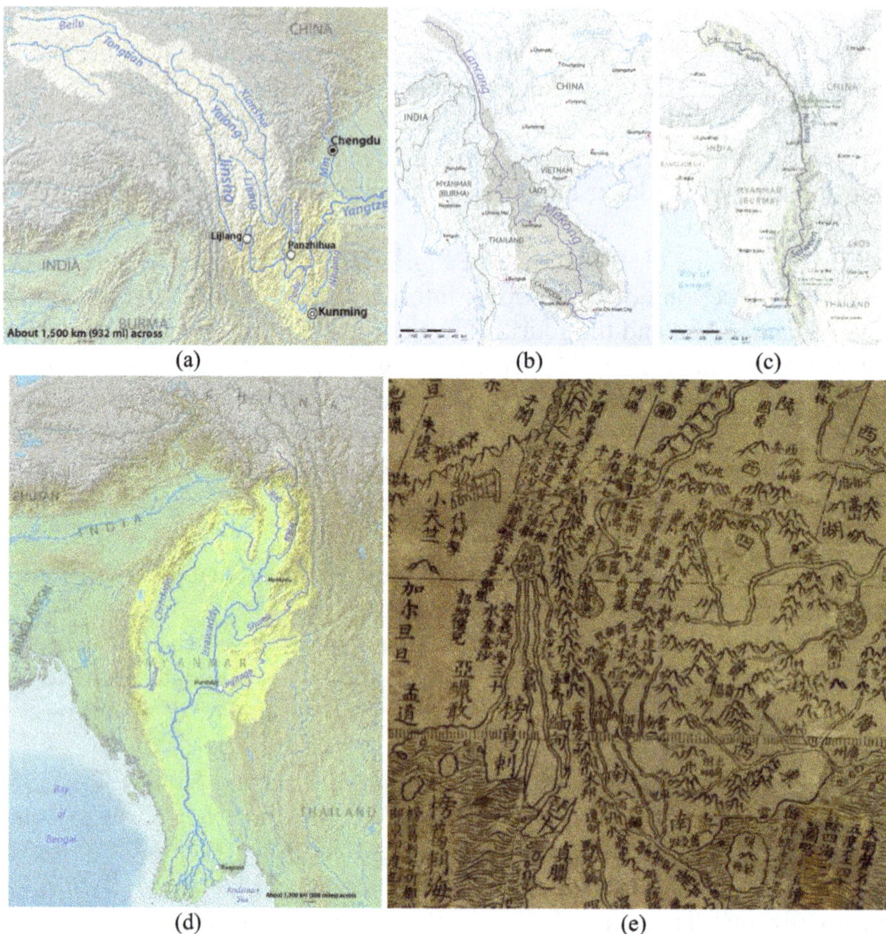

(a)   (b)   (c)

(d)   (e)

Fig. 5.8.   (a) shows the Jinsha River drainage basin in southwestern China on a modern map (*Source*: Shannon, under CC BY-SA 4.0 International, 3.0 Unported, 2.5 Generic, 2.0 Generic and 1.0 Generic license, https://commons.wikimedia.org/wiki/File:Jinsharivermap.jpg);[36] the Min River joins the Jinsha River to form the Yangtze River. (b) shows the Mekong River watershed on a modern map (*Source*: Shannon 1, under CC BY-SA 4.0 International, 3.0 Unported, 2.5 Generic, 2.0 Generic and 1.0 Generic license, https://commons.wikimedia.org/wiki/File:Mekong_river_basin.png).[37] (c) shows the Salween River basin on a modern map (*Source*: Shannon 1, under CC BY-SA 4.0 International, 3.0 Unported, 2.5 Generic, 2.0 Generic and 1.0 Generic license, https://commons.wikimedia.org/wiki/File:Salween_river_basin_map.png).[38] (d) shows the course, watershed, cities, and major tributaries of the Irrawaddy River (*Source*: Shannon, under CC BY-SA 4.0 International, 3.0 Unported, 2.5 Generic, 2.0 Generic and 1.0 Generic license, https://commons.wikimedia.org/wiki/File:Irrawaddyrivermap.jpg).[39] (e) shows the ISC&EP-KWQ extracted from the KWQ (public domain)[40] to compare with (a)–(d).

in Figs. 5.1(a)–5.1(e) depicts the Jinsha River correctly. The ISC&EP-Ortelius (1570; #1) only depicts two Asian rivers.

(3) The Salween River (薩爾溫江上游/萨尔温江上游), see Fig. 5.8(c): The river originates in the Tanggula Mountain Range (唐古拉山脈/唐古拉山脉) in the central Tibetan Plateau (青藏高原). Its headwaters are located near Dengka Peak (登卡峰), east of Tanggula Pass

(唐古拉山口),[30] and it flows into the Andaman Sea. The river is known by various names along its course, including its Chinese name Nu Jiang (怒江). The commonly used spelling "Salween" is an anglicisation of the Burmese name, Thanlwin,[31] dating from nineteenth-century British maps. Three of the five ancient maps, Figs. 5.1(a), (b) and (d), each has an unnamed river flowing into the Andaman Sea, but the ISC&EP-Ortelius (1570; #1) omits this river entirely.

Parts of the Jin Sha, Lancang and Salween rivers make up the Three Parallel Rivers of the Yunnan Protected Areas. But they *do not* originate from a common lake.

(4) The Irrawaddy River (伊洛瓦底江), see Fig. 5.8(d): The river originates from the confluence of the N'mai River (恩梅開江/恩梅开江; originate from Tibet, China) and Mali River (邁立開江/迈立开江; originated in northeastern Myanmar/Burma); both rivers find their sources in the Himalayan glaciers. The Irrawaddy River flows from north to south through Myanmar (Burma) before emptying through the Irrawaddy Delta (伊洛瓦底江三角洲) in the Ayeyarwady Region (伊洛瓦底地區/伊洛瓦底地区) into the Andaman Sea.

This fourth major Asian river also does not originate from a common lake with the previous three rivers. The ISC&EP-Ortelius (1570; #1) only depicts two major Asian rivers in Southeast Asia.

It is in fact quite surprising that on the ISC&CP-KWQ, there are four major Asian rivers originating from a common lake named 加湖 which is the Chinese transliteration of Lake Chiamay. The reason is that early Chinese maps do not show this feature. For example, the Huayi Tu (華夷圖/华夷图) or Map of China and Barbaric Countries,[32] carved in stone in 1136/1137 (perhaps dating as early as 1014) and showing China with its nearby regions, has accurate depictions of great Chinese rivers, but does not show Lake Chiamay, see Fig. 5.9(a).

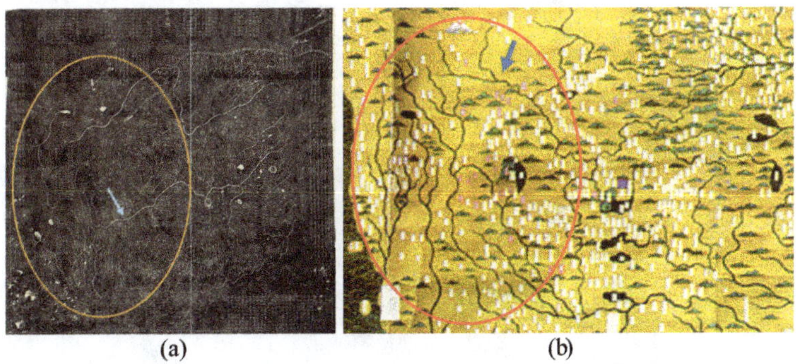

(a)   (b)

Fig. 5.9. (a) shows a 1933 (the date is not firm) rubbing of a 1136/1137 map (on a stele) of China — Huayi Tu Stele — from the US Library of Congress (public domain).[41] The stele is housed in the Forest of Stone Steles Museum in Xian, China. (b) shows the great rivers in Southeast China extracted from Da Ming Hunyi Tu (大明混一圖/大明混一图) or Composite Map of the Ming Empire (public domain)[42] around 1389 in the early Ming Dynasty. In each figure, the blue arrow points to the Jin Sha River which later connects to the Yangtze River in the eastern direction. Other than Jin Sha River, each of the figures has several more great rivers flowing generally from north to south in parallel directions until they empty into the seas, respectively. The river system shown in (b) is more complicated than in (a). But none of these rivers originates from a common lake, hence, 加湖 (Lake Chiamay) does not exist.

The Da Ming Hunyi Tu (大明混一圖/大明混一图) or Composite Map of the Ming Empire[33] around 1389 in the early Ming Dynasty also does not depict 加湖 (Lake Chiamay), see Fig. 5.9(b).

Furthermore, the Guang Yu Tu (廣輿圖/广舆图) or Enlarged Terrestrial Atlas published in 1579 (during the Wanli's reign) is the oldest extant comprehensive atlas of China by the famous Ming cartographer Luo Hongxian (羅洪先/罗洪先; 1504–64), and it also does not show that the great Asian rivers originate from a common lake; 加湖 (Lake Chiamay) is not depicted on this map shown in Fig. 5.10.

The map is based on the Yu Di Tu (輿地圖/輿地图) or Terrestrial Map by Zhu Siben (朱思本; 1273–1333)[34] in the Yuan Dynasty. Zhu Siben's map is no longer extant. Using Zhu's earlier work as a blueprint,[35] the maps made by Luo Hongxian (羅洪先/罗洪先; 1504–1564) were drawn on a grid scale, then bound in book form. Juan (Volume) 1 contains a comprehensive atlas of China in the mid-Ming era, with sixteen maps of the southern and northern provinces, two provinces — Jingshi Shuntianfu (京師順天府/京师顺天府; Jingshi, lit. "capital") in today's Beijing, parts of Hebei Province (河北省), and Tianjin City (天津市); and Nanjing Yingtianfu (南京應天府/南京应天府) in today's Naijing and some neighbouring cities — under direct rule, and thirteen provincial administrative governments. The atlas had a significant and lasting influence both in China and in foreign countries: until the late seventeenth century, maps of China published in Europe were, without exception, drawn based on this work.

The mystical Lake Chiamay first appeared in 1554 in Europe on a new map of Southeast Asia and re-appeared on subsequent European maps with various modifications and under different names,

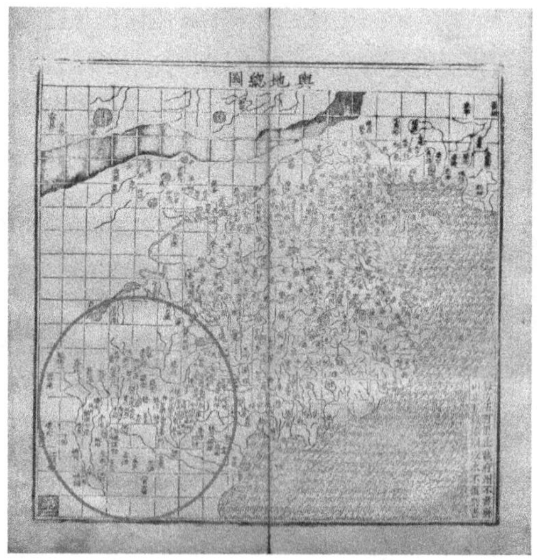

Fig. 5.10. This map is the Guang Yu Tu • Yudi Zongtu (廣輿圖•輿地總圖)/(广舆图•舆地总图) or General Geographical Map in Enlarged Terrestrial Atlas (public domain)[43] by Luo Hongxian (羅洪先/罗洪先; 1504–1564). The oval highlights the major Asian rivers which do not originate from a common lake.

until 1751, nearly 200 years after its invention,[44] even when the lake was proven to not exist.[45] It would be illogical to assume that Matteo Ricci did not know the famous and important 1579 Chinese map — Guang Yu Tu • Yudi Zongtu (廣輿圖 • 輿地總圖)/(广舆图 • 舆地总图) or General Geographical Map in Enlarged Terrestrial Atlas — while he was in the Forbidden City of Beijing in 1601 invited by the Wanli Emperor (萬曆皇帝/万历皇帝), just a year before he published the KWQ. It is also obvious that his Chinese colleagues and close collaborators like Li Zhizao (李之藻) should have known the content of Guang Yu Tu • Yudi Zongtu (廣輿圖 • 輿地總圖)/(广舆图 • 舆地总图) or General Geographical Map in Enlarged Terrestrial Atlas, Huayi Tu (華夷圖/华夷图) or Map of China and Barbaric Countries, and Da Ming Hunyi Tu (大明混一圖/大明混一图) or Composite Map of the Ming Empire. But why did they still decide to draw the mystical Lake Chiamay on the KWQ to be the common origin of the four great Southeast Asian rivers?

Perhaps they were deeply affected by the amazingly influential comment[47] made by the Portuguese geographer/historian Joao de Barros (1496–1570) in his book *Decadas da Asia*, and the follow-up maps such as the one made by Luiz Jorge de Barbuda (c. 1564–c.1613), see Fig. 5.11. Or perhaps copying their mystical Lake Chiamay onto the ISC&EP-KWQ would naturally make the map look more European to cover up its Chinese origin?

Even though the ISC&EP-KWQ depicts four great Asian rivers deriving from the same lake and made it look more like the other four European maps, the map does contain the Chinese river names 金沙江 (Item 40) and 安義河 (An Yi River; in annotation @5) and reveals its Chinese origin. It appears that Matteo Ricci was repeating what he did in depicting the *Terra Australis Incognita* on the KWQ by adopting an Euro-centric view: he extended the southern unknown land in Europeans' mind into a huge and expanded region with its coastal features remarkably looking like those in northern Australia, New Zealand, *Terra del Fuego*/Land of Fire in South America, and the Antarctic Sound,[48] while at the same time filling its interior void with all the exploratory data collected in the 1420s by the Chinese mariners under the command of Admiral Zheng He during their sixth voyage to the Western Ocean. Four hundred years after its publication, this portion of the map's Chinese origin is finally analysed and exposed in my previous book.[49]

Fig. 5.11.   The map is "A new description of China, once called the region of Sina, by Ludovicus Georgius" (public domain);[46] Ludovicus Georgius is the Latinized name of the Portuguese cartographer Luiz Jorge de Barbuda.

2.4   *Out of the 44 geographical items analysed on the ISC&EP-KWQ, the names of 18 of them (c. 41%) are not present on any of the four European maps; while there are seven annotations on the ISC&EP-KWQ, none of the four European maps provides any*

Hence, it is incorrect to say that the ISC&EP-KWQ is a direct or adapted copy of the three European maps. The following is a list of these 18 geographical names and their numbered positions in the Appendix.

---

1. Jaunpur (詔納僕兒/诏纳仆儿; Item 4); 2. *Varnu* (伐剌拏 or 伐剌拏; Item 5); 3. Xiao Tianzhu (小天竺; Item 6); 4. Iliyastan (阿利沙彈/阿利沙弹; Item 10); 5. Palavas (巴羅襪斯/巴罗袜斯; Item 12); 6. Coulam (哥爛/哥烂; Item 15); 7. Paravat (巴辣瓦得; Item 20); 8. Idalcan (伊達爾幹/伊達尔幹; Item 22); 9. Mughal (莫臥爾/莫卧尔; Item 29); 10. Gor (敖羅/敖罗; Item 30); 11. Xi Tianzhu or Western Tianzhou (西天竺國/西天竺国; Item 31); 12. Small Western Ocean (小西洋; Item 33); 13. One Hundred Thousand Islands (萬島/万岛; Item 35); 14. The Big Quicksand (大流沙; Item 37); 15. Jinsha River (金沙江; Item 40); 16. Mengyang (孟養/孟养; Item 41); 17. Dagula (大古剌; Item 42); 18. Region under the jurisdictions of the three Xuanwei Divisions (三宣地方; Item 44).

2.5   *There are seven geographical names of Chinese origin on the ISC&EP-KWQ, revealing explicitly the map's Chinese identity*

These items are (1) Xiao Tianzhu (Small Tiamzhu; 小天竺; Item 6); (2) Anyi River (安義河/安义河; @5); (3) Xi Tianzhu (Western Tianzhu; 西天竺國/西天竺国; Item 31); (4) One Hundred Thousand Islands (萬島/万岛; Item 35); (5) Big Quicksand (大流沙; Item 37); (6) Jinsha River (金沙江; Item 40); and (7) Region under the jurisdictions of the three Xuanwei Divisions (三宣地方; Item 44).

2.6   *There are only three geographical items of European origin on the ISC&EP-KWQ*

(1) Narasinga Kingdom (那心瓦國/那心瓦国; Item 13); (2) Small Western Ocean (小西洋; Item 33); and (3) the non-existent Lake Chiamay (加湖; Item 39).

3. Determining the Era of the ISC&EP-KWQ

Here are the four steps taken for determining the true era revealed by the ISC&EP-KWQ.

3.1 *The era revealed by the ISC&EP-KWQ is initially limited to the period 1394–1493, because the Muslim kingdom — Jaunpur Sultanate (詔納僕兒/诏纳仆儿; 1394–1493) — is depicted on the map*

Min Shi (明史) or *History of Ming* records that during the Yongle period (reigning from 1402 to 1424) of the Ming Dynasty, Jaunpur was Ming's tributary country. This is why the kingdom is not depicted on the ISC&EP-Mercator (1569), the ISC&EP-Ortelius (1570; #1), the ISC&EP-Plancius (1594), and the ISC&EP-Ortelius (1570; #2), because it was no longer in existence in the sixteenth century.

3.2 *The era revealed by ISC&EP-KWQ can be narrowed to 1430–1493*

On the ISC&EP-KWQ, Arakan (亞臘敢/亚腊敢; Item 3) faces the Bay of Bengal and is located to the west of Bengal (榜葛剌). This Arakan Kingdom seems to refer to the Mrauk-U Dynasty of Arakan established in the Bengal region from 1430 to 1785, hence, the era revealed by the ISC&EP-KWQ can be narrowed to 1430–1493.

3.3 *The last geographical data collected by the Chinese mariners, before being depicted on the ISC&EP-KWQ, must stem from the period 1432–1433, which falls within 1430–1493*

The Ming Treasure Fleet arrived at Calicut during its seventh (and last) voyage on its way out in 1432, and returned to Calicut in 1433 on its journey back home. After this voyage, the Ming Court stopped the treasure voyages abruptly and imposed the sea ban until the middle of the sixteenth century when the restrictions were lifted.

In my previous book titled *The Last Journey of the San Bai Eunuch, Admiral Zheng He*,[50] I have extracted the entire navigational routes, timelines, and evidence of Ming Treasure Fleet's seventh (and last) voyage from the epic work titled *San Bao Tai Jian Xi Yang Gi* (三寶太監西洋記/三宝太监西洋记)[51] or *An Account of the Western World Voyage of the San Bao Eunuch* written by Luo Maodeng (罗懋登/罗懋登) in 1597. I have shown in Chapter 3 of this book[52] that the treasure fleet led by Principal Envoy Zheng He (鄭和/郑和; 1371–1433 or 1435) arrived at Calicut (Kozhikode; 乜利客; Item 18; also known by Chinese as 古里; Gu-li) on 19 December 1432 and departed on 24 December 1432, while the fleet led by Deputy Envoy Hong Bao (洪保; 1370–unknown year) arrived earlier at Calicut on 10 December 1432 and left there on 14 December 1432. Although Zheng He seems to have disappeared from the Chinese official historical record after his last voyage, in my book I show that Zheng He might have sailed with his main fleet for Mecca, then North America, and then perhaps New Zealand (he could have passed away either in North America in an unknown year or in New Zealand in c. 1435).[53] Although the mainstream historians have proposed that he died of illness at Calicut in 1433 on his way returning to China, there has been neither hard nor soft evidence to support this claim thus far. Hong Bao's squadrons not only visited Calicut for four days in 1432, but his squadrons also returned to Calicut on 31 March 1433 after they visited Hormuz. There he regrouped his squadrons with the fleet led by the other Principal Envoy Wang Jinghong (王景弘; fourteenth century–after 1437) who joined him later at Calicut. Their joint fleet led by Wang Jinghong left on 9 April 1433 for China.

I point out in my book that Hong Bao's itinerary[54] in fact seems to match what is recorded in *Qian Wen Ji • Xia Xi Yang* (前闻记 • 下西洋)[55] or *A Record of History Once Heard: Down to the Western Ocean*, written by the Ming historian Zhu Yunming (祝允明; 1461–1527) about a portion of the treasure fleet's navigational routes during their seventh voyage.

Since both Hong Bao and Wang Jinhong were in Calicut and they returned to China together in July 1433, they were most able and likely to update their map (source map) of the ISC&EP-KWQ in 1432 and/or 1433 with their newly acquired geographical data. The years 1432–1433 fall in the range of 1430–1493. The historical facts of the rest of the geographical items depicted on the ISC&EP-KWQ and analysed in the Appendix also have no contradiction with this time range.

3.4 *The years 1432–1433 are also consistent with the drawing date in the period 1428–1433 of the KWQ's source map*

This map drawing date is deduced from Section 3 in Chapter 1 of my book titled *Chinese Global Exploration in the Pre-Columbian Era: Evidence from an Ancient World Map*, based on the annotation written next to the location of the Iberian Peninsula, the name of 安南 (Annan Kingdom) and the political condition revealed by the African continent on the KWQ.[56]

Combining 3.1 to 3.4 gives the true era of the ISC&EP-KWQ as 1432–1433.

4. Conclusions

This chapter aims to determine the origin and true era of the portion of the map which covers the Indian subcontinent and its eastern peripheral region on the map Kunyu Wanguo Quantu (坤輿萬國全圖/坤輿万国全图) (abbreviated as KWQ) or Complete Geographical Map of All the Kingdoms of the World. This portion of the map is abbreviated as ISC&EP-KWQ. The KWQ was published by Italian missionary Matteo Ricci in 1602 when he lived in China and was often regarded as of European origin.

In the Introduction, I have applied my previous methodology used for analysing the geographical information and histories of all the written texts on the different regions of the KWQ to determine the eras when these regions were last explored before the cartographers drew them onto the source map of the KWQ.

In Section 2, I have shown that the ISC&EP-KWQ is of Chinese origin in five main aspects in comparison with four major European maps in the sixteenth century: (1) the general geographical shape of the regions depicted on the map; (2) the river system depicted on the map; (3) the explanation of what led to the mystical Lake Chiamay on these five maps; (4) the name list of about 41% of the geographical items on the ISC&EP-KWQ being not depicted on any of the four European maps, and (5) seven geographical names (c. 16%) being of Chinese origin on the ISC&EP-KWQ, revealing explicitly the map's Chinese identity. The four European maps are abbreviated as ISC&EP-Mercator (1569), the ISC&EP-Ortelius (1570; #1), the ISC&EP-Plancius (1594) and the ISC&EP-Ortelius (1570; #2); they are extracted respectively from the 1569 World Map by Gerardus Mercator, the 1570 World Map by Abraham Ortelius, the 1594 World Map by Petrus Plancius, and the 1570 Ortelius Map of Asia (first edition)-*Asiae Nova Descriptio.*

In Section 3, I have deduced, based on analysing the histories of all the geographical items (44 in total) on the ISC&EP-KWQ, the era of the map as 1432–1433, based on the four facts: (1) the political era revealed by the map is 1394–1493 due to the depiction of the Muslim kingdom Jaunpur Sultanate (詔納僕兒/诏纳仆儿; 1394–1493); (2) the Arakan Kingdom on the ISC&EP-KWQ refers to the Mrauk-U Dynasty of Arakan established in the Bengal region from 1430 to 1785, hence, the era of the ISC&EP-KWQ can be narrowed to 1430–1493; (3) the dates of Ming Treasure Fleets' visits of Calicut were in 1432 during their seventh voyage on their way out in 1432 and during the fleets' seventh voyage on their way returning to China in 1433; the Ming mariners were most able and likely to update their map (source map) of the ISC&EP-KWQ in 1432 and/or 1433 with their newly acquired survey data; hence the era of the ISC&EP-KWQ can be further barrowed to 1432–1433, which falls within 1430–1493; and (4) the drawing date of the entire source map of Kunyu Wanguo Quantu was 1428–1433, deduced from the annotation next to the location of the Iberian Peninsula on the KWQ; but he political era of Africa on the KWQ is deduced as 1433; hence, the entire source map of the KWQ was drawn in 1433. The era, 1432–1433, for the Indian Subcontinent and its Eastern periphery being explored by the Chinese is no later than 1433 and is consistent with the drawing date of the source map of the KWQ.

What Matteo Ricci contributed to the ISC&EP-KWQ is adding several European names–Narasinga Kingdom (那心瓦國/那心瓦国; Item 13), Mughal (莫卧尔; 1526–1857; Item 29), Small Western Ocean (小西洋; Item 35)–and introducing the mystical and non-existent Lake Chiamay (加湖; Item 39).

Appendix

This appendix analyses all the geographical items and annotations on the ISC&EP-KWQ. These geographical items and annotations are categorised into two zones, Zone I: the Indian subcontinent and Zone II: the eastern periphery to the Indian subcontinent. On the map, they are almost all written in Chinese traditional characters. Since many readers use the Chinese simplified characters, this book uses both: all the Chinese characters in the main text and in the Appendix of this chapter are expressed first by the modern Chinese traditional characters, then followed by the Chinese simplified characters, separated by a slash "/" between them. If both characters are the same, that expression will be presented only once without repetition. Moreover, all the geographical items are numbered in sequence in this Appendix. The annotations are numbered as well with an "@" sign in front of the number. The "•" in front of an item number denotes that this item does not appear on any of the four sixteenth-century European maps. The Chinese "pinyin" version of each item is given right after it is introduced. My comments are given in the last column of Appendix.

	Indian Subcontinent and Its Eastern Periphery			
Item #	Name	Pinyin	Etymology	History, Geography and My Comments

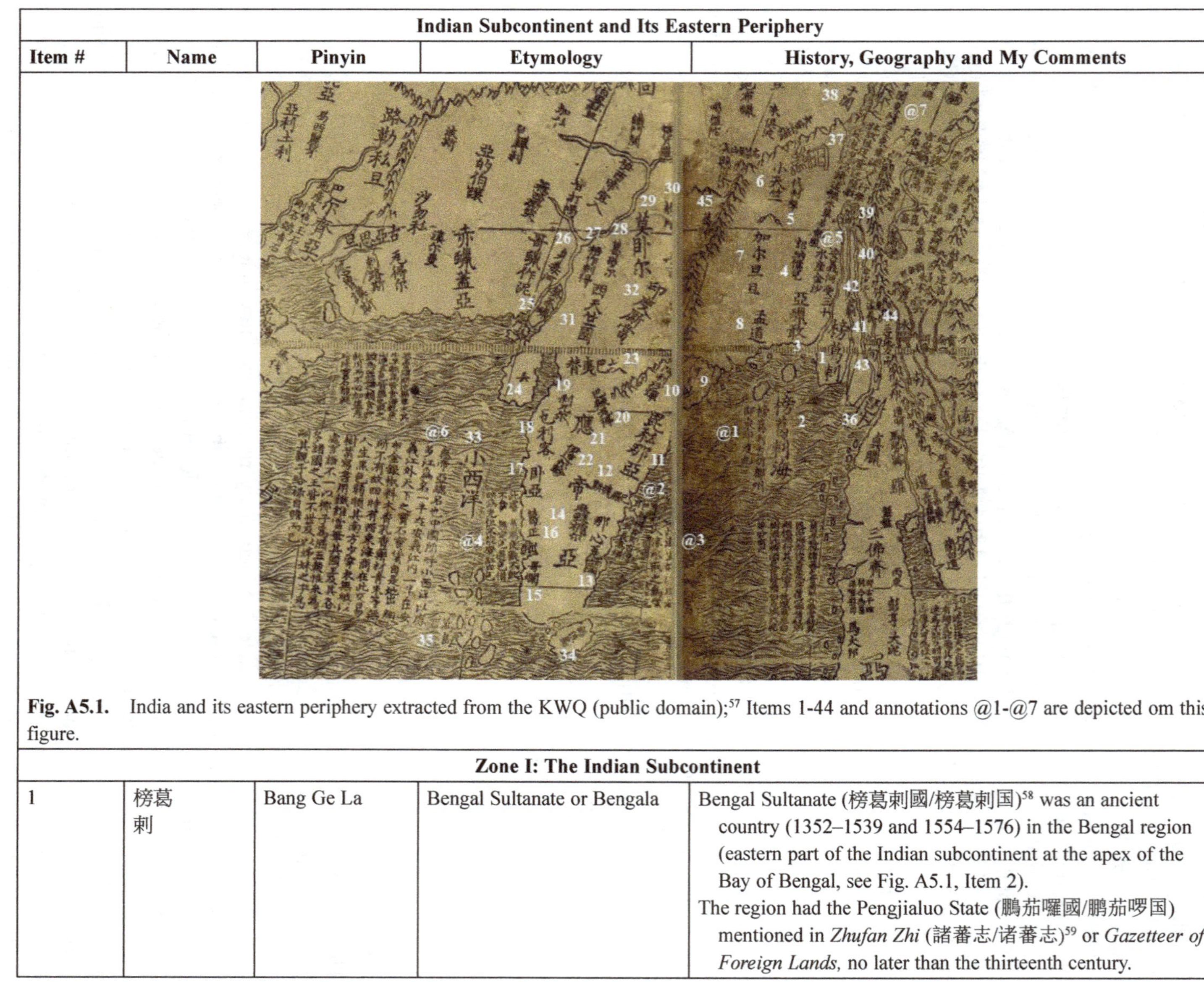

Fig. A5.1. India and its eastern periphery extracted from the KWQ (public domain);[57] Items 1-44 and annotations @1-@7 are depicted om this figure.

	Zone I: The Indian Subcontinent			
1	榜葛剌	Bang Ge La	Bengal Sultanate or Bengala	Bengal Sultanate (榜葛剌國/榜葛剌国)[58] was an ancient country (1352–1539 and 1554–1576) in the Bengal region (eastern part of the Indian subcontinent at the apex of the Bay of Bengal, see Fig. A5.1, Item 2). The region had the Pengjialuo State (鵬茄囉國/鹏茄啰国) mentioned in *Zhufan Zhi* (諸蕃志/诸蕃志)[59] or *Gazetteer of Foreign Lands,* no later than the thirteenth century.

In 1408 (the sixth year of Yongle's reign), a Bengal envoy was sent to establish friendly relations with China, and China also sent an envoy to pay a return visit. *Yingya Shenglan* (瀛涯勝覽/瀛涯胜览) or *The Overall Survey of the Ocean's Shores* written by Ma Huan (馬歡/马欢; c. 1380–1460) in 1451, *Xingcha Shenglan* (星槎勝覽/星槎胜览) or *The Overall Survey of the Star Raft* written by Fei Xin (費信/费信; c. 1385/1388–after 1436) in 1436 and *Ming Shi* (明史) or *History of Ming* written by Zhang Tingyu (張廷玉/张廷玉; 1672–1755), *et al.* in 1739, all have records about the country. Here is the record from the *History of Ming*:[60]

榜葛剌, 即漢身毒國, 東漢曰天竺。。。唐亦分五天竺, 又名五印度。宋仍名天竺。榜葛剌則東印度也。自蘇門答剌順風二十晝夜可至。

My translation and comments (between square brackets) are as follows:

Bengala was the Country of Shendu [身毒國/身毒国] in the Han Dynasty [漢朝/汉朝; 202 B.C.–9 A.D. and 25–220 A.D.; 9–23 A.D.; Xin Dynasty, also known as Xin Mang or 新莽], and it was called Tianzhu [天竺] in the Eastern Han Dynasty [東漢/东汉; 25–220 A.D.] ... In the Tang Dynasty, it [the Bengal region] was divided into five Tianzhus, known as the Five Regions of India [五印度]. In the Song Dynasty, it was still named Tianzhu and was regarded as Eastern India [東印度/东印度]. Bengal can be reached by ship from Sumatra [蘇門答剌/苏门答剌] in twenty days and nights with favourable winds.

In the Yongle era, Admiral Zheng He (鄭和/郑和) and Eunuch Wang Jinghong (王景弘) visited Bengal several times during their voyages to the Western Ocean. The country presented

(*Continued*)

Indian Subcontinent and Its Eastern Periphery				
Item #	Name	Pinyin	Etymology	History, Geography and My Comments
				famous horses and qilins (referring to giraffes) to the Ming Court. During their seventh (and last voyage) in Xuande's reign, the Ming Treasure Fleets visited Bengala in 1433.[61] On the ISC&EP-Mercator (1569), "Bengala" is depicted. On the ISC&EP-Ortelius (1570; #1), the Bengal region is not denoted. But on the ISC&EP-Ortelius (1570; #2), there is "BENGALA". On the ISC&EP-Plancius (1594), "Bengala" is depicted.
@ 1	榜葛剌古忻都州即東印度也	Bang Ge La Gu Xin Du Zhou Ji Dong Yin Du Ye		My translation of the left annotation and my comment (between square brackets) are as follows: Bengala was the ancient Xindu [忻都] state, and that is Eastern India. The above record can be seen on *Xu Wenxian Tongkao • Di Shi Jiu Juan* (續文獻通考 • 第十九 卷/续文献通考 • 第十九 卷) or *Continued Comprehensive Investigations Based on Literary and Documentary Sources, Juan (Volume) 19.*[62]
2	榜葛剌海	Ban Ge La Hai	Bay of Bengal	The Bay of Bengal (榜葛剌海) is the northeastern part of the Indian Ocean as shown in Fig. A5.2. In the sixteenth century the Portuguese built trading posts in the north of the Bay of Bengal.[63] However, from Item 1, we know that in the early fifteenth century the Bengal Sultanate and the Ming Court already sent envoys to each other, and Ma Huan and Fei Xin described Bengal in their books, revealing that the Chinese mariners knew the Bay of Bengal well, because they had to sail through it to reach Bengal during their voyages. The ISC&EP-Mercator (1569) and the ISC&EP-Ortelius (15970; #1) depict a bay where the present-day Bay of Bengal should be, but without giving it a name. But on the ISC&EP-Ortelius (15970; #2), there is "GOLFO DI BENGALA". The ISC&EP-Plancius (1594) depicts the Bay of Bengal as "Golfo Bengala".

| 3 | 亞臘敢 | Ya La Gan | Arakan[65] | 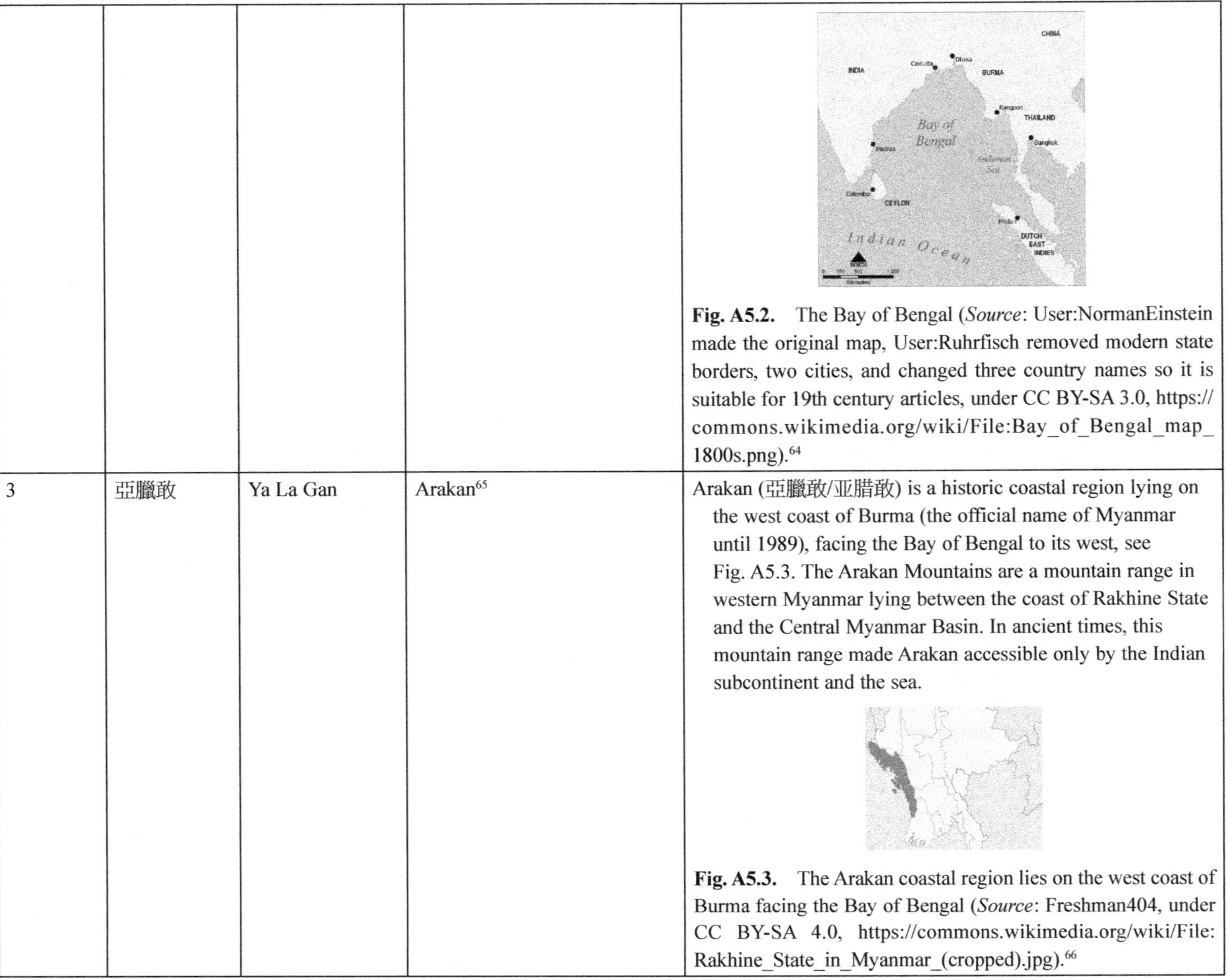

Fig. A5.2. The Bay of Bengal (*Source*: User:NormanEinstein made the original map, User:Ruhrfisch removed modern state borders, two cities, and changed three country names so it is suitable for 19th century articles, under CC BY-SA 3.0, https://commons.wikimedia.org/wiki/File:Bay_of_Bengal_map_1800s.png).[64]

Arakan (亞臘敢/亚腊敢) is a historic coastal region lying on the west coast of Burma (the official name of Myanmar until 1989), facing the Bay of Bengal to its west, see Fig. A5.3. The Arakan Mountains are a mountain range in western Myanmar lying between the coast of Rakhine State and the Central Myanmar Basin. In ancient times, this mountain range made Arakan accessible only by the Indian subcontinent and the sea.

Fig. A5.3. The Arakan coastal region lies on the west coast of Burma facing the Bay of Bengal (*Source*: Freshman404, under CC BY-SA 4.0, https://commons.wikimedia.org/wiki/File:Rakhine_State_in_Myanmar_(cropped).jpg).[66] |

(*Continued*)

		Indian Subcontinent and Its Eastern Periphery		
Item #	Name	Pinyin	Etymology	History, Geography and My Comments
				The Burmese invaded the Arakan region in 1406. After the invasion, the last king of the Launggyet Dynasty (Laymro Kingdom of Arakan; 1237/1251– 1429) in the region fled to Bengal and founded the Mrauk-U Dynasty of Arakan (Kingdom of Mrauk-U; 1430–1785).[67] Finally, the Konbaung Dynasty also known as the Third Burmese Empire (1752–1885) conquered Arakan in 1785.[68] On the ISC&EP-KWQ, Arakan faces the Bay of Bengal and is located to the west of the Bengal (榜葛 剌) region. This Arakan seems to refer to the Mrauk-U Dynasty of Arakan established in the Bengal region from 1430 to 1785. The ISC&EP-Mercator (1569) depicts Arakan facing the Bay of Bengal, but the ISC&EP-Plancius (1594) depicts Arakan as "Aracam" in the inland area. Neither the ISC&EP-Ortelius (1570; #1) nor the ISC&EP-Ortelius (1570; #2) depicts Arakan.
• 4	詔納僕兒	Zhao Na Pu Er	Jaunpur[69]	The Jaunpur Sultanate (詔納僕兒蘇丹國/诏纳仆儿苏丹国; 1394–1493)[70] was a Persianate Muslim kingdom in northern India ruled by the Sharqi Dynasty.[71] The *Xu Wenxian Tongkao* (續文獻通考/续文献通考) or *Continued Comprehensive Investigations Based on Literary and Documentary Sources* has this record:[72] 沼纳朴儿在榜葛迤西, 古天竺国也。居印度之中，又名金刚宝座国。乃释迦得道之所, 明永乐中遣使诏谕国王, 一不剌金遣人来朝贡。 My translation and comments (between square brackets) are as follows: Jaunpur is to the west of Bengal, which was the ancient Country of Tianzhu [古天竺國/古天竺国]. It is in the middle of India and known as the Vajra Throne Kingdom [金剛寶座國/金刚宝座国]. This

				was the place where Sakyamuni [The Buddha] attained his enlightenment. During the Ming Yongle's era, the Ming emperor sent envoys to edict the king [to become a tributary kingdom to Ming]. The ruler Shams ud-din Ibrahim Shah Sharqi [沙姆斯·乌丁·一不剌金·沙阿·沙尔奇; ruled 1401–1440][73] started to send envoys to pay tribute. In *Min Shi* (明史) or *History of Ming*, it records that during the Yongle's reign of the Ming Dynasty, Jaunpur was Ming's tributary country.[74] Since The ISC&EP-KWQ depicts the Jaunpur Sultanate which existed about 100 years from 1394 to 1493, the political era revealed by the map can be determined initially as 1394–1493. This era will be narrowed afterwards by other considerations as explained in Section 3 in the main text of this chapter.
● 5	伐剌拏	Fa La Na	伐剌拏 is the transliteration of Sanskrit *Varnu*.[75]	Varnu (伐剌拏; also known as 伐剌拏) was an ancient kingdom (Country of Fa-la-na or Varana or Varnu)[76] in India. It is mentioned *in Da Tang Da Ci En Si San Cang Fa Shi Chuan* (大唐大慈恩寺三藏法師傳/大唐大慈恩寺三藏法师传)[77] or *A Biography of the Tripitaka Master of the Great Ci'en Monastery of the Great Tang Dynasty.* The location of the kingdom may be called Vaneh in the middle reaches of the Gumal River (古瑪爾河/古玛尔河), a tributary of the Indus River in present-day Pakistan; or it may be called Banna (伐奈) on the banks of the Kuram River (庫臘河/库腊河).
● 6	小天竺	Xiao Tian Zhu	Xiao Tianzhu	Tianzhu (天竺) was the old name in China and Eastern Asia for India. The name is described in detail as Shendu (身毒; the transliteration of Sanskrit *Sindhu* for Indus River)[78] in *Hou Han Shu · Xi Yu Chuan* (後漢書·西域傳/后汉书·西域传) or *The Book of the Later Han · Biography of the Western Regions*.[79] The name also referred to the entire South Asian subcontinent

(Continued)

Indian Subcontinent and Its Eastern Periphery				
Item #	Name	Pinyin	Etymology	History, Geography and My Comments
				in a broad sense. 身毒國/身毒国 is also mentioned multiple times in *Shi Ji • Juan Yi Bai Er Shi San • Da Wan Lie Chuan* (史記 • 卷一百二十三 • 大宛列傳/史记 • 卷一百二十三 • 大宛列传) or *Records of the Grand Historian, Juan (Volume) 123: Biography of Da Wan.* 小 means "small", 小天竺 means "a small part of the entire Tianzhu.". Here 小天竺 (Xiao Tianzhu) on the ISC&EP-KWQ seems to refer to the Northern Tianzhu based on its geographical location in the northern part of India. The ancient Indians divided the South Asian subcontinent into five parts: east, west, south, north, and central regions according to five directions. The ancient Chinese, even before the Eastern Jin Dynasty (東晉/东晋; 316–420), agreed with this classification and put forward the concept of "Five Tianzhus or *Wutianzhu* (五天竺)"[80] for the South Asian subcontinent. Around that time, Chinese monk and traveller Fa Xian (法顯/法显; 337–422) went to India for the first time to inspect some areas of the *Wutianzhu* (五天竺) and wrote records about them. The Indians themselves did not directly propose the phrase *Wutianzhu*, and in the Buddhist scriptures there is also no mention of *Wutianzhu*.[81] However, since the ancients of the Indian subcontinent had already divided their land into east, west, south, north, and central regions according to five directions, then the appearance of *Wutianzhu* or "Five Tianzhus" is reasonable. According to Indian Buddhists, Shakyamuni was born in Central Tianzhu (中天竺).

				The word *Wutianzhu* (五天竺) is not explicitly mentioned in *Fa Xian Zhuan* (法顯傳/法显传) or *A Record of Buddhistic Kingdoms, Being an Account by the Chinese Monk Fa Xian.* But the South Asian subcontinent is already collectively called Tianzhu in the book.[82] Since Northern Tianzhu, Central Tianzhu and Southern Tianzhu are mentioned in the biography of *Fa Xian Zhuan*, this clearly shows that Fa Xian already had in his mind the concept of Tianzhu being divided into five parts. But in *Fa Xian Zhuan*, there is only one mention of Southern Tianzhu and Western Tianzhu, and no record of Eastern Tianzhu.[83] Later Xuanzang (玄奘; 602–664) was the founder of Chinese Buddhist Dharma, one of the four great translators in China, and one of the greatest translators of Chinese Buddhism. He was the only person who went to all five regions in India.[84] The famous *Da Tang Xiyu Ji* (大唐西域記/大唐西域记) or *Great Tang Records on the Western Regions* was dictated by Xuanzang and written by his disciples. It is an important document for the study of ancient Indian history. While traveling around *Wutianzhu* (五天竺) or "Five Tianzhus", Xuanzang changed the "Five Tianzhus" to the "Five Regions of India (五印度)" and recorded in detail the divisions of India.[85] From the mentioning of "Five Tianzhus (五天竺)" to "Five Regions of India (五印度)", we know that the ancient "India", before the British rule, was not a country, but a region of a thoroughly divided landmass.
7	加尔旦旦	Jia Er Dan Dan	Cardandan[86]	The place cannot be identified on a modern map of India. On the ISC&EP-Ortelius (1570; #2), there is CARDANDAN.
To the west of 加尔旦旦, there is a place name starting with "葛", but since the remaining part is missing. This item will not be discussed in this Appendix.				
8	孟道	Meng Dao	Mandao[87]	Mandao (孟道; 22.7°N, 74.1°E) is a place in the region of Gujarat in India.

(Continued)

Indian Subcontinent and Its Eastern Periphery				
Item #	Name	Pinyin	Etymology	History, Geography and My Comments
				On the ISC&EP-KWQ, 孟道 is depicted at c. 25.0°N (KWQ latitude → modern latitude), c. 86°E. The latitudinal reading is reasonably good, but the longitudinal reading is too far eastward by almost 12°. On the ISC&EP-Mercator (1569), Mandao is depicted at c. 24.3°N, c. 100.5°E. The latitudinal reading is reasonably good, but the longitudinal reading is too far eastward by more than 26°. On the ISC&EP-Ortelius (1570; #1), Mandao is depicted at c. 26.0°N, c. 93.5°E. The latitudinal reading is less good, and the longitudinal reading is also too far eastward by almost 20°. On the ISC&EP-Ortelius (1570; #2), there is "MENDAO" depicted at c. 31.5°N, c. 102.5°E. On the ISC&EP-Plancius (1594), Mandao is depicted at c. 23.5°N, c. 96.8°E. The latitudinal reading is very good, but the longitudinal reading is too far eastward by almost 23°.
9	何里沙	He Li Sha	Orissa was the former name of Odisha;[88] both terms *Odisha* and *Orissa* derive from the ancient Prakrit word and can be dated to 1025.[89]	Orissa (何里沙) is the former name of the state of Odisha in Eastern India. The year 1568 is considered a pivotal point in the region's history. In 1568, the region was conquered by the armies of the Sultanate of Bengal and integrated into the Mughal Empire[90] (莫臥兒帝國/莫卧儿帝国). After 1568, the region lost its political identity. On the ISC&EP-Mercator (1569), Orissa is depicted as "Orixa". On the ISC&EP-Ortelius (1570; #1), Orissa (何里沙) is depicted as "Orixa" and on the ISC&EP-Ortelius (1570; #2) there is "ORISSA". However, on the ISC&EP-Plancius (1594), "Orixa" is depicted in BENGALA, best reflecting the political change there after 1568.
● 10	阿利沙彈	A Li Sha Dan	Iliyastan[91]	The place cannot be identified on a modern map.

11	毘私那亞	Pi Si Na Ya	Bisnaga Kingdom[92]	The Bisnaga (毘私那亞/毗私那亚) Kingdom was the old name of the Vijayanagara Empire[93] (1336–1646; also known as the Karnata Kingdom)[94] which occupied much of Southern India.[95] In 1520, Portuguese traveller Domingo Paes visited Vijayanagara and wrote his memoir as *Chronica dos reis de Bisnaga.*[96] Italian merchant, explorer, and writer Niccolò de' Conti (c. 1395–c. 1469) visited this empire for its wealth and fame. He was one of the sources used to create the Fra Mauro World Map in the 1450s. The map shows a sea route from Europe around Africa to India in the 1420s.[97] De' Conti's travel overlapped with Zheng He's voyages in the early fifteenth century. The ISC&EP-Plancius (1594) depicts Bisnaga as "Bissnagar".
● 12	巴羅襪斯	Ba Luo Wa Si	Palavas[98]	Palavas (巴羅襪斯/巴罗袜斯) was in the region of the Coromandel Coast on the east coast of the Indian Peninsula.[99]
@ 2	巴羅襪斯即古儋耳人以耳長爲媚	Ba Luo Wa Si Ji Gu Dan Er Ren Yi Er Chang Wei Mei		My translation and comment (between square brackets) are as follows: Palavas was the ancient Dan Er [儋耳]. The people there were proud of having long ears. According to the record of *Han Shu* (漢書/汉书) or *Book of Han*: "儋耳者，大耳种也", the meaning is "people with Dan'er (儋耳) or have long/big ears or ears with long/big lobes". The record of *Shan Hai Jing* (山海經/山海经) or *Classic of Mountains and Seas* also mentions "南荒之外，有离耳国，其人耳长及肩，每逆风走，则将耳反搭". This is translated by Sheng-Wei Wang as "Beyond the Southern Wilderness, there was the Country of Li'er (離耳/离耳), where people had long ears reaching their shoulders; whenever they walked against the wind, their ears will turn backward falling on their shoulders".

(Continued)

Indian Subcontinent and Its Eastern Periphery				
Item #	Name	Pinyin	Etymology	History, Geography and My Comments
				The meaning of Li Er or Li'er (離耳/离耳) is like that of Dan Er or Dan'er (儋耳). However, historical records show that the ancient Country of Dan'er (儋耳) was in today's Hainan Province of China, not in India. Besides, did people really have such long ears? On Hainan Island, Dan'er was the common name of Danzhou in ancient times. There was an ancient custom in Danzhou: people liked to use shells as currency for exchange and circulation. After the Qin Dynasty (秦朝; 221–206 B.C.) unified the Chinese currency, shells became ornaments for earrings. Residents often wore long earrings hanging down to their shoulders as ornaments, which were called "Dan". The earrings were called "Er", collectively known as "Dan'er". Later "Dan'er" was transformed into the name of the tribe and the name of the place where they lived. The shells worn on both ears hang down to the shoulders and, seen from a distance, it seemed that the ears fell on their shoulders.[100]
@ 3	此海生好珍珠海人沬水取之爲業	Ci Hai Sheng Hao Zhen Zhu Hai Ren Mei Shui Qu Zhi Wei Ye		My translation and comment (between square brackets) are as follows: This sea [Bay of Bengal] produces good pearls, and people who depend on sea for living would dive into the water to take pearls as their occupation. "This sea" refers to the Bay of Bengal. One of the most famous types of pearls found in the Bay of Bengal is known as the "Basra Pearl".[101] It is highly valued for its luster, color, and size, and has been harvested for centuries from the waters of the Bay of Bengal.
13	那心瓦國	Na Xin Wa Guo	Narasinga Kingdom[102]	Europeans referred to the Vijayanagara Empire (1336–1646) as "The Kingdom of Narasinga (那心瓦國/那心瓦国)",[103] a name derived from "Narasimha" by the Portuguese.[104] The kingdom was ruled by Narasingha Raya (1503–1505) when

				the Portuguese first arrived in India. Notice the small difference between Narasingha and Narasinga. The two geographical terms Bisnaga Kingdom (毘私那亞/毗私那亚; Item 11) and Narasinga Kingdom (那心瓦國/那心瓦国; Item 13) in fact correspond to the same Vijayanagara Empire (1336–1646); Bisnaga was the old name of the Vijayanagara Empire and Narasinga was the European name for the empire. On the ISC&EP-KWQ, both names are used to denote the same empire which are depicted at two different locations. On the ISC&EP-Mercator (1569), there is only "Bisnagar", no Narasinga. On the ISC&EP-Ortelius (1570; #1), there is only *Narsiga*, no Bisnaga; but on the ISC&EP-Ortelius (1570; #2), there are NARSINGA and BISNAGA. On the ISC&EP-Plancius (1594), there are both "Bisnnagar" and "Narsinga".
14	麻辣襪尔	Ma La Wa Er	Malabar[105] was a general name for Kerala in foreign trade circles until the arrival of the British.[106]	Malabar (麻辣襪尔/麻辣袜尔) is a coastal region in southwestern India, principally the present state of Kerala. The area is famous for its history as a major spice trade center. This region was the port of first call for Vasco da Gama on his first voyage in 1497 to India.[107] **Fig. A5.4.** Map of the region of the Malabar Coast, India (*Source*: w:user:Planemad, under CC BY-SA 3.0, https://commons.wikimedia.org/wiki/File:India_Malabar_Coast_locator_map.svg).[108]

(*Continued*)

Indian Subcontinent and Its Eastern Periphery				
Item #	Name	Pinyin	Etymology	History, Geography and My Comments
				The ISC&EP-KWQ erroneously depicts Malabar (麻辣襪尔/麻辣袜尔) as an inland region on the side of the Gulf of Bengal. On the ISC&EP-Ortelius (1570; #2), there is "MALABAR" like what is shown on a modern map in Fig. A5.4. On the ISC&EP-Plancius (1594), there is a "Maliapor", but it is facing the Bay of Bengal, not the Indian Ocean. Hence, it is unlikely to be the Malabar region.
● 15	哥爛	Ge Lan	Coulam[109]	The place Coulam (哥爛/哥烂) cannot be identified on a modern map.
16	葛正	Ge Zheng	Cochin or Kochi	Cochin or Kochi (葛正), see Fig. A5.5, is a major port city on the Malabar Coast of India bordering the Laccadive Sea (拉克代夫海), which is a part of the Arabian Sea (阿拉伯海). Cochin was an important spice trading center on the west coast of India from the fourteenth century onward. It maintained a trade network with Arab merchants from the pre-Islamic era. Cochin was occupied by the Portuguese in 1503;[110] it was the first of the European colonies in colonial India. In 1505, the Portuguese established the Portuguese India (1505–1961). **Fig. A5.5.** Old Kochi (*Source*: Vyacheslav Argenberg /http://www.vascoplanet.com/, under CC BY-SA 4.0, https://commons.wikimedia.org/wiki/File:Kochi,_Fishing_nets_at_sunset,_Kerala,_India.jpg).[111]

				The earliest documented references to Cochin occurred in the 1451 *Yingya Shenglan* (瀛涯勝覽/瀛涯胜览) or *The Overall Survey of the Ocean's Shores*[112] written by Chinese voyager Ma Huan (馬歡/马欢; c. 1380–1460). Ma Huan visited Cochin in the early fifteenth century during his voyages as part of Admiral Zheng He's treasure fleet (Ma Huan accompanied Zheng He to sail the Western Ocean during the fourth, sixth and seventh voyages). There are also references to Cochin in accounts written by Italian traveller Niccolò de' Conti (c. 1396–1469), who visited Cochin in 1440. In 1411 (the ninth year of Yongle's reign), Cochin sent envoys and tributes to Ming[113] and Admiral Zheng He visited the country in 1412 (the tenth year of Yongle's reign).[114] Cochin sent in tribute for two consecutive years after that. Per Cochin envoy's request, the Ming Yongle Emperor instructed Zheng He to confer a seal upon Keyili of Cochin's ruler and to enfeoff a mountain in Cochin's kingdom as the Zhenguo Zhi Shan (鎮國之山/镇国之山, lit. "Mountain Which Protects the Country").[115] Zheng He delivered a stone tablet, inscribed with a proclamation composed by the Yongle Emperor himself, to Cochin.[116] In 1430 (the fifth year of Xuande's reign), Zheng He was also sent back to give instructions to Cochin. In 1433 (the eighth year of Xuande's reign), Cochin sent envoys together with envoys of Sri Lanka to China to pay tribute.[117] On the Malabar Coast during the early fifteenth century, the dominant port-city Calicut and the emerging port-city Cochin were in an intense rivalry. Since Cochin was under the protection of Ming China, Calicut was unable to invade Cochin and a military conflict was averted.

(*Continued*)

Indian Subcontinent and Its Eastern Periphery				
Item #	Name	Pinyin	Etymology	History, Geography and My Comments
				However, the cessation of the Ming treasure voyages after the seventh voyage around the middle of 1430s had negative impacts on Cochin, as Calicut would eventually launch an invasion of Cochin. In the late fifteenth century, the ruler — Zamorin of Calicut (1124–1806) — occupied Cochin and installed his representative as the king of the port-city.[118] Both the ISC&EP-Mercator (1569) and the ISC&EP-Plancius (1594) depict Cochin, the ISC&EP-Ortelius (1570; #1) does not, but the ISC&EP-Ortelius (1570; #2) does.
@ 4	此處有革馬良獸不飲不食身無定色遇色借映為光但不能變紅白色	Ci Chu You Ge Ma Liang Shou Bu Yin Bu Shi Shen Wu Ding Se Yu Se Jie Ying Wei Guang Dan Bu Neng Bian Hong Bai Se	My translation and comment (between square brackets) are as follows: There is a kind of lizard called Chameleon or chamaeleon [革馬良/革马良]. It does not need to drink or eat, and its body has no fixed colour. When its body shows a colour, it reflects the colour as light, but it cannot reflect red and white colour. The English word "chameleon" is a simplified spelling of Latin *chamaeleōn*,[119] a borrowing of the Greek χαμαιλέων (*khamailéōn*).[120]	My translation and comment (between square brackets) are as follows: There is a kind of lizard called Chameleon or chamaeleon [革馬良/革马良]. It does not need to drink or eat, and its body has no fixed colour. When its body shows a colour, it reflects the colour as light, but it cannot reflect red and white colour. The English word "chameleon" is a simplified spelling of Latin *chamaeleōn*,[119] a borrowing of the Greek χαμαιλέων (*khamailéōn*).[120] Chameleon or chamaeleon are a distinctive and highly specialized clade of Old-World lizards.[121] They have distinct range of colours, being capable of shifting to different hues and degrees of brightness. They climb trees and are good at wrapping around tree trunks.
17	卧亞	Wo Ya	Goa; the origin of the city name "Goa" is unclear.	Goa (卧亞/卧亚) is a state in India, comprising a mainland district on the country's southwestern coast and an offshore island. In 1312, Goa was under the governance of the Delhi Sultanate (1206–1526).[122] But by 1370 it was forced to surrender to the Vijayanagara Empire (1336–1646)[123] until 1469 when this territory was appropriated by the Bahmani

				sultans of Gulbarga. After that, the area fell into the hands of the Adil Shahis of Bijapur (1490–1686),[124] and a city was founded and served as the capital of Portuguese Indian possessions to replace Govapuri (a port lies a few kilometres to the south). This city was known under the Portuguese as *Velha Goa* (or Old Goa which was the name first used in the 1960s).
				In 1510, the Portuguese set up a permanent settlement in *Velha Goa*, marking the beginning of Portuguese colonial rule in Goa that would last until its annexation by India in 1961.[125] After 1510 the city gave its name to the contiguous territories.
				On the ISC&EP-KWQ, Goa (卧亞; Item 17) is depicted to the south of Calicut (乜利客; Item 18); in fact, it is North of Calicut.
				"Goa" is depicted on all the four European maps.
18	乜利客	Mie Li Ke	Calicut or Kozhikode; the exact origin of the name Kozhikode is uncertain.[126] It is called "Guli (古里)" in the Ming Dynasty books.	Kozhikode (乜利客), or Chinese Guli (古里), also known as Calicut, was the kingdom of the Zamorin of Calicut (1124–1806), in the present-day Indian state of Kerala. Present-day Kozhikode is the second largest city in Kerala, as well as the headquarters of the Kozhikode district. In 1498, Vasco da Gama (c. 1460s–1524) landed in Kappad,[127] near Calicut.
				The king of Calicut started to send tribute to Ming China in 1403 (the first year of Yongle's reign), then in 1405 (the third year of Yongle's reign), and then paid tribute every year.[128]
				Admiral Zheng He's fleet sailed to the Western Ocean seven times and visited Guli (古里) seven times. During his first voyage in the winter of 1405 (the third year of Yongle's reign), he visited Guli for the first time. In 1407 (the fifth year of Yongle's reign) during his second voyage, the Guli

(*Continued*)

Indian Subcontinent and Its Eastern Periphery				
Item #	Name	Pinyin	Etymology	History, Geography and My Comments
				king accepted the Ming imperial edict, and the silver seal of the imperial edict sent by the Yongle Emperor of the Ming Dynasty conferred him the title of King of Guli.[129] Zheng He also built a pavilion with stone tablet in Guli to commemorate the event.[130] Zheng He's fleet used Guli as a base for supplying fresh water and food for the fleet's westward voyage into the Arabian Sea, the African coast, and the regions beyond. Calicut was dubbed the "City of Spices", and its port acted as gateway to the medieval South Indian coast for the Persians, the Arabs, the Chinese, and finally the Europeans. In the fifteenth century, Calicut was the most powerful kingdom on the medieval Malabar Coast, reducing the status of Cochin to a vassal state of Calicut. On the ISC&EP-Ortelius (1570; #1), the ISC&EP-Ortelius (1570; #2), and the ISC&EP-Plancius (1594), Calicut is depicted as "Calecut".
19	刹兀尔	Cha Wu Er	Kannur; the pronunciation of Kannur is closer to 刹兀尔 than "Caul" suggested by other scholars.[131] But Caul and Chual are depicted at where Kannur should be on three of the four European maps discussed in this chapter.	Kannur (刹兀爾/刹兀尔), formerly known in English as Cannanore, is a city in the state of Kerala, India. Cannanore was the headquarters of Kolathunadu (c. sixth century B.C.–modern era),[132] one of the four most important dynasties on the Malabar Coast, along with the Zamorin of Calicut (1124–1806),[133] Kingdom of Cochin (c. twelfth century–1947)[134] and Kingdom of Quilon or Venad (c. eighth, ninth century or twelfth century–1729).[135] On the medieval (465–1453) Kerala coast, the port at Calicut held the superior economic and political position over several important but secondary trading ports connecting Persia, Arabia and China, including Kannur, Kollam, and Cochin. On the ISC&EP-Mercator (1569) and the ISC&EP-Plancius (1594), "Chaul" is depicted at where Kannur (刹兀爾/刹兀尔) should be. On the ISC&EP-Ortelius (1570; #2), "Caul" is depicted at where Kannur (刹兀爾/刹兀尔) should be.

• 20	巴辣瓦得	Ba La Wa De	Paravat[136]	The place may be today's Bhadravati in India.
21	應帝亞	Ying Di Ya	India; the name is derived from the Classical Latin *India* which was a reference to South Asia and an uncertain region to its east.[137]	From 600 to 1200, the Indian early medieval age was defined by regional kingdoms and cultural diversity.[138] In the fifteenth century, the Vijayanagara Empire (毗奢耶那伽羅帝國/毗奢耶那伽罗帝国; 1336–1646) created a composite Hindu culture in south India.[139] Starting in the sixteenth century, the Mughal Empire (莫臥兒帝國/莫卧儿帝国) maintained about two centuries of relative peace from 1526.[140] In the seventeenth century, the British East India Company (英國東印度公司/英国东印度公司; 1600–1874) gradually turned India into a colonial economy, and initiated the beginnings of the British Empire in India.[141] India (應帝亞/应帝亚) is now a country in South Asia. Present-day India is bounded by the Indian Ocean on the south, the Arabian Sea on the southwest, and the Bay of Bengal on the southeast. It shares land borders with Pakistan to the west; China, Nepal (尼泊爾/尼泊尔) and Bhutan (不丹) to the north; and Bangladesh (孟加拉) and Myanmar (緬甸/缅甸) to the east. In the Indian Ocean, India's neighbouring countries are Sri Lanka (斯里蘭卡/斯里兰卡) and the Maldives (馬爾地夫/马尔代夫); its union territories of the Andaman (安達曼/安达曼) and Nicobar (尼科巴) Islands share a maritime border with Thailand, (泰國/泰国) Myanmar, and Indonesia (印尼). On the ISC&EP-KWQ, "India" (應帝亞/应帝亚) matches closely the present-day India. On the ISC&EP-Mercator (1569), the name "India" is not depicted. On the ISC&EP-Ortelius (1570; #1), there is an "India orientalis" which extends from India to Western and Southern China. On the ISC&EP-Ortelius (1570; #2), there is "India intra Gangem".

(Continued)

Indian Subcontinent and Its Eastern Periphery				
Item #	Name	Pinyin	Etymology	History, Geography and My Comments
				On the ISC&EP-Plancius (1594), the name "India" is not depicted.
@ 5	安義河受三十水产金沙		An Yi He Shou San Shi Shui Chan Jin Sha	My translation and comment (between square brackets) are shown as follows: The Anyi River [安義河/安义河] receives water from 30 tributaries and produces gold sand. Although this annotation is written next to an unnamed river, the unnamed river must be the Irrawaddy River or the Ayeyarwady River (meaning "abounding in riches"; 伊洛瓦底江 is the transliteration of "Irrawaddy River"; however, the Chinese pronunciation of "安義" or "Anyi" is closer to "Ayey" in the name "Ayeyarwady"). The river flows relatively straight north–south through Myanmar (Burma) and empties into the Andaman Sea (安達曼海/安达曼海). It is the country's longest river. Here the annotation says that the river is fed by 30 tributaries. In fact, there are five major tributaries: Taping, Shweli, Myitnge, Mu and Chindwin. The Irrawaddy River empties through the Irrawaddy Delta (伊洛瓦底江三角洲) in the Ayeyarwady Region into the Andaman Sea, see Fig. A5.6.

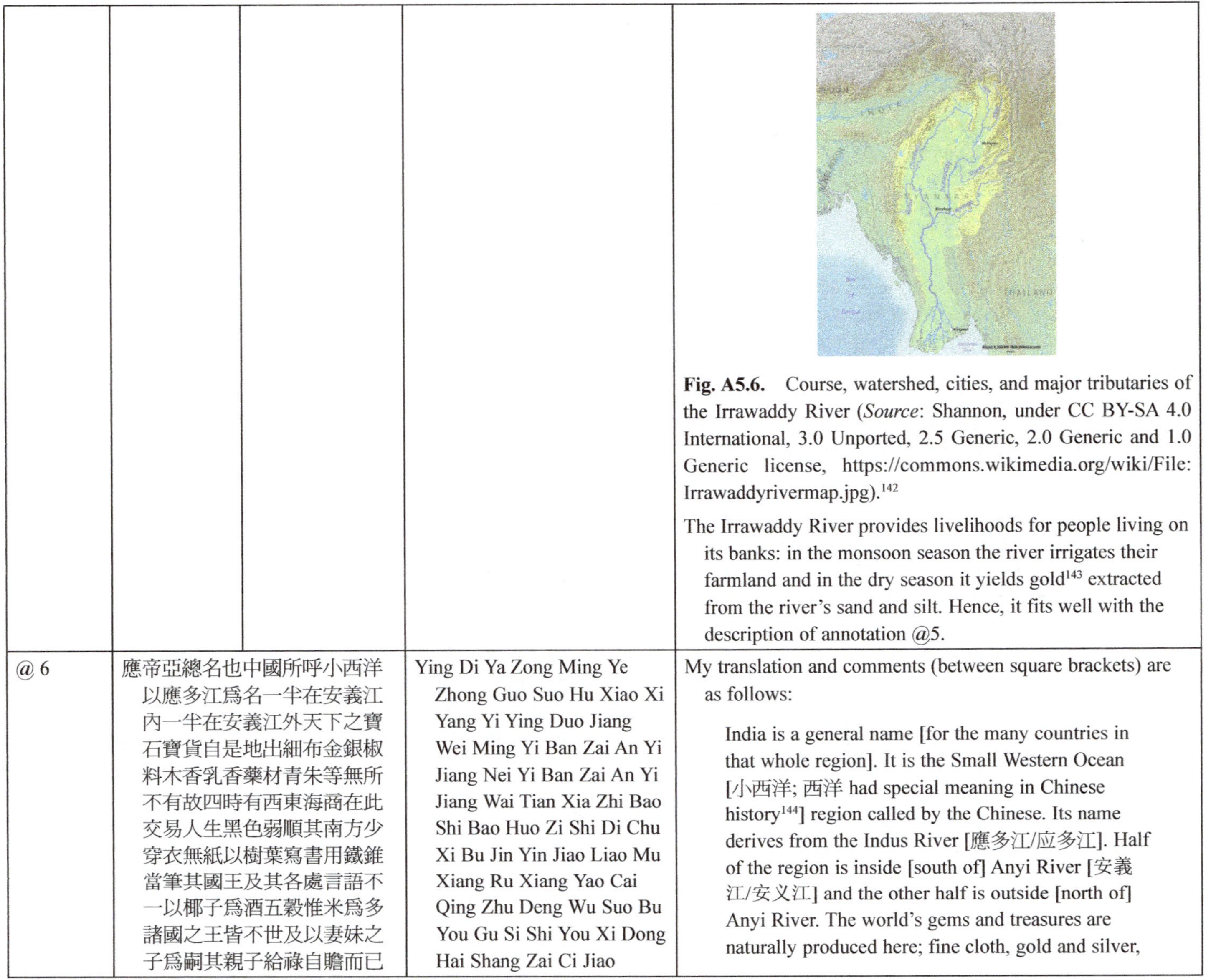

Fig. A5.6. Course, watershed, cities, and major tributaries of the Irrawaddy River (*Source*: Shannon, under CC BY-SA 4.0 International, 3.0 Unported, 2.5 Generic, 2.0 Generic and 1.0 Generic license, https://commons.wikimedia.org/wiki/File:Irrawaddyrivermap.jpg).[142]

The Irrawaddy River provides livelihoods for people living on its banks: in the monsoon season the river irrigates their farmland and in the dry season it yields gold[143] extracted from the river's sand and silt. Hence, it fits well with the description of annotation @5.

| @ 6 | 應帝亞總名也中國所呼小西洋以應多江爲名一半在安義江內一半在安義江外天下之寶石寶貨自是地出細布金銀椒料木香乳香藥材青朱等無所不有故四時有西東海商在此交易人生黑色弱順其南方少穿衣無紙以樹葉寫書用鐵錐當筆其國王及其各處言語不一以椰子爲酒五穀惟米爲多諸國之王皆不世及以妻妹之子爲嗣其親子給祿自贍而已 | Ying Di Ya Zong Ming Ye Zhong Guo Suo Hu Xiao Xi Yang Yi Ying Duo Jiang Wei Ming Yi Ban Zai An Yi Jiang Nei Yi Ban Zai An Yi Jiang Wai Tian Xia Zhi Bao Shi Bao Huo Zi Shi Di Chu Xi Bu Jin Yin Jiao Liao Mu Xiang Ru Xiang Yao Cai Qing Zhu Deng Wu Suo Bu You Gu Si Shi You Xi Dong Hai Shang Zai Ci Jiao | My translation and comments (between square brackets) are as follows:

India is a general name [for the many countries in that whole region]. It is the Small Western Ocean [小西洋; 西洋 had special meaning in Chinese history[144]] region called by the Chinese. Its name derives from the Indus River [應多江/应多江]. Half of the region is inside [south of] Anyi River [安義江/安义江] and the other half is outside [north of] Anyi River. The world's gems and treasures are naturally produced here; fine cloth, gold and silver, |

(*Continued*)

Indian Subcontinent and Its Eastern Periphery				
Item #	Name	Pinyin	Etymology	History, Geography and My Comments
			Yi Ren Sheng Hei Se Ruo Shin Qi Nan Fang Shao Chuan Yi Wu Zhi Yi Shu Xie Shu Xie Shu Yong Tie Shui Dang Bi Qi Guo Wang Ji Qi Ge Chu Yan Yu Bu Yi Yi Ye Zi Wei Jiu Wu Gu Wei Mi Wei Duo Zhu Guo Zhi Wang Jie Bu Shi Ji Yi Qi Mei Zhi Zi Wei Si Qi Qin Zi Gei Lu Zi Shan Er Yi	peppers, woody incense, frankincense, medicinal materials, green vermilion, etc., are all produced here; hence, merchants from the West and the East trade here in all seasons. The people here are dark-skinned; they appear mild and obedient. In the south, people wear less clothes. They have no paper but use iron drills as pens to write on leaves. They have kings and languages in different regions. They use coconuts as wine, and their grains are mostly rice. The kings of all their countries are not hereditary. They take the sons of their wives' sisters as heirs and give their own children salaries for supporting themselves. The only river which can separate the country into the north and the south two parts to fit the description of @6 is the Ganga/Ganges River, not the Anyi River (the Irrawaddy River or Ayeyarwady River), because about 60% of India is south of the Ganga/Ganges and 40% north of the Ganga/Ganges. The Anyi River (the Irrawaddy River or Ayeyarwady River) flow in north-south direction. The Ganga/Ganges is a great river of the plains of the northern Indian subcontinent and a trans-boundary river flowing through India and Bangladesh; it flows in roughly east-west direction and forms the boundary between the two countries.[145] The river flows into the Bay of Bengal. Here the ISC&EP-KWQ makes a serious mistake denoting the Anyi River (安義江/安义江; Irrawaddy River or Ayeyarwady River)", instead of the Ganges, as the river separating India into two parts. It surprisingly omits the Ganges.

				We know that it is the Ganga/Ganges which flows south and east from the Himalayas, then winds its way through northern India, eventually emptying into the Bay of Bengal, see Fig. A5.7. 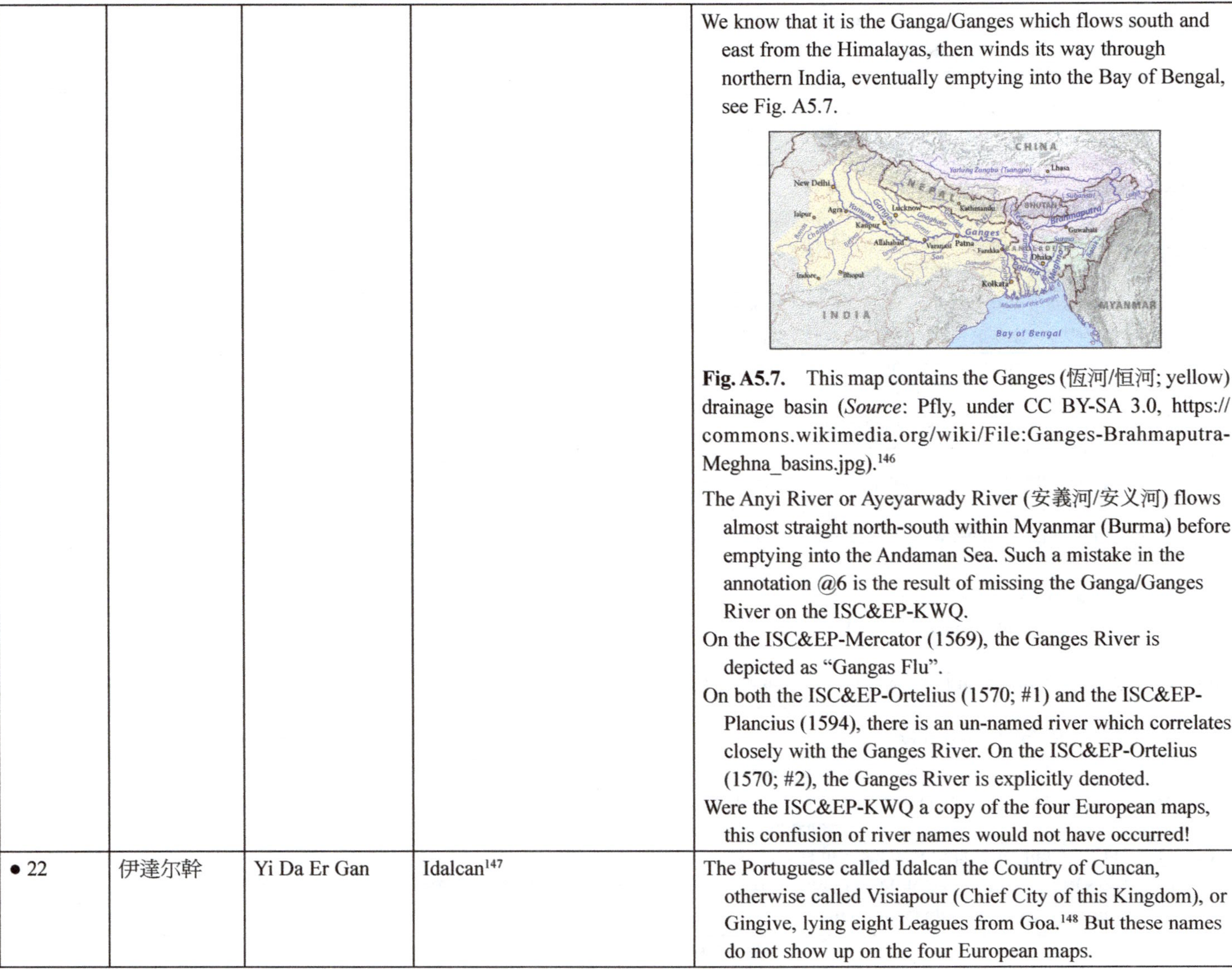 **Fig. A5.7.** This map contains the Ganges (恆河/恒河; yellow) drainage basin (*Source*: Pfly, under CC BY-SA 3.0, https://commons.wikimedia.org/wiki/File:Ganges-Brahmaputra-Meghna_basins.jpg).[146] The Anyi River or Ayeyarwady River (安義河/安义河) flows almost straight north-south within Myanmar (Burma) before emptying into the Andaman Sea. Such a mistake in the annotation @6 is the result of missing the Ganga/Ganges River on the ISC&EP-KWQ. On the ISC&EP-Mercator (1569), the Ganges River is depicted as "Gangas Flu". On both the ISC&EP-Ortelius (1570; #1) and the ISC&EP-Plancius (1594), there is an un-named river which correlates closely with the Ganges River. On the ISC&EP-Ortelius (1570; #2), the Ganges River is explicitly denoted. Were the ISC&EP-KWQ a copy of the four European maps, this confusion of river names would not have occurred!
● 22	伊達尔幹	Yi Da Er Gan	Idalcan[147]	The Portuguese called Idalcan the Country of Cuncan, otherwise called Visiapour (Chief City of this Kingdom), or Gingive, lying eight Leagues from Goa.[148] But these names do not show up on the four European maps.

	Indian Subcontinent and Its Eastern Periphery			
Item #	Name	Pinyin	Etymology	History, Geography and My Comments
23	坎巴夷替	Kan Ba Yi Ti	Koyampadi was the old name of Coimbatore.	坎巴夷替 should be corrected as 坎夷巴替 (Koyampadi) which was in western Kerala state, India; its present name is Coimbatore.[149] On the ISC&EP-Ortelius (1570; #2), there is "CAMBAIA" depicted at where Koyampadi should be.
24	丢	Diu	Diu	Diu (丢) is an island off the southern tip of the Kathiawar Peninsula, see Fig. 5.5(b). Diu remained a possession of the Portuguese from 1535 until 1961.[150] For detailed comparisons of Diu on the ISC&EP-KWQ with the four Europeans maps, please see Subsections 2.1 and 2.2 in the main text.
25	吴茶蠟	Wu Cha La	Gujarat; the name is derived from the Gurjara-Pratihara Dynasty, who ruled Gujarat in the eighth and ninth centuries.[151]	Gujarat is a state along the western coast of India, most of which lies on the Kathiawar peninsula. Fragments of printed cotton from Gujarat have been discovered in Egypt, providing evidence for medieval trade in the western Indian Ocean.[152] Please refer to the detailed discussion of Gujarat (吴茶蠟/吴茶蜡) in Subsection 2.2 in the main text.
26	身毒河	Shen Du He	Indus River	The Indus River is discussed in detail in Subsection 2.2 of Section 2 in the main text, which will not be repeated here.
27	梧作剌得	Wu Zuo La De	Guzarate[153]	Guzarate (梧作剌得) was an alternate form of Gujarat (吴茶蠟/吴茶蜡), but they are depicted separately as Gujarat (吴茶蠟/吴茶蜡; Item 25; an island) and Guzarate (梧作剌得; Item 27; an inland region) on the ISC&EP-KWQ. Perhaps the mistake was caused by the complicated shape of the state of Guzarate. On the ISC&EP-Ortelius (1570; #1), Gujarat is depicted as the "Guzarate" region, not as an island. The two other European maps do not show Guzarate or Gujarat. On the ISC&EP-Ortelius (1570; #2), there is "GVZARATE".

| 28 | 葛步尔 | Ge Bu Er | Kabul or Cabul; it is unknown when the name "Kabul" was first applied to the city. The city became prominent in the thirteenth century.[154] | Kabul or Cabul (葛步爾/葛步尔) is the capital and largest city of present-day Afghanistan. It is in the eastern half of the country.
Afghanistan is a landlocked country located at the crossroads of Central Asia and South Asia. The country is bordered by Pakistan to the east and south, Iran to the west, Turkmenistan to the northwest, Uzbekistan to the north, Tajikistan to the northeast, and China to the northeast and east.
On the ISC&EP-KWQ, the geographical location of Kabul or Cabul is incorrectly depicted in India and to the south of the Indus River. Kabul is to the northwest of the Indus River on a modern map, see Fig. A5.8.
On the ISC&EP-Ortelius (1570; #2), there is "CABVL". |

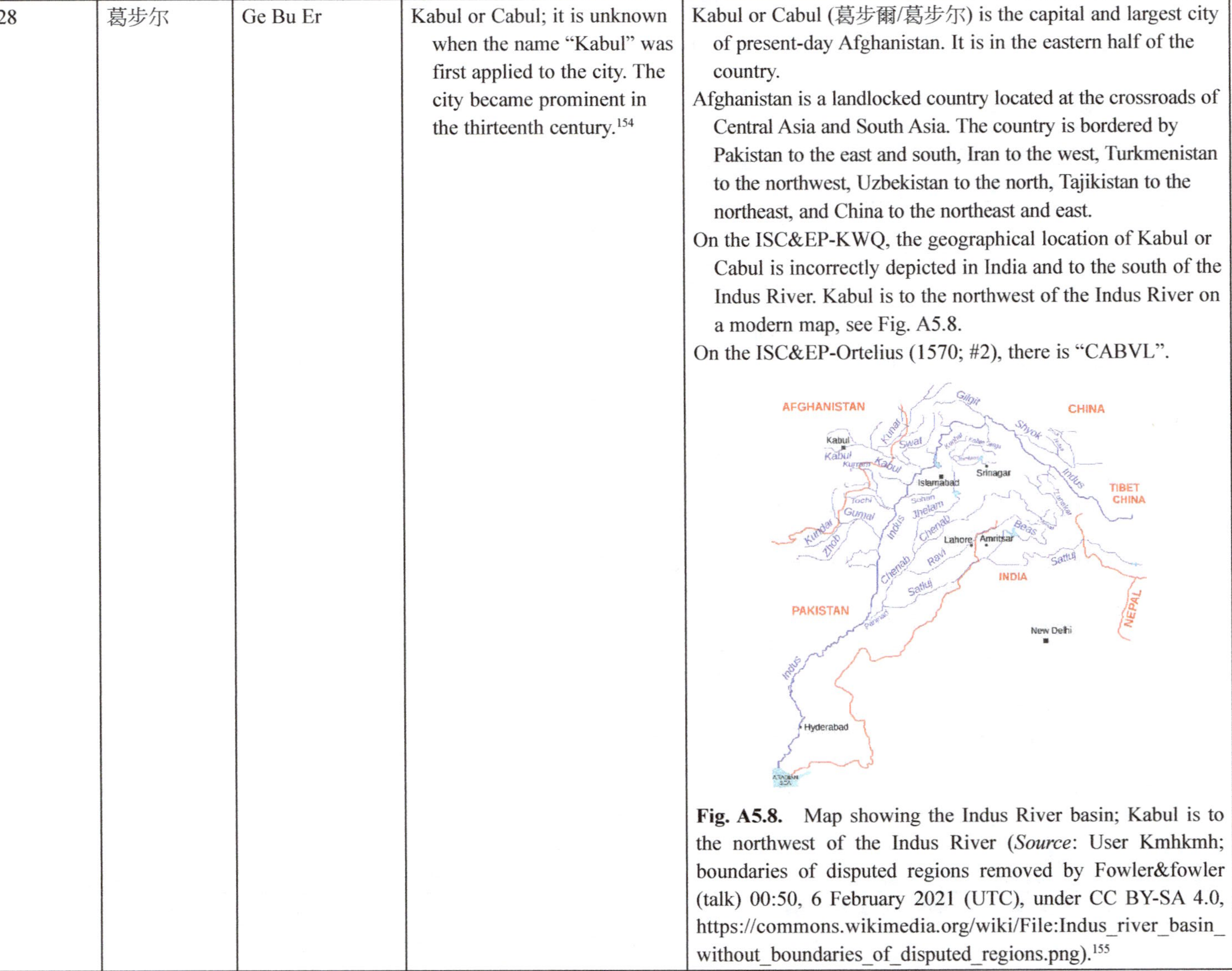

Fig. A5.8.   Map showing the Indus River basin; Kabul is to the northwest of the Indus River (*Source*: User Kmhkmh; boundaries of disputed regions removed by Fowler&fowler (talk) 00:50, 6 February 2021 (UTC), under CC BY-SA 4.0, https://commons.wikimedia.org/wiki/File:Indus_river_basin_without_boundaries_of_disputed_regions.png).[155]

(*Continued*)

Indian Subcontinent and Its Eastern Periphery				
Item #	Name	Pinyin	Etymology	History, Geography and My Comments
● 29	莫臥尔	Mo Wo Er	Mughal or Moghul is from the Arabic and Persian corruption of "Mongol".	The Mughal Empire (莫臥兒帝國/莫卧儿帝国; 1526–1857) was an empire in South Asia,[156] founded in 1526 by Babur, a chieftain from where it was today's Uzbekistan; it occupied the plains of North India. The closest to an official name for the empire was Hindustan (see Item 32), which was documented in the Ain-i-Akbari.[157] The Mughals claimed they were descendants of Genghis Khan of the Mongol Empire.[158] This geographical item must have been added by Matteo Ricci when he published the ISC&EP-KWQ, because none of the four European maps depicts this empire on their maps. According to the annotation next to the Iberian Peninsula on the KWQ, as explained in Chapter 1 of *Chinese Global Exploration in the Pre-Columbian Era: Evidence from an Ancient World Map*,[159] the entire KWQ was drawn in 1433. Hence, the Mughal Empire founded in 1526 was added by Matteo Ricci onto the ISC&EP-KWQ before he published the map.
● 30	敖羅	Ao Luo	Gor[160]	Gor (敖羅/敖罗) is a place in the region of Madhya Pradesh in India.[161] The distance from Gor to India's capital New Delhi in North India is approximately 439 km /273 mi (as the crow flies).[162]
● 31	西天竺國	Xi Tian Zhu Guo	Xi Tianzhu or Western Tianzhu	See detailed explanation in Item 6.
32	印度厮當	Yin Du Si Dang	Indostan[163] is the obsolete form of Hindustan. Hindustan is derived from the Persian word *Hindū* cognate with the Sanskrit *Sindhu*.[164]	Indostan (印度厮當/印度厮当; lit. Indo-land) broadly referred to the Indian subcontinent; also known as Hind[165] or *Sindh*.[166] Hindustan was introduced during the Mughal Empire (1526–1857) and was closest to an official name for the empire. On the ISC&EP-Mercator (1569), there is "INDOSTAN" expanding deep into the Western China region.

				On the ISC&EP-Ortelius (1570; #1), there is no "Indostan" or "Hindustan", but on the ISC&EP-Ortelius (1570; #2), there is "INDOSTAN". On the ISC&EP-Plancius (1594), there is "INDOSTAN" near the northeastern part of the Indian subcontinent.
● 33	小西洋	Xiao Xi Yang	Small Western Ocean; it is within the range of the Arabian Sea.	On the ISC&EP-KWQ, today's Indian Ocean is depicted as Xiao Xi Yang (小西洋, lit. "Small Western Ocean"). On the KWQ, there is a Da Xi Yang (大西洋, lit. "Great Western Ocean"; it is today's Atlantic Ocean, see Fig. A5.9) depicted to the west of the North African continent. None of the four European maps depicts "Small Western Ocean (小西洋)". **Fig. A5.9.** Location of the Arabian Sea in the Indian Ocean. Created by Norman Einstein, 19 July 2005 (*Source*: NormanEinstein, Ras67, under CC BY-SA 3.0, https://commons.wikimedia.org/wiki/File:Arabian_Sea_map.png).[167]
34	錫狼島	Xi Lang Dao	Island of Mount Ceylon	Here 錫狼島/锡狼岛 refers to the Island of Mount Ceylon where the Kingdom of Mount Ceylon (錫蘭山國/锡兰山国) was located. The Kingdom is recorded in *Ming Shi* (明史) or *History of Ming*[168] as follows: 永樂中, 鄭和使西洋至其地, 其王亞烈苦奈兒欲害和, 和覺, 去之他國。。。及和歸, 復經其地, 乃誘和至國中, 發兵五萬劫和, 塞歸路。 和乃率步卒二千, 由間道乘虛攻拔其城, 生擒亞烈苦奈兒及妻子, 頭目, 獻俘於朝。 廷臣請行戮,

(*Continued*)

Indian Subcontinent and Its Eastern Periphery				
Item #	Name	Pinyin	Etymology	History, Geography and My Comments
				帝憫其無知, 并妻子皆釋, 且給以衣食。命擇其族之賢者立之。。。其舊王亦遣歸。自是海外諸蕃益服天子威德, 貢使載道, 王遂屢入貢。 宣德五年, 鄭和撫諭其國。八年, 王不剌葛麻巴忽剌批遣使來貢。。。 My translation and comments (between square brackets) are as follows: During the Yongle period [it was during Zheng He's third voyage to the Western Ocean starting from 1409], Zheng He reached this land [Island of Mount Ceylon].[169] But the king of the country, Ariel Kunel, wanted to harm Zheng He. Zheng He found out about it and went to another country. . . When he returned [from his journey], he passed through this land again. The king lured him to enter the country but sent fifty thousand soldiers to seize Zheng He and block his return. Zheng He then led 2,000 infantries, took a side road, attacked the emptied capital, and captured Ariel Kunel, his wife, and the leaders alive. They were brought back to China and presented as captives to the Ming Court [this happened in 1411, after Zheng He's fleet returned to China]. The court officials demanded them to be killed, but the emperor [Yongle] pitied the king's ignorance, released him and his wife, gave them food and clothes, and issued an edict to choose another virtuous monarch from his clan to be the new king. . . The captured king was also sent back to his country. Since then, all the overseas tribes were convinced of the majesty and virtue of the Yongle Emperor, and their tribute envoys all praised the emperor's endeavour. Since then, the king repeatedly paid tribute.

In 1430 (the fifth year of Xuande's reign), Zheng He, [see Fig. A5.10] gave appeasing instructions to this country. In 1433 (the eighth year of Xuande's reign), King Bu La Ge Ma Ba Hu La Pi (transliteration of 不剌葛麻巴忽剌批) sent envoys to pay tribute. . .

From the journey of Zheng He's seventh voyage, we know that the last identifiable location of Zheng He's entire fleet on their way to the Western Ocean was near present-day Ceylon/Sri Lanka on 28 November 1432 (November 6 in the seventh year of Xuande's reign). Then, this fleet was split into a main fleet led by Zheng He and Wang Jinghong (王景弘), and a squadron led by Hong Bao (洪保).[170] This last location is also recorded in *Qian Wen Ji • Xia Xi Yang* (前聞記 • 下西洋)[171] or *A Record of History Once Heard: Down to the Western Ocean* by the Ming historian Zhu Yunming (祝允明; 1461–1527). Hong Bao's squadron arrived at Calicut on 10 December 1432 and left there on 14 December 1432.[172] The fleet led by Zheng He and Wang Jinghong arrived at Calicut around 19–24 December 1432; they had a banquet with the king and left the same day.[173]

Fig. A5.10. The figure shows the Zheng He wax statue in the Quanzhou Maritime Museum (*Source*: Jonjanego, under CC BY-SA 2.0, https://commons.wikimedia.org/wiki/File:Admiral_Zhenghe.jpg).[174]

The ISC&EP-Mercator (1569) depicts *Zeilam insula.*
The ISC&EP-Ortelius (1570; #1) depicts "Zeilan". On the ISC&EP-Ortelius (1570; #2), there is ZEILAN.

(*Continued*)

Indian Subcontinent and Its Eastern Periphery				
Item #	Name	Pinyin	Etymology	History, Geography and My Comments
				The ISC&EP-Plancius (1594) depicts *Zeilam insula*. All these names denote the Island of Mount Ceylon.
● 35	萬島	Wan Dao	One Hundred Thousand Islands; the classical Sanskrit texts dating back to the Vedic period (1500 B.C.–600 B.C.) mention the "one lakh islands" (*Lakshadweepa;* meaning, "one hundred thousand islands") which would include not only the present-day Maldives, but also the Laccadives, Aminidivi Islands, Minicoy Island, and the Chagos island groups.[175]	One Hundred Thousand Islands (萬島/万岛) was a generic name which would include not only the present-day Maldives, but also the Laccadives, Aminidivi Islands, Minicoy Island, and the Chagos island groups as shown, respectively, in Fig. A5.11. **Fig. A5.11.** Location of Maldives in the Indian Ocean (top left; highlighted by the small circle; *Source*: Parvez gsm at English Wikipedia, under CC BY-SA 3.0, https://commons.wikimedia.org/wiki/File:Maldives_(orthographic_projection).svg);[176] Laccadive Islands (top right; highlighted by a small circle; *Source*: Uwe Dedering, under CC BY-SA 3.0, https://commons.wikimedia.org/wiki/File:India_location_map_3.png);[177] Aminidivi Islands (bottom left; highlighted by a small circle; *Source*: Uwe Dedering, under CC BY-SA 3.0, https://commons.wikimedia.org/wiki/File:India_location_map_3.png);[178] Minicoy Island (bottom centre; highlighted by a small circle; *Source*: Uwe Dedering, under CC BY-SA 3.0, https://commons.wikimedia.org/wiki/File:India_location_map_3.png);[179] and Chagos Archipelago (bottom right; *Source*: Tentotwo, under CC BY-SA 3.0, https://commons.wikimedia.org/wiki/File:Indian_Ocean_laea_location_map.svg).[180]

				The ISC&EP-Mercator (1569) depicts *Maldiuar* and many other named island groups nearby. The ISC&EP-Ortelius (1570; #1) depicts *Yedi Maldiuar* and many unnamed island groups nearby. On the ISC&EP-Ortelius (1570; #2), there are many separately named islands. The ISC&EP-Plancius (1594) depicts *Maldiua* and many unnamed island groups nearby. However, none of these four European maps uses the name "One Hundred Thousand Islands (萬島/万岛)".
Zone II: Indian subcontinent's eastern periphery				
36	琶牛	Pa Niu	Bago, formally spelled as Pegu; Pegu, formerly also known as Hanthawaddy.	Bago (勃固), formally spelled as Pegu (琶牛), is a city in today's Myanmar. It was founded in 1276/77 A.D.[181] according to the *Zabu Kuncha*, an early fifteenth century Burmese administrative treatise, or even earlier according to a Chinese source.[182] In the fifteenth century, Bago became a centre of commerce and Theravada Buddhism. Pegu is depicted on the ISC&EP-Mercator (1569) and the ISC&EP-Plancius (1594), not on the ISC&EP-Ortelius (1570; #1), but on the ISC&EP-Ortelius (1570; #2), there is PEGV.
● 37	大流沙	Da Liu Sha	Big Quicksand	Quicksand (流沙) refers to water-saturated loose sandy soil (sand, silt, or clay) that loses strength and cannot support weight. The liquified soil can flow under the action of external forces and becomes dangerous. Tibet has a unique landscape which is called the Big Quicksand (大流沙). It is located on the edge of the Nu River (怒江) in the Bingchacha (丙察察) section of the 219 National Highway from Yunnan to Tibet.[183] It is unique because there is no other natural landscape like this in China or even in the world. It is very dangerous to pass this area because once trapped in quicksand, it is difficult to leave by one's own effort and the situation can cause death.

(*Continued*)

Indian Subcontinent and Its Eastern Periphery				
Item #	Name	Pinyin	Etymology	History, Geography and My Comments
38	于闐	Yu Tian	Khotan/Cotan	The Kingdom of Khotan/Cotan was an ancient Buddhist Saka Kingdom located on the branch of the Silk Road that ran along the southern edge of the Taklamakan Desert (塔克拉瑪乾沙漠/塔克拉玛干沙漠) in the Tarim Basin (塔里木盆地; in modern-day Xinjiang, China). From the Han Dynasty until at least the Tang Dynasty, the kingdom was known in Chinese as Yutian (Chinese: 于闐/于阗, 于寘, or於闐/於阗; c. 300 B.C.–1006). It was conquered by the Muslim Kara-Khanid Khanate in 1006, during the Islamization and Turkicisation of Xinjiang.[184] But the Yutian city continued to exist as a significant city on the Silk Road, see Fig. A5.12. **Fig. A5.12.** Map of the Kingdom of Khotan c. 1000 (*Source*: SY, under CC BY-SA 3.0, https://commons.wikimedia.org/wiki/File:Kingdom_of_Khotan.png).[185] The ISC&EP-Ortelius (1570; #1) depicts Khotan as "Cotam".
@ 7	于闐東磧石又東為流沙人行無跡故往返皆迷聚骸以識道無水草多熱風	Yu Tian Dong Qi Shi You Dong Wei Liu Sha Ren Xing Wu Ji Gu Wang Fan Jie Mi Ju Hai Yi Shi Dao Wu Shui Cao Duo He Feng		My translation and comments (between square brackets) are as follows: To the east of Yutian [于闐/于阗] is Qishi [磧石/碛石], and then to the east of Qishi is an area of quicksand [flowing sands or moving sands] where there is no trace of people walking through. So, it is easy to

				confuse coming and going directions. Only guided by the gathered corpses [of the previously deceased travellers] can one know the way. There is no water, no plants, but a lot of hot wind. Khotan was located on the branch of the Silk Road that ran along the southern edge of the Taklamakan Desert, see Fig. A5.13, in the Tarim Basin. The annotation is a vivid description of traveling through this desert in Southwestern Xinjiang (新疆), Northwest China. Fig. A5.13 shows that the Taklamakan Desert is bounded by the Kunlun Mountains (崑崙山/昆仑山) to the south, the Pamir Mountains (帕米爾高原/帕米尔高原) to the west, the Tian Shan (天山; Fig. A5.13 denotes "Tien Shan") range to the north, and the Gobi Desert (戈壁) to the east (the name "Gobi Desert" is not marked in this figure). **Fig. A5.13.** This is a map of the Tarim River drainage basin; the map shows crossing the Taklamakan Desert which is usually dry (*Source*: SY, under CC BY-SA 3.0, https://commons.wikimedia.org/wiki/File:Tarimrivermap.png).[186]
39	加湖	Jia Hu	A fictitious Lake Cha: *Lago de Chiamay*	In a map published in 1554, there appeared a lake, *Lago de Chiamay*,[187] which was depicted as the source of four of the great rivers of Southeast Asia. This lake continued to show up on maps for almost two hundred years, until it was shown to be non-existent; then it disappeared from the maps.

(*Continued*)

			Indian Subcontinent and Its Eastern Periphery	
Item #	Name	Pinyin	Etymology	History, Geography and My Comments
				On the ISC&EP-KWQ, this fictitious lake has a Chinese name 加湖 (Lake Cha) and is depicted as the source of four of the great rivers in Southeast Asia. Please see the detailed discussion of 加湖 in Subsection 2.3 of Section 2 in the main text. For brevity, it will not be repeated here.
● 40	金沙江	Jin Sha Jiang	Jinsha River; the name "Jinsha" originates in the Song Dynasty when the river attracted large numbers of gold prospectors. Gold prospecting along the Jinsha continues to this day.[188]	金 means "gold", 沙 "sand" and 江 "jiang or river". Jinsha River (金沙江), lit. "Gold Sand River". The name describes that actual placer gold, alluvial gold powder sometimes still panned from the river's waters. Jinsha River is the Chinese name for the upper stretches of Changjiang (長江/长江; "Chang" means "long" and "jiang" means "river") or Yangtze River/Yangzijiang[189] (揚子江/扬子江) from Yushu in Qinghai Province to Yibin in Sichuan. The Jinsha River runs through the three provinces of Tibet, Sichuan, and Yunnan. Along the way, its largest tributary is the Yalong Jiang (雅礱江/雅砻江). When it reaches Yibin (宜賓/宜宾) in Sichuan Province (四川), where the Min River (岷江) merges into it, it was named the Changjiang or Yangtze River or Yangzijiang (長江/长江 or 揚子江/扬子江), see Fig. A5.14. The Jinsha River originates from Qinghai Province, then flows south through a deep gorge parallel to the similar gorges of the upper Mekong River (湄公河上游 or Lancang River; 瀾滄江/澜沧江) and upper Salween River (薩爾溫江上游/萨尔温江上游 or Nu River; 怒江). A part of the Jinsha River belongs to the "Three Parallel Rivers of the Yunnan Protected Areas".[190] The three rivers do not originate from a common lake.

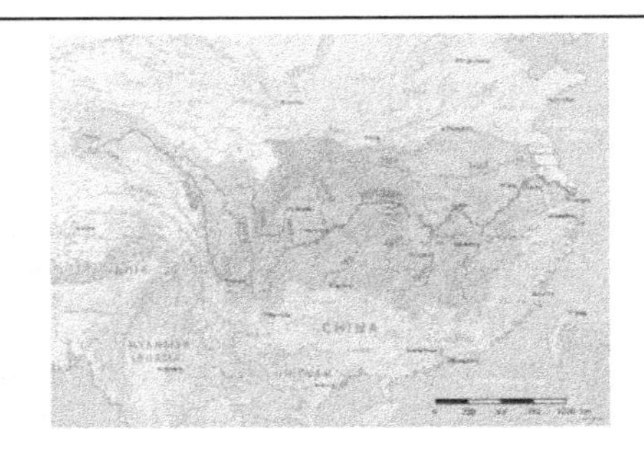

Fig. A5.14.   Map of the Yangtze River (Changjiang) drainage basin, with major tributaries and cities (*Source*: User:Shannon 1, under CC BY-SA 4.0 International, 3.0 Unported, 2.5 Generic, 2.0 Generic and 1.0 Generic license, https://commons. wikimedia.org/wiki/File:Yangtze_river_map.png).[191]

On a modern map, see Fig. A5.14, the Jinsha River forms the western border of Sichuan and then flows into Yunnan Province. After a large and long loop to the north of Dali Bai Autonomous Prefecture (大理白族自治州), the Jinsha swings northeast, forming the Sichuan-Yunnan provincial boundary until it joins the Min River (岷江) at Yibin (宜賓/宜宾) in Sichuan to form the Yangtze.

However, on the ISC&EP-KWQ, 金沙江 (Jinsha River) is depicted not only as flowing from an erroneous source 加湖 (Lake Cha) with three other major Asian rivers, but also mistakenly flowing southward into the territory of Myanmar (緬甸/缅甸); from there it empties into the Bay of Bengal (榜葛剌海 or 孟加拉灣/孟加拉湾); it does not join the Min River to form the Yangtze River. This is incorrect. Jinsha River and the three other major Asian rivers all have their separate sources, as follows: Jinsha River: the source is the Tongtian River (通天河; in China's Qinghai Province); Lancang River: the source is the Lasaigongma Spring (拉賽貢瑪泉/拉赛贡玛泉; in Zadoi, Yushu Tibetan Autonomous Prefecture, Qinghai, China); Salween River: the source is the Tanggula Mountains

(*Continued*)

Indian Subcontinent and Its Eastern Periphery				
Item #	Name	Pinyin	Etymology	History, Geography and My Comments
				(唐古拉山; in Nagqu, Tibet, China); and Irrawaddy River: the river originates from the confluence of the N'mai River (恩梅開江/恩梅开江; originate from Tibet, China) and Mali River (邁立開江/迈立开江; originated in northeastern Myanmar/Burma); both rivers find their sources in the Himalayan glaciers.
● 41	孟養	Meng Yang	Mengyang	Mengyang (孟養/孟养) is now a place with a very small population in Shan State, Myanmar. In 1404 (the second year of Yongle's reign), the Yongle Emperor made Mengyang a Xuanwei Division (宣慰司).[192] The political status of Mengyang has changed several times after that. In 1417 (the fifteenth year of Yongle's reign), Mengyang's leader sent envoys to present horses and local products to the Ming Court.[193]
● 42	大古剌	Da Gu La	Dagula	Dagula (大古剌) was the Hanthawaddy (漢達瓦底/汉达瓦底) Kingdom (1287–1551)[194] in southern Myanmar in ancient times. During the Ming period of 1402 to 1425, the kingdom was a tributary country to the Ming Court. In 1406 (the fourth year of Yongle's reign), the Ming Court established the Dagula Regional Military and Civilian Commission (an appeasement department; 大古剌軍民宣慰使司/大古剌军民宣慰使司).[195] But starting from the Xuande period (1425–1435), the kingdom stopped sending in tribute. In 1551, the kingdom was destroyed by Myanmar.[196] The kingdom is not depicted on any of the European maps, because it no longer existed after 1551. If ISC&EP-KWQ were drawn by Matteo Ricci in 1602, Dagula (大古剌) should not show up on the ISC&EP-KWQ.

| 43 | 緬甸 | Mian Dian | Today's Myanmar; also known as Burma (the official name until 1989) | Myanmar (緬甸/缅甸) is a country in Southeast Asia. It borders the Andaman Sea in the southwest, India and Bangladesh in the northwest, China in the northeast, Thailand, and Laos in the southeast.
Myanmar formed a unified country in 1044 and was successively ruled by the Bagan Dynasty (蒲甘王朝; 849–1297), Taungoo Dynasty (東籲王朝/东吁王朝; 1531–1752) and Konbaung Dynasty (貢榜王朝/贡榜王朝; 1752–1885).[197]
In 1287, the Mongolian ruler Kublai Khan led the Yuan Dynasty army to invade Myanmar massively (the Yuan-Burma War), which caused the Bagan Dynasty to lose its independence and perish. After the fall of the Bagan Dynasty, the Mongols left the Irrawaddy basin, and the Bagan Dynasty was divided into several small kingdoms.[198]
Despite its internal fights, Myanmar maintained a tributary relationship with the Ming Court in the Yongle and Xuande periods in the fifteenth century. In 1424 (the 22nd year of Yongle's reign), the Ming Dynasty established an appeasement department in Diwu La (i.e., Taungoo). In 1425 (the first year of Hongxi's reign) and 1430 (the fifth year of Xuande's reign), Taungoo sent envoys to China twice to pay tribute.[199]
The ISC&EP-Mercator (1569) depicts Burma as "Verma".
The ISC&EP-Ortelius (1570; #1) depicts Burma as "Brema". On the ISC&EP-Ortelius (1570; #2), there is VERMA.
The ISC&EP-Plancius (1594) also depicts Burma as "Verma". |
| ● 44 | 三宣地方 | San Xuan Di Fang | Region under the jurisdictions of the three Xuanwei Divisions: they were the Dagula, the Dewulash and the Demasasa. These three divisions were in the continental portion of Southeast Asia. | In 1406 (the fourth year of Yongle's reign), the Ming Court established the Dagula Xuanwei Division (大古刺宣慰司).[200] Since the local chiefs in Dagula and other places "begged for orders to serve", the Ming Court in the same year added the Demasasa Xuanwei Division (底馬撒宣慰司/底马撒宣慰司);[201] in 1424 (the 22nd year of Yongle's reign) the Dewulash Xuanwei Division (底兀刺宣慰司) was also added.[202] |

(Continued)

Indian Subcontinent and Its Eastern Periphery				
Item #	Name	Pinyin	Etymology	History, Geography and My Comments
				"司", pronounced as "si", means "division" in the Tusi (土司) system. Tusi, also known as Tuguan (土官), was a hereditary official position held by ethnic minority leaders in the Northwest and Southwest regions during the Yuan, Ming, and Qing dynasties.[203] The "si" in Tusi has the meaning of "presiding over management". In fact, the Xuanwei Division was a local administrative agency set up in the Ming Dynasty to manage the border areas. The Ming Court also has the responsibility to mediate and protect the Xuanwei Division. Since the Xuanwei Division has jurisdiction over the territory of the Ming Dynasty, the chief Xuanwei envoy had to be canonized by the emperor.

Endnotes

[1]Library of Congress. "Kun yu wan guo quan tu." www.loc.gov, Library of Congress, 2025, loc.gov/item/2010585650

[2]Ronnie Po-Chia Hsia. *Matteo Ricci and the Catholic Mission to China, 1583–1610: A Short History with Documents (Passages: Key Moments in History)*. Indianapolis, IN, USA: Hackett Publishing Company, Inc., 2016

[3]Sheng-Wei Wang. *Chinese Global Exploration in the Pre-Columbian Era: Evidence from an Ancient World Map*. Singapore: World Scientific Publishing Co., 2023.

[4]"File:Kunyu Wanguo Quantu by Matteo Ricci Plate 1-3.jpg." *Wikimedia Commons: The Free Media Repository,* Wikimedia Foundation, 13 June 2023, upload.wikimedia.org/wikipedia/commons/b/b8/Kunyu_Wanguo_Quantu_by_Matteo_Ricci_Plate_1-3.jpg

[5]"File:Mercator 1569 world map composite.jpg." *Wikimedia Commons*, 26 Nov. 2016, upload.wikimedia.org/wikipedia/commons/4/4b/Mercator_1569_world_map_composite.jpg

[6]"File:OrteliusWorldMap1570.jpg." *Wikimedia Commons*, 12 July 2022, upload.wikimedia.org/wikipedia/commons/e/e2/OrteliusWorldMap1570.jpg

[7]"File:1594 double hemisphere world map by Petrus Plancius.jpg." *Wikimedia Commons*, 2 Sept. 2022, upload.wikimedia.org/wikipedia/commons/0/0d/1594_double_hemisphere_world_map_by_Petrus_Plancius.jpg

[8]"File:1570 Ortelius Map of Asia (first edition) - Geographicus - AsiaeNovaDescriptio-ortelius.jpg." *Wikimedia Commons*, 22 Mar. 2011, upload.wikimedia.org/wikipedia/commons/a/a3/1570_Ortelius_Map_of_Asia_(first_edition)_-_Geographicus_-_AsiaeNovaDescriptio-ortelius.jpg

[9]*Ibid.*

[10]"File:South Asia non political, with rivers.jpg." *Wikimedia Commons*, 11 Sept. 2023, upload.wikimedia.org/wikipedia/commons/5/52/South_Asia_non_political%2C_with_rivers.jpg

[11]"File:Ganges-Brahmaputra-Meghna basins.jpg." *Wikimedia Commons*, 10 Nov. 2023, upload.wikimedia.org/wikipedia/commons/3/34/Ganges-Brahmaputra-Meghna_basins.jpg

[12]Stephen Alter. *Sacred Waters: A Pilgrimage Up the Ganges River to the Source of Hindu Culture*. Telangana, India: Penguin Books India, 2001.

[13]Sukumari Bhattacharji. *Legends of Devi*. Hyderabad, Telangana, India: Orient Longman/Disha, 1995.

[14]Sally Wriggins. *Xuanzang: A Buddhist Pilgrim On The Silk Road*. Boulder, CO, USA: Westview Press, 1996.

[15]Li Rongxi. *The Great Tang Dynasty Record of the Western Regions*. Berkeley, USA: Bukkyo Dendo Kyokai and Numata Center for Buddhist Translation and Research, 1996, pp. 109–115.

[16]W. W. Tarn. "Alexander and the Ganges." *The Journal of Hellenic Studies*, vol. 43, no. 2, 1923, pp. 93–101. https://doi.org/10.2307/625798

[17]J. Lennart Berggren and Alexander Jones. *Ptolemy's Geography: An Annotated Translation of the Theoretical Chapters*. Princeton, NJ, USA: Princeton University Press, 2000.

[18]Ernest George Ravenstein. "Map." In Hugh Chisholm, ed., *Encyclopædia Britannica*, vol. 17, 11th ed. Cambridge University Press, 1911, p. 637.

[19]"File:Peutinger India.png." *Wikimedia Commons*, 23 Nov. 2022, upload.wikimedia.org/wikipedia/commons/8/8c/Peutinger_India.png

[20]Arun B. Shrestha, *et al. The Himalayan Climate and Water Atlas: Impact of Climate Change on Water Resources in Five of Asia's Major River Basins*, Lalitpur, Nepal: International Centre for Integrated Mountain Development, 2015, p. 58.

[21]DK and Smithsonian. *Natural Wonders of the World*, New York, NY, USA: Penguin Random House, 2017, p. 240.

[22]Sima Qian (司馬遷; 134 B.C.–c. 86 B.C.). *Shi Ji • Juan Yi Bai Yi Shi Liu • Xi Nan Yi Lie Chuan* (史记 • 卷一百一十六 • 西南夷列傳) or *Records of the Grand Historian, Juan (Volume) 116, Biography of Southwest Yi*; Sima Qian (司馬遷; 134 B.C. –c. 86 B.C.). *Shi Ji • Juan Yi Bai Er Shi San • Da Yuan Lie Chuan* (史记 • 卷一百二十三 • 大宛列傳) or *Records of the Grand Historian, Juan (Volume) 123, Biography of Da Wan*; an electronic file can be found at https://ctext.org/shiji/zh

[23]"File:IN-GJ.svg." *Wikimedia Commons*, 5 April 2023, upload.wikimedia.org/wikipedia/commons/thumb/1/1e/IN-GJ.svg/1799px-IN-GJ.svg.png

[24]"File:Map GujDist North.png." *Wikimedia Commons*, 21 July 2022, upload.wikimedia.org/wikipedia/commons/b/b9/Map_GujDist_North.png?uselang=zh

[25]"File:Course and major tributaries of the Indus.jpg." *Wikimedia Commons*, 6 Feb. 2021, upload.wikimedia.org/wikipedia/commons/d/d2/Course_and_major_tributaries_of_the_Indus.jpg

[26]Michael Witzel. "Early Indian history: Linguistic and textual parameters." In George Erdosy, ed. *The Indo-Aryans of Ancient South Asia: Language, Material Culture and Ethnicity*, Berlin, Germany: Walter de Gruyter, 1995, pp. 85–125; P. Thieme. "Sanskrit sindu-/Sindhu- and Old Iranian hindu-/Hindu-" In Mary Boyce and Ilya Gershevitch, eds. *W. B. Henning Memorial Volume*, London, UK: Lund Humphries, 1970, p. 450.

[27]Bratindra Nath Mukherjee. *Nationhood and Statehood in India: A Historical Survey*, Washington, D.C., USA: Regency Publications, 2001, p. 3.

[28]P. Thieme. *Op. cit.*, pp. 447–450.

[29]Peter Bauer-Gottwein. "Lancang River: A brief introduction." www.eoforchina.env.dtu.dk/, EOForChina (funded by the Danish Ministry of Foreign Affairs), 2025, https://www.eoforchina.env.dtu.dk/field-sites/lancang

[30]Helgi Björnsson, *et al. Encyclopedia of Snow, Ice and Glaciers*. Springer, 2011.

[31]Francis Mason. *Tenasserim: Or Notes on the Fauna, Flora, Minerals, and Nations of British Burmah and Pegu*. Charleston, SC, USA: American Mission Press, 2011.

[32]"File:Rubbing of Huayi tu map.jpg." *Wikimedia Commons*, 26 Mar. 2023, upload.wikimedia.org/wikipedia/commons/8/86/Rubbing_of_Huayi_tu_map.jpg

[33]"File:Da-ming-hun-yi-tu.jpg." *Wikimedia Commons*, 17 Oct. 2016, upload.wikimedia.org/wikipedia/commons/c/cd/Da-ming-hun-yi-tu.jpg

[34]Luo Hongxian (羅洪先; 1504–1564). "Guang yu tu (廣輿圖)." www.loc.gov, Library of Congress, 2025, loc.gov/item/2021666327/

[35]W. Huttmann. "On Chinese and European maps of China." *Journal of the Royal Geog. Soc.*, vol. 14, 1844, pp. 117–127; W. Fuchs. *The "Mongol Atlas" of China by Chu Ssu-pen and the Kuang-yu-t'u. Monumenta Serica, Monograph no. 8*. Peiping, China: Fu Jen University Press, 1946.

[36]"File:Jinsharivermap.jpg." *Wikimedia Commons*, 27 July 2022, upload.wikimedia.org/wikipedia/commons/a/a7/Jinsharivermap.jpg

[37]"File:Mekong river basin.png." *Wikimedia Commons*, 27 July 2022, upload.wikimedia.org/wikipedia/commons/7/7c/Mekong_river_basin.png

[38]"File:Salween river basin map.png." *Wikimedia Commons*, 15 Nov. 2020, upload.wikimedia.org/wikipedia/commons/3/3c/Salween_river_basin_map.png

[39]"File:Irrawaddyrivermap.jpg." *Wikimedia Commons*, 27 July 2022, upload.wikimedia.org/wikipedia/commons/c/c1/Irrawaddyrivermap.jpg

[40]"File:Kunyu Wanguo Quantu by Matteo Ricci Plate 1-3.jpg. *Op. cit.*

[41]"File:Rubbing of Huayi tu map.jpg." *Op. cit.*

[42]"File:Da-ming-hun-yi-tu.jpg." *Op. cit.*

[43]"File:廣輿圖輿地總圖.jpg." *Wikimedia Common*, 4 Apr. 2023, upload.wikimedia.org/wikipedia/commons/5/55/廣輿圖輿地總圖.jpg

[44]Michael Pearson. "Lake Chiamay: Asia's mythical mother of rivers." *The Globe, Journal of the Australian and New Zealand Map Society*, vol. 83, 2018, pp. 43–62; https://www.researchgate.net/publication/327155061_%27Lake_Chiamay_Asia%27s_mythical_mother_of_rivers%27_The_Globe_Journal_of_the_Australian_and_New_Zealand_Map_Society_83_43-62_2018

[45]Michael Pearson. *Op. cit.*

[46]"File:CEM-11-Chinae-nova-descriptio-2521.jpg." *Wikimedia Commons*, 21 June 2020, upload.wikimedia.org/wikipedia/commons/9/94/CEM-11-Chinae-nova-descriptio-2521.jpg

[47]*Ibid.*

[48]Sheng-Wei Wang. "Chapter 2: Chinese Explored Australia, New Zealand, Land of Fire and Antarctica Long Before the Europeans." *Chinese Global Exploration in the Pre-Columbian Era: Evidence from an Ancient World Map, op. cit.*, pp. 31–69.

[49]*Ibid.*

[50]Sheng-Wei Wang. *The Last Journey of the San Bao Eunuch, Admiral Zheng He.* Hong Kong, China: Ptoverse Hong Kong, 2019.

[51]Luo Maodeng (羅懋登/罗懋登; lived c. 1597). *San Bao Tai Jian Xi Yang Gi* (三寶太監西洋記/三宝太监西洋记) or *An Account of the Western World Voyage of the San Bao Eunuch*; an electronic version can be found at https://www.diancang.xyz/wenxueyishu/sanbaotaijianxiyangji/

[52]Sheng-Wei Wang. "Chapter 3: Luo Maodeng's Tianfang/Yun Chong was Mecca in Saudi Arabia." *The Last Journey of the San Bao Eunuch, Admiral Zheng He, op. cit.*, pp. 91–138.

[53]Sheng-Wei Wang. "Chapter 2: Chinese Explored Australia, New Zealand, Land of Fire and Antarctica Long Before the Europeans." *Chinese Global Exploration in the Pre-Columbian Era: Evidence from an Ancient World Map, op. cit.*, pp. 52–61.

[54]Sheng-Wei Wang. "Chapter 3: Luo Maodeng's Tianfang/Yun Chong was Mecca in Saudi Arabia." *The Last Journey of the San Bao Eunuch, Admiral Zheng He, op. cit.*, pp. 91–138.

[55]Zhu Yunmeng (祝允明; 1461–1527). *Qian Wen Ji • Xia Xi Yang* (前闻记 • 下西洋) or *A Record of History Once Heard: Down to the Western Ocean*; an electronic version can be found at https://ctext.org/wiki.pl?if=gb&chapter=203549

[56]Sheng-Wei Wang. "Chapter 1: Chinese Explored Cape Breton Island Long Before the Europeans." *Chinese Global Exploration in the Pre-Columbian Era: Evidence from an Ancient World Map, op. cit.*, pp. 7–8.

[57]"File:Kunyu Wanguo Quantu by Matteo Ricci Plate 1-3.jpg." *Op.cit.*

[58]Zhang Tingyu (張廷玉; 1672–1755), *et al. Ming Shi • Juan San Bai Er Shi Liu • Lie Chuan Di Er Bai Shi Si • Wai Guo Qi* (明史 • 卷三百二十六 • 列傳第二百十四 • 外國七) or *History of Ming, Juan (Volume) 326, Liezhuan (US type of biography) 214, Foreign Country 7*; an electronic version can be found at https://ctext.org/wiki.pl?if=gb&res=410835

[59]Zhao Rugua (趙汝适; 1170–1231). *Zhufan Zhi • Xi Tian Peng Qie Luo Guo* (諸蕃志 • 西天鵬茄囉國) or *Gazetteer of Foreign Lands, Pengjialuo State in the Western Tianzhu*; an electronic version can be found at https://ctext.org/wiki.pl?if=gb&res=520299

[60]Zhang Tingyu (張廷玉; 1672–1755), *et al. Ming Shi • Juan San Bai Er Shi Liu • Lie Chuan Di Er Bai Shi Si • Wai Guo Qi* (明史 • 卷三百二十六 • 列傳第二百十四 • 外國七) or *History of Ming, Juan (Volume) 326, Liezhuan (US type of biography) 214, Foreign Country 7*; *op. cit.*

[61]Sheng-Wei Wang. "Chapter Three: Luo Maodeng's *Tianfang*/Yun Chong was Mecca in Saudi Arabia." *The last journey of the San Bao Eunuch, Admiral Zheng He, op. cit.*, pp. 105–106.

[62]Wang Qi (王圻; 1530–1615). *Xu Wenxian Tongkao • Di Shi Jiu Juan* (續文獻通考 • 第十九卷）or *Continued Comprehensive Investigations Based on Literary and Documentary Sources, Juan (Volume) 19*, the electronic version can be found at https://books.google.com.hk/books?id=ue5bAAAAcAAJ&pg=PP448&lpg=PP448&dq=%E6%A6%9C%E8%91%9B%E5%89%8C%E5%8F%A4%E5%BF%BB%E9%83%BD%E5%B7%9E&source=bl&ots=1PFxdEgx1G&sig=ACfU3U0judomE6U9Xlfax8fInDR0bB3NSA&hl=zh-TW&sa=X&ved=2ahUKEwiMzOmK3vCCAxU2lFYBHRDnDJ4Q6AF6BAgTEAM#v=onepage&q=%-E6%A6%9C%E8%91%9B%E5%89%8C%E5%8F%A4%E5%BF%BB%E9%83%BD%E5%B7%9E&f=false

[63]J.J.A. Campos. *History of the Portuguese in Bengal*. Open Library: Gyan Publishing House, 2020.

[64]"File:Bay of Bengal map 1800s.png." *Wikimedia Commons*, 6 Nov. 2023, upload.wikimedia.org/wikipedia/commons/a/a0/Bay_of_Bengal_map_1800s.png

[65]Huang Shijian (黄时鉴) and Gong Yingyan (龚缨晏). *Li Ma Dou Shi Jie Di Tu Yan Jiu* (利玛窦世界地图研究) or *Research on Matteo Ricci's World Map*, Shanghai, China: Shanghai Chinese Classics Publishing House (上海古籍出版社), 2004, p. 194.

[66]"File:Rakhine State in Myanmar (cropped).jpg." *Wikimedia Commons*, 26 Jan. 2021, upload.wikimedia.org/wikipedia/commons/8/8c/Rakhine_State_in_Myanmar_(cropped).jpg

[67]William J. Topich and Keith A. Leitich. *The History of Myanmar*. Santa Barbara, CA, USA: ABC-CLIO, LLC, 2013.

[68]*Ibid.*

[69]Huang Shijian (黄时鉴) and Gong Yingyan (龚缨晏). Op. cit., p. 204.

[70]N. Jayapalan. *History of India, from 1206 to 1773, Volume II*, New Delhi, India: Atlantic Publishers and Distributors Pvt Ltd., 2001, p. 76.

[71]Joseph E. Schwartzberg, ed. *A Historical Atlas of South Asia (The Association for Asian Studies Reference Series, No. 2)*, Chicago, IL, USA: 1978, p. 147, map XIV.4.

[72]Wang Qi (王圻; 1530–1615). *Xu Wenxian Tongkao* (續文獻通考）or *Continued Comprehensive Investigations Based on Literary and Documentary Sources*; the electronic version can be found at https://ctext.org/wiki.pl?if=gb&res=110962

[73]Mian Muhammad Saeed. *The Sharqi of Jaunpur: A Political & Cultural History*, Karachi, Sindh, Pakistan: University of Karachi, 1972, p. 61.

[74]Zhang Tingyu (張廷玉; 1672–1755), *et al. Ming Shi • Juan San Bai Er Shi Liu • Lie Chuan Di Er Bai Shi Si • Wai Guo Qi* (明史 • 卷三百二十六 • 列傳第二百十四 • 外國七) or *History of Ming, Juan (Volume) 326, Liezhuan (US type of biography) 214, Foreign Country 7*; an electronic version can be found at https://ctext.org/wiki.pl?if=gb&res=410835

[75]Huang Shijian (黄时鉴) and Gong Yingyan (龚缨晏). Op. cit., p. 191.

[76]Samuel Beal. *Buddhist records of the Western world (Xuanzang)*. Darya Ganj, India: Motilal Banarsidass Publishers Pvt. Ltd., 1884; an electronic version can be found at https://www.wisdomlib.org/south-asia/book/buddhist-records-of-the-western-world-xuanzang/d/doc220288.html

[77]"大唐大慈恩寺三藏法師傳/卷05 (*A Biography of the Tripitaka Master of the Great Ci'en Monastery of the Great Tang Dynasty/Juan 5*)." *Wikipedia*, 9 Jan. 2020, zh.wikisource.org/wiki/大唐大慈恩寺三藏法師傳/卷05

[78]Si Maqian (司馬遷; 134 B.C. –c. 86 B.C.). *Shi Ji • Juan Yi Bai Er Shi San • Da Wan Lie Chuan* (史记 • 卷一百二十三 • 大宛列傳) or *Records of the Grand Historian, Juan (Volume) 123, Biography of Da Wan* an electronic file can be found at https://ctext.org/shiji/zh

[79]Fan Ye (范曄; 398–445). Hou Han Shu • Xi Yu Chuan (後漢書 • 西域傳) *or Book of the Later Han: Biography of the Western Regions*; an electronic version can be found at https://ctext.org/hou-han-shu/xi-yu-zhuan/zh

[80]Xue Keqiao (薛克翹). "From Faxian's Wu Tianzhus to Xuanzang's Wu Yindu (從法顯的'五天竺'到玄奘的'五印度')." *The Hualin International Journal of Buddhist Studies (華林國際佛學學刊/华林国际佛学学刊)*,   vol. 2, no. 1, 2019, pp. 151–166. https://dx.doi.org/10.6939/HIJBS.201904_2(1).0006

[81]*Ibid.*

[82]*Ibid.*

[83]*Ibid.*

[84]*Ibid.*

[85]*Ibid.*

[86]Huang Shijian (黄时鉴) and Gong Yingyan (龚缨晏). Op. cit., p. 190.

[87]Huang Shijian (黄时鉴) and Gong Yingyan (龚缨晏). Op. cit., p. 197.

[88]Huang Shijian (黄时鉴) and Gong Yingyan (龚缨晏). *Op. cit.*, p. 193.

[89]C.B. Patel. *Origin and Evolution of the Name ODISA*, Bhubaneswar, India: I&PR Department, Government of Odisha, 2015, pp. 28–30.

[90]Sailendra Sen. *A Textbook of Medieval Indian History*, Delhi, India: Primus Books. 2013, pp. 121–122.

[91]Huang Shijian (黄时鉴) and Gong Yingyan (龚缨晏). Op. cit., p. 197.

[92]Huang Shijian (黄时鉴) and Gong Yingyan (龚缨晏). *Op. cit.*, p. 199.

[93]Robert Sewell. *A Forgotten Empire (Vijayanagar)*. New Delhi, India: Asian Educational Services, 2011.

[94]Burton Stein. *The New Cambridge History of India: Vijayanagara*. Cambridge, UK: Cambridge University Press, 1989.

[95]Robert Sewell. *Op. cit.*

[96]Domingos Paes and Fernpo Nunes. *A Forgotten Empire: Vijayanagar A Contribution to the History of India "Chronica Dos Reis De Bisnaga"*. Open Library, 2006.

[97]Sheng-Wei Wang. "A Chinese-based Word Map Depicts Africa in 1433." *Chinese Global Exploration in the Pre-Columbian Era: Evidence from an Ancient World Map, op. cit.*, pp. 147–148.

[98]Huang Shijian (黄时鉴) and Gong Yingyan (龚缨晏). Op. cit., p. 187.

[99]Chen Jiarong, et al (陳佳榮等). *Gu Dai Nan Hai Di Ming Hui Shi* (古代南海地名匯釋) or *Explanation of Place Names in the Ancient South China Sea*. Beijing, China: Zhonghua Book Company (中华书局), 1984; an electronic file can be found at https://www.world10k.com/blog/?p=1083

[100]Yue Shi (樂史; 930–1007). *Tai Ping Huan Yu Ji* (太平寰宇記) or *Universal Geography of the Taiping Era [976–983]*; an electronic version can be found at https://ctext.org/wiki.pl?if=gb&res=637263

[101]Harish Johari. *The Healing Power of Gemstones: In Tantra, Ayurveda, and Astrology*, Rochester, VT, USA: Destiny Books, 1886, p. 72. ISBN 9780892816088.

[102]Huang Shijian (黄时鉴) and Gong Yingyan (龚缨晏). Op. cit., p. 194.

[103]DHNS. "The Narasinga Kingdom." www.deccanherald.com, Deccan Herald, 2025, www.deccanherald.com/india/karnataka/narasinga-kingdom-2123698; Angelo Paratico. "Andrea Corsali. L'esploratore Amico Di Leonardo Da Vinci, Scomparso In Oriente." https://archive.org/, Internet Archive, 2025, archive.org/details/andrea-corsali.-lesploratore-amico-di-leonardo-da-vinci-scomparso-in-oriente

[104]Anjana. "The Rayas of Vijayanagar." www.notesonindianhistory.com, Notes on Indian History, 2025, https://www.notesonindianhistory.com/2013/10/vijayanagara-city-of-victory.html

[105]Huang Shijian (黄时鉴) and Gong Yingyan (龚缨晏). *Op. cit.*, p. 202.

[106]A. Sreedhara Menon. *Keralacharithram (Malayalam)*. Kerala, India: DC Books, 2016.

[107]Nigel Cliff. *The Last Crusade: The Epic Voyages of Vasco da Gama*. New York, NY, USA: Harper Perennial, 2012.

[108]"File:India Malabar Coast locator map.svg." *Wikimedia Commons*, 4 Nov. 2021, upload.wikimedia.org/ wikipedia/commons/thumb/5/50/India_Malabar_Coast_locator_map.svg/1812px-India_Malabar_Coast_locator_map.svg.png

[109]Huang Shijian (黄时鉴) and Gong Yingyan (龚缨晏). Op. cit., p. 200.

[110]Philip M. Parker. *Cochin: Webster's Timeline History, 1330—2007*. Las Vegas, NV, USA: ICON Group International, Inc., 2008.

[111]"File:Kochi, Fishing nets at sunset, Kerala, India.jpg." *Wikimedia Commons*, 14 July 2024, upload.wiki-media.org/wikipedia/commons/6/6e/Kochi%2C_Fishing_nets_at_sunset%2C_Kerala%2C_India.jpg

[112]Ma Huan (c. 1380–1460). J.V.G. Mills, trans. *Ying-yai Sheng-lan: The Overall Survey of the Ocean's Shores (1433)*. London, UK: Hakluyt Society, 1970.

[113]Zhang Tingyu (張廷玉; 1672–1755), *et al. Ming Shi • Juan San Bai Er Shi Liu • Lie Chuan Di Er Bai Shi Si • Wai Guo Qi* (明史 • 卷三百二十六 • 列傳第二百十四 • 外國七) or *History of Ming, Juan (Volume) 326, Liezhuan (US type of biography) 214, Foreign Country 7*; an electronic version can be found at https:// ctext.org/wiki.pl?if=gb&res=410835

[114]*Ibid.*

[115]*Ibid.*

[116]*Ibid.*

[117]*Ibid.*

[118]Philip M. Parker. *Cochin: Webster's Timeline History, 1330–2007*. Las Vegas, USA: ICON Group International, Inc., 2008.

[119]Charlton T Lewis and Charles Short. *A Latin Dictionary*. Open Library: Nigel Gourlay, 2020.

[120]Henry George Liddell and Robert Scott, Compiler. Henry Stuart Jones, ed. Roderick McKenzie, Assistant. *Greek-English Lexicon*. Oxford, UK: Clarendon Press, 1996.

[121]F. Glaw, "Taxonomic checklist of chameleons (Squamata: Chamaeleonidae)." *Vertebrate Zoology*, vol. 65, no. 2, 2015, pp. 167–246.

[122]Michael Shally-Jensen and Anthony Vivian. *A Cultural Encyclopedia of Lost Cities and Civilizations*. Santa Barbara, CA, USA: ABC-CLIO, 2022.

[123]Burton Stein. *Op. cit.*

[124]Sailendra Sen. *A Textbook of Medieval Indian History*, Delhi, India: Primus Books, 2013, p. 119.

[125]Jagan Pillarisetti. "The Liberation of Goa: 1961." https://web.archive.org/, Bharat Rakshak, 2025, web. archive.org/web/20070809202539/http://www.bharat-rakshak.com/IAF/History/1960s/Goa01.html

[126]A. Sreedhara Menon. *Kerala District Gazetteers, Vol. 5*. Kerala, India: Superintendent of Government Presses, 1962.

[127]Nigel Cliff. *The Last Crusade: The Epic Voyages of Vasco da Gama*. New York, NY, USA: 2012.

[128]Zhang Tingyu (張廷玉; 1672–1755), *et al. Ming Shi • Juan San Bai Er Shi Liu • Lie Chuan Di Er Bai Shi Si • Wai Guo Qi* (明史 • 卷三百二十六 • 列傳第二百十四 • 外國七) or *History of Ming, Juan (Volume) 326, Liezhuan (US type of biography) 214, Foreign Country 7*; an electronic version can be found at https:// ctext.org/wiki.pl?if=gb&res=410835

[129]*Ibid.*

[130]*Ibid.*

[131]Huang Shijian (黄时鉴) and Gong Yingyan (龚缨晏). *Op. cit.*, p. 196; Nurlan Kenzheakhmet and Alpamys Abu. "Some Medieval and Post-Golden Horde's Towns of the Itil (Volga) and Syr-Darya Basins According

to the Arabic and Chinese Maps." *Golden Horde Review*, vol. 9, no. 3, 2021, pp. 611–653. https://www.researchgate.net/publication/354971343_Some_Medieval_and_Post-Golden_Horde's_Towns_of_the_Itil_Volga_and_Syr-Darya_Basins_According_to_the_Arabic_and_Chinese_Maps

[132]A. Sreedhara Menon. *Kerala History and its Makers*. Kerala, India: D C Books, 2011.

[133]K.V. Krishna Ayyar. "The Zamorins Of Calicut." https://archive.org/details/, Internet Archive, 2025, archive.org/details/TheZamorinsOfCalicut

[134]A Sreedhara Menon. *A Survey of Kerala History*. Kerala, India: DC Books, 2008.

[135]Noburu Karashmia, ed. *A Concise History of South India: Issues and Interpretations*, New Delhi, India: Oxford University Press, 2014. pp. 143–44.

[136]Huang Shijian (黄时鉴) and Gong Yingyan (龚缨晏). *Op. cit.*, p. 187.

[137]Sir James Murray, ed. *The Oxford English Dictionary*. Oxford, UK: Oxford University Press, 1989.

[138]B. Stein. *A History of India*. Oxford, UK: Wiley-Blackwell, 1998.

[139]C. B. Asher and C. Talbot. *India Before Europe*. Cambridge, UK: Cambridge University Press, 2006.

[140]*Ibid.*

[141]Anthony Farrington. *Trading Places: The East India Company and Asia 1600-1834*. London, UK: The British Library, 2002.

[142]"File:Irrawaddyrivermap.jpg." *Wikmedia Commons*, 27 July 2022, upload.wikimedia.org/wikipedia/commons/c/c1/Irrawaddyrivermap.jpg

[143]Patrick Blanche. "River of Gold." www2.irrawaddy.com, Irrawaddy Publishing Group, https://www2.irrawaddy.com/article.php?art_id=19667

[144]Sheng-Wei Wang. *Chinese Global Exploration in the Pre-Columbian Era: Evidence from an Ancient World Map, op. cit.*, p. 125.

[145]Tahmina Ahmed. "Ichamati River." In Sirajul Islam and Ahmed A. Jamal, eds. *Banglapedia: National Encyclopedia of Bangladesh (Second ed.)*. Nimtali, Dhaka, Bangladesh: Asiatic Society of Bangladesh, 2012.

[146]"File:Ganges-Brahmaputra-Meghna basins.jpg." *Wikimedia Commons*, 10 Nov. 2023, upload.wikimedia.org/wikipedia/commons/3/34/Ganges-Brahmaputra-Meghna_basins.jpg

[147]Huang Shijian (黄时鉴) and Gong Yingyan (龚缨晏). *Op. cit.*, p. 191.

[148]John Ogilby. "The Kingdom of Decan." *The Empire of the Great Mogol and India*. London, UK: Early English Books Online Text Creation Partnership, 2011; an electronic version can be found at https://quod.lib.umich.edu/e/eebo2/A53223.0001.001/1:12.48?rgn=div2;view=fulltext

[149]Chen Jiarong (陳佳榮), *et al.* "Section 4: The Jews settled in China (第四節 猶太人之入居中國)." www.world10k.com, Chen Jiarong (陳佳榮), 2025, www.world10k.com/blog/?p=699

[150]M. N. Pearson. *The Portuguese in India (The New Cambridge History of India)*. Cambridge, UK: Cambridge University Press, 2006.

[151]Official Gujarat State Portal. "History of Gujarat." web.archive.org, Gujarat State, 2025, https://web.archive.org/web/20100203081044/http://www.gujaratindia.com/about-gujarat/history-1.htm

[152]Ruth Bames. "Indian Cotton for Cairo: The Royal Ontario Museum's Gujarati Textiles and the Early Western Indian Ocean Trade." *Textile History*, vol. 48, no. 1, 2017, pp. 15–30. https://www.tandfonline.com/doi/full/10.1080/00404969.2017.1294814

[153]Huang Shijian (黄时鉴) and Gong Yingyan (龚缨晏). Op. cit., p. 201.

[154]M. Hassan Kakar. "Kabul." In Peter N. Stearns, ed. *Oxford Encyclopedia of the Modern World*. Oxford, UK: Oxford University Press, 2008.

[155]"File:Indus river basin without boundaries of disputed regions.png." *Wikimedia Commons*, 20 May 2023, upload.wikimedia.org/wikipedia/commons/8/8f/Indus_river_basin_without_boundaries_of_disputed_regions.png

[156]John F. Richards. *The Mughal Empire*. Cambridge, UK: Cambridge University Press, 1995.

[157]Evgenia Vanina. *Medieval Indian Mindscapes: Space, Time, Society, Man*. Open Library, 2012.

[158]Robert L. Canfield. *Turko-Persia in Historical Perspective*, Cambridge, UK: Cambridge University Press, 2002, p. 20.

[159]Sheng-Wei Wang. "Chapter 1: Chinese Explored Cape Breton Island Long before the Europeans." *Chinese Global Exploration in the Pre-Columbian Era: Evidence from an Ancient World Map, op. cit.*, pp. 7–8.

[160]Huang Shijian (黄时鉴) and Gong Yingyan (龚缨晏). Op. cit., p. 200.

[161]Places in the World. "Gor, India." https://india.places-in-the-world.com/, Places in the World, 2025, india.places-in-the-world.com/7018625-place-gor.html

[162]*Ibid.*

[163]Huang Shijian (黄时鉴) and Gong Yingyan (龚缨晏). *Op. cit.*, p. 189.

[164]Arvind Sharma. "On Hindu, Hindustan, Hinduism and Hindutva." *Numen.* vol. 49, no. 1, 2002, pp. 1–36. https://www.jstor.org/stable/3270470

[165]Anu Kapur. *Mapping Place Names of India*. Oxfordshire, U.K.: Taylor & Francis, 2019.

[166]Bratindra Nath Mukherjee. *The Foreign Names of the Indian Subcontinent: Place Names Society of India*, Google Book, 1989, p. 154.

[167]"File:Arabian Sea map.png." *Wikimedia Commons, The Free Media Depository*, Wikimedia Foundation, 31 Dec. 2023, upload.wikimedia.org/wikipedia/commons/4/41/Arabian_Sea_map.png

[168]Zhang Tingyu (張廷玉; 1672–1755), *et al. Ming Shi • Juan San Bai Er Shi Liu • Lie Chuan Di Er Bai Shi Si • Wai Guo Qi* (明史 • 卷三百二十六 • 列傳第二百十四 • 外國七) or *History of Ming, Juan (Volume) 326, Liezhuan (US type of biography) 214, Foreign Country 7*; an electronic version can be found at https://ctext.org/wiki.pl?if=gb&res=410835

[169]Edward L. Dreyer. *Zheng He: China and the Oceans in the Early Ming Dynasty, 1405–1433*. New York, NY, USA: Pearson Longman, 2007.

[170]Sheng-Wei Wang. "Chapter Three: Luo Maodeng's *Tianfang*/Yun Chong was Mecca in Saudi Arabia." *Op. cit.*, pp. 98–102.

[171]Zhu Yunming (祝允明; 1461–1527). *Qian Wen Ji • Xia Xi Yang* (前闻記 •下西洋) or *A Record of History Once Heard: Down to the Western Ocean*; an electronic version can be found at https://ctext.org/wiki.pl?if=gb&chapter=203549&remap=gb

[172]*Ibid.*

[173]Sheng-Wei Wang. "Chapter Three: Luo Maodeng's *Tianfang*/Yun Chong was Mecca in Saudi Arabia." *The Last Journey of the San Bao Eunuch, Admiral Zheng He, Op. cit.*, p. 105.

[174]"File:Admiral Zhenghe.jpg." *Wikimedia Commons*, 5 Sept. 2023, upload.wikimedia.org/wikipedia/commons/2/28/Admiral_Zhenghe.jpg

[175]Vaman Shivram Apte. *Sanskrit–English Dictionary*. New Delhi, India: Motilal Banarsidass Publishing House, 1985.

[176]"File:Maldives (orthographic projection).svg." *Wikimedia Commons*, 6 June 2023, upload.wikimedia.org/wikipedia/commons/thumb/4/4b/Maldives_(orthographic_projection).svg/2048px-Maldives_(orthographic_projection).svg.png

[177]"File:India location map 3.png." *Wikimedia Commons*, 4 Oct. 2023, upload.wikimedia.org/wikipedia/commons/c/ca/India_location_map_3.png

[178]*Ibid.*

[179]*Ibid.*

[180]"File:Indian Ocean laea location map.svg." *Wikimedia Commons*, 29 July 2023, upload.wikimedia.org/wikipedia/commons/thumb/c/c7/Indian_Ocean_laea_location_map.svg/2180px-Indian_Ocean_laea_location_map.svg.png

[181]Michael A. Aung-Thwin. *Myanmar in the Fifteenth Century.* Honolulu, Hawaii, USA: University of Hawaii Press, 2017.

[182]Bijan Raj Chatterji. "Jayavarman VII (1181–1201 A.D.) (The last of the great monarchs of Cambodia)." *Proceedings of the Indian History Congress*, vol. 3, 1939, pp. 377–385. https://www.jstor.org/stable/44252387

[183]Western Legend Vision (西部传奇视界). "(Traveling through Bingchacha, how dangerous and beautiful the road to Tibet is, Bingchacha carpooling (穿越丙察察，丙察察进藏路有多险有多美，丙察察拼车)." www.sohu.com, Sohu, 2025, https://www.sohu.com/a/483183332_100111728

[184]James A. Millward. *Eurasian Crossroads: A History of Xinjiang.* New York, NY, USA: Columbia University Press, 2007.

[185]"File:Kingdom of Khotan.png." *Wikimedia Commons*, 21 Dec. 2022, upload.wikimedia.org/wikipedia/commons/3/3e/Kingdom_of_Khotan.png

[186]"File:Tarimrivermap.png." *Wikimedia Commons*, 31 May 2022, upload.wikimedia.org/wikipedia/commons/7/7e/Tarimrivermap.png

[187]Michael Pearson. *Op. cit.* https://www.researchgate.net/publication/327155061_%27Lake_Chiamay_Asia%27s_mythical_mother_of_rivers%27_The_Globe_Journal_of_the_Australian_and_New_Zealand_Map_Society_83_43-62_2018

[188]She Youji (佘由基). "The Lijiang government is finally determined to crack down on illegal gold mining in the Jinsha River (丽江政府部门终下决心 打击金沙江非法淘金)." www.cngaosu.com, Lijiang government, 2025, https://web.archive.org/web/20150402140618/http://wajueji.cngaosu.com/peijian/yunnan/lijiang/2012-04-17/11612.html

[189]Yangtze River was the old name of the downstream section from Nanjing to the estuary of the Changjiang. Why the Changjiang was called the Yangtze River is generally explained as follows: Because the Western missionaries who came to China first encountered the Yangtze section of the Changjiang and heard the name "Yangtze River", they generally referred Changjiang as the "Yangtze River", and "Yangtze River" has also become the name of Changjiang in English.

[190]World Heritage Centre. "Three Parallel Rivers of Yunnan Protected Areas." https://whc.unesco.org/, UNESCO World Heritage convention, 2025, whc.unesco.org/en/list/1083/

[191]"File:Yangtze river map.png." *Wikimedia Commons*, 24 Oct. 2022, upload.wikimedia.org/wikipedia/commons/f/fc/Yangtze_river_map.png

[192]Zhang Tingyu (張廷玉; 1672–1755), *et al. Ming Shi • Juan San Bai Yi Shi Wu • Lie Chuan Di Er Bai San • Yun Nan Tu Si San* (明史 • 卷三百一十五 • 列傳第二百三 • 雲南土司三) or *History of Ming, Juan (Volume) 315, Liezhuan (US type of biography) 203, Yunnan Tusi 3*; an electronic version can be found at https://ctext.org/wiki.pl?if=gb&res=410835

[193]*Ibid.*

[194]He Shengda (賀聖達). *Mian Dian Shi* (緬甸史) or *History of Myanmar.* Beijing, China: People's Publishing House (人民出版社), 1992.

[195]Zhang Tingyu (張廷玉; 1672–1755), *et al. Ming Shi • Juan Si Shi Liu • Zhi Di Er Shi Er • Di Li Qi* (明史 • 卷四十六 • 志第二十二 • 地理七) or *History of Ming, Juan (Volume) 46, Record 22, Geography 7*; an electronic version can be found at https://ctext.org/wiki.pl?if=gb&res=410835

[196]Li Mou (李謀), *et al.*, trans. *Liu Li Gong Shi* (琉璃宮史) or *History of the Liu Li Palace.* Hong Kong, China: 5he Commercial Press (商務印書館), 2007.

[197]G. E. Harvey. *History of Burma.* Open Library: Gyan Publishing House, 2020.

[198]*Ibid.*

[199]He Shengda (賀聖達). *Op. cit.*

[200]*Ibid.*

[201]Zhang Tingyu (張廷玉; 1672–1755), *et al. Ming Shi • Juan Si Shi Liu • Zhi Di Er Shi Er • Di Li Qi* (明史 • 卷四十 六 • 志第二十 二 • 地理七) or *History of Ming, Juan (Volume) 46, Record 22, Geography 7*; an electronic version can be found at https://ctext.org/wiki.pl?if=gb&res=410835

[202]*Ibid.*

[203]Zhang Tingyu (張廷玉; 1672–1755), *et al. Ming Shi • Juan Qi Shi Liu • Zhi Di Wu Shi Er • Zhi Guan Wu* (明史 • 卷七十 六 • 志第五十 二 • 職官五) or *History of Ming, Juan (Volume) 76, Record 52, Official 5*; an electronic version can be found at https://ctext.org/wiki.pl?if=gb&res=410835

Chapter 6

An Ancient World Map Depicts Chinese Exploration of West and Central Asia in 1433

Abstract

The West and Central Asia portions of Kunyu Wanguo Quantu (坤輿萬國全圖/坤輿万国全图) (abbreviated as KWQ), or Complete Geographical Map of All the Kingdoms of the World, are fully analysed in this chapter. These portions of the map are abbreviated as West&Central Asia-KWQ. The KWQ was the first world map written in Chinese and published in China by Italian missionary Matteo Ricci in 1602. The map has been generally considered as a European map.

However, after analysing the complete 93 geographical items and their histories, and the nine annotations on the West&Central Asia-KWQ, and comparing them with seven major sixteenth-century European maps, Ricci's map is shown to be of Chinese origin.

The political era revealed directly by the West&Central Asia-KWQ is deduced step by step in Section 3 as being within the period 1192–1464. After verifying the Chinese historical records on Admiral Zheng He's seventh (and last) voyage to the Western Ocean, such a wide period can be narrowed down to the year 1433 which falls in the range of 1192–1464.

Keywords: Kunyu Wanguo Quantu, Matteo Ricci, Ming treasure voyages, West and Central Asia, Zheng He

1. Introduction

For over four hundred years, the ancient world map — Kunyu Wanguo Quantu (坤輿萬國全圖/坤輿万国全图),[1] abbreviated here as KWQ — drawn with Chinese characters and having latitudinal and longitudinal lines, has been generally regarded as a European map. The map was published in 1602 by Matteo Ricci (an Italian Jesuit priest and one of the founding figures of the Jesuit China missions) and was believed to be based on the European maps which Ricci brought with him to China in 1582.[2]

In my previous book titled *Chinese Global Exploration in the Pre-Columbian Era: Evidence from an Ancient World Map*,[3] and in Chapters 1–5 of this book, I have made thorough analyses to show that the true origin and the true eras of many major portions of the KWQ can be deduced from their geographical and historical information, and from the annotations on the map.

359

Most importantly, these analyses have shown that the afore-mentioned portions of the KWQ must have used Chinese maps as source maps (they no longer exist). They are not based on contemporary European maps.

In this chapter, I shall continue my effort for the West and Central Asia regions on the KWQ, abbreviated here as West&Central Asia-KWQ, using the same approach: a thorough analysis of the nine annotations, and of the geographical locations and histories of the 93 geographical items depicted on the West&Central Asia-KWQ, and comparison of the information obtained with that on the seven major sixteenth century European maps.

First, we need to know the geographical regions covered by West Asia and Central Asia:

West Asia (西亞/西亚): It is a subregion of Asia. The term *West Asia* is used pragmatically and has no "correct" or generally accepted definition. In this chapter, I adopt the definition that it is the westernmost region of Asia and the subregion, which consists of Anatolia (安那托利亞/安那托利亚; also known as Asia Minor or 小亞細亞/小亚细亚), the Arabian Peninsula (阿拉伯半島/阿拉伯半岛), Iran, Mesopotamia (美索不達米亞/美索不达米亚), the Armenian Highlands (亞美尼亞高原/亚美尼亚高原), the Levant (黎凡特; a large area in the Eastern Mediterranean region of West Asia), the Island of Cyprus (塞浦路斯島/塞浦路斯岛), the Sinai Peninsula (西奈半島/西奈半岛), and the southern part of the Caucasus Region (南高加索地區/南高加索地区).

Figure 6.1(a) shows that the West Asia region is separated from Africa by the Isthmus of Suez (蘇伊士地峽/苏伊士地峡) in Egypt, from Europe by the waterways of the Turkish Straits

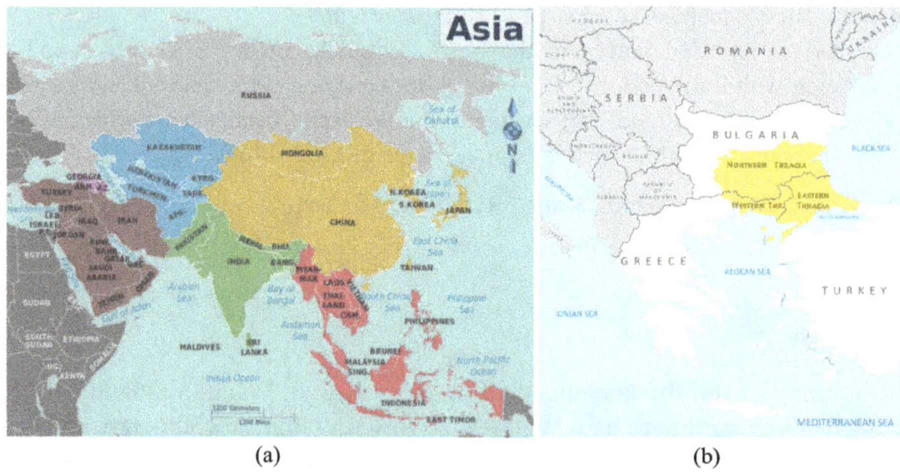

(a)   (b)

Fig. 6.1. (a) is a modern map of Asia (*Source*: Cacahuate, adapted by Peter Fitzgerald, Globe-trotter, Joelf, Texugo, Piet-c and Bennylin, under CC BY-SA 4.0 International, 3.0 Unported, 2.5 Generic, 2.0 Generic and 1.0 Generic license, https://commons.wikimedia.org/wiki/File:Map_of_Asia.svg);[6] West and Central Asia cover the regions in brown and blue colors; for Central Asia (the blue-colored region), the northern and western areas of Pakistan and the Kashmir Valley of India, as well as the Tibetans and Ladakhis, (all in the green-colored region) may also be included. (b) shows the Sea of Marmara connects the Black Sea with the Aegean Sea through the Bosporus Strait and the Dardanelles Strait and separates Anatolia from Thrace in the Balkan Peninsula of Southeastern Europe (*Source*: Пакко, under CC BY-SA 3.0, https://commons.wikimedia.org/wiki/File:Thrace_and_present-day_state_borderlines.png).[7]

(土耳其海峽/土耳其海峡), and the watershed of the Greater Caucasus (大高加索地區/大高加索地区), with Central Asia (中亞/中亚) lies to its northeast and South Asia (南亞/南亚) to its east. Twelve seas surround the region (clockwise): the Aegean Sea (愛琴海/爱琴海), the Sea of Marmara (馬摩拉海/马摩拉海), the Black Sea (黑海), the Caspian Sea (里海), the Persian Gulf (波斯灣/波斯湾), the Gulf of Oman (阿曼灣/阿曼湾), the Arabian Sea (阿拉伯海), the Gulf of Aden (亞丁灣/亚丁湾), the Red Sea (红海), the Gulf of Aqaba (亞喀巴灣/亚喀巴湾), the Gulf of Suez (蘇伊士灣/苏伊士湾), and the Mediterranean Sea (地中海).

Among the subregions of West Asia, Anatolia is a large peninsula that forms the western-most extension of continental Asia, as it borders European Turkey (歐洲土耳其/欧洲土耳其). The land mass of Anatolia constitutes most of the territory of contemporary Turkey. Geographically, the Anatolian region is bounded by the Turkish Straits (土耳其海峽/土耳其海峡) to the north-west, the Black Sea (黑海) to the north, the Armenian Plateau (亞美尼亞高原/亚美尼亚高原) to the east, the Mediterranean Sea (地中海) to the south, and the Aegean Sea to the west. Topographically, the Sea of Marmara connects the Black Sea with the Aegean Sea through the Bosporus Strait (博斯普魯斯海峽/博斯普鲁斯海峡) and the Dardanelles Strait (達達尼爾海峽/达达尼尔海峡) and separates Anatolia from Thrace (色雷斯) in the Balkan Peninsula (巴爾幹半島/巴尔干半岛) of Southeastern Europe, see Fig. 6.1(b).

Central Asia: It is a subregion of Asia. The borders of Central Asia are subject to multiple definitions. In this chapter, I adopt the definition that it stretches from the Caspian Sea in the southwest, and Eastern Europe in the northwest, to Western China and Mongolia in the east, and from Afghanistan and Iran in the south to Russia in the north. The region includes in part the present-day Kazakhstan, Kyrgyzstan, Tajikistan, Turkmenistan, and Uzbekistan; the northern and western areas of Pakistan and the Kashmir Valley of India; the Tibetans and Ladakhis (Ladakh is one of the most sparsely populated regions in India; its culture and history are closely related to those of Tibet) may also be included.

Historically Central Asia was closely tied to the Silk Road trade routes, connecting Europe and the Far East. Through this connection, the movement of people, goods, and ideas between East and West enhanced the exchanges of culture and commerce.[4]

In the pre-Islamic and early Islamic eras (c. 1000 and earlier), Central Asia was inhabited predominantly by Iranian people.[5] After expansion by Turkic people, Central Asia also became the homeland for the Uzbeks, Kazakhs, Tatars, Turkmens, Kyrgyz, and Uyghurs.

The geographical information on the West&Central Asia-KWQ will be compared with that on seven major European maps of the sixteenth century; they are abbreviated as: (1) the West&Central Asia-Mercator (1569), extracted from the 1569 World Map by Gerardus Mercator; (2) the West&Central Asia-Ortelius (1570; #1), extracted from the 1570 World Map by Abraham Ortelius; (3) the West&Central Asia-Plancius (1594), extracted from the 1594 World Map by Petrus Plancius; (4) the West&Central Asia-Ortelius (1570; #2), which is the 1570 Ortelius Map of Asia (first edition) — *Asiae Nova Descriptio*; (5) the West&Central Asia-Ortelius (1570; #3), which is the 1570 representation of the Turnish Empire by Ortelius; (6) the West&Central Asia-Ortelius (1570; #4) which is the Ortelius Map of Persia, Antwerp, 1570; and (7) the West&Central Asia-Ortelius (1570; #5) which is the Ortelius Tartariae sive Magni Chami Regni tÿpus, 1570.

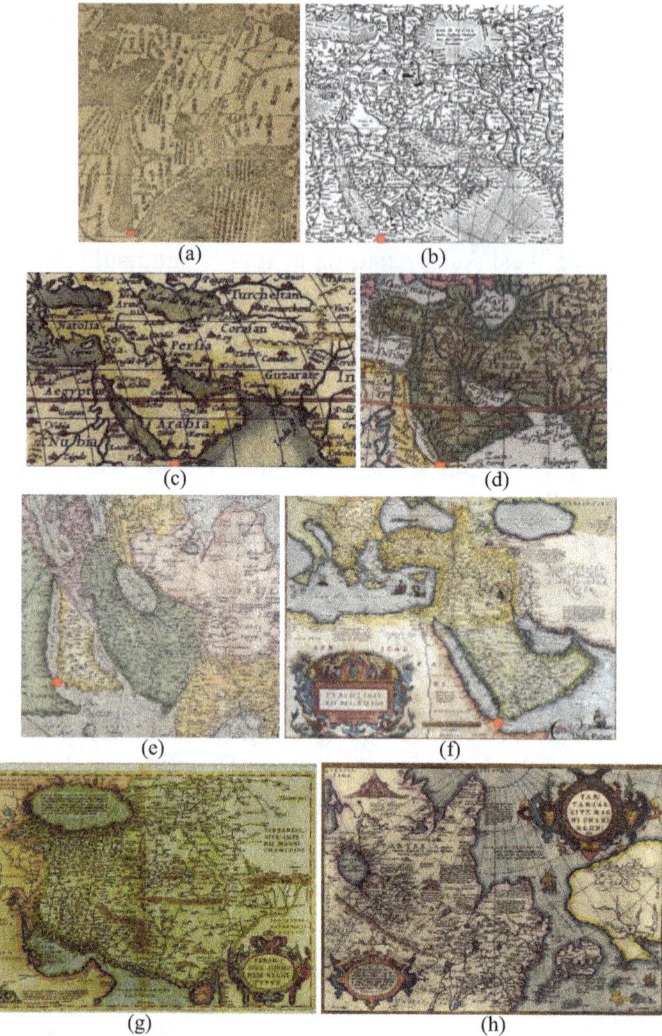

Fig. 6.2   (a) shows the West&Central Asia-KWQ extracted from the KWQ (public domain);[8] (b) shows the West&Central Asia-Mercator (1569) extracted from the 1569 World Map by Gerardus Mercator (public domain);[9] (c) shows the West&Central Asia-Ortelius (1570; #1) extracted from the 1570 World Map by Abraham Ortelius (public domain);[10] (d) shows the West&Central Asia-Plancius (1594) extracted from the 1594 World Map by Petrus Plancius (public domain);[11] (e) shows the West&Central Asia-Ortelius (1570; #2) extracted from *the* 1570 Ortelius Map of Asia (first edition) — *Asiae Nova Descriptio* (public domain).[12] (f) shows the West&Central Asia-Ortelius (1570; #3) which is the representation of the Turkish Empire (public domain);[13] it is part of Abraham Ortelius's atlas Teatrum Orbis Terrarum, Antwerp, 1570, map no. 50; the map is the best known of all the sixteenth-century maps of the Ottoman Empire;[14] it covers the Middle East, from Egypt in the west to the Caspian Sea in the east, prominently featuring the Arabian Peninsula, and extending north of the Mediterranean from Turkey across Greece to Italy; (g) shows the West&Central Asia-Ortelius (1570; #4) which is the Ortelius Map of Persia, Antwerp, 1570 (public domain);[15] and (h) shows the West&Central Asia-Ortelius (1570, #5) which covers North Asia and is the Ortelius Tartariae sive Magni Chami Regni tÿpus, 1570 (public domain).[16]

The several Ortelius maps other than his world map are included for comparison because they give more geographical data than his world map.

In the Appendix, I have analysed the geographical information and histories of all 93 geographical items, and the nine original annotations depicted on the West&Central-KWQ shown in Fig. 6.2(a) and compared them with those on the above-mentioned seven major European maps shown in Figs. 6.2(b)–6.2(h) whenever possible. The results of these comparisons are used as the basis for determining the origin and era of the West&Central-KWQ.

2. The West&Central Asia-KWQ is of Chinese Origin

We can analyse the origin of the West&Central Asia-KWQ from three aspects as follows:

2.1 *The geographical shapes of the Arabian Peninsula on the European maps are similar but differ from that on the West&Central Asia-KWQ*

The geographical shapes of the Arabian Peninsula on European maps look like a rain boot, see Figs. 6.2(b)–6.2(f), unlike the triangular shape observed on the West&Central Asia-KWQ. This is an obvious distinction. Figures 6.2(g) and 6.2(h) do not show the complete Arabian Peninsula.

2.2 *Comparisons made between the coordinates of the southernmost tip of the Arabian Peninsula on a modern map, the West&Central Asia-KWQ and the European maps in Figs. 6.2(a)–6.2(f) (see the small red dot) show that the West&Central Asia-KWQ cannot be a direct or adapted copy of the European maps*

See Table 6.1.

Table 6.1.   shows the coordinates of the southernmost tip of the Arabian Peninsula on a modern map and on the six relevant ancient maps.

#	Source map	Coordinates of the Southernmost tip on Arabian Peninsula	Location on the source map
1	Arabian Peninsula on Google Map	12.7°N, 43.5°E	Samudera, Yemen
2	West&Central Asia-KWQ	10.8°N (KWQ latitude → modern latitude), 50.0°E	Aden Kingdom
3	West&Central Asia-Mercator (1569)	11.5°N, 61.5°E	Aden
4	West&Central Asia-Ortelius (1570; #1)	12.3°N, 53.0°E	Aden
5	West&Central Asia-Plancius (1594)	12.0°N, 52.5°E	Aden
6	West&Central Asia-Ortelius (1570; #2)	12.5°N, 57.0°E	Ara in Aden
7	West&Central Asia-Ortelius (1570; #3)	12.5°N, 61.0°E	Ara in Aden

Table 6.1 shows that the coordinates of the southernmost tip of the Arabian Peninsula on the West&Central Asia-KWQ give the worst latitude, but the best longitude than those on the other five European maps. If the West&Central Asia-KWQ were a direct or adapted copy of these European maps, it should show a similar trend as the European maps, namely, a good latitudinal value and a poor longitudinal value. But Table 1 shows the opposite. Hence, we can conclude that the West&Central Asia-KWQ has its own source map which is not of European origin.

2.3 *Out of the 93 geographical items analysed on the West&Central Asia-KWQ, the names of 21 of them (c. 23%) are not present on any of the seven European maps; while there are nine annotations on the West&Central Asia-KWQ, the European maps provide far fewer or none*

The following lists the 21 geographical items and their numbered positions in the Appendix.

---

1. Arabian Sea (曷剌比海 or 阿拉伯海; Item 1); 2. Upemoin (五伯麻尹; Item 6); 3. Hehemo (赫曷抹; Item 15); 4. Irmin (挹爾名/挹尔名; Item 16); 5. *Ngo, tribu di* (厄人雜野/厄人杂野; Item 18); 6. El Yemanah (扼落野; Item 19); 7. *Hogiuhoie* (曷入曷野; Item 21); 8. Caldea (磕爾突牙/磕尔突牙; Item 24); 9. Licia (利细亞/利细亚; Item 29); 10. Cappadocia (葛八多齊亞/葛八多齐亚; Item 31); 11. Menteshe district (悶突色/闷突色; Item 32); 12. Anadole (曷捹多勒; Item 34); 13. Hexiliya (曷西麗牙/曷西丽牙; Item 46); 14. The Kingdom of Women (女人國/女人国; Item 48); 15. Lurestan (路勒私旦; Item 50); 16. Laluse (剌路斯; Item 53); 17. Kesh (加私; Item 66); 18. Huihui (回回; Item 74); 18. Hepantuo (喝盤陀/喝盘陀; Item 76); 19. Zhujubo (朱俱波; Item 80); 20. Huozhou (火州; Item 90); 21. Suyab River or Chu River (素葉河 or 楚河 ; Item 92)

Moreover, on the West&Central Asia-Ortelius (1570; #5), there is "Calis". How did Abraham Ortelius in 1570 get to know this place even before Damião de Góis and his traveling companions became the first Europeans to reach the Kingdom of Cialis in 1605? Did he in fact copy this geographical item from the Chinese map but used a European name?

The nine annotations on the West&Central Asia-KWQ cover local precious stones, produce, livestock, wild beasts, religious hub, mountains, volcano, lakes, city, and kingdoms, whereas the few annotations, see Table 6.2, on the European maps are narrowly focused on the Caspian Sea and Hormoz.

Moreover, annotation @8 mentions the ancient Jiaohe City (交河) where the capital of the Nearer Jushi/Anterior Jushi (車師前國/车师前国) was located. The Jushi Kingdom was separated into the Nearer Jushi/Anterior Jushi (a southern area controlled by the Han) and the Further Jushi/ Posterior Jushi (車師後國/车师后国; a northern area dominated by the Xiongnu; the capital appears to have been called Yuli or Yulai).[17] The separation occurred in c. 60 B.C. in the Han Dynasty (202 B.C.–9 A.D., 25–220 A.D.). This record is based on *Shi Ji • Juan Yi Bai Er Shi San • Da Yuan Lie Chuan* (史记 • 卷一百二十三 • 大宛列傳) or *Records of the Grand Historian, Juan (Volume) 123, Biography of Da Wan*.[18] Hence, it is apparent that Matteo Ricci has used Chinese geographical and historical information (not seen on the European maps discussed here) to depict geographical items on the West&Central Asia-KWQ.

Table 6.2 shows the number of annotations on the West&Central Asia-KWQ and on the seven relevant European maps.

#	Source	Number of relevant annotations
1	West&Central Asia-KWQ	9
2	West&Central Asia-Mercator (1569)	0
3	West&Central Asia-Ortelius (1570; #1)	0
4	West&Central Asia-Plancius (1594)	0
5	West&Central Asia-Ortelius (1570; #2)	0
6	West&Central Asia-Ortelius (1570; #3)	2
7	West&Central Asia-Ortelius (1570; #4)	2
8	West&Central Asia-Ortelius (1570; #5)	1

2.4 *On the West&Central Asia-KWQ, six geographical items are of Chinese origin; they are not transliterated geographical names*

The following lists the six geographical names of Chinese origin on the West&Central Asia-KWQ.

> 1. The Kingdom of Women (女人國/女人国; Item 48); 2. Iron Gate Pass (鐵門関/铁门关; Item 72); 3. Huihui (回回; Item 74); 4. Pamir Mountains (大葱嶺/大葱岭; Item 81); 5. Huozhou (火州; Item 90); 6. Bei Gao Hai/Caspian Sea (北高海; Item 93)

In summary, the evidence contained in Sections 2.1–2.4 strongly shows that the West&Central Asia-KWQ is of Chinese origin. The map is not a direct or adapted copy of the seven major sixteenth-century European maps.

3. The West&Central Asia-KWQ Depicts an Era of the Year 1433

Here are the six steps for determining the political era of the West&Central Asia-KWQ:

3.1 *The era revealed by the West&Central Asia-KWQ is initially limited to the period c. 1008–1490 because the map depicts the Georgiani or Kingdom of Georgia (熱阿爾入亞那/热阿尔入亚那; Item 49), which lasted from c. 1008 to1490*

3.2 *The era revealed by the West&Central Asia-KWQ can be narrowed to 1184–1490*

On the West&Central Asia-KWQ, Lurestan (路勒私旦; 1184–1597: Item 50) is depicted. Combining Subsections 3.1 and 3.2 narrows the political era of the map to the period 1184–1490.

3.3 *The era revealed by the West&Central Asia-KWQ can be further narrowed to 1192–1489*

On the West&Central Asia-KWQ, Cyprus (止波里; 1192–1489: Item 71) is depicted. Combining Subsections 3.2 and 3.3 narrows the political era of the map to the period 1192–1489.

3.4 *The era revealed by the West&Central Asia-KWQ can be again narrowed to 1192–1464*

On the West&Central Asia-KWQ, Huozhou (火州: Item 90) is depicted, implying that the political era revealed by the map is before 1464 or before the end of the Tianshun Period (1457–1464) of the Ming Yingzong, when the Huozhou Kingdom was annexed by Turpan. Combining Subsections 3.3 and 3.4 narrows the political era of the map to the period 1192–1464.

3.5 *The era revealed by the West&Central Asia-KWQ can be further narrowed to the thirteenth century to 1464 because the city El Tor (黑入多; Item 22) depicted on the West&Central Asia-KWQ was founded in the thirteenth century*

3.6 *The last geographical data collected by the Chinese mariners, before being depicted on the West&Central Asia-KWQ, most likely stem from the year 1433, which falls within the thirteenth century–1464*

The Ming Treasure Fleet arrived at Calicut during its seventh (and last) voyage on its outbound journey in 1432 and returned to Calicut in 1433 on its journey back home. After this voyage, the Ming Court abruptly stopped the treasure voyages and imposed a sea ban until the middle of the sixteenth century when the restrictions were lifted.

In my previous book titled *The Last Journey of the San Bao Eunuch, Admiral Zheng He,*[19] I have extracted the entire navigational routes, timelines, and evidence of Ming Treasure Fleet's seventh (and last) voyage from the epic work titled *San Bao Tai Jian Xi Yang Gi* (三寶太監西洋記/三宝太监西洋记)[20] or *An Account of the Western World Voyage of the San Bao Eunuch* written by Luo Maodeng (罗懋登/罗懋登) in 1597. I have shown in Chapter 3 of that book[21] that the treasure fleet led by Principal Envoy Zheng He (鄭和/郑和; 1371–1433 or 1435) arrived at Calicut (Kozhikode; 乜利客; also known by Chinese as 古里; Gu-li) on 19 December 1432 and departed on 24 December 1432, while the fleet led by Deputy Envoy Hong Bao (洪保; 1370–Year od death unknown) arrived earlier at Calicut on 10 December 1432 and left there on 14 December 1432.

The squadron led by Hong Bao made only a short stay in Calicut, then continued north to reach Dandi Bandar (near latitude 16.2°N and longitude 73.4°E), a small city in the middle of the west coast of India. Later Hong Bao's squadron crossed the Arabian Sea to pass the Jabal Khamis Mountain (22.4°N, 59.5°E) in today's northeast corner of Oman. A few days later, the fleet continued heading towards the Persian Gulf in today's southern Iran, and on 16 January 1433,[22] Hong

Bao's squadron arrived at Bandar Abbas, the Gulf port of Hormuz. Hong Bao and his men stayed in the neighbourhood of Hormuz for almost two months before traveling homeward on 9 March 1433. During these two months, Hong Bao's squadron was most able and likely to update their map of the West and Central Asia in 1433 with their newly acquired geographical data. The year 1433 falls in the range of the thirteenth century to 1464.

On 31 March 1433, the squadron led by Hong Bao returned to Calicut to meet the squadron led by Principal Envoy Wang Jinghong (王景弘; 1369–1437) who had sailed with Zheng He to the east coast of Africa in 1433, then spent about two and a half months visiting and exploring the Arabian Peninsula, and Hormuz; they sailed back from Hormuz to Aden and Wang Jinghong departed with Zheng He from Aden later in 1433.[23] Wang Jinghong's squadron also had the great opportunity to update their map of West and Central Asia. On 9 April 1433, all the squadrons were returned to China, led by Wang Jinghong, and they arrived at Nanjing in July of that year.

Hong Bao's voyage seems to match the navigational routes and timelines recorded in *Qian Wen Ji • Xia Xi Yang* (前闻记•下西洋)[24] or *A Record of History Once Heard: Down to the Western Ocean* by the Ming historian Zhu Yunming (祝允明; 1461–1527)[25] for the treasure fleet's seventh voyage.

I have extracted Wang Jinghong's voyage and timelines from *San Bao Tai Jian Xi Yang Gi* (三寶太監西洋記/三宝太监西洋记)[26] or *An Account of the Western World Voyage of the San Bao Eunuch* written by Luo Maodeng (罗懋登/罗懋登) and published in my previous book.[27]

However, Zheng He seems to have disappeared from the Chinese official historical records after his last voyage. In my previous book, I show that the fleet led by Zheng He continued to sail from Aden to Mecca, and then left Mecca to round the southernmost corner of the African continent to reach North America; after staying for a few months there, some of the mariners (they might include Zheng He himself) sailed to China en route New Zealand, but met with a huge natural disaster and vanished there around 1435/36.[28] The mainstream historians have proposed that he died of illness at Calicut in 1433 on his way returning to China, but there has been neither hard nor soft evidence to support this claim.

From the above historical records, we can see that the Ming mariners most likely explored West and Central Asia in 1433 to produce the source map which was later used by Matteo Ricci to make the West&Central Asia-KWQ, because this map is of Chinese origin as shown in Section 2 of this chapter.

4. Conclusions

As mentioned in the Introduction of this chapter, I have aimed to use the same methodology which I used in the previous five chapters to analyse the West and Central Asia portions of the map Kunyu Wanguo Quantu (坤輿萬國全圖/坤舆万国全图) (abbreviated as KWQ) or Complete Geographical Map of All the Kingdoms of the World. These portions are abbreviated as West&Central Asia-KWQ. The KWQ was the first world map written in Chinese and published in China, by Italian missionary Matteo Ricci in 1602; it has been considered a European map.

However, after analysing the geographical information and histories of the complete 93 geographical items and the nine annotations on the West&Central Asia-KWQ in the Appendix and comparing the map with seven major sixteenth-century European maps in Section 2, I show that the map is of Chinese origin.

The era revealed directly by the West&Central Asia-KWQ is deduced step by step in Section 3 as in the period thirteenth century–1464. But, after verifying the Chinese historical records on Admiral Zheng He's seventh (and last) voyage to the Western Ocean, such a wide period can be narrowed to the year 1433.

Appendix

This Appendix analyses all the geographical items (93) and annotations (9) on the West&Central Asia-KWQ. These geographical items and annotations are separated into two parts. On the West&Central Asia -KWQ, Chinese traditional characters are used, except for a handful of cases where one or two Chinese simplified characters appear since they already existed in ancient Chinese writings. I use Chinese traditional characters followed by the corresponding simplified characters separated by a "/" sign in the main text and the Appendix. If they are identical, then only one expression is given to avoid repetition. Items preceded by "●" are names of the geographical items which do not appear on any of the seven European maps. The sign "@" means "annotation", and all the geographical items and annotations are numbered in sequence in this Appendix. The Chinese "pinyin" spelling of each item is given right after it is introduced.

West and Central Asia: Part I

Fig. A6.1.   Extracted from the West&Central Asia-KWQ (public domain); Items 1-75 and annotations @1-@7 are depicted on this Figure.

Item #	Name	Pinyin	Etymology	Analysis of history, politics, geography: comparison or comment
● 1	曷剌比海	He La Bi Hai	Arabian Sea	The Arabian Sea (曷剌比海 or 阿拉伯海), see Fig. A6.2, is a region of the northern Indian Ocean bounded on the north by Pakistan, Iran and the Gulf of Oman, (阿曼灣/阿曼湾), on the west by the Gulf of Aden (亞丁灣/亚丁湾) and the Arabian Peninsula (阿拉伯半島/阿拉伯半岛), on the southeast by

(*Continued*)

Item #	Name	Pinyin	Etymology	Analysis of history, politics, geography: comparison or comment
				the Laccadive Sea (拉克代夫海) and the Maldives (馬爾地夫/马尔代夫), on the southwest by Somalia (索馬利亞/索马里), and on the east by India. The Gulf of Aden in the west connects the Arabian Sea to the Red Sea (紅海) through the Bab-el-Mandeb Strait (曼德海峽/曼德海峽), and the Gulf of Oman is in the northwest, connecting it to the Persian Gulf (波斯灣/波斯湾). The Arabian Sea has been an important marine trade route since Antiquity and through the so-called Age of Sail, a period that lasted from the mid-fifteenth[29] or mid-sixteenth (at the latest) to the mid-nineteenth centuries. **Fig. A6.2.** Location of the Arabian Sea in the Indian Ocean. Created by Norman Einstein, 19 July 2005 (*Source*: Norman Einstein, Ras67, under CC BY-SA 3.0, https://commons.wikimedia.org/wiki/File:Arabian_Sea_map.png).[30] On the West&Central Asia-KWQ, see Fig. A6.1, in addition to the Arabian Sea (曷剌比海 or 阿拉伯海) which is depicted close to the Arabian Peninsula, to its east there is the Xiao Xi Yang/Small Western Ocean (小西洋 or the Small Western Ocean) near the west coast of the Indian subcontinent, but still within the Indian Ocean. I have explained Xiao Xi Yang in Item 33 in the Appendix of Chapter 5. On the West&Central Asia-Mercator (1569), there is the name MARE RV. On the West&Central Asia-Ortelius (1570; #1), no name is given to this sea; only MAR DI INDI is denoted south of the Tropic of Capricorn.

				On the West&Central Asia-Plancius (1594), the name given to this sea is OCEANUS INDICUS, north of the Equator. On the West&Central Asia-Ortelius (1570; #2), the name of this sea is given as INDIA, olim Rubrum, north of the Equator. On the West&Central Asia-Ortelius (1570; #3), the name of this sea is given as MARE INDICVM, ol. RVBRVM.
2	馬伐	Ma Fa	Maffa[31]	The place cannot be identified on a modern map. It may be in today's Oman. Oman is one of the six countries Bahrain, Kuwait, Oman, Qatar, Saudi Arabia, and the UAE, which form the Gulf Cooperation Council (GCC), see Fig. A6.3. **Fig. A6.3.** The countries of the Gulf Cooperation Council (*Source*: Derivative work: UA3 (talk); Map_of_the_Arabic_peninsula_-_en.svg: Original: Lokal Profil et al., under CC BY-SA 4.0, https://commons.wikimedia.org/wiki/File:Arabian_Peninsula_Map.svg).[32] Maffa is depicted as "Haffa" on the West&Central Asia-Ortelius (1570; #2) and as MAFFA on the West&Central Asia-Ortelius (1570; #3).
3	亞依漫	Ya Yi Man	Ayman[33] or Oman	On the West&Central-KWQ, 亞依漫 is depicted at where today's Oman is. Oman is a country in West Asia and has existed since Antiquity.[34] It is located on the southeastern coast of the Arabian Peninsula and overlooks the mouth of the Persian Gulf. In the Ming Dynasty, the West&Central Asia-KWQ depicts it as 亞依漫 and the pronunciation of Ayman (亞依漫) is close to 阿曼 or today's Oman. On the West&Central Asia-Mercator (1569), there is "AYAMA", but it is landlocked; it is in the western side of the Peninsula and occupies a large area.

(*Continued*)

Item #	Name	Pinyin	Etymology	Analysis of history, politics, geography: comparison or comment
				On the West&Central Asia-Ortelius (1570; #2), there is a land locked AYMAN, but its location is also in the centre of the Peninsula and occupies a large area. On the West&Central Asia-Ortelius (1570; #3), there is "AYAMAN", landlocked and in the north of the Arabian Peninsula. On the West&Central Asia-Plancius (1594), there is "Ayaman", but its location is in the center of the Peninsula and occupies a large area. The European cartographers seem to have misplaced the geographical location of Ayman.
4	默生丁海	Mo Shen-g Ding Hai	Persian Gulf; before being given its present name, the Persian Gulf was called many different names. In the Greek sources, the body of water that bordered this province came to be known as the "Persian Gulf".[35]	默生丁海 (*Mare di Mesendin*) is the present-day Persian Gulf (the Arabian Gulf) in Western Asia, see Fig. A6.4. **Fig. A6.4.** Map of the Persian Gulf. The Gulf of Oman leads to the Arabian Sea (public domain).[36] The body of water is an extension of the Indian Ocean located between Iran and the Arabian Peninsula. It is connected to the Gulf of Oman in the east by the Strait of Hormuz, see Fig. A6.4.

				The Persian Gulf, along with the Silk Road, were important trade routes. Many of the trading ports of the Persian empires were in or around the Persian Gulf. The West&Central Asia-Mercator (1569) uses *Mare di Mesendin* for this body of water. On the West&Central Asia-Ortelius (1570; #1), the Persian Gulf is unnamed. On the West&Central Asia-Ortelius (1570; #2), there is MARE MESENDIN. On the West&Central Asia-Ortelius (1570; #3) and the West&Central Asia-Ortelius (1570; #4), the name is *Mare El Catif, olim Sinus Persicus.* On the West&Central Asia-Plancius (1594), the Persian Gulf is depicted as *Mar Mesendin.*
5	曷剌比亞	He La Bi Ya	Arabia or Araby or Arabian Peninsula; during the Hellenistic period, the Arabian Peninsula was known as *Arabia* or *Aravia* (Greek: Αραβία).	曷剌比亞 is the Arabian Peninsula or Arabia or Araby. Geographically, the Arabian Peninsula includes present-day Bahrain, Kuwait, Oman, Qatar, Saudi Arabia, the United Arab Emirates (UAE) and Yemen, as well as southern Iraq and Jordan,[37] together with offshore islands, located in the extreme southwestern corner of Asia, see Fig. A6.4. Figure A6.4 also shows that the Arabian Peninsula is bounded by (clockwise) the Persian Gulf, the Strait of Hormuz, the Gulf of Oman, the Arabian Sea, the Gulf of Aden, the Guardafui Channel, the Bab-el-Mandeb Strait and the Red Sea. The northern portion of the peninsula merges with the Syrian Desert with no clear borderline, although the northern boundary of the peninsula is generally considered to be the northern borders of Saudi Arabia and Kuwait, also southern regions of Iraq and Jordan. This is what we know today. The West&Central Asia-Mercator (1569) depicts "Arabia de ferra" in the northern region of the Peninsula. The West&Central Asia-Ortelius (1570; #1) and the West&Central Asia-Ortelius (1570; #3) depict "Arabia" and ARABIA, respectively, but not the West&Central Asia-Plancius (1594).

(*Continued*)

Item #	Name	Pinyin	Etymology	Analysis of history, politics, geography: comparison or comment
@ 1	乳香產於此地其樹甚小他處則無又產一藥名乜尔刺涂尸不敗	Ru Xiang Chan Yu Ci Di Qi Shu Shen Xiao Ta Chu Ze Wu You Chan Yi Yao Ming Mie Er La Tu Shi Bu Bai	My translation and comment (between square brackets) are as follows: Frankincense is produced here. The incense trees are very small, and do not grow in other places. The place also produces another medicine named Mie Er La [乜爾剌/乜尔刺]; when applied on a corpse, it can prevent rotting. Frankincense is the resinous dried sap harvested from the olibanum tree. The frankincense tree grows in an Arabian Desert wadi where water occasionally flows in the winter.[38] Summers in the central region are extremely hot and dry, ranging from 27°C to 43°C in the inland areas and 27°C to 38°C in coastal areas; hence, preventing dead bodies from rotting relies on chemicals applied on corpses.
● 6	五伯麻尹	Wu Bo Ma Yin	Upemoin[39]	The place cannot be identified on a modern map. It may be in present-day Oman or Yemen, see Fig. A6.4.
7	赫力	He Li	Herit[40]	The place cannot be identified on a modern map. It may be in today's Yemen, see Fig. A6.4. On the West&Central Asia-Ortelius (1570; #3), there is HERIT.
8	曷禡西利非丁	He Ma Xi Li Fei Ding	Amansirisdi or Amansirisdin	The place cannot be identified on a modern map. It may be in present-day Saudi Arabia. On the West&Central Asia-Ortelius (1570; #2), there is "Amansirisdi". On the West&Central Asia-Ortelius (1570; #3), there is "Amansirisdin".
9	嚕密	Lu Mi	Brum or Rum[41]	The place cannot be identified on a modern map. It may be in today's Saudi Arabia. On both the West&Central Asia-Ortelius (1570; #2) and the West&Central Asia-Ortelius (1570; #3), there is "Brum".

10	法勤達	Fa Qin Da	Fartach[42]	The place cannot be identified on a modern map. It may be in today's Yemen. "Fartach" or "Fartache" is depicted on the West&Central Asia-Ortelius (1570; #1), the West&Central Asia-Ortelius (1570; #2) and the West&Central Asia-Plancius (1594).
11	妻畢	Qi Bi	Sebid[43]	The place cannot be identified on a modern map. It may be in today's Yemen. Both the West&Central Asia-Ortelius (1570; #1) and the West&Central Asia-Ortelius (1570; #3) depict "Zibit". The West&Central Asia-Plancius (1594) depicts "Tybi".
12	墨加	Mo Jia	Mecca	Mecca (墨加 or 麥加/麦加, commonly known as Makkah) is the holiest city in Islam and the capital of Mecca Province in western Saudi Arabia. The name "Mecca" was first seen as Majia Guo (麻嘉國/麻嘉国) in *Ling Wai Dai Da* (嶺外代答/岭外代答) or *Representative Answers from the Region beyond the Mountains* written by Zhou Qufei (周去非) in 1178 of the Southern Song Dynasty; in *Song Huiyao* (宋會要/宋会要) or *Song Government Manuscript Compendium*, it is called Mojia (摩迦); in *Zhu Fan Zhi* (諸蕃志/诸蕃志) or *A Description of Barbarian Nations* it is called Majia (麻嘉); and in *Ming Shi* (明史) or *History of Ming*, it is called Mojia (默加) which is depicted on the West&Central Asia-KWQ. Mecca is generally considered the fountainhead and cradle of Islam. It is revered in Islam as the birthplace of the Islamic prophet Muhammad. Visiting Mecca for the *Hajj* is an obligation upon all able Muslims. The Great Mosque of Mecca, known as the *Masjid al-Haram*, is believed by Muslims to have been built by Abraham and Ishmael. It is one of Islam's holiest sites and the direction of prayer for all Muslims (*qibla*).[44]

(*Continued*)

Item #	Name	Pinyin	Etymology	Analysis of history, politics, geography: comparison or comment
				Admiral Zheng He and many of his mariners were devoted Muslims. They too went to Mecca for pilgrimage during their seventh (and last) voyage.[45] They arrived at Port Jeddah (吉達港/吉达港) in the period 25 March 25–30 April 1433, visited *Tianfang*/Yun Chong (Mecca; 天方國/筠冲/麥加 or 天方国/筠冲/麦加), and left Port Jeddah four days later in the period 29 March–4 May 1433, before sailing towards the Americas. They finally landed on the north Atlantic coast in late autumn of 1433.[46] During the same voyage, Ma Huan (馬歡/马欢; c. 1380–1460) and seven senior officers together were sent by Assistant Envoy Hong Bao (洪保) from Calicut to conduct trade and pilgrimage by joining a ship of Calicut's merchants and went to *Tianfang*/Mo-jia in ancient Tunisia, North Africa, on 14 December 1432. There the seven-men delegation visited the Great Mosque or the Mosque of Uqba. They returned to Calicut a year later.[47] All the European maps depict "Mecha" or "Mecca".
13	亞登國	Ya Deng Guo	Kingdom of Aden	The Kingdom of Aden (亞登/亚登) was a city state in the eastern approach to the Red Sea (the Gulf of Aden), some 170 km (177 km/110 mi) east of the Bab-el-Mandeb Strait. Islam arrived in Aden (today's Yemen) in 630 A.D. and Yemen became part of the Muslim realm. During the first part of the fourteenth century, from 1325 to 1354, the Moroccan traveler Ibn Battuta visited Aden.[48] In 1421, the Yongle Emperor of the Ming Dynasty ordered Principal Envoy Grand Eunuch Li Xing (李興/李兴) and Grand Eunuch Zhou Man (周滿/周满) of Zheng He's fleet to convey an imperial edict with hats and robes to bestow on the King of Aden. The envoys boarded three treasure ships and set sail from Sumatra to the port of Aden. The event was recorded in the book *Yingyai Shenglan*

				(瀛涯勝覽/瀛涯胜览) or *The Overall Survey of the Ocean's Shores* written in 1451 by Ma Huan who accompanied the imperial envoy. The Ming Treasure Fleets visited Aden during their fifth and sixth voyages to the Western Ocean, The Treasure Fleet led by Zheng He and Wang Jinghong (王景弘; fourteenth century–after 1437) also visited Aden in the period 15 March 1433–19 April 1433 during their seventh voyage; they left there in the period 17 March–11 April 1433.[49] All the European maps depict ADEN or ADEM.
14	默德那	Mo De Na	Medina	Medina (默德那) officially Al Madinah Al Munawwarah, lit. "The Enlightened City", is the second-holiest city in Islam and the capital of Medina Province in western Saudi Arabia. Medina is generally considered the "cradle of Islamic culture and civilization".[50] The *Al-Masjid al-Nabawi*, lit. "The Prophet's Mosque", is the burial site of the last Islamic prophet, Muhammad, by whom the mosque was built in 622 A.D.
● 15	赫曷抹	He He Mo	Hehemo[51]	The place cannot be identified on a modern map. It may be in today's Saudi Arabia, see Fig. A6.4.
● 16	挹尔名	Yi Er Ming	Irmin[52]	The place cannot be identified on a modern map. It may be in today's Saudi Arabia, see Fig. A6.4.
17	耶惹曷尸	Ye Re He Shi	Egias[53]	The place cannot be identified on a modern map. It may be in today's Saudi Arabia, see Fig. A6.4. On the West&Central Asia-Ortelius (1570; #3), there is EGIAS.
● 18	戹人雜野	E Ren Za Ye	*Ngo, tribu di*[54]	The name cannot be identified on a modern map.
● 19	扼落野	E Luo Ye	El Yemanah[55]	The place cannot be identified on a modern map. It may be in today's Saudi Arabia, see Fig. A6.4.

(*Continued*)

Item #	Name	Pinyin	Etymology	Analysis of history, politics, geography: comparison or comment
20	兒葛諦夫	Er Ge Di Fu	Elcaif or El Katif[56]	The place cannot be identified on a modern map. It may be in today's Saudi Arabia, see Fig. A6.4. On the West&Central Asia-Ortelius (1570; #2), there is "Eltaif", and on the West&Central Asia-Ortelius (1570; #3) there is ELCATIF.
● 21	曷入曷野	He Ru He Ye	*Hogiuhoie*[57]	The place cannot be identified on a modern map. It may be in today's Saudi Arabia, see Fig. A6.4.
22	黑入多	Hei Ru Duo	El Tor; also Romanised as *Al-Tur* and *At-Tur*; the name comes from the Arabic term for the mountain *Jabal Al Tor*, where the prophet Moses received the Tablets of the Law from God.	El Tor (黑入多), see Fig. A6.5, is a small city on the Sinai Peninsula (西奈半島/西奈半岛) and the capital of the South Sinai Governorate of Egypt on the northern edge of Africa. The city itself appears to have been founded in the thirteenth century.[58] On the West&Central Asia-KWQ, El Tor (黑入多) is mistakenly depicted as part of the Arabian Peninsula. Both the West&Central Asia-Ortelius (1570; #2) and the West&Central Asia-Ortelius (1570; #3) also mistakenly depict "Eltor" on the Arabian Peninsula.

Fig. A6.5. Location (dark dot) of El Tor in Egypt;[59] it is separated by the Red Sea from the Arabian Peninsula; adapted from "Location map of Egypt" (https://commons.wikimedia.org/wiki/File:Egypt_adm_location_map.svg) by NordNordWest, under license https://creativecommons.org/licenses/by-sa/3.0/de/legalcode

23	死海	Si Hai	Dead Sea; this English name is in reference to the scarcity of aquatic life caused by the lake's extreme salinity (9.6 times as salty as the ocean), hence its name.[60] The name "Dead Sea" occasionally appears in Hebrew literature to mean "Sea of Death".[61]	The Dead Sea (死海) is a salt lake bordered by Jordan to the east and the West Bank and Israel to the west. It lies in the Jordan Rift Valley. Fig. A6.6. Location of the Dead Sea (*Source*: NordNordWest, under CC BY-SA 3.0, https://commons.wikimedia.org/wiki/File:Israel_location_map.svg).[62] On the West&Central Asia-Ortelius (1570; #2), there is an unnamed lake at the approximate location of the Dead Sea. On the West&Central Asia-Ortelius (1570; #3), there is *M. Mortuu* (Dead Sea in Latin: *Mare Mortuum*).
@ 2	此海無所產名為死海然水性常浮人溺其中不沉	Ci Hai Wu Suo Chan Ming Wei Si Hai Ran Shui Xing Chang Fu Ren Ni Qi Zhong Bu Chen		My translation is as follows: This sea produces nothing; it is called the Dead Sea. However, the water is always floating, and people who drown in it will not sink. The water in the Dead Sea is about 9.6 times as salty as the ocean — and has a density of 1.24 kg/l, which makes swimming like floating.[63]
• 24	磕尔突牙	Ke Er Tu Ya	Caldea[64]	The place cannot be identified on a modern map in today's Saudi Arabia or Iraq, see Fig. A6.4.
25	西利牙	Xi Li Ya	Syria (region)	Syria (西利牙 or 敍利亞/叙利亚) historically referred to a wide region located east of the Mediterranean Sea in West Asia, broadly synonymous with the Levant.[65] The region's boundaries have changed throughout history.

(Continued)

Item #	Name	Pinyin	Etymology	Analysis of history, politics, geography: comparison or comment
				In 1400, the Muslim Turco-Mongol conqueror Tamerlane invaded Syria[66] and by the end of the fifteenth century, the discovery of a sea route from Europe to the Far East ended the need for an overland trade route through Syria. In 1516, the Ottoman Empire conquered Syria and incorporated it into its empire as Ottoman Syria.[67] The West&Central Asia-Mercator (1569) depicts "Syria". The West&Central Asia-Ortelius (1570; #1) depicts "Soria". The West&Central Asia-Plancius (1594) depicts "Syria". The West&Central Asia-Ortelius (1570; #2) depicts "Soria". The West&Central Asia-Ortelius (1570; #3) depicts "Soria".
@ 3	天主降生於是地故人謂之聖土	Tian Zhu Jiang Sheng Yu Shi Di Gu Ren Wei Zhi Sheng Tu	My translation is as follows: God was born at this place, so people call it the holy land. This annotation refers to如德亞 (Judea or Judaea) to be discussed in Item 26.
26	如德亞	Ru De Ya	Iudea,[68] Judea or Judaea; *Judea* is a Greek and Roman adaptation of the name "Judah".	Judea or Judaea (如德亞/如德亚) is a mountainous region of the Levant. Traditionally dominated by the city of Jerusalem, it is now part of Palestine and Israel. In 132 A.D., the Roman province of Judaea was merged with Galilee to form the enlarged province of Syria Palaestina.[69] Later Judaea went through the Byzantine period (330–1453) and Jerusalem became a major Christian pilgrimage and ecclesiastical hub.[70] During the Crusader period, the Franks tended to settle in the southern half of the region between Jerusalem and Nablus since there was a sizable Christian population there.[71] Most of the people living in the northern portion of Judaea in the late sixteenth century were Muslims; some of them resided in towns that today have significant Christian populations.

Ming Shi (明史) or *History of Ming* has the following record:[72]

拂菻, 即漢大秦。。。晉及魏皆曰大秦, 嘗入貢。
唐曰拂菻, 宋仍之, 亦數入貢。元末, 入市中國, 元
亡不能歸。太祖聞之, 以洪武四年八月召見。。。
乃遣使入貢, 後不復至。萬曆時, 大西洋人至京師,
言天主生於如德亞, 即古大秦國也。

My translation and comments (between square brackets) are as follows:

Fu Lin [拂菻] was Da Qin [大秦] in the Han Dynasty . . . it was called Da Qin in both the Jin [晉/晋] and Wei [魏] dynasties, and they paid tribute [to these Chinese dynasties]. In the Tang [唐] Dynasty, the country was also called Fu Lin, and it was the same in the Song [宋] Dynasty; they sent in tribute several times. At the end of the Yuan [元] Dynasty, the country was incorporated into China. But after the Yuan Dynasty collapsed, the kingdom could not return to China. The Taizu Emperor [太祖] of the Ming Dynasty heard about it and summoned their king in August of the fourth year of the Hongwu era (1371) . . . The king then sent envoys to pay tribute, but they never came again. During the Wanli [萬曆/万历] period [1572–1620], people from the Atlantic Ocean [Europeans] came to the capital [of China] and said that God was born in Judea [or Judaea; 如德亞/如德亚]; that was where the ancient Da Qin Kingdom was located.

The West&Central Asia-Ortelius (1570; #3) depicts IVDEA.

(*Continued*)

Item #	Name	Pinyin	Etymology	Analysis of history, politics, geography: comparison or comment
27	歐法獵得河	Ou Fa Lie De He	Euphrates River	The Euphrates River (歐法獵得河/欧法猎得河) is the longest and one of the most historically important rivers of Western Asia. Together with the Tigris River (底格里斯河), they define Mesopotamia (lit. "the land between the rivers"; 美索不達米亞/美索不达米亚) in modern-day Iraq, see Fig. A6.7. The region was home to the ancient Sumerians (蘇美人/苏美人), Babylonians (巴比倫人/巴比伦人), and Assyrians (亞述人/亚述人). The river originates in Turkey, flows through Syria and Iraq and joins the Tigris at Al-Qurnah (c. 31°N, c. 47.4°E) in the Shatt al-Arab, then empties into the Persian Gulf. On the West&Central Asia-KWQ, the confluence of the Euphrates and an unnamed river (which must be the Tigris) is at c. 30.6°N (KWQ latitude → modern latitude), c. 52.5°E, in good agreement with the coordinates on a modern map. On the West&Central Asia-Mercator (1569), the Euphrates and Tigris rivers are depicted; their confluence is at (c. 31.5°N, c. 61.5°E). On the West&Central Asia-Ortelius (1570; #1), the Euphrates and Tigris rivers are not named; their confluence is at c. 32.5°N, c. 57.0°E. On the West&Central Asia-Ortelius (1570; #2), the Euphrates *fl.* meets an unnamed river (presumably the Tigris River) at c. 33°N, c. 64.0°E. On the West&Central Asia-Ortelius (1570; #3), there is Euphrates, but its estuary has complicated river structure and hard to determine its confluence with the Tigris. On the West&Central Asia-Plancius (1594), the Euphrates and Tigris rivers are not named; their confluence is at (c. 31.5°N, c. 55.0°E). The longitudinal values of four of the European maps show larger errors than that on the West&Central Asia-KWQ.

				Fig. A6.7. This is a map of the Tigris-Euphrates Watershed. Karl Musser, created it based on USGS data (*Source*: No machine-readable author provided. Kmusser assumed (based on copyright claims), under CC BY-SA 2.5, https://commons.wikimedia.org/wiki/File:Tigr-euph.png).[73]
28	沙尔加龍	Sha Er Jia Long	Sarcum	The place cannot be identified on a modern map. It should be in the western side of Asia Minor. Asia Minor, see Fig. A6.8, is a historical term for the westernmost part of Asia, roughly corresponding to modern-day Turkey. It has been known by various names throughout history, including Anatolia, see Fig. A6.8 in Item 29. On the West&Central Asia-Ortelius (1570; #3), there is SARCVM.
● 29	利细亞	Li Xi Ya	Licia[74]	Lycia (利细亞/利细亚) was an ancient maritime district of southwestern Anatolia (now Turkey), see Fig. A6.8. After the fall of the Byzantine Empire (330/395–1453) in the fifteenth century, Lycia was under the Ottoman Empire and was inherited by the Turkish Republic[75] on the Dissolution of the Ottoman Empire. On the West&Central Asia-KWQ, Licia is mistakenly depicted at the northwestern side of Asia Minor/Anatolia.

(Continued)

Item #	Name	Pinyin	Etymology	Analysis of history, politics, geography: comparison or comment
				Fig. A6.8. Region of Licia (in dark gray; *Source*: Caliniuc, under CC BY-SA 4.0, https://commons.wikimedia.org/wiki/File:Map_of_Lycia.jpg)[76] in Asia Minor/Anatolia.
30	區大亦	Qu Da Yi	Chiutaio[77]	The place cannot be identified on a modern map. It should be in today's Turkey, see Fig. A6.8. On the West&Central Asia-Ortelius (1570; #2), there is "Chiutaio". On the West&Central Asia-Ortelius (1570; #3), there is CHIV TAIE.
● 31	葛八多齊亞	Ge Ba Duo Qi Ya	Cappadocia[78]	Cappadocia (葛八多齊亞/葛八多齐亚) is a historical region in Central Anatolia. Starting from the second half of the thirteenth century, Cappadocia was ruled by successive Turkic states: the Karaman-based Beylik of Karaman and then the Ottoman Empire (奧斯曼帝國/奥斯曼帝国).[79] Cappadocia remained part of the Ottoman Empire until 1922, when it became part of the modern state of Turkey.[80] Cappadocia is at the center right in Fig. A6.8, among the classical regions of Anatolia/Asia Minor. On the West&Central Asia-KWQ, Cappadocia is mistakenly depicted on the western side of Asia Minor.

| @ 4 | 有山名為嵇沒辣山頂吐火頂傍出獅子中多豐草產羊甚廣山腳有龍蛇舊無人住後一異人率眾門山以居世傳嵇沒辣之獸獅子首羊身龙尾吐火有聖人除之蓋寓言也 | You Shan Ming Wei Ji Mei La Shan Ding Tu Huo Ding Bang Chu Shi Zi Zhong Duo Feng Cao Chan Yang Shen Guang Shan Jiao You Long She Jiu Wu Ren Zhu Hou Yi Yi Ren Shuai Zhong Men Shan Yi Ju Shi Chuan Ji Mei La Zhi Shou Shi Zi Shou Yang Shen Long Wei Tu Huo You Sheng Ren Chu Zhi Gai Yu Yan Ye | My translation and comment (between square brackets) are as follows:

There is a mountain called Ji Mei La [嵇沒辣 or Qi Mei La (奇美拉) or Chimaera] which spits fire from its top, and near the area of the mountain top there were lions. In the middle of the mountain area, there are many grasslands to feed sheep raised in a large area. Since there were dragons and snakes at the foot of the mountain, no one lived there in the old days. Later a strange man led a group of people to live on the foot of the mountain and guarded its entrance. Legend says that the beast in the Ji Mei La Mountain had a lion's head, sheep's body, dragon's tail, and spit fire, until a sage got rid of it. This is what the fables generally tell.

In Greek mythology, there is indeed a story about Bellerophon beheading Chimaera. Chimaera originally meant a female goat with a lion's head, a sheep's body, and a snake's tail. This annotation on the West&Central Asia-KWQ specially emphasises that the story of Ji Mei La (嵇沒辣 or 奇美拉 or Qi Mei La or Chimaera) is a fable. We can imagine that when the mariners of Zheng He's fleet heard this Greek mythology hundreds of years ago when they visited Asia Minor, they must have found the story very interesting, so they specifically recorded the fable on their map which later became the source map of the West&Central Asia-KWQ.
The fire spewing from the top of Ji Mei La [嵇沒辣] Mountain indicates that it was a volcano. It is said that hundreds of millions of years ago, a volcano named Erciyes erupted in Cappadocia, Turkey, which changed the surrounding terrain and accumulated volcanic ash and lava. Since then, the slowly weathered rocks have gradually turned into a unique geological wonder in the world—the forest of magic stone pillars—known as the "fairy chimneys (精靈煙囱/精灵烟囱岩)". |

(Continued)

Item #	Name	Pinyin	Etymology	Analysis of history, politics, geography: comparison or comment
				Turkey is one of the most seismically active regions on the planet.[81] Turkey's major volcanic mountains are in Eastern and Central Anatolia regions including Cappadocia.[82] As said, the volcanic character of the area was recognised in Antiquity, which was mostly suited for viticulture.
● 32	悶突色	Men Tu Se	Menteshe district[83]	Menteshe Dynasty (1260/1290–1426; 悶突色/闷突色 or 門泰謝/门泰谢) ruled in the Muğla-Milas region of southwestern Anatolia. Menteshe became a sanjak (an administrative district) of the Ottoman Empire in 1426.[84] On the West&Central Asia-KWQ, 悶突色 (Menteshe) is depicted in central Asia Minor and to the east of Cappadocia; this is incorrect. It should be in Southwestern Asia Minor, see Fig. A6.9. **Fig. A6.9.** The map shows the Beylik of Menteshe (near the center) in 1300 (*Source*: Gabagool, under CC BY-SA 3.0, https://commons.wikimedia.org/wiki/File:Anatolia1300.png).[85]
33	亞尔的溺里	Ya Er De Ni Li	Aidinelli or Aldenelli[86]	Aidinelli or Aldenelli (亞爾的溺裡/亞尔的溺里) should be in Anatolia, but it cannot be located on a modern map. On the West&Central Asia-Ortelius (1570; #3), there is ALDINELLI.
● 34	曷捺多勒	He Na Duo Le	Anadole[87]	The name refers most likely to 那多里亞 (Natolia or Anatolia) which is also depicted on the West&Central Asia-KWQ and correctly designates the entire Anatolia Peninsula. It is known as Asian Minor which is now a major part of the nation of Turkey. On the European maps, Anatolia is depicted as NATOLIA.

35	加臘馬溺	Jia La Ma Ni	Carmania[88] or Caramania	Carmania (加臘馬溺/加腊马溺) is a historical region that approximately corresponds to the current province of Kerman, Iran. Carmania was considered part of Ariana (a general geographical term used by some Greek and Roman authors of the ancient period for a district of wide extent between Central Asia and the Indus River).[89] But the West&Central Asia-KWQ mistakenly depicts it in Asia Minor/Anatolia. On the West&Central Asia-Ortelius (1570; #3), there is CARAMANIA. The geographical location is better depicted than that on the West&Central Asia-KWQ.
36	阿禡西亞	A Ma Xi Ya	Amasya; it was called Amaseia or Amasia in Antiquity.[90]	Amasya (阿禡西亞/阿禡西亚) is a city in northern Turkey, in the Black Sea Region. On the West&Central Asia-Ortelius (1570; #3), there is AMASIA.
37	波里六泥	Bo Li Liu Ni	Bolli Roni[91]	The place cannot be identified on a modern map. It should be in today's Turkey, see Fig. A6.4. On the West&Central Asia-Ortelius (1570; #3), there is only BOLLI RONI.
38	蘇襪斯	Su Wa Si	Suvas[92]	The place cannot be identified on a modern map. It should be today's Sivas in Turkey, see Fig. A6.4. On the West&Central Asia-Ortelius (1570; #2), there is "Suuas". On the West&Central Asia-Ortelius (1570; #3), there is SVVAS and "Suuas".
39	波促	Bo Cu	Bozoch[93]	The place cannot be identified on a modern map. It should be in today's eastern part of Turkey, see Fig. A6.4. On the West&Central Asia-Ortelius (1570; #3), there is Bozoc.
40	入耨	Ru Nou	Gino poli or Giul[94] or Genic[95]	The place cannot be identified on a modern map. It should be in today's eastern part of Turkey, see Fig. A6.4. On the West&Central Asia-Ortelius (1570; #3), there is GENIC.[96]

(*Continued*)

Item #	Name	Pinyin	Etymology	Analysis of history, politics, geography: comparison or comment
41	撥壞	Bo Rang	Pegian[97]	The place cannot be identified on a modern map. It should be in today's eastern part of Turkey, see Fig. A6.4. On the West&Central Asia-Ortelius (1570; #3), there is PEGIAN.
42	亞馬西亞	Ya Ma Xi Ya	Amasia[98]	Amasia or Amasya (亞馬西亞/亚马西亚) is a repetition of 阿㼮西亞 (Item 36). On the West&Central Asia-Ortelius (1570; #3), there is AMASIA.
43	西入厮儋	Xi Ru Si Dan	Sijistan[99]	Sakastan, also known as Sistan or Sijistan (西入厮儋), is a historical region in present-day south-eastern Iran, south-western Afghanistan, and north-western Pakistan,[100] see Fig. A6.10. Fig. A6.10. Map of Sakastan (also known as Sijistan) around 100 B.C. (public domain).[101]

				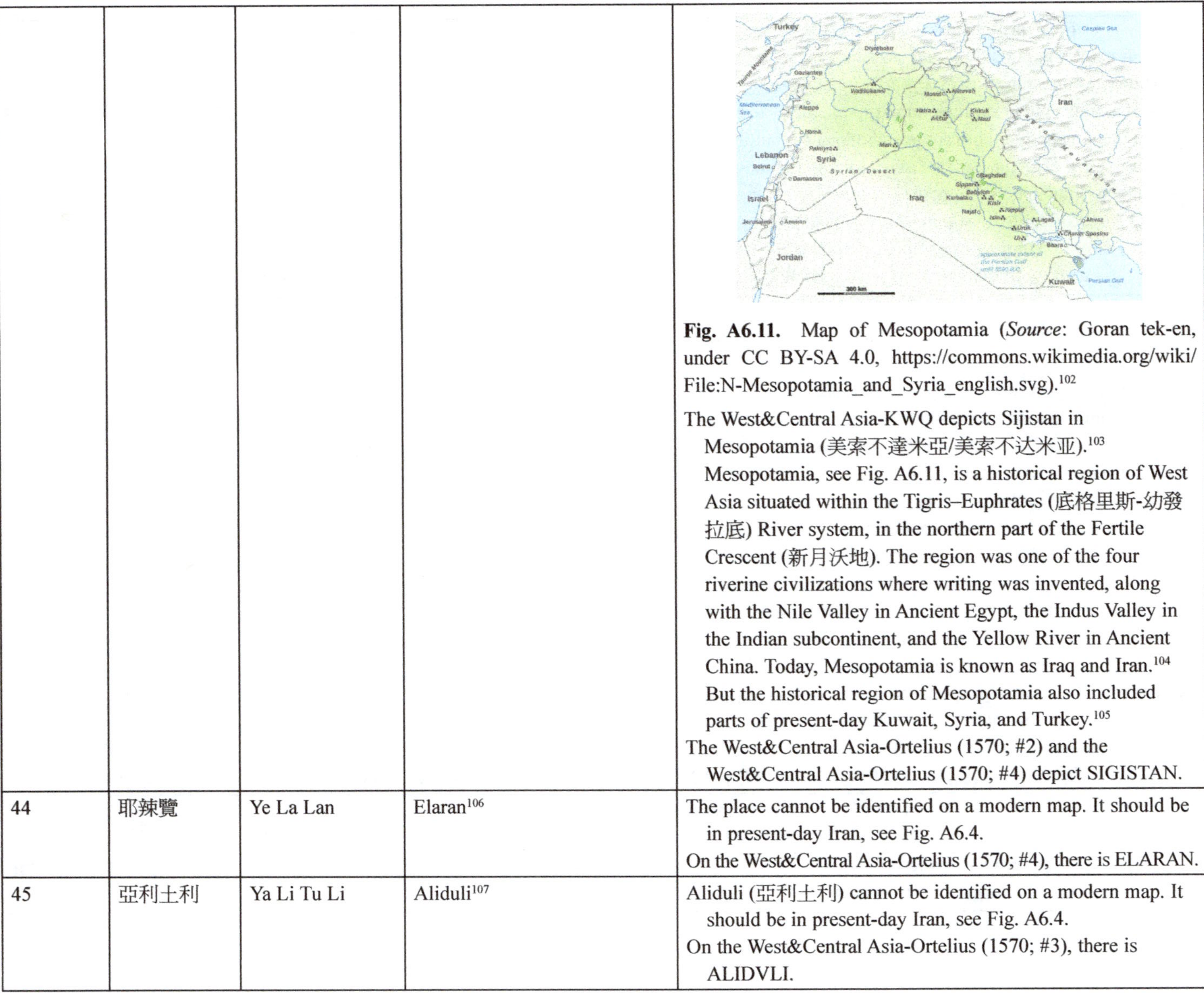 **Fig. A6.11.** Map of Mesopotamia (*Source*: Goran tek-en, under CC BY-SA 4.0, https://commons.wikimedia.org/wiki/File:N-Mesopotamia_and_Syria_english.svg).[102] The West&Central Asia-KWQ depicts Sijistan in Mesopotamia (美索不達米亞/美索不达米亚).[103] Mesopotamia, see Fig. A6.11, is a historical region of West Asia situated within the Tigris–Euphrates (底格里斯-幼發拉底) River system, in the northern part of the Fertile Crescent (新月沃地). The region was one of the four riverine civilizations where writing was invented, along with the Nile Valley in Ancient Egypt, the Indus Valley in the Indian subcontinent, and the Yellow River in Ancient China. Today, Mesopotamia is known as Iraq and Iran.[104] But the historical region of Mesopotamia also included parts of present-day Kuwait, Syria, and Turkey.[105] The West&Central Asia-Ortelius (1570; #2) and the West&Central Asia-Ortelius (1570; #4) depict SIGISTAN.
44	耶辣覽	Ye La Lan	Elaran[106]	The place cannot be identified on a modern map. It should be in present-day Iran, see Fig. A6.4. On the West&Central Asia-Ortelius (1570; #4), there is ELARAN.
45	亞利土利	Ya Li Tu Li	Aliduli[107]	Aliduli (亞利土利) cannot be identified on a modern map. It should be in present-day Iran, see Fig. A6.4. On the West&Central Asia-Ortelius (1570; #3), there is ALIDVLI.

(*Continued*)

Item #	Name	Pinyin	Etymology	Analysis of history, politics, geography: comparison or comment
● 46	曷西麗牙	He Xi Li Ya	Hexiliya[108]	The place cannot be identified on a modern map. It should be in present-day Iran, see Fig. A6.4.
47	亞尔墨尼亞	Ya Er Mo Ni Ya	Armenia[109] region	The Kingdom of Armenia (亞爾墨尼亞/亚尔墨尼亚; 331 B.C.–428 A.D.) was a landlocked country in the Armenian Highlands of Western Asia.[110] In the early fourth century (301 A.D.), Armenia became the first country in the world to regard Christianity as its official religion.[111] The kingdom lasted until the fifth century. The Armenians later fell under Byzantine (330–1453), Sassanid Persian (224–651) and Islamic hegemony, but reinstated their independence. In 1454, shortly after Byzantium was conquered by the Ottoman Turks, Armenia lost the last bit of its sovereignty and was annexed by the Ottoman Empire (c. 1299–1922). In the early sixteenth century, much of Armenia came under Safavid Persian (1501–1736) rule; however, the vast Western Armenian region has been under the rule of the Turks for a long time,[112] while Eastern Armenia remained under Persian rule.[113] Armenia is depicted on all the European maps in Figs. 6.1(b)–6.1(f).
● 48	女人國	Nu Ren Guo	The Kingdom of Women	Regarding the records of the Kingdom of Women (女人國/女人国), the earliest records can be found in *Ling Wai Dai Da* (岭外代答) or *Representative Answers from the Region beyond the Mountains* written by Zhou Qufei (周去非) in the Southern Song Dynasty (南宋; 1127–1279). However, regarding the geographical location of the Kingdom of Women, there are different opinions in the Chinese records. On the West&Central Asia-KWQ, the Kingdom of Women is depicted in the Caucasus (高加索) region between the Black Sea (黑海) and the Caspian Sea (里海).

@ 5	舊有此國亦有男子但多生男即殺之今又爲男子所併徒存其名耳。		Jiu You Ci Guo Yi You Nan Zi Dan Duo Sheng Nan Ji Sha Zhi Jin You Wei Nan Zi Suo Bing Tu Cun Qi Ming Er	My translation and comments (between square brackets) are as follows: In the ancient days, there was this country [the Kingdom of Women] where also lived men. But when more male babies were born, they would be killed. Now the kingdom is occupied by men again but keeps its old name [the Kingdom of Women].
49	熱阿尔入亞那	Re A Er Ru Ya Na	Georgiani;[114] the Kingdom of Georgia	The Kingdom of Georgia (Georgiani; c. 1008–1490), also known as the Georgian Empire,[115] was a medieval Eurasian monarchy which reached its Golden Age of political and economic strength during the reign of King David IV and Queen Tamar the Great from the eleventh to the thirteenth centuries,[116] see Fig. A6.12. On the West&Central Asia-Ortelius (1570; #2), there is GEORGIANI. On the West&Central Asia-Ortelius (1570; #3) and the West&Central Asia-Ortelius (1570; #5), there is GEORGIANL But none of the West&Central Asia-Mercator (1569), the West&Central Asia-Ortelius (1570; #1) and the West&Central Asia-Plancius (1594) depicts this kingdom. The reason is that it was no longer in existence. Here we see that some of Ortelius's maps give conflicting geographical data due to a lack of true knowledge of the history of this kingdom. The Kingdom of Georgia (Georgiani; c. 1008–1490) was no longer in existence in the sixteenth century, hence, should not be depicted on the sixteenth-century European maps.

(Continued)

Item #	Name	Pinyin	Etymology	Analysis of history, politics, geography: comparison or comment
				Fig. A6.12. Kingdom of Georgia in c. 1220, at the peak of its territorial expansion (*Source*: Lazicum, under CC BY-SA 4.0, https://commons.wikimedia.org/wiki/File:Kingdom_of_ Georgia_(1210s)_Sphere_of_influence.png).[117]
● 50	路勒私旦	Lu Le Si Dan	Lurestan,[118] meaning "land of the Lurs".[119]	Lurestan (路勒私旦) Province, also known as Luristan, Lorestan, or Loristan, is one of the 31 provinces of Iran. It is in the western part of the country in the Zagros Mountains (扎格羅斯山脈/扎格羅斯山脈).[120] The Khorshidi Dynasty, Abbasi Dynasty or Shahs of Little Lorestan (1184–1597) was a Lur Dynasty that ruled Little Lorestan which was a region roughly corresponding to the Ilam and Lorestan provinces in the present-day Iran. From the sixteenth century onwards, it was also referred to as Lorestan. In the nineteenth century, the area was split into Ilam and Lorestan.[121]
51	巴尔齊亞	Ba Er Qi Ya	Parthia;[122] the name is a continuation from Latin *Parthia*, from Old Persian *Parthava*, meaning "of the Parthians (安息人的)" who were an Iranian people. In the Western world, Persia was historically the common name used for Iran.[123]	Parthia (巴爾齊亞/巴尔齐亚) was a historical region, see Fig. A6.13, located in northeastern Greater Iran (a region having been long ruled by the dynasties of various Iranian empires, comprising parts of West Asia, the Caucasus, Central Asia, South Asia, and East Asia, specifically Xinjiang).[124] Ancient Iran, historically known as Persia, was the dominant nation of Western Asia for over twelve centuries, with three successive native dynasties — the Achaemenid (Persia Empire; 550–330 B.C.), the Parthian (247 B.C.–224 A.D.), and the Sasanian (224–651).

However, both the Roman (later the Byzantine Empire) and Persian empires (ancient Iran) competed for influence in Arabia. The Roman–Persian Wars (54 B.C.–628 A.D.)[125] gave opportunities for Arabia's rise, and the early Muslim conquests brought about the collapse of the Sasanian Empire and great territorial losses for the Byzantine Empire (330–1453).

The Islamic Golden Age is traditionally dated from the eighth century to the thirteenth century[126] but ended due to Mongol invasions and the Siege of Baghdad in 1258.[127]

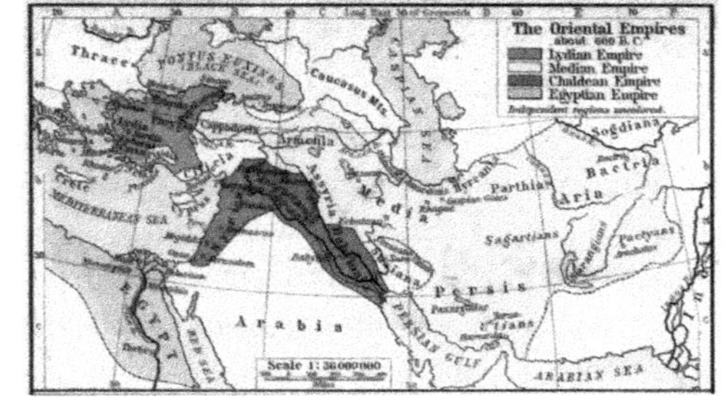

Fig. A6.13.   Median Empire (c. 678 B.C.–c. 550 B.C.) about 600 B.C. showing Parthia and Persis (public domain).[128]

The West&Central Asia-Mercator (1569) and the West West&Central Asia-Ortelius (1570; #1) depict "Persia", not "Parthia".

The West&Central Asia-Ortelius (1570; #2) depicts "Persia", not "Parthia".

The West&Central Asia-Plancius (1594) depicts both "Parthia" and "Persis". The latter is the origin of the name "Persia",[129] which is located to the southwest of modern-day Iran, now a province. The Persians initially migrated either from Central Asia or, more probably, from the north through the Caucasus.[130] They would then have migrated to the current region of Persis in the early first millennium B.C.[131]

(*Continued*)

Item #	Name	Pinyin	Etymology	Analysis of history, politics, geography: comparison or comment
@ 6	地產各色玉石金剛石雞青石	Di Chan Ge Se Yu Shi Jin Gang Shi Ji Qing Shi		My translation and comment (between the square brackets) are as follows: The place produces all kinds of jade, diamond, and bluestones [in China, this kind of stone is used to carve chicken-shaped statues].
52	忽魯謨斯	Hu Lu Mo Si	Ormus Kingdom or Hormoz Kingdom	The Kingdom of Ormus (also known as Hormoz; 忽魯謨斯國/忽鲁谟斯国; eleventh century–1622) was in the eastern side of the Persian Gulf. The kingdom was a vassal state of the Portuguese Empire in the East from 1515 to 1622. The kingdom's capital was a fortified port city which had the same name. It was one of the most important ports in the Middle East at the time as it controlled seaway trading routes through the Persian Gulf to China, India, and East Africa. This port was originally located on the southern coast of Iran to the east of the Strait of Hormuz, near the modern city of Minab (denoted by "2" in Fig. A6.14), and was relocated around the year 1300[132] to the island of Jarun which came to be known as Hormuz Island (denoted by "1"), which is located near the modern city of Bandar-e Abbas (denoted by "3"), see Fig. A6.14. On the West&Central Asia-KWQ, Hormuz Island is unnamed, the Kingdom of Ormus (忽魯謨斯/忽鲁谟斯) lying on the mainland is denoted. On the West&Central Asia-Mercator (1569), Hormuz Island is denoted. On the West&Central Asia-Ortelius (1570; #1), Hormuz Island is denoted. On the West&Central Asia-Ortelius (1570; #2), Hormuz Island is denoted, but the Kingdom of Ormus is also depicted on the Arabian Peninsula. On the West&Central Asia-Ortelius (1570; #3), the Kingdom of Ormvs is denoted on the mainland and on the Arabian Peninsula.

			On the West&Central Asia-Ortelius (1570; #4), the Kingdom of Ormvs is denoted on the mainland and on the Arabian Peninsula. On the West&Central Asia-Plancius (1594), the Kingdom of Ormus is denoted on the mainland. 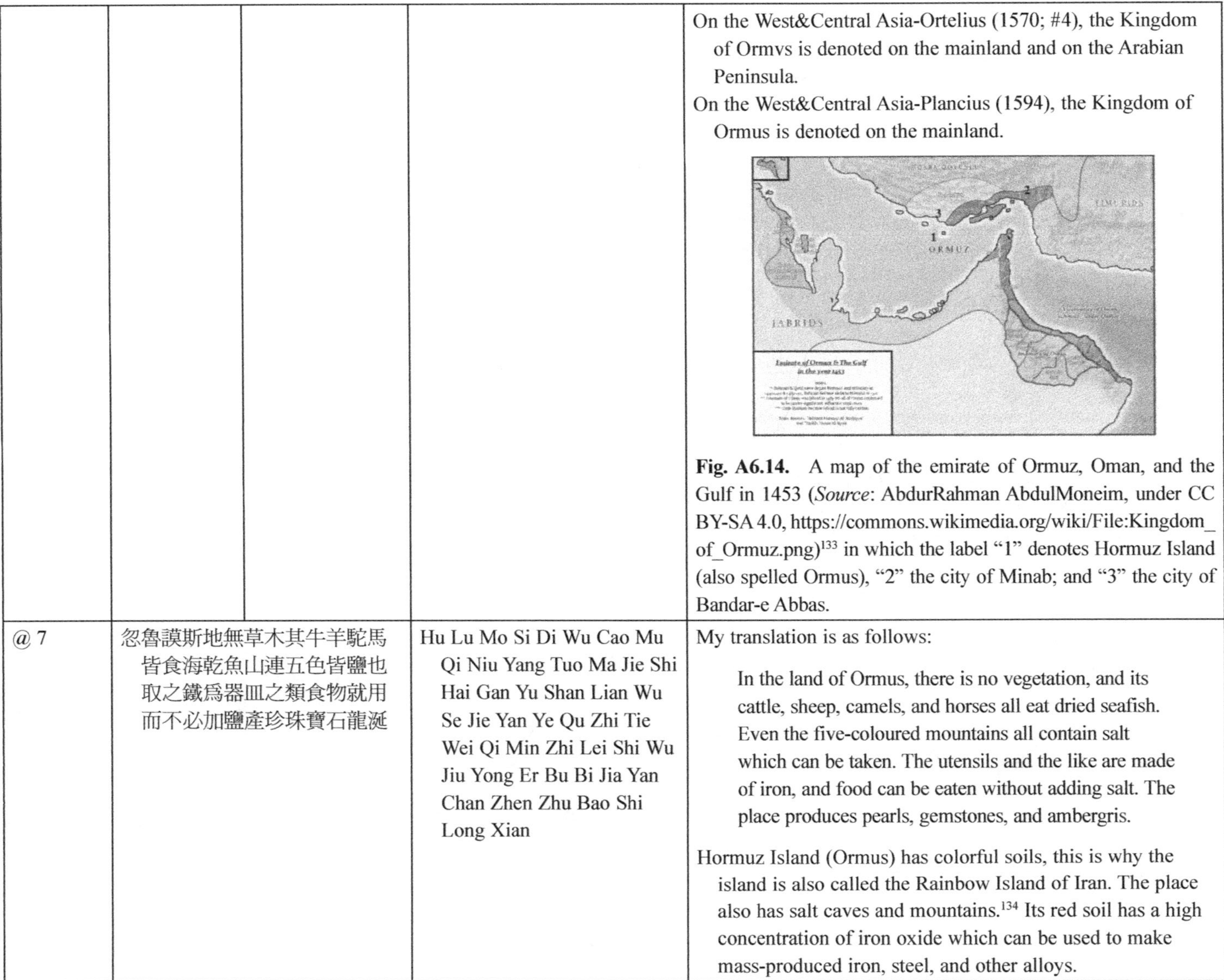 **Fig. A6.14.** A map of the emirate of Ormuz, Oman, and the Gulf in 1453 (*Source*: AbdurRahman AbdulMoneim, under CC BY-SA 4.0, https://commons.wikimedia.org/wiki/File:Kingdom_of_Ormuz.png)[133] in which the label "1" denotes Hormuz Island (also spelled Ormus), "2" the city of Minab; and "3" the city of Bandar-e Abbas.
@ 7	忽魯謨斯地無草木其牛羊駝馬皆食海乾魚山連五色皆鹽也取之鐵爲器皿之類食物就用而不必加鹽產珍珠寶石龍涎	Hu Lu Mo Si Di Wu Cao Mu Qi Niu Yang Tuo Ma Jie Shi Hai Gan Yu Shan Lian Wu Se Jie Yan Ye Qu Zhi Tie Wei Qi Min Zhi Lei Shi Wu Jiu Yong Er Bu Bi Jia Yan Chan Zhen Zhu Bao Shi Long Xian	My translation is as follows: In the land of Ormus, there is no vegetation, and its cattle, sheep, camels, and horses all eat dried seafish. Even the five-coloured mountains all contain salt which can be taken. The utensils and the like are made of iron, and food can be eaten without adding salt. The place produces pearls, gemstones, and ambergris. Hormuz Island (Ormus) has colorful soils, this is why the island is also called the Rainbow Island of Iran. The place also has salt caves and mountains.[134] Its red soil has a high concentration of iron oxide which can be used to make mass-produced iron, steel, and other alloys.

(Continued)

Item #	Name	Pinyin	Etymology	Analysis of history, politics, geography: comparison or comment
● 53	刺路斯	La Lu Si	Laluse[135]	The place cannot be identified on a modern map. It should be in present-day Iran, see Fig. A6.4.
54	瓦得尔	Wa De Er	Guadel[136]	The place cannot be identified on a modern map. It should be in present-day Iran, see Fig. A6.4. "Guadel" is depicted on the West&Central Asia-Ortelius (1570; #2), the West&Central Asia-Ortelius (1570; #3), and the West&Central Asia-Ortelius (1570; #4).
55	古亞思旦	Gu Ya Si Dan	Cusistan[137]	古亞思旦 should be corrected as 古西思旦 (Cusistan). Cusistan (古西思旦) in Persis was called Cutha, or the land of Cuth. The place cannot be identified on a modern map. It should be in present-day Iran, see Fig. A6.4. The West&Central Asia-Mercator (1569) depicts Cusistan, but the West&Central Asia-Ortelius (1570; #1) and the West&Central Asia-Plancius (1594) do not. On the West&Central Asia-Ortelius (1570; #2), there is CVSISTAN.
56	沙勿私	Sha Wu Si	Savas[138]	The place cannot be identified on a modern map. It should be in present-day Iran, see Fig. A6.4. Savas is depicted on the West&Central Asia-Ortelius (1570; #3).
57	溪尔曼	Xi Er Man	Chirman[139]	Chirman (溪尔曼) is Kerman in Iran. Kerman was founded as a defense outpost in the third century.[140] On the West&Central Asia-Mercator (1569), there is "Chirman". On the West&Central Asia-Ortelius (1570; #2), there is CHIRMAN. On the West&Central Asia-Ortelius (1570; #4), there is CHIRMAN. On the West&Central Asia-Plancius (1594), there is "Chirma".
58	赤蠟蓋亞	Chi La Gai Ya	Erachaian[141]	The place cannot be identified on a modern map. It should be in present-day Iran or Pakistain, see Fig. A6.4

				On the West&Central Asia-Ortelius (1570; #2), there is ERACHAIAN. On the West&Central Asia-Ortelius (1570; #4), there is ERACHAIAN.
59	哥蠟作泥	Ge La Zuo Ni	Chorasan[142] or Khorasan or Khurasan or Khorassan	Chorasan/Khorasan (哥蠟作泥/哥蜡作泥) is a historical eastern region in the Iranian Plateau in West and Central Asia. The extent of the region referred to as *Khorasan* varied over time. **Fig. A6.15.** Greater Khorasan (Chorasan) and neighbouring historical regions (*Source*: Don-kun, *Source*: Lencer, User:Виктор B, under CC BY-SA 3.0, https://commons.wikimedia.org/wiki/File:Chorasan-Transoxanien-Choresmien_neu.svg).[143] Chorasan is depicted as "Corassan" on the West&Central Asia-Ortelius (1570; #1) and the West&Central Asia-Ortelius (1570; #2).
60	伊西帝宜入	Yi Xi Di Yi Ru	Istigias[144]	One of the most ample and magnificent cities of Persia was Istigias (伊西帝宜入) which was the chief seat of Bactria (Bactria[145] was a province of the Persian Empire located in modern Afghanistan, Uzbekistan, and Tajikistan). The West&Central Asia-Ortelius (1570; #2), the West&Central Asia-Ortelius (1570; #4) and the West&Central Asia-Ortelius (1570; #5), all depict ISTIGIAS.

(*Continued*)

Item #	Name	Pinyin	Etymology	Analysis of history, politics, geography: comparison or comment
61	甘打喝	Gan Da He	Kandahar or Kandahu	Kandahar (甘打喝 or 坎大哈) is a city in Afghanistan, located in the south of the country. Kandahar has been a frequent target for conquest because of its strategic location in Asia, controlling the main trade route linking the Indian subcontinent with the Middle East and Central Asia.[146] On the West&Central Asia-Ortelius (1570; #1), there is "Candahar". On the West&Central Asia-Plancius (1594), there is also "Candahar". On the West&Central Asia-Ortelius (1570; #4), there is CANDAHAR.
62	惹西厮突	Re Xi Si Tu	Sirisitvr[147]	The place cannot be identified on a modern map. It should be in present-day Afghanistan. On the West&Central Asia-Ortelius (1570; #4), there is SIRISITVR. However, the geographical location of Sirisitvr on the West&Central Asia-KWQ is very different from that on the West&Central Asia-Ortelius (1570; #4).
63	色耨利	Se Nu Li	Sables	The place cannot be identified on a modern map. It should be in present-day Afghanistan. On the West&Central Asia-Ortelius (1570; #4), there is SABLES.
64	亞的伯讓	Ya De Bo Rang	Adilbegian[148] region	It was the Province of Aderbeitzan, or Aderbajon, formerly Great Media, or Satrapene.[149] The Country of Aderbeitzan, or Adherbaigan, or Aderbajon, was anciently called South, or Great Media.[150] Castald named it Adilbegian.[151] Aderbeitzan borders Schirwan in the North and North-West; Persia in the South; the Country of Kilan and Parthia in the East; and Great Armenia in the West.[152] The territory was first ruled by Caucasian Albania (second century B.C.–eighth century A.D.) and later various Persian empires. Until the nineteenth century, it remained part of Qajar Iran (1789–1925).[153] On the West&Central Asia-Ortelius (1570; #4), there is ADILBEGIAN.

| 65 | 波斯 | Bo Si | Persia; Iran has a long history and is a well-known ancient country in the world. The ancient Chinese history and ancient European history called this country "Persia". But Iranians have never called their country "Persia". It was not until 1935 that the international community followed the wishes of the country's government and renamed it "Iran".[154] | Iran, located in West Asia and officially the Islamic Republic of Iran, was called Persia (波斯) before 1935.[155] The present Iran borders thirteen land and maritime countries and faces the Caspian Sea to the north and the Persian Gulf to the south, see Fig. A6.16.
The country is home to one of the world's oldest civilizations, beginning with the formation of the Elamite kingdoms[156] in the fourth millennium B.C. Arab Muslims conquered the empire in the seventh century A.D., which led to the Islamization of Iran.[157] In the fifteenth century, the native Safavids re-established a unified Iranian state and national identity, and converted the country to Shia Islam.[158] |

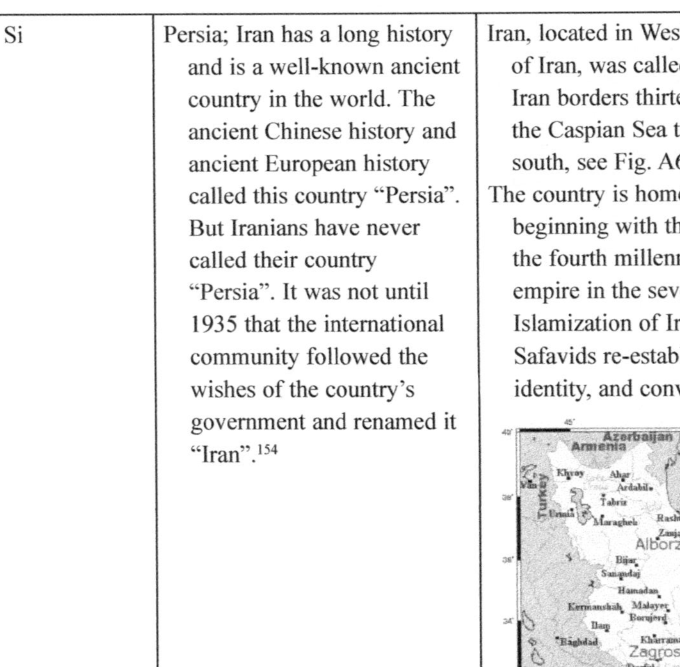

Fig. A6.16. Borders of Iran (*Source*: Kelisi at English Wikipedia, under CC BY-SA 3.0, https://commons.wikimedia.org/wiki/File:IranOMC.png):[159] land borders with seven countries (clockwise: Turkmenistan, Afghanistan, Pakistan, Iraq, Turkey, Armenia and Azerbaijan) and maritime borders with six other countries (counterclockwise: Kuwait, Saudi Arabia, Bahrain, Qatar, the United Arab Emirates, and Oman).

(*Continued*)

Item #	Name	Pinyin	Etymology	Analysis of history, politics, geography: comparison or comment
				From 1381 to 1467, Persia became part of the Timurid Empire (帖木兒帝國/帖木儿帝国; 1370–1507), a Turkic (突厥) empire[160] that arose in the river area.[161] Shortly after the death of Timur (1336–1405; a Turco-Mongol conqueror who founded the Timurid Empire), Western Persia was successively ruled by the Black Sheep Dynasty (黑羊王朝; 1374–1468) established by the Turkmen (土庫曼人/土库曼人; culturally Persianate and Muslim)[162] from c. 1374 to 1468[163] and the White Sheep Dynasty (1378–1503; a culturally Persianate, Sunni Turkoman tribal confederation) from 1378 to 1503.[164] From 1405 to 1433, Emperor Yongle of the Ming Dynasty sent Zheng He (鄭和/郑和) to organised a fleet to sail to the Western Ocean, reaching the Timurid Empire many times.[165] The stone tablet erected by Zheng He in Sri Lanka (Ceylon/Island of Mount Ceylon) is written in Chinese, Tamil, and Persian (Stele of Giving to the Buddhist Temple on Mount Ceylon). Arabic and Persian were widely used in the Ming Dynasty Muslim scripture education.[166] At the end of the sixteenth century and the beginning of the seventeenth century, during the reign of Emperor Abbas the Great (阿巴斯大帝), the Safavid Dynasty (薩非王朝/萨非王朝) reached its peak of national power. It could compete with the two great powers of the East and the West: the Ottoman Empire (also known as the Turkish Empire, which was an imperial state that controlled much of Southeast Europe, West Asia, and North Africa between the fourteenth and early 20th centuries; 奧斯曼帝國/奥斯曼帝国; c. 1299–1922) and the Mughal Empire (1526–1858; a Muslim empire that ruled most of India and Pakistan; 莫臥兒帝國/莫卧儿帝国). Both the West&Central Asia-Mercator (1569) and the West&Central Asia-Ortelius (1570; #1) depict Persia.

				On the West&Central Asia-Ortelius (1570; #2), there is also "Persia". The West&Central Asia-Plancius (1594) depicts "Persis" and "Parthia", not Persia. Persis, also called Persia proper, is the Fars region, located to the southwest of modern-day Iran, now a province.
● 66	加私	Jia Si	Kesh[167]	The place is in present-day Beruniy, Karakalpakstan, Uzbekistan.
67	得力利大伯里私旦	De Li Li Da Bo Li Si Dan	Tares Taperistan[168] or Tabaristan. Tabaristan was named after the Tapurians, who had been deported there from Parthia by the Parthian king Phraates I (r. 176–171 B.C.).[169]	Tabaristan or Tares Taperistan (得力利大伯里私旦), see Fig. A6.17, was a mountainous region located on the Caspian coast of northern Iran. It corresponded to the present-day province of Mazandaran, which became the predominant name of the area from the eleventh century onwards.[170] On the West&Central Asia-Ortelius (1570; #2), there is TEREA TAPERISTAN. **Fig. A6.17.** Northern Iran and its surroundings during the Iranian intermezzo or Persian Renaissance after the seventh-century Muslim conquest of Persia and the fall of the Sasanian Empire. (*Source*: HistoryofIran, under CC BY-SA 4.0, https://commons.wikimedia.org/wiki/File:Northern_Iran_and_its_surroundings_during_the_Iranian_intermezzo.svg).[171]
68	入蘭	Ru Lan	Gilan[172]	Gilan (入蘭/入兰) is an Iranian province along the southwestern coast of the Caspian Sea. In 1307 the Ilkhan Öljeitü conquered the region.[173] This was the first time the region came under the rule of the Mongols. After 1336, the region seemed to be independent again. Gilan was ruled over by two local dynasties in the late fifteenth-early sixteenth century; it also recognised twice, for short periods of time, the suzerainty of the Ottoman Empire in 1534 and 1591.[174]

(Continued)

Item #	Name	Pinyin	Etymology	Analysis of history, politics, geography: comparison or comment
				The West&Central Asia-Ortelius (1570; #2) and the West&Central Asia-Ortelius (1570; #4) depict GILAN.
69	陀拔斯單	Tuo Ba Si Dan	Tabaristan	Tabaristan (陀拔斯單/陀拔斯单) was the name of an ancient region in Central Asia, see Fig. A6.17. This is a repetition of Tares Taperistan (得力利大伯里私旦; Item 67). The region is recorded in the *Xin Tang Shu • Bo Si Chuan* (新唐書 • 波斯傳/新唐书 • 波斯传) or *New Book of Tang, Persian Biography*. Between 744 and 754, the people there maintained friendly relations with the Tang Dynasty of China.[175] Around 765, it was absorbed by the Abbasid (750–1258) caliphates, creating one of the largest land empires in history.[176] On the West&Central Asia-Ortelius (1570; #4), there is TAPERISTAN.
70	葛尔打闌	Ge Er Da Lan	Caldarian[177]	The place cannot be identified on a modern map. It should be in present-day Iran or Afghanistan, see Fig. A6.16. On the West&Central Asia-Ortelius (1570; #4), there is *Caldaran.*
71	止波里	Zhi Bo Li	Cyprus; the earliest attested reference to *Cyprus* is the fifteenth century B.C. Mycenaean Greek.[178]	Cyprus (止波里) is the third-largest island in the Mediterranean Sea.[179] It is in the eastern Mediterranean Sea, north of the Sinai Peninsula, south of the Anatolian Peninsula, and west of the Levant. The Kingdom of Cyprus was one of Crusader states that existed between 1192 and 1489.[180] As a strategic location in the Eastern Mediterranean, it was subsequently occupied by several major powers. The Island of Cyprus was an overseas possession of the Republic of Venice (697–1797) from 1489, when the independent Kingdom of Cyprus ended, until 1571,[181] when the island was conquered by the Ottoman Empire.[182]

			The West&Central Asia-Mercator (1569) and the West&Central Asia-Plancius (1594) depict the island as "Cyprus", while the West&Central Asia-Ortelius (1570; #1) depict it as "Ciprus". On the West&Central Asia-Ortelius (1570; #2), there is *Cipro*. On the West&Central Asia-Ortelius (1570; #3), there is CIPRVS.
72	鐵門関	Tie Men Guan	Iron Gate Pass or Tiemen Guan[183]	The Iron Gate Pass (鐵門関/铁门关; 41.8°N, 86.2°E) is a checkpoint in Korla, Bayingolin Mongolian Autonomous Prefecture, Xinjiang Uyghur Autonomous Region (新疆維吾爾自治區巴音郭楞蒙古自治州庫爾勒市/新疆维吾尔自治区巴音郭楞蒙古自治州库尔勒市), China. The Iron Gate Pass was of historical strategical significance because it formed a vulnerable bottleneck on the Silk Road. At present, the pass is no longer part of the road infrastructure of the region and is preserved as a scenic and historical area. On the West&Central Asia-KWQ, the Iron Gate Pass (鐵門関) is depicted at c. 36.5°N (KWQ latitude → modern latitude), 75°E, being too southward and westward than its correct location on a modern map. On the West&Central Asia-Ortelius (1570; #2), there is *Termet*.
73	覩貨羅	Du Huo Luo	Tocharian Kingdom: Sanskrit name: *Tukhāra, Tuhkhāra* or *Tusāra*; Greek name *Tokharoi*; on the European maps, it is depicted as Tacalistan.	The Tocharian Kingdom (Tukhāra, Tuhkhāra or Tusāra; 覩貨羅國/覩货罗国 or睹貨邏國/覩货逻国 or 吐火羅國/吐火罗国)[184] was in today's northern region of Afghanistan near the upper reaches of the Oxus River (烏滸河/乌浒河）or today's Amu Darya River (阿姆河). It was located to the southwest of the Pamir Plateau (帕米爾高原/帕米尔高原) or Congling (蔥嶺/葱岭) and to the south of the Wuhu River, being at the crossroads from East Turkestan to Persia and India. It was once the land of ancient Bactria (大夏; 2500/2000 B.C.–900/1000 A.D.).[185] By the seventh century, it was attached to the Tang Dynasty,[186] and later was under the rule of Arab Muslims.[187]

(Continued)

Item #	Name	Pinyin	Etymology	Analysis of history, politics, geography: comparison or comment
				The West&Central Asia-Ortelius (1570; #2), the West&Central Asia-Ortelius (1570; #4), and the West&Central Asia-Ortelius (1570; #5), all depict Tacalistan.
● 74	回回	Hui Hui	Huihui[188]	Huihui (回回) was the alternative name for 回鶻/回鶻 (*Huíhú*); it was a tribal alliance. Around the thirteenth century, the Mongol Empire rose, and the descendants of the Uighurs who migrated westward sought refuge in Mongolia. According to the Mongolian transliteration, they were called Uighurs in Chinese historical records. The *Huíhú* people is a minority tribe in China. They were the ancestors of the Uighurs[189] and were renamed from Uighurs.[190] They are mainly distributed in Xinjiang, but also scattered in Inner Mongolia, Gansu, Mongolia, and some areas in Central Asia.
75	蠟大私	La Da Si	Tares[191]	The place cannot be identified on a modern map. It should be in today's Northern Iran. On the West&Central Asia-Ortelius (1570; #5), there is TARSE.

West and Central Asia: Part II

Fig. A6.18. Extracted from West&Central Asia-KWQ (public domain), Items 76-93 and annotations @8-@9 are depicted on this Figure.

● 76	喝盤陀	He Pan Tuo	Hepantuo	Hepantuo (喝盤陀/喝盘陀) was a small country on the Pamir Plateau (帕米爾高原/帕米尔高原) in the Western Regions (西域) during the Northern and Southern Dynasties (南北朝; 420–589). Geographically, it controlled key points on the China-India Silk Roads. The capital of Hepantuo was in Tashkurghan (塔什庫爾干鎮/塔什库尔干镇) in Xinjiang in the far west of China, close to the country's border with Tajikistan. Tashkurgan was a significant stop on the Silk Road, with roads leading to major centres of trade such as Kashgar (喀什; one of the westernmost cities of China in the Tarim Basin region of southern Xinjiang, China, near the country's border with Kyrgyzstan and Tajikistan). During the Kaiyuan period (開元年間/开元年间; 713–734) of Emperor Xuanzong (玄宗) in the Tang Dynasty (唐朝), the Hepantuo Kingdom perished. Later, the Tang Dynasty set up Congling Guards (蔥嶺守捉/葱岭守捉) in Hepantuo and implemented effective jurisdiction. Although the small kingdom is long gone, the Tashkurghan Town still exists till today and the West&Central Asia-KWQ depicts it using the name of the kingdom.
77	地布蠟	Di Bu La	Tibura[192]	The place cannot be identified on a modern map. But Tibura (地布蠟/地布蜡) should be in northeast India. The West&Central Asia-KWQ erroneously depicts it in the present Xingjiang area. On the West&Central Asia-Ortelius (1570; #5), there is TIPVRA.
78	懸度山	Xuan Du Shan	Xuandu Shan	Xuandu Shan (懸度山/悬度山) was the name of an ancient mountain range, the present-day Hindu Kush Mountains[193] at the Afghanistan-Pakistan border. On the West&Central Asia-Ortelius (1570; #2), there is a huge mountain range named *Imaus mons nunc Dalanguer* (大葱嶺/大葱岭) and *Vsonte* (懸度山/悬度山).

(*Continued*)

Item #	Name	Pinyin	Etymology	Analysis of history, politics, geography: comparison or comment
				On the West&Central Asia-Ortelius (1570; #4), there are *Dalanguer M.* and *Naugracot mons.* On the West&Central Asia-Ortelius (1570; #5), there are *Monte Dalanguer* and *Monte Vssonte* as part of *Imaus mons qui & Caucasus*[194]
79	高葛娑山	Gao Ge Po Shan	Caucasus Mountains	The Caucasus Mountains (高葛娑山 or 高加索山脉) are a mountain range at the intersection of Asia and Europe, stretching between the Black Sea (黑海) and the Caspian Sea (里海), see Fig. A6.19. 

Fig. A6.19. Topographic map of the Caucasus Mountains (*Source*: Bourrichon — fr:Bourrichon with English translations, additions, and corrections by Ketone16 (partly following Yuri Koryakov), under CC BY-SA 4.0 International, 3.0 Unported, 2.5 Generic, 2.0 Generic and 1.0 Generic license, https://commons.wikimedia.org/wiki/File:Caucasus_topographic_map-en.svg).[195]

The Caucasus Mountains include the Greater Caucasus (大高加索) in the north and Lesser Caucasus (小高加索) in the south. The Greater Caucasus runs west-northwest to east-southeast, from the Caucasian Natural Reserve

				(高加索自然保護區/高加索自然保护区) in the vicinity of Sochi (索契), Russia, on the northeastern shore of the Black Sea, to Baku (巴庫/巴库), Azerbaijan (阿塞拜疆), on the Caspian Sea. The Lesser Caucasus runs parallel to the Greater about 100 km/62 mi south.[196] The Greater and Lesser Caucasus ranges are connected by the Likhi Range (利克山脈/利克山脉). The Caucasus is a 1200 km/750 mi stretch of land. It lies between Europe and Asia. It stretches from the Caspian to the Black Sea. The Caucasus helped connect Europe and Asia. The old Silk Road passed through the region. However, on the West&Central Asia Asia-KWQ, 高葛娑山 (Caucasus Mountains) are misplaced to the far southeast of the Caspian Sea, instead of between the Black Sea and the Caspian Sea. On the West&Central Asia-Mercator (1569), between the Black Sea and the Caspian Sea, there are several mountain ranges, but none of them are named. On the West&Central Asia-Ortelius (1570; #1), the West&Central Asia-Ortelius (1570; #2), the West&Central Asia-Ortelius (1570; #3), the West&Central Asia-Ortelius (1570; #4), and the West&Central Asia-Plancius (1594), no mountain range is depicted between the Black Sea and the Caspian Sea. On the West&Central Asia-Ortelius (1570; #5), there are *Monte Dalanguer* and *Monte Vssonte* as part of the *Imaus mons qui & Caucasus*.[197] But their geographical locations are not between the Black Sea and the Caspian Sea.
• 80	朱俱波	Zhu Ju Bo	Zhujubo or Karghalik[198] or Kargilik	Zhujubo (朱俱波) was a small town where the ancient Zhujubo Kingdom was in the foothills of the Kunlun Mountains (崑崙山/昆仑山). The town is in the present-day Qipan Township, southwest of Yecheng County and southeast of Taxkorgan County, Xinjiang Uygur Autonomous Region (新疆維吾爾自治區葉城縣西南棋盤鄉及塔什庫爾幹縣東南境/新疆维吾尔自治区叶城县西南棋盘乡及塔什库尔干县东南境).

(*Continued*)

Item #	Name	Pinyin	Etymology	Analysis of history, politics, geography: comparison or comment
				In the *Xin Tang Shu • Xi Yu Chuan* (新唐書 • 西域傳) or *New Book of Tang, Biography of the Western Region*, it states that "朱俱波亦名朱俱槃, 漢子合國也". It is translated by Sheng-Wei Wang as: Zhujubo (朱俱波) is also known as Zhujupan (朱俱槃) which was the Kingdom of Zihe (子合國/子合国) in the Han Dynasty.
81	大葱嶺	Da Cong Ling	Monte Dalanguer[199] or Pamir Mountains; given that Allium Giganteum (大葱) is a kind of wild green onion which grows on the plateau of the Pamir Mountains,[200] hence, its Chinese name "大葱嶺", meaning "Onion Range", here "嶺/岭" means "mountains or mountain range".	The Pamir Mountains (帕米爾高原/帕米尔高原 or 蔥嶺/葱岭, lit. "Onion Range") are a mountain range between Central Asia and Pakistan. It is located at a junction with other notable mountains, namely the Tian Shan (天山), Karakoram (喀喇崑崙山脈/喀喇昆仑山脉), Kunlun (崑崙/昆仑), Hindu Kush (興都庫什/兴都库什) and the Himalaya Mountain Ranges (喜馬拉雅山脈/喜马拉雅山脉). They are among the world's highest mountains.[201] Much of the Pamir Mountains lie in the Gorno-Badakhshan Province (戈爾諾-巴達赫尚省/戈尔诺-巴达赫尚省) in Tajikistan (塔吉克斯坦). To the south, they border the Hindu Kush Mountains (庫什山脈/库什山脉) along Afghanistan's Wakhan Corridor (阿富汗瓦罕走廊) and Gilgit-Baltistan (吉爾吉特-巴爾蒂斯坦/吉尔吉特-巴尔蒂斯坦) regions in Pakistan (巴基斯坦). To the north, they join the Tian Shan Mountains (天山山脈/天山山脉 or祁連山/祁连山) along the Alay Valley (阿萊谷) in Kyrgyzstan (吉爾吉斯斯坦/吉尔吉斯斯坦). To the east, they extend to the range that includes China's Kongur Tagh (公格爾峰/公格尔峰), in the Eastern Pamirs (東帕米爾高原/东帕米尔高原), separated by the Yarkand Valley (莎車谷/莎车谷) from the Kunlun Mountains (崑崙山/昆仑山).[202] On top of the Pamir plateau, there are eight U-shaped River valleys including Great Pamir (大帕米爾/大帕米尔）and Little Pamir (小帕米爾/小帕米尔). The Pamir River is in the south-west of the Pamirs.

				The seven European maps all show mountain ranges in the general neighbourhood of Central Asia, Pakistan, India, and areas of China, but their names are not often depicted. On the West&Central Asia-Ortelius (1570; #2), there is a huge mountain range named "Imaus mons nunc Dalanguer" (大葱嶺/大葱岭) and Vsonte (懸度山/悬度山). On the West&Central Asia-Ortelius (1570; #4), there are "Dalanguer M." and "Naugracot mons". On the West&Central Asia-Ortelius (1570; #5), there are "Monte Dalanguer" and "Monte Vssonte".
82	貌力南客尔	Mao Li Nan Ke Er	Mavrenacher or Mavrenaher	The place cannot be identified on a modern map. It should be in Transoxiana (in lower Central Asia roughly corresponding to modern-day eastern Uzbekistan, western Tajikistan, parts of southern Kazakhstan, parts of Turkmenistan and southern Kyrgyzstan). On the West&Central Asia-Ortelius (1570; #2), there is MAVRENACHER. On the West&Central Asia-Ortelius (1570; #4), there is also MAVRENACHER. On the West&Central Asia-Ortelius (1570; #5), there is MAVRENAHER.
83	没善土	Mei Shan Tu	Mesandaran	The place cannot be identified on a modern map. It should be in present-day Iran. On the West&Central Asia-Ortelius (1570; #4), there is MESANDARAN.
84	大革里思旦	Da Ge Li Si Dan	Tacalistan[203]	Tacalistan (大革里思旦) was a region to the east of the Caspian Sea in Central Asia. This item seems to be a repetition of the Tocharian Kingdom (Tukhāra, Tuhkhāra or Tusāra; 覩貨羅國/覩货罗国 or 睹貨邏國/覩货逻国 or 吐火羅國/吐火罗国; Item 73). The West&Central Asia-Ortelius (1570; #2), the West&Central Asia-Ortelius (1570; #4), and the West&Central Asia-Ortelius (1570; #5), all depict TACALISTAN.

(Continued)

Item #	Name	Pinyin	Etymology	Analysis of history, politics, geography: comparison or comment
85	耶塞拨	Ye Sai Bo	Iesel Bas	Here 耶塞拨 may refer to the region where Iesel Bas had their settlement. The Qizilbash, or "Red Heads" (they had red-coloured headwear), were a diverse array of mainly Turkoman[204] Shia militant groups and their origin can be dated from the fifteenth century onward. They flourished in Azerbaijan (亞塞拜然/阿塞拜疆),[205] Anatolia (安那托利亞/安纳托利亚), the Armenian Highlands (亞美尼亞高原/亚美尼亚高原), the Caucasus (高加索), and Kurdistan (庫德斯坦/库尔德斯坦) from the late fifteenth century onwards. They contributed to the foundation of the Safavid Dynasty (薩法維王朝/萨法维王朝; 1501–1736) in early modern Iran (伊朗).[206] The West&Central Asia-Ortelius (1570; #2), the West&Central Asia-Ortelius (1570; #4), and the West&Central Asia-Ortelius (1570; #5), all depict IESEL BAS.
86	何加入	He Jia Ru	Ocrage[207]	The place cannot be identified on a modern map. It should be in present-day Iran or Turkmenistan, see Fig. A6.16. The West&Central Asia-Ortelius (1570; #2), the West&Central Asia-Ortelius (1570; #4) depict OCRAGE.
87	打喇巴	Da La Ba	Starabat	The place cannot be identified on a modern map. It should be in present-day Iran, see Fig. A6.16. On the West&Central Asia-Ortelius (1570; #4), there is STARABAT.
88	察瓦泰	Cha Wa Tai	Zagatai[208]	The place cannot be identified on a modern map. It should be in present-day Kazakhstan or Turkmenistan, see Fig. A6.16. The West&Central Asia-Ortelius (1570; #2), the West&Central Asia-Ortelius (1570; #4), and the West&Central Asia-Ortelius (1570; #5) depict ZAGATAI.

| 89 | 焉耆 | Yan Qí | Yanqi[209] or Karasahr or Karashar or Karaxahr | Karasahr or Karashar was originally known in the Tocharian languages as *Ārśi* (or Arshi), Qarašähär, Agni or the Chinese derivative Yanqi (焉耆). It is now an ancient town on the Silk Road and the capital of Yanqi Hui Autonomous County in the Bayingolin Mongol Autonomous Prefecture, Xinjiang (新疆巴音郭楞蒙古自治州焉耆回族自治縣/新疆巴音郭楞蒙古自治州焉耆回族自治县).
Between the mid-thirteenth century and the eighteenth century, Karasahr was part of the Mongol Chagatai Khanate (察合台汗國/察合台汗国).
Fig. A6.20. The Tarim Basin in the third century, showing the so-called Tocharian and related states; Karasahr is the green-coloured area at top right (*Source*: Schreiber, under CC BY-SA 3.0, https://commons.wikimedia.org/wiki/File:Tarimbecken_3._Jahrhundert.png).[210]
Karashahr was known to late medieval Europeans as *Calis, Cialis, Chalis,* or *Chialis*: in the early seventeenth century, because the Portuguese Jesuit lay brother Bento de Góis visited the Tarim Basin on his way from India to China (via Kabul in Afghanistan and Kashgar on the western side of the Tarim Basin of southern Xinjiang, China, see Fig. A6.20). De Góis and his traveling companions spent several months in the "Kingdom of Cialis" in 1605 before reaching the Ming border at Jiayuguan.[211] But this was three years after Matteo Ricci published Kunyu Wanguo Quantu. |

(*Continued*)

Item #	Name	Pinyin	Etymology	Analysis of history, politics, geography: comparison or comment
				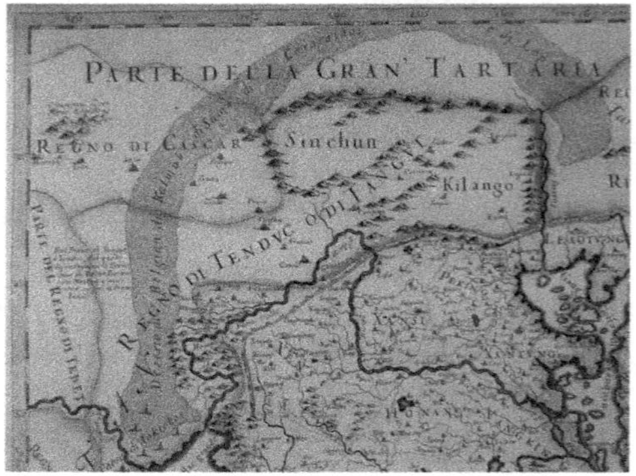 Fig. A6.21. "Cilia" (enclosed by the circle) is one of the cities in the chain stretching from Hiarcan to Sucieu (originally published 1682; public domain).[212] Surprisingly, on the West&Central Asia-Ortelius (1570; #5), there is "Calis". How did Abraham Ortelius get to know this place even before De Góis and his traveling companions reached the Kingdom of Cialis in 1605? He must have copied the location from the Chinese map but used the European name! Yanqi (Karasahr or Karashar or Karaxahr) is also depicted on this seventeenth century European map as "Cilia", see Fig. A6.21, as of one of the cities in the chain stretching from Hiarcan to Sucieu (originally published 1682).
● 90	火州	Huo Zhou	Huozhou; Qoco, Karakhoja[213]	Huozhou (火州) is 112.7 km/70 mi west of Liucheng (柳城) and 48.3 km /30 mi east of Turpan (吐魯番/吐鲁番)[214] in the Xinjiang Uygur Autonomous Region (新疆維吾爾自治區/新疆维吾尔自治区).

				In 840 A. D., after the Mobei Uyghurs (漠北迴鶻/漠北回鶻) collapsed, one of the tribes moved to Huozhou (火州) and established the Gaochang Uyghur Kingdom (高昌回鶻王國/高昌回鹘王国).[215] During the Yongle period of the Ming Dynasty, the king sent envoys to Ming.[216] During the Ming Yingzong period (明英宗; reigning during the Zhengtong period 1436–1449, and the Tianshun (天順/天顺) period 1457–1464), the kingdom still sent tributes in the thirteenth year (2/03/1448–1/22/1449) of the Zhengtong period, but later discontinued, because it was annexed by Turpan during the Tianshun period.[217]
91	土鲁番	Tu Lu Fan	Turfan; the name "Turfan" is older than Turpan, but was not used until the middle of the second millennium A.D. and the name became widespread only in the post-Mongol period.[218]	Turpan (also known as Turfan or Tulufan; 吐鲁番/吐鲁番) is a prefecture-level city located in the east of the autonomous region of Xinjiang, China. Turpan was an important town on the ancient Silk Road. At that time, other kingdoms of the region included Korla (庫爾勒/库尔勒) and Yanqi (焉耆). Along with city-states such as Krorän (Loulan; 樓蘭/楼兰) and Kucha (龜茲/龟兹), Turfan appears to have been inhabited by people speaking the Indo-European Tocharian languages in prehistory.[219] Later the Uyghurs established the Kara-Khoja Kingdom or Gaochang Kingdom[220] (Uyghuria Idikut state; 856–1389) in the Turpan region. The Uyghur state later became a vassal state of the Kara-Khitans (喀喇契丹人) and then became a vassal of the Mongol Empire. The Ming Dynasty established local government, and the place was named Turpan. At the beginning of the fifteenth century, tribute began to be paid to the Ming Dynasty. In 1414 (the twelfth year of Yongle's reign), Chen Cheng (陳誠/陈诚) visited this place when he was sent as an envoy to the Western Regions and there were constant exchanges of tribute envoys in the early fifteenth century.[221]

(*Continued*)

Item #	Name	Pinyin	Etymology	Analysis of history, politics, geography: comparison or comment
				The Moghul ruler of Turpan Yunus Khan, also known as Ḥājjī 'Ali (ruled 1462–1478), unified Moghulistan (roughly corresponding to today's Eastern Xinjiang) under his authority in 1472. Around that time, conflicts with the Ming China started over the issues of tribute, trade, borders, and internal succession to the throne of Moghulistan (or Turpan). These conflicts were called the Ming-Turpan conflict (Chinese: 哈密之爭).[222] The West&Central Asia-Ortelius (1570; #1) depicts "Turfvn". The West&Central Asia and the 1570; #2) and West&Central Asia-Ortelius (1570; #5) depict "Turfan". But the West&Central Asia-Mercator (1569) and the West&Central Asia-Plancius (1594) do not depict Turpan or Turfan.
@ 8	漢車師地古之交河	Han Ju Shī Dì Gu Zhī Jiao He		My translation and comment (between square brackets) are as follows: This was the ancient Jiaohe City [交河] where the Jushi Kingdom was in the Han Dynasty. Jushi was a kingdom located in the Western Regions. The kingdom was separated into two parts in c. 60 B.C. in the Han Dynasty (202 B.C.–9 A.D., 25–220 A.D.). One kingdom was the Nearer Jushi/Anterior Jushi (車師前國/车师前国). Its capital was called Jiaohe and the site was in the Yarzigou Valley to the northwest of present-day Turpan, Xinjiang. The kingdom occupied a southern area controlled by the Han. The other kingdom was the Further Jushi/Posterior Jushi (車師後國/车师后国). Its capital was called Yuli or Yulai. This kingdom held a northern area dominated by the Xiongnu. The Jushi Kingdom's original name was Gushi (姑師/姑师), which first appeared in *Shi Ji • Juan Yi Bai Er Shi San • Da Yuan Lie Chuan* (史記 • 卷一百二十三 • 大宛列傳/史记 • 卷一百二十三 • 大宛列传) or

| | | | | *Records of the Grand Historian, Juan (Volume) 123, Biography of Da Yuan.*[223] The kingdom's southeast was connected to Dunhuang (敦煌), the south was connected to Loulan (樓蘭/楼兰) and Shanshan (鄯善; c. 77 B.C.–c. 630 A.D.), the west was connected to Yanqi (焉耆), the northwest was connected to Wusun (烏孫/乌孙), and the northeast was connected to the Xiongnu (匈奴). The kingdom was the key passage of the Silk Road. The Jiaohe or Yarkhoto ruins (10 km west of the city of Turpan in Xinjiang Uyghur Autonomous Region, China),[224] along with other archaeological sites associated with the Silk Road, were inscribed in 2014 on the UNESCO World Heritage List as the "Silk Roads: the Routes Network of Chang'an-Tianshan Corridor World Heritage Site. (絲綢之路：長安－天山廊道的路網/世界遺產)."[225] |
| ● 92 | 素葉河 | Su Ye He | The old name was Suyab (碎葉/碎叶); the present name is Chu River (楚河).[226] | The Chu River (楚河; Shu or Chüy or Chwu River) is in Northern Kyrgyzstan (北吉爾吉斯斯坦/北吉尔吉斯斯坦) and Southern Kazakhstan (南哈薩克/南哈萨克斯坦), see Fig. A6.23. The Chu River is neither the Syr Darya River nor the Amu Darya River shown in Fig. A6.22. The Chu River is formed by the confluence of the rivers Joon Aryk and Kochkor near the village Kochkor in Kyrgyzstan (吉爾吉斯斯坦/吉尔吉斯斯坦).[227] As the Chu River flows through the Chüy Valley (楚伊谷), it forms the border between Kyrgyzstan and Kazakhstan for more than a hundred kilometres, but then it leaves Kyrgyzstan and flows into Kazakhstan, see Fig. A6.22, where it flows at the northern edge of the Moiynkum Desert (莫因庫姆沙漠/莫因库姆沙漠), with Lake Kokuydynkol (科庫丁科湖/科库丁科湖) close to its channel. During the Middle Ages, the area was strategically important, because an ancient Silk Road city — Suyab (碎葉/碎叶) — the capital of several dynasties in the Western Regions, was |

(Continued)

Item #	Name	Pinyin	Etymology	Analysis of history, politics, geography: comparison or comment
				in the Chu River Valley (present-day Kyrgyzstan). The ruins of this city (Ak-Beshim; 阿克-貝希姆/阿克-贝希姆), along with other archaeological sites associated with the Silk Road, were inscribed in 2014 on the UNESCO World Heritage List as the "Silk Roads: the Routes Network of Chang'an-Tianshan Corridor World Heritage Site (絲綢之路: 長安－天山廊道的路網/世界遺產/丝绸之路: 长安－天山廊道的路网/世界遗产).[228] On the West&Central Asia-KWQ, the Chu River is erroneously depicted as flowing into 北高海 (the present-day Caspian Sea). On the West&Central Asia-Ortelius (1570; #4), there is *Oxus fl.*, which is the Amu Darya, *not* the Chu River. However, Amu Darya is a major river in Central Asia and Afghanistan, see Fig. A6.23. In this figure, the Chu River is not depicted (see Fig. A6.22 for comparison). **Fig. A6.22.** Chu River in the Syr Darya basin (*Source*: Background layer attributed to DEMIS Mapserver, map created by Shannon1, under CC BY-SA 4.0 International, 3.0 Unported, 2.5 Generic, 2.0 Generic and 1.0 Generic license, https://commons.wikimedia.org/wiki/File:Syrdaryamap.png).[229]

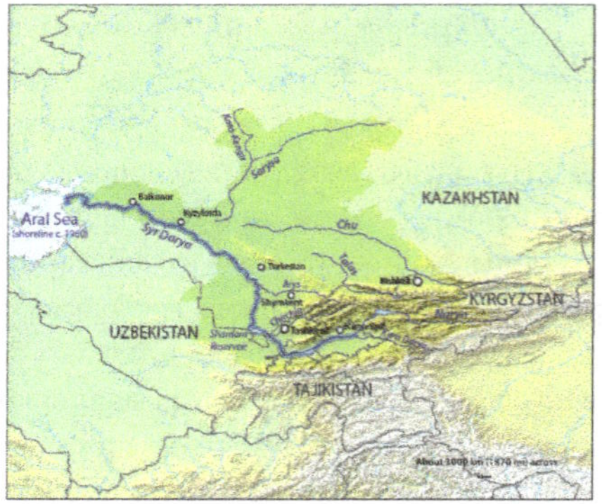

				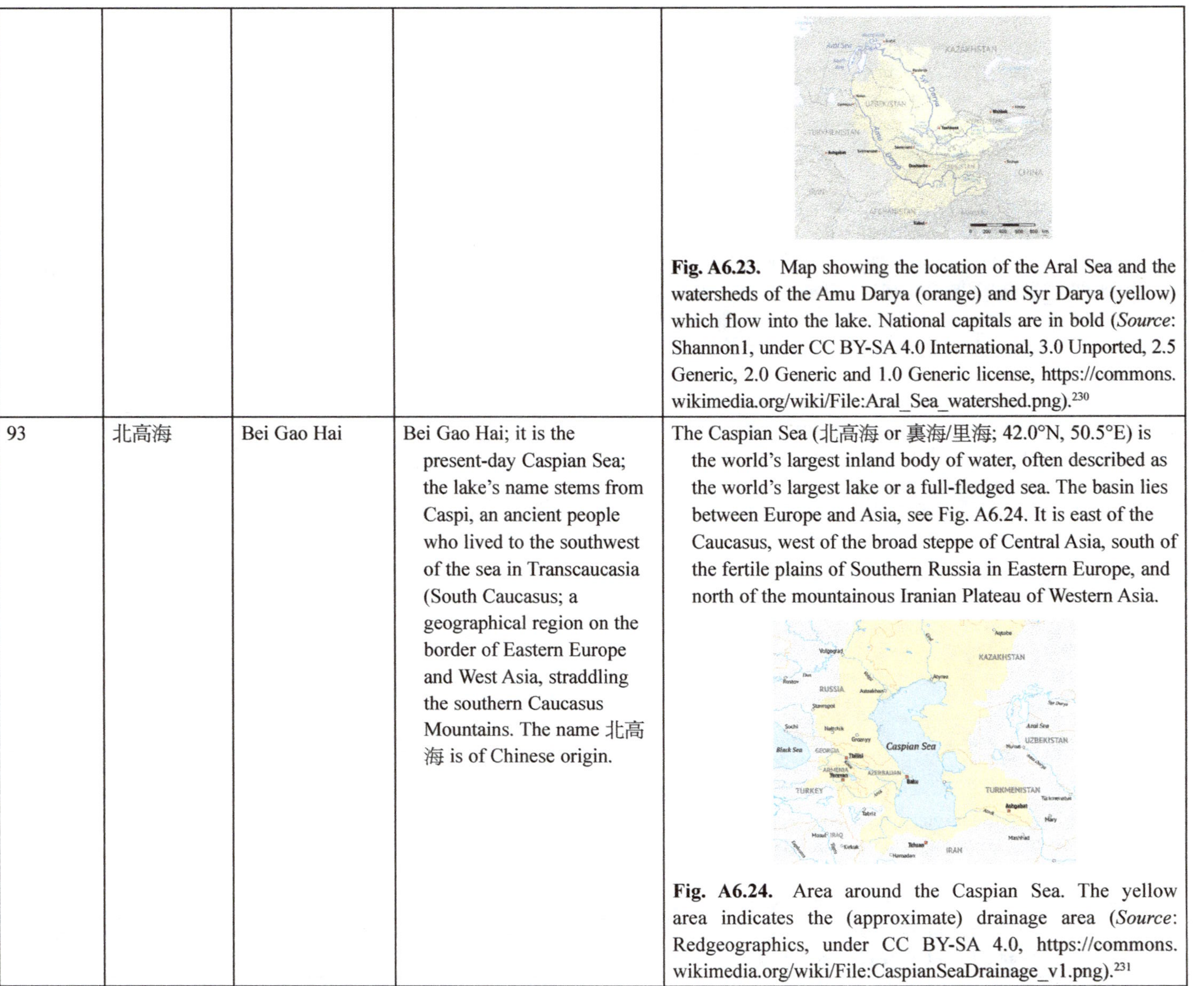Fig. A6.23. Map showing the location of the Aral Sea and the watersheds of the Amu Darya (orange) and Syr Darya (yellow) which flow into the lake. National capitals are in bold (*Source*: Shannon1, under CC BY-SA 4.0 International, 3.0 Unported, 2.5 Generic, 2.0 Generic and 1.0 Generic license, https://commons.wikimedia.org/wiki/File:Aral_Sea_watershed.png).[230]
93	北高海	Bei Gao Hai	Bei Gao Hai; it is the present-day Caspian Sea; the lake's name stems from Caspi, an ancient people who lived to the southwest of the sea in Transcaucasia (South Caucasus; a geographical region on the border of Eastern Europe and West Asia, straddling the southern Caucasus Mountains. The name 北高海 is of Chinese origin.	The Caspian Sea (北高海 or 裏海/里海; 42.0°N, 50.5°E) is the world's largest inland body of water, often described as the world's largest lake or a full-fledged sea. The basin lies between Europe and Asia, see Fig. A6.24. It is east of the Caucasus, west of the broad steppe of Central Asia, south of the fertile plains of Southern Russia in Eastern Europe, and north of the mountainous Iranian Plateau of Western Asia. Fig. A6.24. Area around the Caspian Sea. The yellow area indicates the (approximate) drainage area (*Source*: Redgeographics, under CC BY-SA 4.0, https://commons.wikimedia.org/wiki/File:CaspianSeaDrainage_v1.png).[231]

(*Continued*)

Item #	Name	Pinyin	Etymology	Analysis of history, politics, geography: comparison or comment
				Figure A6.25 shows the topographic map of the Caucasus. The South Caucasus roughly corresponds to modern Armenia, Georgia, and Azerbaijan (see Fig. A6.25), which are sometimes collectively known as the Caucasian States. 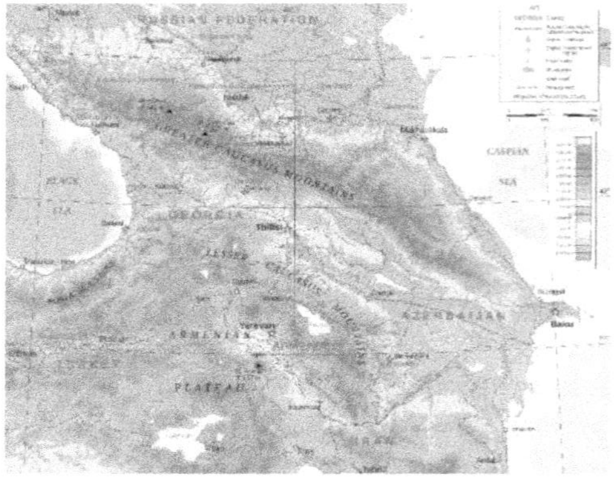 **Fig. A6.25.** Topographic map of the Caucasus (*Source*: Bourrichon — fr:Bourrichon with English translations, additions, and corrections by Ketone16 (partly following Yuri Koryakov), under CC BY-SA 4.0 International, 3.0 Unported, 2.5 Generic, 2.0 Generic and 1.0 Generic license, https://commons.wikimedia.org/wiki/File:Caucasus_topographic_map-en.svg).[232] On the West&Central Asia-KWQ, this lake has a name of Chinese origin: 北高海. 北 means "north", 高 "high or tall", and 海 "sea", literally, a sea in the far north (at a high latitude). The coordinates of the Caspian Sea on a modern map are 42.0°N, 50.5°E in comparison with the following four ancient maps as examples: On the West&Central Asia-KWQ, the coordinates of the Caspian Sea (北高海) are c. 43.4.0°N (KWQ latitude → modern latitude), c. 70.0°E. The longitudinal coordinate has a quite large error.

On the West&Central Asia-Mercator (1569), the Caspian Sea is depicted as *Mare de Sala Vel de Bachu* at c. 44.0°N, c. 72.5°E)

On the West&Central Asia-Ortelius (1570; #1), the Caspian Sea is depicted as *Mare de Bachu* at c. 43.5°N, c. 66.5°E.

On the West&Central Asia-Plancius (1594), the Caspian Sea is depicted as *Mare de Sala Bachu* at c. 44.0°N, c. 64.5°E.

The Caspian Sea is long and narrow from north to south, and its shape resembles an "S". The Caspian Sea depicted on a modern map, spans in the longitudinal direction no more than c. 6°, but in the latitudinal direction as much as c. 11°. But the eight maps discussed below all show this sea being long and narrow from east to west with an elliptical shape:

The Caspian Sea depicted on the West&Central Asia-KWQ spans in the longitudinal direction by almost 20°, but in the latitudinal direction by less than c. 7°.

The Caspian Sea depicted on the West&Central Asia-Mercator (1569) spans in the longitudinal direction by c. 17°, but in the latitudinal direction by only c. 7°.

The Caspian Sea depicted on the West&Central Asia-Ortelius (1570; #1) spans in the longitudinal direction by c. 18°, but in the latitudinal direction by only c. 8°.

The Caspian Sea depicted on the West&Central Asia-Plancius (1594) spans in the longitudinal direction by c. 13°, but in the latitudinal direction by c. 7°.

The Caspian Sea depicted on the West&Central Asia-Ortelius (1570; #2) spans in the longitudinal direction by c. 18°, but in the latitudinal direction only c. 7°.

The Caspian Sea depicted on the West&Central Asia-Ortelius (1570; #3) spans in the longitudinal direction by c. 20°, but in the latitudinal direction by only c. 6.5°.

The Caspian Sea depicted on the West&Central Asia-Ortelius (1570; #4) spans in the longitudinal direction by c. 21°, but in the latitudinal direction only c. 6.5°.

(Continued)

Item #	Name	Pinyin	Etymology	Analysis of history, politics, geography: comparison or comment
				The Caspian Sea depicted on the West&Central Asia-Ortelius (1570; #5) spans in the longitudinal direction by c. 20°, but in the latitudinal direction by only c. 7°. In summary, we can see that none of these ancient maps produces the correct shape of the Caspian Sea, unless their latitudinal and longitudinal widths are exchanged.
@ 9	此水甚浩蕩不通大海故疑爲海爲湖然其水鹹則姑謂之海	Ci Shui Shen Hao Dang Bu Tong Da Hai Gu Yi Wei Hai Wei Hu Ran Qi Shui Xian Ze Gu Wei Shi Hai		My translation and comments (between the square brackets) are as follows: This water [北高海; Caspian Sea] is so vast but is not connected with a sea, so it is suspected as a sea or a lake. However, since the water is salty,[233] we can call it a sea. This is the Caspian Sea which was known by Chinese as 北高海 (Item 93) because it is considered as a sea in the far north at a high latitude.

Endnotes

[1] Library of Congress. "Kun yu wan guo quan tu." www.loc.gov, Library of Congress, 2025, loc.gov/item/201058565

[2] Ronnie Po-Chia Hsia. *Matteo Ricci and the Catholic Mission to China, 1583–1610: A Short History with Documents (Passages: Key Moments in History)*. Indianapolis, IN, USA: Hackett Publishing Company, Inc., 2016.

[3] Sheng-Wei Wang. *Chinese Global Exploration in the Pre-Columbian Era: Evidence from an Ancient World Map*. Singapore: World Scientific Publishing Co., 2023.

[4] Peter B. Golden. *Central Asia in World History*. Oxford, UK: Oxford University Press, 2011.

[5] C. E. Bosworth. "Central Asia iv. In the Islamic Period up to the Mongols." In Ehsan Yarshater, ed. *Encyclopædia Iranica, V/2: Čehel Sotūn, Isfahan–Central Asia XIII*, Oxfordshire, UK: Routledge & Kegan Paul. 1990, pp. 169–172; an electronic version can be found at https://www.iranicaonline.org/articles/central-asia-iv/.

[6] "File:Map of Asia.svg." *Wikimedia Commons, The Free Media Depository*, Wikimedia Foundation, 22 Nov. 2021, upload.wikimedia.org/wikipedia/commons/thumb/4/40/Map_of_Asia.svg/2738px-Map_of_Asia.svg.png

[7] "File:Thrace and present-day state borderlines.png." *Wikimedia Commons,* 13 Mar. 2024, upload.wikimedia.org/wikipedia/commons/2/2b/Thrace_and_present-day_state_borderlines.png

[8] "File:Kunyu Wanguo Quantu by Matteo Ricci Plate 1-3.jpg." *Wikimedia Commons,* 13 June 2023, upload.wikimedia.org/wikipedia/commons/b/b8/Kunyu_Wanguo_Quantu_by_Matteo_Ricci_Plate_1-3.jpg

[9] "File:Mercator 1569 world map composite.jpg." *Wikimedia Commons*, 26 Nov. 2016, upload.wikimedia.org/wikipedia/commons/4/4b/Mercator_1569_world_map_composite.jpg

[10] "File:OrteliusWorldMap1570.jpg." *Wikipedia Commons*, 12 July 2022, upload.wikimedia.org/wikipedia/commons/e/e2/OrteliusWorldMap1570.jpg

[11] "File:1594 double hemisphere world map by Petrus Plancius.jpg." *Wikimedia Commons*, 2 Sept. 2022, upload.wikimedia.org/wikipedia/commons/0/0d/1594_double_hemisphere_world_map_by_Petrus_Plancius.jpg

[12] "File:1570 Ortelius Map of Asia (first edition) - Geographicus - AsiaeNovaDescriptio-ortelius.jpg." *Wikimedia Commons*, 22 Mar. 2011, upload.wikimedia.org/wikipedia/commons/a/a3/1570_Ortelius_Map_of_Asia_(first_edition)_-_Geographicus_-_AsiaeNovaDescriptio-ortelius.jpg

[13] "File:Abraham Ortelius - Tvrcici imperii descriptio.jpg." *Wikimedia Commons*, 4 July 2023, upload.wikimedia.org/wikipedia/commons/9/94/Abraham_Ortelius_-_Tvrcici_imperii_descriptio.jpg

[14] Abraham Ortelius. "Representation of the Turkish Empire. Turcici Imperii descriptio." www.loc.gov, Library of Congress, 2025, loc.gov/item/2021668700/

[15] "File:Persici Sive Sopho Rvm Regni Typvs.The Ortelius map of Persia, Antwerp, 1570.jpg." *Wikimedia Commons*, 6 Mar. 2024, upload.wikimedia.org/wikipedia/commons/1/1c/Persici_Sive_Sopho_Rvm_Regni_Typvs.The_Ortelius_map_of_Persia%2C_Antwerp%2C_1570.jpg

[16] "File:Tartariae sive Magni Chami Regni tÿpus. LOC 83690086.jpg." *Wikimedia Commons*, 5 June 2022, upload.wikimedia.org/wikipedia/commons/6/67/Tartariae_sive_Magni_Chami_Regni_tÿpus._LOC_83690086.jpg

[17] Fan Ye (398-446 A.D). John E. Hill, trans. "Chronicle on the Western Regions from the Hou Hanshu." https://depts.washington.edu/, University of Washington in Seattle, 2025. https://depts.washington.edu/silkroad/texts/hhshu/hou_han_shu.html

[18]Sima Qian (司馬遷; 134 B.C. –c. 86 B.C.). *Shi Ji • Juan Yi Bai Er Shi San • Da Yuan Lie Chuan* (史记 • 卷一百二十三 • 大宛列傳) or *Records of the Grand Historian, Juan (Volume) 123, Biography of Da Yuan*; an electronic file can be found at https://ctext.org/shiji/zh

[19]Sheng-Wei Wang. *The Last Journey of the San Bao Eunuch, Admiral Zheng He*. Hong Kong, China: Proverse Hong Kong, 2019.

[20]Luo Maodeng (羅懋登/罗懋登; lived c. 1597). *San Bao Tai Jian Xi Yang Gi* (三寶太監西洋記/三宝太监西洋记) or *An Account of the Western World Voyage of the San Bao Eunuch*; an electronic version can be found at https://www.diancang.xyz/wenxueyishu/sanbaotaijianxiyangji/

[21]Sheng-Wei Wang. "Chapter 3: Luo Maodeng's Tianfang/Yun Chong was Mecca in Saudi Arabia." *The Last Journey of the San Bao Eunuch, Admiral Zheng He, op. cit.*, pp. 91–138.

[22]Edward L. Dreyer has the date as January 17, 1433. The one-day difference is due to conversion from the Lunar calendar into Gregorian calendar; Dreyer, Edward L. "Zheng He's Seventh and Final Voyage, 1431–1433." *Zheng He: China and the Oceans in the Early Ming Dynasty, 1405–1433*, New York, NY, USA: Pearson Longman, 2007, pp. 150–163.

[23]Sheng-Wei Wang. "Chapter 3: Luo Maodeng's Tianfang/Yun Chong was Mecca in Saudi Arabia." *The Last Journey of the San Bao Eunuch, Admiral Zheng He, op. cit.*, 2019, pp. 91–138.

[24]Zhu Yunmeng (祝允明; 1461–1527). *Qian Wen Ji • Xia Xi Yang* (前闻記 •下西洋) or *A Record of History Once Heard: Down to the Western Ocean*; an electronic version can be found at https://ctext.org/wiki.pl?if=gb&chapter=203549

[25]*Ibid.*

[26]Luo Maodeng (羅懋登/罗懋登; lived c. 1597). *San Bao Tai Jian Xi Yang Gi* (三寶太監西洋記/三宝太监西洋记) or *An Account of the Western World Voyage of the San Bao Eunuch*; an electronic version can be found at https://www.diancang.xyz/wenxueyishu/sanbaotaijianxiyangji/

[27]Sheng-Wei Wang. "Chapter 3: Luo Maodeng's Tianfang/Yun Chong was Mecca in Saudi Arabia." *The Last Journey of the San Bao Eunuch, Admiral Zheng He, op. cit.*, pp. 91–138.

[28]Sheng-Wei Wang. "Chapter 2: Chinese Explored Australia, New Zealand, Land of Fire and Antarctica Long Before the Europeans." *Chinese Global Exploration in the Pre-Columbian Era: Evidence from an Ancient World Map, op. cit.*, pp. 52–61.

[29]Jennifer L. Gaynor. "Ages of Sail, Ocean Basins, and Southeast Asia." *Journal of World History*, vol. 24, no. 2, June 2013, pp. 309–333. Project MUSE, https://doi.org/10.1353/jwh.2013.0059.

[30]"File:Arabian Sea map.png." *Wikimedia Commons*, 31 Dec. 2023, upload.wikimedia.org/wikipedia/commons/4/41/Arabian_Sea_map.png

[31]Huang Shijian (黄时鉴) and Gong Yingyan (龚缨晏). *Li Ma Dou Shi Jie Di Tu Yan Jiu* (利玛窦世界地图研究) or *Research on Matteo Ricci's World Map*, Shanghai, China: Shanghai Chinese Classics Publishing House (上海古籍出版社), 2004, p. 197.

[32]"File:Arabian Peninsula Map.svg." *Wikimedia Commons*, 7 Nov. 2022, upload.wikimedia.org/wikipedia/commons/thumb/0/0c/Arabian_Peninsula_Map.svg/1854px-Arabian_Peninsula_Map.svg.png

[33]Huang Shijian (黄时鉴) and Gong Yingyan (龚缨晏). *Op. cit.*, p. 194.

[34]Government of Sharjah. "Pre-Islamic – SAA." https://saa.shj.ae/, Sharjah Archaeology Authority, 2025, saa.shj.ae/en/age-exc/pre-islamic/

[35]William Vincent, *et al.*, *The Voyage Of Nearchus From The Indus To The Euphrates: Collected From The Original Journal Preserved By Arrian, And Illustrated By Authorities Ancient And Modern*. Open Library: Legare Street Press, 2022.

[36]"File:Middle east.jpg." *Wikimedia Commons*, 10 Dec. 2023, upload.wikimedia.org/wikipedia/commons/1/13/Middle_east.jpg

[37]Saul Bernard Cohen. *Geopolitics of the World System*, Lanham, Maryland, USA: Rowman & Littlefield, 2003, p. 337.

[38]Elise Vernon Pearlstine. "A Brief History of Frankincense: Following a scent trail." www.laphamsquarterly.org, Lapham's Quarterly, 2025, laphamsquarterly.org/roundtable/brief-history-frankincense

[39]Huang Shijian (黄时鉴) and Gong Yingyan (龚缨晏). *Op. cit.*, p. 186.

[40]Huang Shijian (黄时鉴) and Gong Yingyan (龚缨晏). *Op. cit.*, p. 205.

[41]Huang Shijian (黄时鉴) and Gong Yingyan (龚缨晏). *Op. cit.*, p. 208.

[42]Huang Shijian (黄时鉴) and Gong Yingyan (龚缨晏). *Op. cit.*, p. 196.

[43]Huang Shijian (黄时鉴) and Gong Yingyan (龚缨晏). *Op. cit.*, p. 192.

[44]Seyyed Hossein Nasr. *Mecca the Blessed, Medina the Radiant: The Holiest Cities of Islam.* North Clarendon, VT, USA: Tuttle Publishing, 2013.

[45]Sheng-Wei Wang. "Chapter 3: Luo Maodeng's Tianfang/Yun Chong was Mecca in Saudi Arabia." *The Last Journey of the San Bao Eunuch, Admiral Zheng He, op. cit.*, pp. 91–138.

[46]Sheng-Wei Wang. "Chapter 3: Luo Maodeng's Tianfang/Yun Chong was Mecca in Saudi Arabia." *The Last Journey of the San Bao Eunuch, Admiral Zheng He, op. cit.*, p. 138.

[47]Sheng-Wei Wang. "Chapter 2: Ma Huan's *Tianfang*/Mo-jia was ancient Tunisia in North Africa." *The Last Journey of the San Bao Eunuch, Admiral Zheng He, op. cit.*, pp. 53–90.

[48]Ibn Battuta. Samuel Lee, trans. *The Travels of Ibn Battuta: In the Near East, Asia and Africa.* Mineola, New York, USA: Dover Publications, 2004.

[49]Sheng-Wei Wang. "Chapter 3: Luo Maodeng's Tianfang/Yun Chong was Mecca in Saudi Arabia." *The Last Journey of the San Bao Eunuch, Admiral Zheng He, op. cit.*, p. 138.

[50]H. Lammens. *Islam: Beliefs and Institutions*, Oxfordshire, England, UK: Routledge, 2013, p. 5.

[51]Huang Shijian (黄时鉴) and Gong Yingyan (龚缨晏). *Op. cit.*, p. 205.

[52]Huang Shijian (黄时鉴) and Gong Yingyan (龚缨晏). *Op. cit.*, p. 200.

[53]Huang Shijian (黄时鉴) and Gong Yingyan (龚缨晏). *Op. cit.*, p. 198.

[54]Huang Shijian (黄时鉴) and Gong Yingyan (龚缨晏). *Op. cit.*, p. 189

[55]Huang Shijian (黄时鉴) and Gong Yingyan (龚缨晏). *Op. cit.*, p. 192.

[56]Huang Shijian (黄时鉴) and Gong Yingyan (龚缨晏). *Op. cit.*, p. 196.

[57]Huang Shijian (黄时鉴) and Gong Yingyan (龚缨晏). *Op. cit.*, p. 199.

[58]David Peacock and Andrew Peacock. "The Enigma of 'Aydhab: a Medieval Islamic Port on the Red Sea Coast." *International Journal of Nautical Archaeology*, vol. 37, no. 1, 2008, pp. 32–48. https://www.tandfonline.com/doi/full/10.1111/j.1095-9270.2007.00172.x

[59]"File:Egypt adm location map.svg." *Wikimedia Commons*, 20 May 2023, upload.wikimedia.org/wikipedia/commons/thumb/e/e7/Egypt_adm_location_map.svg/2221px-Egypt_adm_location_map.svg.png

[60]Lawrence A. Brown and R. W. McColl, eds. *Encyclopedia of World Geography*, New York, NY, USA: Facts on File, 2005, p. 237.

[61]David Bridger and Samuel Wolk. *The New Jewish Encyclopedia*, Millburn, NJ, USA: Behrman House, Inc., 1976, p. 109.

[62]"File:Israel location map.svg." *Wikimedia Commons*, 9 Oct. 2023, upload.wikimedia.org/wikipedia/commons/thumb/1/17/Israel_location_map.svg/1102px-Israel_location_map.svg.png

[63]R. W. Mccoll, ed. *Encyclopedia Of World Geography*, Alphen aan den Rijn, Netherlands: Facts on File, 2005.

[64]Huang Shijian (黄时鉴) and Gong Yingyan (龚缨晏). *Op. cit.*, p. 206.

[65]M. L. Steiner and A. E. Killebrew. *The Oxford Handbook of the Archaeology of the Levant: C. 8000–332 BCE*, Oxford, England, UK: Oxford University Press, 2014, p. 2.

[66]Trevor Bryce. *Ancient Syria: A Three Thousand Year History*. Oxford, England, UK: Oxford University Press, 2014.

[67]*Ibid.*

[68]Huang Shijian (黃时鉴) and Gong Yingyan (龚缨晏). *Op. cit.*, p. 192.

[69]Gordon Clouser. *Jesus, Joshua, Yeshua of Nazareth Revised and Expanded Paperback*. Bloomington, IN, USA: iUniverse , 2011; H. H. Ben-Sasson. *A History of the Jewish People*. Cambridge, MA, USA: Harvard University Press, 1976, p. 334.

[70]Doron Bar. "Rural Monasticism as a Key Element in the Christianization of Byzantine Palestine." *The Harvard Theological Review*, vol. 98, no. 1, 2005, 49–65. https://www.researchgate.net/publication/ 231951606_Rural_Monasticism_as_a_Key_Element_in_the_Christianization_of_Byzantine_Palestine

[71]Ellenblum Ronnie. *Frankish Rural Settlement in the Latin Kingdom of Jerusalem*. Cambridge, UK: Cambridge University Press, 2010, pp. 225–229; Michael Ehrlich. "Judea and Jerusalem." *The Islamization of the Holy Land, 634–1800*, York, UK.: Arc Humanities Press, 2022, pp. 111–112.

[72]Zhang Tingyu (張廷玉; 1672–1755), *et al.*, *Ming Shi • Juan San Bai Er Shi Liu • Lie Chuan Di Er Bai Shi Si • Wai Guo Qi* (明史 • 卷三百二十六 • 列傳第二百十四 • 外國七) or *History of Ming, Juan (Volume) 326, Liezhuan (US type of biography) 214, Foreign Country 7*; an electronic version can be found at https:// ctext.org/wiki.pl?if=gb&res=410835

[73]"File:Tigr-euph.png." *Wikimedia Commons*, 12 Dec. 2022, upload.wikimedia.org/wikipedia/commons/ 0/06/Tigr-euph.png

[74]Huang Shijian (黃时鉴) and Gong Yingyan (龚缨晏). *Op. cit.*, p. 192.

[75]Alan Palmer. *The Decline and Fall of the Ottoman Empire*. London, England, UK: Faber and Faber, 2011.

[76]"File:Map of Lycia.jpg." *Wikimedia Commons*, 16 Oct. 2020, upload.wikimedia.org/wikipedia/commons/b/ b2/Map_of_Lycia.jpg

[77]Huang Shijian (黃时鉴) and Gong Yingyan (龚缨晏). *Op. cit.*, p. 201.

[78]Huang Shijian (黃时鉴) and Gong Yingyan (龚缨晏). *Op. cit.*, p. 203.

[79]Norman Stone. *Turkey: A Short History*. London, UK: Thames & Hudson, 2014.

[80]*Ibid.*

[81]Richard Kemeny. "The surprising source of Turkey's volcanoes lies more than 1,000 miles away." www. nationalgeographic.com, National Geographic, 2024, nationalgeographic.com/premium/article/surprising- magma-source-turkeys-volcanoes-1000-miles-away

[82]*Ibid.*

[83]Huang Shijian (黃时鉴) and Gong Yingyan (龚缨晏). *Op. cit.*, p. 201.

[84]Stanford J. Shaw. *History of the Ottoman Empire and Modern Turkey: Volume 1, Empire of the Gazis: The Rise and Decline of the Ottoman Empire 1280–1808*. Cambridge, UK: Cambridge University Press, 1976.

[85]"File:Anatolia1300.png." *Wikimedia Commons*, 10 Aug. 2023, upload.wikimedia.org/wikipedia/ commons/c/cc/Anatolia1300.png

[86]Huang Shijian (黃时鉴) and Gong Yingyan (龚缨晏). *Op. cit.*, p. 195.

[87]Huang Shijian (黃时鉴) and Gong Yingyan (龚缨晏). *Op. cit.*, p. 199.

[88]Huang Shijian (黃时鉴) and Gong Yingyan (龚缨晏). *Op. cit.*, p. 190.

[89]William Smith. "Ariana." *Dictionary of Greek and Roman Geography*, Boston, MA, USA: Little, Brown and Company, 1980, pp. 210–211.

[90]Britannica. *The New Encyclopædia Britannica, 15th ed., Vol. 1*. Chicago, USA: Encyclopædia Britannica Inc., 1992, p. 313.

[91]Huang Shijian (黃时鉴) and Gong Yingyan (龚缨晏). *Op. cit.*, p. 196.

[92]Huang Shijian (黃时鉴) and Gong Yingyan (龚缨晏). *Op. cit.*, p. 208.

[93]Huang Shijian (黄时鉴) and Gong Yingyan (龚缨晏). *Op. cit.*, p. 196.
[94]Huang Shijian (黄时鉴) and Gong Yingyan (龚缨晏). *Op. cit.*, p. 184.
[95]"Some Medieval and Post-Golden Horde's Towns of the Itil (Volga) and Syr-Darya Basins According to the Arabic and Chinese Maps." Golden Horde Review, vol. 9, no. 3, 2021, pp. 611–653. DOI: 10.22378/2313-6197.2021-9-3.611–653.
[96]*Ibid.*
[97]Huang Shijian (黄时鉴) and Gong Yingyan (龚缨晏). *Op. cit.*, p. 206.
[98]Huang Shijian (黄时鉴) and Gong Yingyan (龚缨晏). *Op. cit.*, p. 195.
[99]Huang Shijian (黄时鉴) and Gong Yingyan (龚缨晏). *Op. cit.*, p. 190.
[100]Albrecht Classen. *Handbook of Medieval Studies: Terms – Methods – Trends*, Berlin, Germany: Walter de Gruyter, 2010, p. 6.
[101]"File:SakastanMap.jpg." *Wikimedia Commons*, 7 Dec. 2023, upload.wikimedia.org/wikipedia/commons/2/2a/SakastanMap.jpg
[102]"File:N-Mesopotamia and Syria english.svg." *Wikimedia Commons*, 31 Jan. 2024, upload.wikimedia.org/wikipedia/commons/thumb/e/ed/N-Mesopotamia_and_Syria_english.svg/3281px-N-Mesopotamia_and_Syria_english.svg.png
[103]Paul Kriwaczek. *Babylon: Mesopotamia and the Birth of Civilization*. Open Library: St. Martin's Griffin, 2012.
[104]Michael Seymour. "Ancient Mesopotamia and Modern Iraq in the British Press, 1980–2003." *Current Anthropology*, vol. 45, np. 3, 2004, pp. 351–368. https://www.jstor.org/stable/10.1086/383004
[105]Susan Pollock. *Ancient Mesopotamia. The Eden that never was, Case Studies in Early Societies*. Cambridge, UK: Cambridge University Press, 1999, p. 1.
[106]Huang Shijian (黄时鉴) and Gong Yingyan (龚缨晏). *Op. cit.*, p. 197.
[107]Huang Shijian (黄时鉴) and Gong Yingyan (龚缨晏). *Op. cit.*, p. 195.
[108]Huang Shijian (黄时鉴) and Gong Yingyan (龚缨晏). *Op. cit.*, p. 199.
[109]Huang Shijian (黄时鉴) and Gong Yingyan (龚缨晏). *Op. cit.*, p. 195.
[110]Mack Chahin. *The Kingdom of Armenia*. Open Library: 1991.
[111]Martin D. Stringer. *A Sociological History of Christian Worship*. Cambridge, UK: Cambridge University Press, 2005.
[112]Rouben Paul Adalian. *Historical Dictionary of Armenia*. Lanham, MD, USA: Scarecrow Press, 2010; Tom Allen and Deirdre Holding. *Armenia & Nagorno Karabagh*. Chalfont St. Peter, Buckinghamshire, UK: Bradt Travel Guides, 2013.
[113]Europa Publications. Imogen Bell, ed. *Eastern Europe, Russia and Central Asia 2003*. Milton Park, Abingdon-on-Thames, Oxfordshire, England, UK: Taylor & Francis, 2002.
[114]Huang Shijian (黄时鉴) and Gong Yingyan (龚缨晏). *Op. cit.*, p. 206.
[115]Gennady Chufrin, ed. *The Security of the Caspian Sea Region*. Stockholm, Sweden: Stockholm International Peace Research Institute, 2002.
[116]Stephen H. Rapp Jr. "Georgia, Georgians, until 1300." In Kate Fleet, *et al.*, eds. *Encyclopaedia of Islam, THREE*. Wang. Consulted online on 21 February 2024; http://dx.doi.org/10.1163/1573-3912_ei3_COM_32056; First published online: 2020; First print edition: 9789004413481, 2020, 2020-6 Brill Online, 2002.
[117]"File:Kingdom of Georgia (1210s) Sphere of influence.png." *Wikimedia Commons*, 26 Nov. 2022, upload.wikimedia.org/wikipedia/commons/f/f7/Kingdom_of_Georgia_(1210s)_Sphere_of_influence.png
[118]Huang Shijian (黄时鉴) and Gong Yingyan (龚缨晏). *Op. cit.*, p. 204.
[119]David Blow. Shah Abbas: *The Ruthless King Who Became an Iranian Legend*. London, UK: I.B.Tauris, 2009.

[120]Richard L. Scheffel and Susan J. Wernert. *Natural Wonders of the World.* New York, NY, USA: Reader's Digest, 1980.

[121]V. Minorsky. "Luristān." In *Encyclopaedia of Islam, Second Edition.* P. Bearman, *et al.*, eds. Consulted online on 22 February 2024 <http://dx.doi.org/10.1163/1573-3912_islam_COM_0588>

[122]Huang Shijian (黄时鉴) and Gong Yingyan (龚缨晏). *Op. cit.*, p. 187.

[123]Joshua A. Fishman. *Handbook of Language and Ethnic Identity: Disciplinary and Regional Perspectives. Vol. 1*, Oxford, UK: Oxford University Press, 2010, p. 266.

[124]RICHARD N. FRYE. "Greater Iran: A 20th-Century Odyssey." *Inter39, no. Journal of Middle East Studies*, vol. 39, no. 2, 2007, pp 307–309. DOI: https://doi.org/10.1017/S0020743807070195; Mehrdad Ghodrat-Dizaji. "Remarks on the Location of the Province of Parthia in the Sasanian Period." In Vesta Sarkhosh Curtis, *et al.*, *The Parthian and Early Sasanian Empires: adaptation and expansion*, Barnsley, UK: Oxbow Books, 2016, pp. 42–46. https://www.academia.edu/7326142/_Remarks_on_the_Location_of_the_Province_of_Parthia_in_the_Sasanian_Period_The_Parthian_and_Early_Sasanian_Empires_Adaptation_and_Expansion_eds_Vesta_Sarkhosh_Curtis_et_al_Oxford_and_Philadelphia_Oxbow_Books_2016_pp_42_46

[125]Vesta Sarkhosh Curtis and Sarah Stewart, eds. *The Age of the Parthians, 2nd edn.* London, UK: I. B. Tauris, 2010.

[126]George Saliba. *A History of Arabic Astronomy: Planetary Theories During the Golden Age of Islam.* New York, NY, USA: New York University Press, 1994, pp. 245, 250, 256–257.

[127]James Chambers. *The Devil's Horsemen: The Mongol Invasion of Europe.* New York, NY, USA: Atheneum, 1979.

[128]"File:Median Empire.jpg." *Wikimedia Commons*, 4 Dec. 2023, upload.wikimedia.org/wikipedia/commons/c/c6/Median_Empire.jpg

[129]Percy Sykes. *A History of Persia*, London, UK: Macmillan and Company, 1921, p. 5.

[130]Muhammad A. Dandamaev and Vladimir G. Lukonin. D. J. Dadson and Philip L. Kohl, translators. *The Culture and Social Institutions of Ancient Iran*, Cambridge, UK: Cambridge University Press, 2004, pp. 1–5.

[131]*Ibid.*

[132]H. Yule. *The Book of Ser Marco Polo. Vol. 1.* Open Library: South Asia Books, 1993.

[133]"File:Kingdom of Ormuz.png." *Wikimedia Commons*, 22 Jan. 2024, upload.wikimedia.org/wikipedia/commons/a/ab/Kingdom_of_Ormuz.png

[134]Vasilĭ Vladimirovich and W. Barthold. Svat Soucek, trans. *An Historical Geography of Iran.* Princeton, NJ, USA: Princeton University Press, 1984. https://www.jstor.org/stable/j.ctt7zvv3z

[135]Huang Shijian (黄时鉴) and Gong Yingyan (龚缨晏). *Op. cit.*, p. 198.

[136]Huang Shijian (黄时鉴) and Gong Yingyan (龚缨晏). *Op. cit.*, p. 186.

[137]Huang Shijian (黄时鉴) and Gong Yingyan (龚缨晏). *Op. cit.*, p. 188.

[138]Huang Shijian (黄时鉴) and Gong Yingyan (龚缨晏). *Op. cit.*, p. 193.

[139]Huang Shijian (黄时鉴) and Gong Yingyan (龚缨晏). *Op. cit.*, p. 205.

[140]Xavier de Planhol and Bernard Hourcade, "KERMAN ii. Historical Geography." *Encyclopædia Iranica*, XVI/3, pp. 251–265. http://www.iranicaonline.org/articles/kerman-historical-geography

[141]Huang Shijian (黄时鉴) and Gong Yingyan (龚缨晏). *Op. cit.*, p. 192.

[142]Huang Shijian (黄时鉴) and Gong Yingyan (龚缨晏). *Op. cit.*, p. 200.

[143]"File:Chorasan-Transoxanien-Choresmien neu.svg." *Wikimedia Commons*, 25 Mar. 2024, upload.wikimedia.org/wikipedia/commons/thumb/b/b4/Chorasan-Transoxanien-Choresmien_neu.svg/2020px-Chorasan-Transoxanien-Choresmien_neu.svg.png

[144]Huang Shijian (黄时鉴) and Gong Yingyan (龚缨晏). *Op. cit.*, p. 191.

[145]Giovanni Botero (1540–1617) and Robert, fl. Johnson (1586–1626). "Persia." *Relations of the World*, Early English Books Online Text Creation Partnership 2011, p. 567, https://quod.lib.umich.edu/e/eebo/A16489.0001.001/1:4.4.10?rgn=div3;view=fulltext

[146]Isidore of Charax. "Parthian Stations." www.parthia.com, Edward C. D. Hopkins, 2025, parthia.com/doc/parthian_stations.htm

[147]Nurlan Kenzheakhmet and Alpamys Abu. "Some Medieval and Post-Golden Horde's Towns of the Itil (Volga) and Syr-Darya Basins According to the Arabic and Chinese Maps." *Golden Horde Review*, vol. 9, no. 3, 2021, p. 622. https://www.researchgate.net/publication/354971343_Some_Medieval_and_Post-Golden_Horde's_Towns_of_the_Itil_Volga_and_Syr-Darya_Basins_According_to_the_Arabic_and_Chinese_Maps

[148]Huang Shijian (黄时鉴) and Gong Yingyan (龚缨晏). *Op. cit.*, p. 195.

[149]John Ogilby (1600–1676). "The Description of the Kingdom of Persia." *Relations of the World*, Early English Books Online Text Creation Partnership 2011, p. 21, https://quod.lib.umich.edu/e/eebo2/A53223.0001.001/1:7.4?rgn=div2;view=fulltext

[150]*Ibid.*

[151]*Ibid.*

[152]*Ibid.*

[153]Assef Ashraf. *Making and Remaking Empire in Early Qajar Iran.* Cambridge, UK: Cambridge University Press, 2024.

[154]Michael Axworthy. A History of Iran: Empire of the Mind. New York, NY, USA: Basic Books, 2016.

[155]Homa Katouzian. *The Persians: Ancient, Mediaeval and Modern Iran.* New Haven, CT, USA: Yale University Press, 2010.

[156]Elizabeth Carter and Matthew W. Stolper. *Elam: surveys of political history and archaeology*, Berkeley, CA, USA: University of California Press, 1984, p. 3.

[157]Homa Katouzian. *Op. cit.*

[158]*Ibid.*

[159]"File:IranOMC.png." *Wikimedia Commons*, 8 Mar. 2023, upload.wikimedia.org/wikipedia/commons/9/94/IranOMC.png

[160]Nile Green. *The Persianate World: The Frontiers of a Eurasian Lingua Franca.* Berkeley, CA, USA: University of California Press, 2019.

[161]Embassy of the Republic of China in the Islamic Republic of Iran. "A Brief History of Iran (伊朗简史)." http://ir.china-embassy.gov.cn/, Embassy of the Republic of China in the Islamic Republic of Iran, 2025, ir.china-embassy.gov.cn/zjyl/gjgk/200412/t20041216_2520588.htm

[162]Beaujard Philippe. "Western Asia: Revival of the Persian Gulf." *The Worlds of the Indian Ocean*, Canbridge, UK: Cambridge University Press, 2019, pp. 515–521.

[163]Dickran Kouymjian. "Armenia from the fall of the Cilician Kingdom (1375) to the forced emigration under Shah Abbas." In Richard G. Hovannisian, ed. *The Armenian People From Ancient to Modern Times, Volume I: The Dynastic Periods: From Antiquity to the Fourteenth Century.* London, UK: Palgrave Macmillan, 2004.

[164]Daniel T. Potts. *Nomadism in Iran: From Antiquity to the Modern Era.* Oxford, UK: Oxford University Press, 2014.

[165]Economic and Commercial Office of the Embassy of the People's Republic of China in the Islamic Republic of Iran. "The History of Iran (伊朗历史沿革)." http://ir.mofcom.gov.cn/, Ministry of Commerce. PRC, 2025, https://ir.mofcom.gov.cn/ylgk/art/2017/art_a9171db367834c6c84e1b889de5d2e58.html

[166]Economic and Commercial Office of the Embassy of the People's Republic of China in the Islamic Republic of Iran. *Op. cit.*

[167]Huang Shijian (黄时鉴) and Gong Yingyan (龚缨晏). *Op. cit.*, p. 190.

[168]Huang Shijian (黄时鉴) and Gong Yingyan (龚缨晏). *Op. cit.*, p. 202.

[169]Christopher Brunner. "Geographical and Administrative divisions: Settlements and Economy." In Ehsan Yarshater, ed. *The Cambridge History of Iran, Volume 3(2): The Seleucid, Parthian and Sasanian Periods*, Cambridge, UK: Cambridge University Press, 1983, pp. 747–778.

[170]Peter Webb. "Tabarestan." In Oliver Nicholson, ed. *The Oxford Dictionary of Late Antiquity*. Oxford, UK: Oxford University Press, 2018.

[171]"File:Northern Iran and its surroundings during the Iranian intermezzo.svg." *Wikimedia Commons*, 14 Oct. 2022, upload.wikimedia.org/wikipedia/commons/thumb/0/08/Northern_Iran_and_its_surroundings_during_the_Iranian_intermezzo.svg/3562px-Northern_Iran_and_its_surroundings_during_the_Iranian_intermezzo.svg.png

[172]Huang Shijian (黄时鉴) and Gong Yingyan (龚缨晏). *Op. cit.*, p. 184.

[173]Charles Melville. "The Ilkhan Öljeitü's conquest of Gilan (1307): rumour and reality." In R. Amitai-Preiss and D.O. Morgan, eds. *The Mongol empire and its legacy*, Leiden, Netherlands: Brill Publishers, 2000, pp. 73–125.

[174]D. E. Pitcher. *An Historical Geography of the Ottoman Empire: From Earliest Times to the End of the Sixteenth Century*, Leiden, Netherlands: Brill, 1972, p. 132.

[175]Heritage Database (遗产数据库). "Tobastan (陀拔斯单)." http://www.silkroads.org.cn/, Silk Roads World Heritage (丝绸之路世界遗产), 2025, http://www.silkroads.org.cn/portal.php?mod=view&aid=45630

[176]History Bits. "The Arabic Empire – 632–1258." www.historybits.com, History Bits, 2025, www.historybits.com/world-history/the-arab-empire/

[177]Huang Shijian (黄时鉴) and Gong Yingyan (龚缨晏). *Op. cit.*, p. 203.

[178]John Strange. Caphtor/Keftiu: A New Investigation. Leiden, Netherlands: Brill, 1980, p. 167.

[179]Staff Writer. "Biggest Islands In The Mediterranean Sea By Area." www.worldatlas.com, WorldAtlas, 2025, worldatlas.com/articles/biggest-islands-in-the-mediterranean-sea.html

[180]Helen J. Nicholson. "Remembering the Crusaders in Cyprus: The Lusignans, the Hospitallers, and the 1191 Conquest of Cyprus in Jean d'Arras's Mélusine." In Simon T. Parsons and Linda M. Paterson, eds. *Literature of the Crusades*, Cambridge, UK: Cambridge University Press, 2018, pp. 158–172. doi:10.1017/9781787441736.011

[181]John Julius Norwich. *A History of Venice.* New York, NY, USA: Alfred A. Knopf, 1982.

[182]*Ibid.*

[183]Huang Shijian (黄时鉴) and Gong Yingyan (龚缨晏). *Op. cit.*, p. 209.

[184]*Fo Guang Dictionary* (佛光大辭典). "Tocharian Kingdom (睹貨邏國)." http://m.fodizi.tw/, Buddhist Disciple Library (佛弟子文庫), 2025, http://m.fodizi.tw/f05/79147.html

[185]Ji Xianlin (季羨林; 1911–2009), ed. *Da Tang Xi Yu Ji xiao Zhu • Juan Yi • Du Huo Luo Guo* (大唐西域記校注) • 卷一 • 睹貨邏國 or *Records of the Western Regions of the Tang Dynasty, Juan (Volume)1: Du Huo Luo Guo.* Beijing, China: Zhonghua Book Company (中华书局有限公司), 2000.

[186]Howard J. Wechsler. "T'ai-tsung (reign 624–49) the consolidator." In Dennis Twitchett, ed. *The Cambridge History of China, Volume 3: Sui and T'ang China, 589–906 AD, Part 1*, Cambridge, UK: Cambridge University Press, 1979, pp. 188–241.

[187]Ji Xianlin (季羨林; 1911-2009), ed. *Op. cit.*

[188]Huang Shijian (黄时鉴) and Gong Yingyan (龚缨晏). *Op. cit.*, p. 190.

[189]Lin Gan (林幹). *History of Turks and Uighurs* (突厥與回紇史). Inner Mongolia, China: Inner Mongolia People's Publishing House (內蒙古人民出版社), 2007.

[190]Zhang Zhao (張昭; 156—236), *et al.*, eds. *Jiu Tang Shu • Juan Yi Bai Jiu Shi Wu • Lie Chuan Di Yi Bai Si Shi Wu* (舊唐書 • 卷一百九十五 • 列傳第一百四十五) or *Old Book of Tang, Juan (Volume) 195, Biography 145.* An electronic version can be found at https://ctext.org/wiki.pl?if=gb&res=456206

[191]Nurlan Kenzheakhmet and Alpamys Abu. *Op. cit.* p. 622.

[192]Huang Shijian (黄时鉴) and Gong Yingyan (龚缨晏). *Op. cit.*, p. 190.

[193]Zheng Fenduo (郑芬多). "Where is the ancient Xuandu Mountain (古代的悬度山在哪里)?" https://zhidao.baidu.com, Baidu, 2025, zhidao.baidu.com/question/1496274625163458459.html

[194]"File:Tartariae sive Magni Chami Regni tÿpus. LOC 83690086.jpg." *Wikimedia Commons*, 5 June 2022, upload.wikimedia.org/wikipedia/commons/6/67/Tartariae_sive_Magni_Chami_Regni_tÿpus._LOC_83690086.jpg

[195]"File:Caucasus topographic map-en.svg." *Wikimedia Commons*, 15 Feb. 2024, upload.wikimedia.org/wikipedia/commons/thumb/1/17/Caucasus_topographic_map-en.svg/2560px-Caucasus_topographic_map-en.svg.png

[196]Chris R. Stokes, *et al.*, eds. *Encyclopedia of Snow, Ice and Glaciers*, Berlin/Heidelberg, Germany: Spring Science & Business Media, 2011, p. 127.

[197]"File:Tartariae sive Magni Chami Regni tÿpus. LOC 83690086.jpg." *Wikimedia Commons*, 5 June 2022, upload.wikimedia.org/wikipedia/commons/6/67/Tartariae_sive_Magni_Chami_Regni_tÿpus._LOC_83690086.jpg

[198]Huang Shijian (黄时鉴) and Gong Yingyan (龚缨晏). *Op. cit.*, p. 190.

[199]Nurlan Kenzheakhmet and Alpamys Abu. *Op. cit.*, p. 622.

[200]cat1208. "The Pamirs are called 'Congling or Gree-Onion Plateau, meaning 葱岭'"; is it because there are many green onions growing there (帕米尔高原被叫做'葱岭', 是因为有很多大葱吗)?" http://www.360doc.com, National Humanity History, 2025, http://www.360doc.com/content/21/0823/23/7442640_992337316.shtml

[201]"File:High Asia Mountain Ranges.jpg." *Wikimedia Commons*, 14 Nov. 2022, upload.wikimedia.org/wikipedia/commons/9/9e/High_Asia_Mountain_Ranges.jpg

[202]*Ibid.*

[203]Huang Shijian (黄时鉴) and Gong Yingyan (龚缨晏). *Op. cit.*, p. 185.

[204]Kathryn Babayan. *The Waning of the Qizilbash: The Spiritual and the Temporal in Seventeenth Century Iran.* Princeton, NJ, USA: Princeton University, 1993, pp. 1–6 and 41–47.

[205]Charles H. Parker. *Global Interactions in the Early Modern Age, 1400–1800*, Cambridge, UK: Cambridge University Press, 2010, p. 53.

[206]Roger M. Savory. "Kizil-Bash." *In Encyclopaedia of Islam, Vol. 5,* 1993, pp. 243–245.

[207]Huang Shijian (黄时鉴) and Gong Yingyan (龚缨晏). *Op. cit.*, p. 192.

[208]Huang Shijian (黄时鉴) and Gong Yingyan (龚缨晏). *Op. cit.*, p. 206.

[209]Huang Shijian (黄时鉴) and Gong Yingyan (龚缨晏). *Op. cit.*, p. 201.

[210]"File:Tarimbecken 3. Jahrhundert.png." *Wikimedia Commons*, 29 July 2023, upload.wikimedia.org/wikipedia/commons/d/d0/Tarimbecken_3._Jahrhundert.png

[211]Luther Carrington Goodrich and Zhaoying Fang. "Bento de Góis." *Dictionary of Ming Biography. Vol. 1: 1368–1644,* New York, NY, USA: Columbia University Press, 1976, pp. 472–473.

[212]"File:CEM-36-NW-corner.jpg." *Wikimedia Commons*, 17 Jan. 2024, upload.wikimedia.org/wikipedia/commons/5/56/CEM-36-NW-corner.jpg

[213]Huang Shijian (黄时鉴) and Gong Yingyan (龚缨晏). *Op. cit.*, p. 187.

[214]Zhang Tingyu (張廷玉; 1672–1755), *et al.*, *Ming Shi • Juan San Bai Yi Shi Wu • Lie Chuan Di Er Bai San • Yun Nan Tu Si San* (明史 • 卷三百二十九 • 列傳第二百十七 • 西域一) or *History of Ming, Juan (Volume)*

329, Liezhuan (US type of biography) 217, Western Regions 1; an electronic version can be found at https://ctext.org/wiki.pl?if=gb&res=410835

[215]*Ibid.*

[216]*Ibid.*

[217]*Ibid.*

[218]Denis Sinor. *Inner Asia*, Oxfordshire, England, UK: Routledge, 1997, p. 121.

[219]J. P. Mallory. "The Problem of Tocharian Origins: An Archaeological Perspective." In Victor H. Mair, ed. *Sino-Platonic Papers, Vol. 259*. Philadelphia, PA. USA: Department of East Asian Languages and Civilizations, University of Pennsylvania, 2015, pp. 1–63.

[220]Charles Eliot. *Hinduism and Buddhism: An Historical Sketch, Vol. 1*. Open Library, Qontro Classic Books, 2010.

[221]Zhang Tingyu (張廷玉; 1672–1755), *et al.*, *Ming Shi • Juan San Bai Yi Shi Wu • Lie Chuan Di Er Bai San • Yun Nan Tu Si San* (明史 • 卷三百二十九 • 列傳第二百十七 • 西域一) or *History of Ming, Juan (Volume) 329, Liezhuan (US type of biography) 217, Western Regions 1*; an electronic version can be found at https://ctext.org/wiki.pl?if=gb&res=410835

[222]Muqi Che. *The Silk Road, Past and Present*, Beijing, China: Foreign Languages Press, 1989, p. 115.

[223]Sima Qian (司馬遷; 134 B.C. –c. 86 B.C.). *Shi Ji • Juan Yi Bai Er Shi San • Da Yuan Lie Chuan* (史记 • 卷一百二十三 • 大宛列傳) or *Records of the Grand Historian, Juan (Volume) 123, Biography of Da Yuan*; an electronic file can be found at https://ctext.org/shiji/zh

[224]Judy Bonavia. *The Silk Road: Xi'an to Kashgar. Odyssey Guides. Revised by Christoph Baumer (Reprint edn.)*. Hong Kong, China: Air Photo International, 2004, p. 236.

[225]UNESCO. "Silk Roads: the Routes Network of Chang'an-Tianshan Corridor." https://whc.unesco.org/, World Heritage Convention, 2025, whc.unesco.org/en/list/1442/multiple=1&unique_number=1985

[226]Huang Shijian (黃时鉴) and Gong Yingyan (龚缨晏). *Op. cit.*, p. 200.

[227]Chief Editorial Board. *Иссык-Куль.Нарын:Энциклопедия [Encyclopedia of Issyk-Kul and Naryn Oblasts] (in Russian)*, Bishkek, Kyrgyzstan: Chief Editorial Board of Kyrgyz Soviet Encyclopedia, 1991. p. 512.

[228]UNESCO. "Silk Roads: the Routes Network of Chang'an-Tianshan Corridor." *Op. cit.*

[229]"File:Syrdaryamap.png." *Wikimedia Commons*, 21 Feb. 2022, upload.wikimedia.org/wikipedia/commons/3/33/Syrdaryamap.png

[230]"File:Aral Sea watershed.png." *Wikimedia Commons*, upload.wikimedia.org/wikipedia/commons/7/73/Aral_Sea_watershed.png

[231]"File:CaspianSeaDrainage v1.png." *Wikimedia Commons*, 7 Feb. 2024, upload.wikimedia.org/wikipedia/commons/e/ee/CaspianSeaDrainage_v1.png

[232]"File:Caucasus topographic map-en.svg." *Wikimedia Commons*, 15 Feb. 2024, upload.wikimedia.org/wikipedia/commons/thumb/1/17/Caucasus_topographic_map-en.svg/2560px-Caucasus_topographic_map-en.svg.png

[233]Nikolay Esin, *et. al.* "On the origin of salt in the Caspian Sea." https://ui.adsabs.harvard.edu/, 19th EGU General Assembly, EGU2017, proceedings from the conference held 23-28 April, 2017 in Vienna, Austria., p.18414, ui.adsabs.harvard.edu/abs/2017EGUGA..1918414E/abstract

Chapter 7

An Ancient World Map Depicts Chinese Exploration of Asia to the North of China in 1428–1433, Long Before the First Europeans Explored the Region

Abstract

This chapter reports that a Chinese-based world map — the Kunyu Wanguo Quantu (坤輿萬國全圖/坤輿万国全图) (abbreviated as KWQ) or Complete Geographical Map of All the Kingdoms of the World published by Matteo Ricci in 1602 in China — reveals that the region of Asia to the north of China (abbreviated as ANC-KWQ) was explored by Chinese in the era 1428–1438 and that the map is of Chinese origin.

Since the entire KWQ has been deduced in my previous book titled *Chinese Global Exploration in the Pre-Columbian Era: Evidence from an Ancient World Map* as a map drawn in 1433 and the town of Tyumen was founded by Taibuga (泰布加) probably sometime between 1405 and 1428, this implies that the region was explored in the era 1428–1433.

These conclusions are based on analysing the locations and histories of the complete 102 geographical items and the fourteen annotations depicted on the ANC-KWQ and on comparing them with six major sixteenth-century European maps discussed in this chapter. Out of the 102 items, 57 of them (c. 56%) are not present on any of these European maps; and among the same 102 geographical items, 21 are of Chinese origin.

Keywords: Asia to the North of China, China, Kunyu Wanguo Quantu, Ming Dynasty, Oirats, Siberia, Tartary

1. Introduction

For over four hundred years, the world map — Kunyu Wanguo Quantu (坤輿萬國全圖/坤輿万国全图) (abbreviated as KWQ) or Complete Geographical Map of All the Kingdoms of the World — written with Chinese characters and having latitudinal and longitudinal lines, has generally been regarded as a map drawn in 1602 by an Italian Jesuit priest Matteo Ricci and his Chinese collaborators, based on the European maps which Ricci brought with him to China in 1582.[1] In the previous chapters of this book and in my earlier book,[2] I have analysed and provided evidence to show that geographical information of most Asia and non-Asia portions of the KWQ is based on the knowledge obtained by the Ming (明代; 1388–1644) mariners in the 1420s and 1430s during their sixth and seventh voyages to the Western Ocean led by Admiral Zheng He (郑和; 1371–1433/1435), whereas the information of the European portion and the Japanese portion is based on Chinese exploratory data gathered in the earlier Southern Song Dynasty (南宋; 1127–1279) and in 897 A.D. (nearly 650 years before the arrival of the first Europeans), respectively. This means that all these portions of the KWQ used Chinese maps as source maps (these no longer exist) and that they were not copied from other contemporary European maps.

In this chapter, I shall continue my analyses for the region of Asia to the north of China on the KWQ (abbreviated as ANC-KWQ) using the same approach: (1) thoroughly analysing the locations and histories of the 102 geographical items and the fourteen original annotations depicted on the map; and (2) comparing the information obtained on the ANC-KWQ with those on the major sixteenth-century European maps to determine the origin and the era of the ANC-KWQ. Note: All the Chinese characters in the main text and in the Appendix are expressed first with the Chinese traditional characters, then followed by the Chinese simplified characters, separated by a slash "/". When both characters are the same, that expression is presented only once.

2. The Map of Asia to the North of Ancient China is of Chinese Origin

First, we need to know the western boundary of the ANC-KWQ. Figure 7.1 shows that the western boundary of the ANC-KWQ starts just to the east of 矮人國/矮人国 (Kingdom of Dwarfs; Item 80 in the book given in endnote 3),[3] 諸勿瓦的亞/诺勿瓦的亚 (Novgorod; Item 81)[4] and 兀失丁入 (Usting: Item 92).[5] In fact, the red-colored area which spans roughly to the east of this western boundary and to the north, northeast, and east of the Caspian Sea is the present-day eastern Europe, but it is part of 亞細亞/亚细亚 (Asia) depicted on the coloured Japanese copy of the

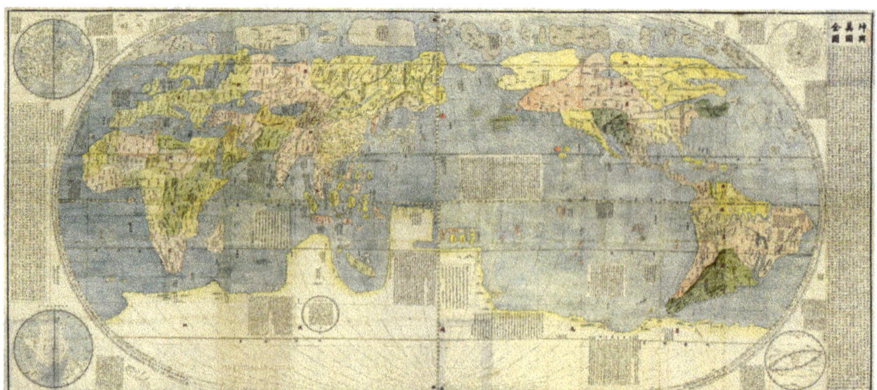

Fig. 7.1.   Asia to the north of ancient China is highlighted in red and yellow colours to the east of 歐邏巴/欧逻巴 (Europe) and to the north of the territory of Unified Great Ming (大明一統/大明一统; as denoted on the KWQ); the map is extracted from the colored Japanese copy of the original Chinese version of the KWQ (public domain).[9]

original Chinese version of the KWQ.[6] The 歐邏巴/欧逻巴 (Europe; Item 117)[7] analysed in my previous book[8] is the Europe defined on the KWQ, and it is smaller than the modern Europe.

My previous book accidentally included 勒贊/勒赞 (Ryazan; Item 93)[10] as part of 歐邏巴//欧逻巴 (Europe; Item 117) on the KWQ. However, since the Principality of Ryazan existed from 1129 to 1521, which overlapped with the era 1157–1166 of the European portion of the KWQ (abbreviated as Europe-KWQ), as I have deduced in my previous book, this accidental inclusion has not affected the determination of the era of the Europe-KWQ. It will also not impact the determination of the era of the ANC-KWQ, as will be seen in the analyses of this chapter.

All the geographical items included in the region of 亞細亞/亚细亚 (Asia) on the KWQ, which are depicted to the north of the Ming China's territory and to the east of the then 歐邏巴//欧逻巴 (Europe) on the KWQ, and if they have not been analysed in the previous chapters of this book, will be included in the ANC-KWQ.

The region on the ANC-KWQ can then be categorized into three zones: Zone I: Eurasia; Zone II: Western Siberia, Central Siberia and parts of Central Asia, parts of Transoxiana (this is the Latin name for the region and civilization located in lower Central Asia roughly corresponding to modern-day eastern Uzbekistan, western Tajikistan, parts of southern Kazakhstan, parts of Turkmenistan and southern Kyrgyzstan), parts of Xingjiang and Mongolia; and Zone III: Eastern Siberia.

The geographical information on the ANC-KWQ shown in Fig. 7.2(a) will be compared with those on the six major European maps of the sixteenth century shown in Figs. 7.2(b)–7.2(g). The first three maps are abbreviated as: (b) the ANC-Mercator (1569), extracted from the 1569 World Map by Gerardus Mercator; (c) the ANC-Ortelius (1570; #1), extracted from the 1570 World Map by Abraham Ortelius; (d) the ANC-Plancius (1594), extracted from the 1594 World Map by Petrus Plancius.

Then, three more Ortelius maps (other than his world map) are brought in for comparison; although they do not cover all the geographical areas to the north of China in comparison to his world map, they do give more geographical data. In Fig. 7.2 caption, they are abbreviated as follows: (e) the Ortelius (1570; #2) which is the 1570 Ortelius Map of Asia (first edition)-*Asiae Nova Descriptio* ("New Map of Asia (亞洲新地圖/亚洲新地图))"; (f) the Ortelius (1570; #3) which is the 1570 Ortelius Tartariae sive Magni Chami Regni tÿpus ("Map of the Tatar or Great Khan Empire (韃靼或大汗帝國地圖/鞑靼或大汗帝国地图))"; and (g) the Jenkinson/Ortelius (1562) which is from a manuscript by Anthony Jenkinson, 1562, in Abraham Ortelius, Theatrum Orbis Terrarum ("Theatre of the Lands of the World (世界地圖/世界地图))".

2.1 *Western knowledge in the sixteenth century of the region to the north of China was extremely limited, but silk goods first produced in Neolithic China were already imported and traded in Siberia from the first millennium B.C., revealing the China — Siberia contacts since Antiquity*

Anthony Jenkinson (1529–1610/1611) was one of the first Englishmen to explore Muscovy/Moscovia and certain regions of the present-day Russia. He was the first English Ambassador to

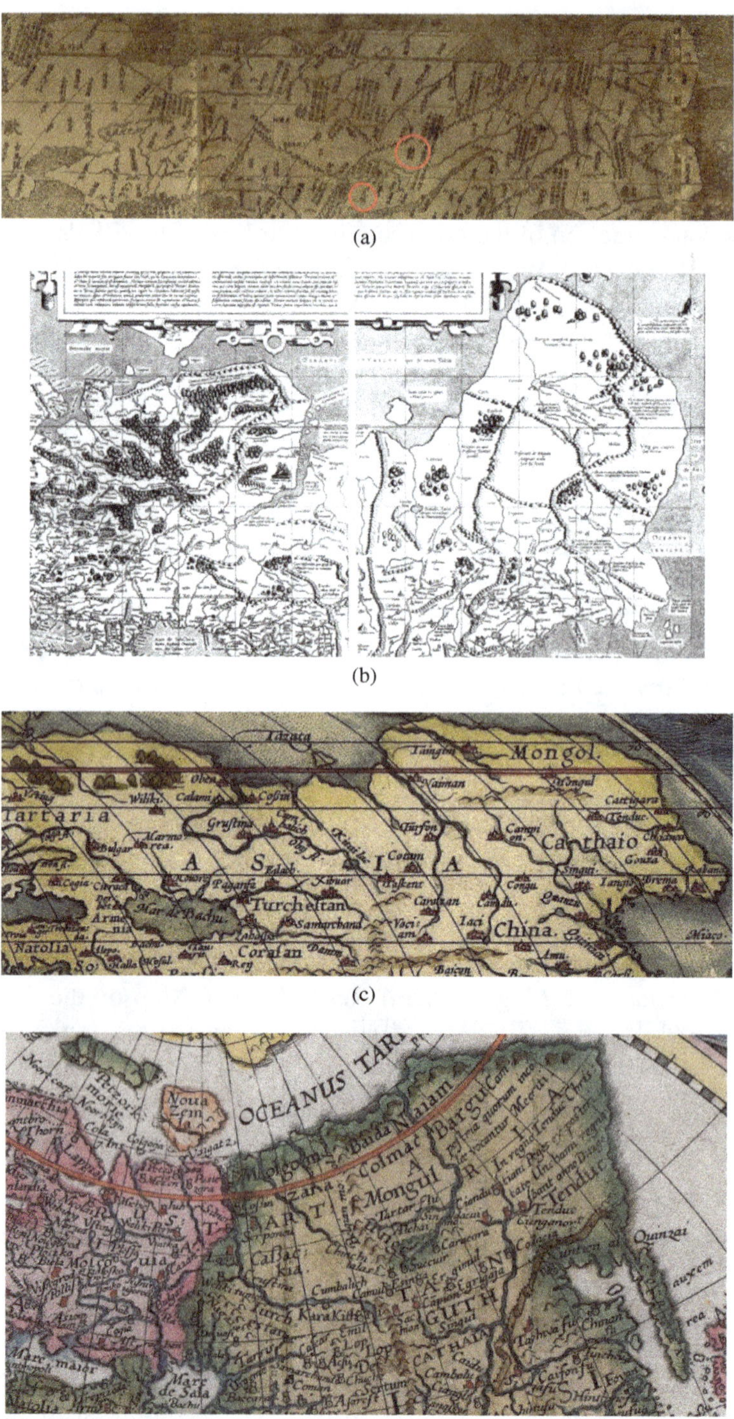

(c)

Idanta
Tangut
Mongol.
Naiman
Oten
Olen
Cossin
Calami
Wilki
Grustina
Turfon
Ongul
Cattigara
Tartaria
Marmo
Bulgar
Cottun
Pentzin
Cangi
Tenduc
A S I A
Cathaio
Sydai
Tuskent
Congi
Singui
Tangut
Brema
Turchestan
Cauxan
Candu
Lanzu
Samarchand
Yaci
Iaci
China.
Natolia
Coralan
Damn
Miaco
Baicon

Noort
Thezora
morte
Nova
Zem
OCEANUS TART
Banda Naiam
Bargu
Colmac
A
Mongul
TART
Casac
Cumbalich
Cathaio
Kara Kitti
GUTH
Cambalu
Quanzai
Tenduc
Mare
de Sala
Natolia

(d)

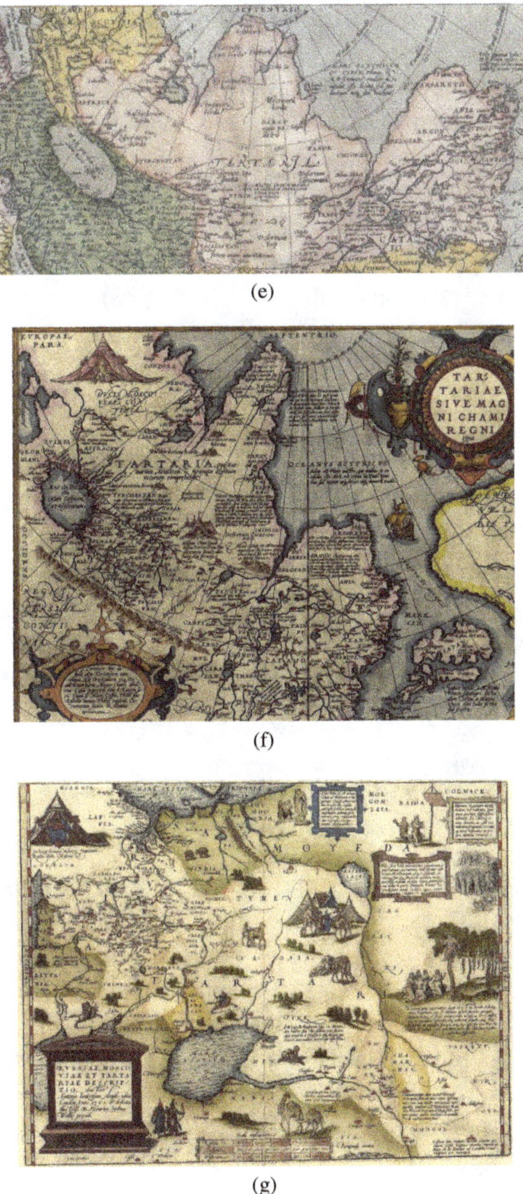

(e)

(f)

(g)

Fig. 7.2.   (a) the ANC-KWQ extracted from the KWQ (public domain);[11] (b) the ANC-Mercator (1569) extracted from the 1569 World Map by Gerardus Mercator (public domain);[12] (c) the ANC-Ortelius (1570; #1) extracted from the 1570 World Map by Abraham Ortelius (public domain);[13] (d) the ANC-Plancius (1594) extracted from the 1594 World Map by Petrus Plancius (public domain);[14] (e) the Ortelius (1570; #2) extracted from the 1570 Ortelius Map of Asia (first edition)-*Asiae Nova Descriptio* (public domain);[15] (f) the Ortelius (1570, #3) which is the Ortelius Tartariae sive Magni Chami Regni tÿpus, 1570 (public domain);[16] and (g) the Jenkinson/Ortelius (1562) which is from a manuscript by Anthony Jenkinson, 1562, in Abraham Ortelius, Theatrum Orbis Terrarum (public domain).[17]

Russia in 1566.[18] But before the eighteenth century, Western knowledge of the region to the north of China, such as the present-day European Russia, Siberia (西伯利亞/西伯利亚), part of Central Asia, and Manchuria (滿洲/满洲), was extremely limited. This huge area was simply known to them as "Tartary (韃靼/鞑靼)"[19] or in Latin as *Tartaria*, and the Mongols who had invaded them as "Tartars (韃靼人/鞑靼人)" (the Tatars formerly also spelled as Tartars).

In fact, the Tatar tribes started in Eastern Mongolia. They were first mentioned in Orkhon inscriptions[20] or Kul Tigin steles (闕特勤碑/阙特勤碑) written in the Old Turkic alphabet around the eighth century (overlapping with the Tang Dynasty: 618–907; with an interregnum between 690 and 705) in the Orkhon Valley (鄂爾渾河谷/鄂尔浑河谷) in present-day Mongolia.

According to historical records, the official Russian incursion into Siberia dates as late as 1581,[21] when the Cossack hetman Ermak Timofeyevich (1532–1585) led a detachment across the Ural Mountains (烏拉爾山脈/乌拉尔山脉). They soon defeated the forces of the Khanate of Sibir (西伯利亞汗國/西伯利亚汗国).[22]

But as early as the first millennium B.C. (the period lasting from 1000 B.C. to 1 B.C.), trade was already underway over the Silk Road. Silk goods were imported and traded in Siberia.[23] Silk is a fabric first produced in Neolithic China from the filaments of the cocoon of the silkworm. The Turkic-Mongols were in southern Siberia around the third century B.C.

Later, the steppes of Siberia were occupied by a succession of nomadic peoples, including the Khitan (契丹) people, various Turkic peoples, and the Mongols, as we shall learn by walking through the histories of the geographical items depicted on the ANC-KWQ, and analysing them in the Appendix at the end of this chapter.

2.2 *The ANC-KWQ reveals a Ming political condition in the fifteenth century, long before the first Europeans' arrival*

First, notice that on the ANC-KWQ in Fig. 7.2(a), there are two names, 韃靼 (right) and 瓦蠟 (left), highlighted by the red circles. These two items are important in telling the political era revealed by the ANC-KWQ and require further explanations.

The name 韃靼/鞑靼 (Dada or Tatar/Tartar) appeared in the Chinese Northern and Southern dynasties (420–589). It was the name used by the Turkic (突厥) tribes to call the Shiwei (室韋/室韦) tribes. In Turkic, "Shiwei" is called "Tatar/Tartar". Shiwei refers to the descendants of the Tribal Khanate Rouran (柔然汗國/柔然汗国; 402–555) which was a branch of the Xianbei (鮮卑/鲜卑 tribe; Shiwei is a homophone of "Xianbei") in the north of Khitan (契丹) or related tribes of the Mongolian language family.

The name 瓦蠟/瓦剌 (Oirats, also formerly known as Eluts and Eleuths) were the westernmost group of the Mongols. Their ancestors lived in the Altai (阿爾泰/阿尔泰) region of Siberia, Xinjiang and western Mongolia. The first documented reference to Elut and Yelut was in the Onginsk "rune" inscriptions dated in the sixth century.[24]

Then, at the beginning of the thirteenth century, different groups of these Mongolian Turkic nomads became part of the troops of the Mongol conqueror Genghis Khan (成吉思汗;

c. 1162–1227). Afterwards, the Mongols and the Turkic peoples were mixed with each other. As a result, the Mongolian troops that invaded Russia and Hungary were collectively called Tartars by the Europeans, and the lands occupied by the Mongolians were called "Tartary" or "Tartaria".

The Yuan Dynasty (元朝; 1271–1368) was a Mongol-led imperial dynasty of China. It was established by Kublai Khan (忽必烈汗; 1215–1294) of the Mongol Empire. In Chinese history, the Yuan Dynasty (1271–1368) followed the Song Dynasty (960–1279) and preceded the Ming Dynasty (1368–1644). In 1368, the Yuan forces were defeated by the Ming army and the Genghisid rulers retreated to the Mongolian Plateau to rule as the Northern Yuan Dynasty (北元; 1368–1635).

But the Ming army continued to pursue the Yuan remnants in the Mongolian steppes. In 1388, the defeat of Uskhal Khan (烏斯哈爾汗/乌斯哈尔汗) effectively shattered Yuan power in the steppes and allowed the four Western Oirat Mongols to rise and split from the Mongolian head-quarters in the east (ruled by Genghisids; the major descendants of Kublai rulers of the Northern Yuan).[25] Later the Oirats even became kingmakers of the Northern Yuan realm for a period of time.[26]

For the Northern Yuan realm, from the end of the fourteenth century, on the one side stood the Western Mongols, and on the other side the Eastern Mongols. The Ming Dynasty (1368–1644) called the political realm in Eastern Mongolia (on the Eastern Mongolian Plateau) "Dada" (韃靼/鞑靼; later the Dada people were known in English as "Tatars/Tartars", and their land as "Tartary or Tartaria"; see Item 52 in the Appendix) and the confederation of the Mongols in Western Mongolia "Wala" (瓦剌/瓦蠟; later these Mongols were known in English as "Oirats"; they formed the Alliance of the Four Oirats or the Oirat Confederation; 1399–1634; Item 47 in the Appendix).[27] This was also mentioned at the beginning of this subsection.

Mongol relations with the Ming Dynasty consisted of sporadic bursts of conflict intermingled with periods of peaceful relations and border trade.

In 1438, Tuo Tuo Buhua (脫脫不花) supported by Tuo Huan (脫歡/脱欢) of the Oirat tribe, defeated the Great Khan Adai (大汗阿岱), and unified the Mongolian tribes.[28]

Esen (也先) Taishi (reigned 1438–1454; son of Tuo Huan) brought the Oirats to the height of their power. In 1449, Esen first invaded the Ming Dynasty and captured the emperor Ming Yingzong Zhu Qizhen (明英宗朱祁鎮/明英宗朱祁镇).[29] In 1453, Esen took the title of not just Khan, but also Northern Yuan (北元) Emperor as the "Da Yuantian Sheng Khan (大元田盛大可汗)".[30] Subsequently, the widespread discontent among the Genghisids resulted in a series of revolts which caused Esen's death in 1454, and the Northern Yuan was once more split into two portions between the Oirats and Eastern Mongols. Esen's death also started the decline of the Oirats, who would not recover until their rise as the Dzungar Khanate (準噶爾汗國/准噶尔汗国; 1634–1758) in the seventeenth century.[31]

In 1479, Dayan Khan (達延汗/达延汗; 1472–1517; the fifteenth grandson of Genghis Khan) became ruler of the Northern Yuan.

In 1487, Dayan Khan eliminated Oirat power and unified the Eastern and the Western Mongolians.

After Dayan Khan's death in 1517, the Mongols began falling apart again due to the two succeeding khans' internal fights.

Under Tümen Zasagt Khan (圖門扎薩格特汗/图门扎萨格特汗; 1539–92), the realm was unified again during his reigning years (1558–1592). He died in 1592. The Northern Yuan (1368–1635) lasted until its conquest by the Jurchen-led Later Jin Dynasty (后金; 1616–1636) in 1635.

From the above history, we know that the unification of the Eastern Mongols and the Western Oirats occurred before 1388, from 1438 to 1454, 1487 to 1517, and after 1557. From 1388 to 1438, 1454 to 1487, and 1517 to 1557, the Northern Yuan was split into two portions between the Oirats and Eastern Mongols. Since on the ANC-KWQ, Dada (韃靼/鞑靼) and Wala (瓦蠟/瓦剌) are depicted as two separate political realms, which era does the map belong to? We shall soon find the answer.

In fact, after Esen's death in 1454, the Oirats started to decline and no longer possessed the power to challenge the Eastern Mongolian regime. Hence, it was mainly during the late fourteenth century and before the mid-fifteenth century that the Eastern Mongolia and Western Mongolia stood as two powerful political realms as shown on the ANC-KWQ in Fig. 7.2(a). This era was long before the Europeans reached the Mongolian steppes.

It is not surprising that the sixteenth-century European cartographers such as the Flemish Gerardus Mercator (1512–1594), the Spanish-Flemish Abraham Ortelius (1527–1598), and the Dutch-Flemish Petrus Plancius (1552–1622) were all eager to incorporate the most recent information about the territory east of Moscow into their maps. That huge area was called Tartary (or Tartaria) on their maps;[32] by then, the Oirat power was already diminished. Hence, the political information revealed by their maps should be very different from that on the ANC-KWQ.

For example, in Fig. 7.2(c) and 7.2(g), Ortelius only depicts Tartaria at the far western region on the ANC-Ortelius (1570; #1) and the Jenkinson/Ortelius (1562); whereas in Figs. 7.2(e) and 7.2(f), it covers the entire western and central regions of the Ortelius (1570; #2) and the Ortelius (1570; #3). In Fig. 7.2(d), Plancius depicts Tartaria throughout the entire ANC-Plancius (1594). But in Fig. 7.2(b), Mercator does not even depict Tartaria on his ANC-Mercator (1569). All these maps reveal the individual European cartographer's knowledge about Asia to the north of China in the sixteenth century. Clearly, none of them depicted the co-existence of Eastern Mongolia and Western Mongolia which occurred mainly in the late fourteenth and before the mid-fifteenth centuries. After 1454, the Oirat power was too weak to compete with Eastern Mongolia.

This is the first evidence and one of the major differences which shows that the ANC-KWQ is not a direct or adapted copy of the major sixteenth-century European maps analysed in this chapter. Matteo Ricci claimed that the KWQ was the map he drew and published in 1602, but we can see clearly that the content of the ANC-KWQ is not of European origin, it belonged to an earlier era than the sixteenth century. Without using the Chinese source maps, it would be impossible for Ricci to draw the ANC-KWQ which reveal this distinct Chinese political information on it.

In the Appendix, I have analysed the geographical information and histories of all the 102 geographical items and the fourteen original annotations depicted on the ANC-KWQ in Fig. 7.2(a).

After comparing them with those on the six major European maps shown in Figs. 7.2(b)–7.2(g) whenever possible, the main results of these comparisons are summarised in the following subsections.

2.3 *Out of the 102 geographical items analysed on the ANC-KWQ, 57 of them (c. 56%) are not present on any of the six major European maps of the sixteenth century; and among the same 102 geographical items, 23 are of Chinese origin*

The details are given in Tables 7.1–7.3.

Here is the list for the 57 geographical items which are not present on any of the six European maps, and their numbered positions in the Appendix.

Table 7.1 Zone I.

1. Boddia (波的亞/波的亚; Item 1); 2. Feirmalechia (非爾馬勒祁亞/非尔马勒祁亚; Item 2); 3. Ualamueico (瓦郎尾可; Item 7); 4. Vladimir (縛羅得抹爾/缚罗得抹尔; Item 8); 5. Miselow (迷色錄/迷色录; Item 11); 6 Yimou Guo/the Country of One-eyed Man (一目國/一目国; Item 16); 7. Esugedaye (額索各苔耶/额索各苔耶; Item 17); 8. Cabardi (加巴爾地亞/加巴尔地亚; Item 18); 9. Kara Kithai (黑其泰; Item 26); 10. Sarmatia Asiatica (亞細亞: 沙爾馬齊亞/亚细亚: 沙尔马齐亚; Item 27); 11. Sarmatia Europea (歐邏巴: 沙爾馬齊亞/欧逻巴: 沙尔马齐亚; Item 28); 12. White Sea (白海; Item 34)
Comment: Thirty-four percent of the geographical items (35 is the total) depicted in Zone I on the ANC-KWQ are not present on the European maps analysed in this chapter (12/35 = 34%).

Table 7.2 Zone II.

13. Alessandria (亞力山的/亚力山的; Item 37); 14. Yilibali (亦力把力; Item 41); 15. Yin Mountains (陰山/阴山; Item 42); 16. Monibei (漠泥彼; Item 43); 17. Kuobaihuadu (濶百花渡; Item 44); 18. Orkhon River (哈剌禾林河; Item 45); 19. Oirats (瓦蠟/瓦蜡 or 瓦剌; Item 47); 20. Sarikil (撒里怯兒/撒里怯儿; Item 48); 21. Tuul River (土剌河; Item 49); 22. Shahuzhen (殺胡鎮/杀胡镇; Item 50); 23. Azhili (阿只里; Item 53); 24. Xia Jia (轄戛/辖戛; Item 54); 25. Lugumoluye (路骨莫路耶; Item 58); 26. Morpa (漠爾拔/漠尔拔; Item 59); 27. Kingdom of Hela (曷剌國/曷剌国; Item 60); 28. Wanzaida (宛在達/宛在达; Item 61); 29. Ghost Country (鬼國/鬼国; Item 63); 30. Niuti Turkic people (牛蹄突厥; Item 65); 31. Heichezi (黑車子/黑车子; Item 66); 32. Tianfang (天方; Item 68); 33. Tiele (tolos) (鐵勒/铁勒; Item 69); 34. Da Kuoye (大壙野/大圹野; Item 71); 35. Calmucchi (哥兒墨/哥儿墨; Item 74)
Comment: 57.5 percent of the geographical items (40 is the total) depicted in Zone II on the ANC-KWQ are not present on the European maps analysed in this chapter (23/40 = 57.5%).

Table 7.3 Zone III.

36. Yujuelü/Wugu (嫗厥律/妪厥律; Item 77); 37. Didouyu (地豆於/地豆于; Item 78); 38. Shiwei (室韋/室韦; Item 79); 39. Mierqi (襪結子/袜结子; Item 80); 40. Zeru (測兒吳/测儿吴; Item 81); 41. Da Shiwei (大室韋/大室韦; Item 82); 42. Zhiheerca (支何兒察/支何儿察; Item 83); 43. Luo Mountains (羅山/罗山; Item 84); 44. Wuluohou (烏洛侯/乌洛侯; Item 85); 45. Qudumei Kingdom (區度寐/区度寐; Item 87); 46. Shenmoda Shiwei (深末怛室韋/深末怛室韦; Item 88); 47. Luo huang ye (羅荒野/罗荒野; Item 90); 48. Bo Shiwei (鉢室韋/钵室韦; Item 93); 49 Hegiuuaia (黑入瓦牙; Item 94); 50. Huangtou Shiwei (黃頭室韋/黄头室韦; Item 95); 51. Northern Shiwei (北室韋/北室韦; Item 96); 52. Lake Khanka (白湖; Item 97); 53. Baode River (包得河; Item 98); 54. Iammasceli (亞馬是里/亚马是里; Item 99); 55. Shou Shiwei (獸室韋/兽室韦; Item 100); 56. Hubu Mountain (胡布山; Item 101); 57. Dog Nation (狗國/狗国; Item 102)
Comment: Eighty-one percent of the geographical items (27 is the total) depicted in Zone III on the ANC-KWQ are not present on the European maps analysed in this chapter (22/27 = 81%).

The above three tables show that from west to east as we move further into eastern Siberia, the ANC-KWQ starts to depict more and more geographical items that are absent from the European maps analysed in this chapter. This shows that these regions were further away from Europe and were less known to the Europeans at the time. However, the Chinese did not have such a problem, since they explored these places long before the Europeans' arrival and were able to depict these places on their maps. These maps (no longer in existence) were then used by Matteo Ricci for making the ANC-KWQ.

Here is the list for the 23 geographical items which are of Chinese origin; their numbered positions in the Appendix are also listed.

Table 7.4 Zone I.

1. Yimou Guo/the Country of One-eyed Man (一目國/一目国; Item 16); 2. White Sea (白海; Item 34)

Table 7.5 Zone II.

3. Yin Mountains (陰山/阴山; Item 42); 4. Hening (和寧/和宁; Item 46); 5. Wala/Oirats (瓦蠟/瓦蜡; Item 47); 6. Shahuzhen (殺胡鎮/杀胡镇; Item 50); 7. Gannan River (幹難河/干难河; Item 51); 8. Xia Jia (轄戛/辖戛; Item 54); 9. Ghost Country (鬼國/鬼国; Item 63); 10. Niuti Turkic peolple (牛蹄突厥; Item 65); 11. Heichezi (黑車子/黑车子; Item 66); 12. Tianfang (天方; Item 68); 13. Da Kuoye (大壙野/大圹野; Item 71)

Table 7.6 Zone III.

<table>
<tr><td>14. Hexi River (賀喜河/贺喜河; Item 76); 15. Yujuelü (嫗厥律/妪厥律; Item 77); 16. Didouyu (地豆於/地豆于; Item 78); 17. Luo Mountains (羅山/罗山; Item 84); 18. Mao Shui (兒水; Item 86); 19. Luo Huang Ye (羅荒野/罗荒野; Item 90); 20. White Lake (白湖; Item 97); 21. Baode River (包得河; Item 98); 22. Hubu Mountain (胡布山; Item 101); 23. Dog Nation (狗國/狗国; Item 102)</td></tr>
</table>

In summary, since out of the 102 geographical items analysed on the ANC-KWQ, 57 of them (c. 56 percent) are not present on any of the six major European maps in the sixteenth century; and among the same 102 geographical items, 23 are of Chinese origin (c. 23 percent), it is clear that the ANC-KWQ is of Chinese origin and not a direct or adapted copy of the major sixteenth-century European maps discussed in this chapter.

3. Determining the Era of the ANC-KWQ

3.1 *The ANC-KWQ reveals the political condition from 1418 to the early sixteenth century, because Yilibali (亦力把力) is depicted on the map*

This requires a more detailed explanation as follows:

Other than the Northern Yuan Dynasty which we have discussed in Subsection 2.2, during the Ming Yongle (1402–1424), Hongxi (1424–1425), Xuande (1425–1435) and Yingzong (1435–1449) periods, there was the Eastern Chagatai Khanate (東察合台汗國/东察合台汗国; 1347–1570[33] or 1347–1705[34]) to the west of the Northern Yuan realm. The Chagatai Khanate (察合台汗國/察合台汗国; 1227–1570 or 1227–1705)[35] was established by Chagatai, the second son of Genghis Khan. The Chagatai Khanate was split into two parts in 1347, the Western Chagatai Khanate and the Eastern Chagatai Khanate (Moghulistan).

When Genghis Khan (成吉思汗; c. 1162–1227) conquered a large area of land, he gave these lands to his four sons, namely the four major Khanates of the Yuan Dynasty: the Golden Horde (金帳汗國/金帐汗国; 1242–1502), the Chagatai Khanate (察合台汗國/察合台汗国; 1227–1570 or 1227–1705), the Wokuotai Khanate (窩闊台汗國/窝阔台汗国; 1251–1309), and the Ilkhanate (伊利汗國/伊利汗国; 1256–1333/1335).[36] Genghis Khan's grandson Kublai was the Khanate of the Great Khan,[37] and this khanate refers to the Yuan Dynasty in China,[38] which was founded by Kublai Khan in 1271. Kublai completed the conquest of China and became the first Yuan ruler of the entire country. But during the Ming Dynasty, only the Golden Horde Khanate and the Eastern Chagatai Khanate were still in existence, and their histories can help in analysing the era of the ANC-KWQ.

In 1418 during the sixteenth year of the Ming Yongle period, the Eastern Chagatai Khanate moved its capital to Yilibali (亦力把力).[39] Because after the fall of the Yuan Dynasty in 1368, the

Eastern Chagatai Khanate took the Ming Dynasty as its suzerain, most of the Ming historians no longer recorded the country as the Eastern Chagatai Khanate but referred to it by the name of its capital, Yilibali (亦力把力). Therefore, on many maps of the Ming Dynasty, Yilibali (亦力把力) replaces the country name of East Chagatai Khanate. This is also what we see on the ANC-KWQ.

However, after 1465, Yilibali came to pay tribute less frequently and finally it ended when Turfan (吐鲁番) conquered Yilibali in the early sixteenth century.[40] Instead, the Turpan City became the capital of the Eastern Chagatai Khanate and the name Yilibali (亦力把力) disappeared from the records of the Ming Dynasty.

Because after the early sixteenth century, the Turpan City became the capital of the Eastern Chagatai Khanate and most of the nomadic tribes in the Yilibali area (Ili River Basin; 今新疆境内伊犁河流域) took refuge in the Kazakh Khanate (哈薩克汗國/哈萨克汗国; 1465–1847) and became its components, Yilibali is not depicted on any of the sixteenth century European maps analysed in the chapter.

3.2 *The ANC-KWQ further reveals the political condition of 1419 to right before 1438, because the Golden Horde (金帳汗國/金帐汗国; 1242–1502) and the Khanate of Kazan (喀山汗國/喀山汗国; 1438–1552) are not depicted on the map*

The detailed analysis is as follows:

The Golden Horde[41] was founded around 1242 by Batu Khan (巴都汗), a grandson of Genghis Khan. It covered regions in Russia and Eastern Europe. The territory of the Golden Horde at its peak in the early fourteenth century extended from Siberia and Central Asia to parts of Eastern Europe, from the Urals to the Danube in the west, and from the Black Sea to the Caspian Sea in the south.

The Golden Horde had shown cracks in the fourteenth century, with periods of chaos within the polity. But the death of Edigu (艾迪古; the last person to ever unite the Horde) in 1419, marked the beginning of the fragmentation and the decay of the Golden Horde. This may explain why the ANC-KWQ does not depict the Golden Horde or Great Horde on the map.

The name "Great Horde" was first mentioned in the 1430s. It might have been used to directly link the now greatly reduced administrative center of the Horde to the original greatness of the Golden Horde.[42] In the 1430s and the 1440s, the Golden Horde already fractured into separate states including the Kazan Khanate (喀山汗國/喀山汗国; 1438–1552), the Nogai Khanate (諾蓋汗國/诺盖汗国; 1440s–1634), and the Kasimov Khanate (卡西莫夫汗國/卡西莫夫汗国; 1445–1552; it had separated itself from Kazan). Each one of these Khanates claimed to be the legitimate successor to the Golden Horde. But the Crimean Khanate (克里米亞汗國/克里米亚汗国; 1441–1783; it was regarded as the direct heir to the Golden Horde and to Desht-i-Kipchak)[43] subjugated the Sarai-based Great Horde (it was centred at the core of the former Golden Horde at Sarai on the lower Volga)[44] in 1502.[45] The Great Horde finally dissipated.[46]

Since the Kazan Khanate (喀山汗國/喀山汗国; 1438–1552) is also not depicted on the ANC-KWQ, the map's political era must be before 1438. This will immediately explain that none of the

Nogai Khanate (諾蓋汗國/诺盖汗国; 1440s–1634), the Kasimov Khanate (卡西莫夫汗國/卡西莫夫汗国; 1445–1552) and the Crimean Khanate (克里米亞汗國/克里米亚汗国; 1441–1783) is depicted on the ANC-KWQ, because they did not exist before 1438.

Combining Subsections 3.1 and 3.2, we can deduce that the ANC-KWQ is a map of 1419 to right before 1438. This time range overlaps with the Yongle (1402–1424), Hongxi (1424–1425), Xuande (1425–1435) and Yingzong (1435–1449) periods.

3.3 *The era of the ANC-KWQ can be finally determined to be from 1428 to 1433*

The reasons are as follows:

My previous analysis[47] has shown that the source map(s) of the KWQ was(were) drawn by the Chinese cartographers in 1433.[48] Then, the era of the ANC-KWQ (the era when these regions were explored by the Chinese before the map was drawn) can be further narrowed to be from 1428 to 1433, because the town of Tyumen (Item 28) was founded by Taibuga (泰布加) probably sometime between 1405 and 1428 and we know that the entire KWQ was drawn in 1433. The era 1428 to 1433 falls nicely in the range from 1419 to right before 1438 deduced from Subsection 3.2.

In summary, from the analysis of Subsection 3.1 we know that the ANC-KWQ reveal a political era from 1418 to the early sixteenth century; from Subsection 3.2 we know that the political era revealed by the ANC-KWQ can be narrowed as from 1419 to right before 1438; and from Subsection 3.3 we know that the era of the ANC-KWQ can finally be narrowed as from 1428 to 1433. During this period, Dada (韃靼/鞑靼) in Eastern Mongolia and the confederation of the Oirats (瓦蠟/瓦蜡 or瓦剌) in Western Mongolia were two separate political realms as shown on the ANC-KWQ.

The detailed analyses presented in the Appendix show that the political landscape revealed by the 102 geographical items on the ANC-KWQ are all consistent with the time range from 1419 to 1433, except for those cases when outdated geographical names are still retained on the map.

4. Conclusions

This chapter is devoted to the analysis of the region of Asia to the north of China depicted on the first Chinese-based world map — the Kunyu Wanguo Quantu (坤輿萬國全圖/坤舆万国全图) (abbreviated as KWQ) or Complete Geographical Map of All the Kingdoms of the World published by Matteo Ricci in 1602 in China. On this map, the Asian portion to the north of China and to the east of Europe is abbreviated as ANC-KWQ in this chapter.

In the Introduction, I reviewed the successful results published in my previous book titled *Chinese Global Exploration in the Pre-Columbian Era: Evidence from an Ancient World Map*. The book has analysed and provided evidence to show that the geographical information of several major portions on the KWQ is based on the knowledge obtained by the Ming (明代; 1388–1644) mariners in the 1420s and 1430s during their sixth and seventh voyages to the Western Ocean led by Admiral Zheng He (郑和; 1371–1433/1435). However, the European portion on the map is based on Chinese exploratory data gathered in the much earlier Southern Song Dynasty (南宋;

1127–1279). This means that all these portions of the KWQ used Chinese maps as source maps (these no longer exist) and that they were not copied from other contemporary European maps.

Following the same methodologies, in the previous six chapters of this book I have shown that none of the eras of the following Asian regions on the KWQ can be dated 1602. Instead, their eras are: (1) Japan in 897 during the Tang Dynasty; (2) Korea in 1433 during the Ming Dynasty; (3) the Great Ming in 1433; (4) Southeast Asia in 1432–1433; (5) the Indian subcontinent and its eastern periphery in 1432–1433; and (6) West and Central Asia also in 1433. Moreover, I have provided evidence to show that the portions of the map depicting all these different regions on the KWQ are of Chinese origin.

Then in Section 2 of the seventh (and last) chapter of this book, I apply the same methodologies to show that the map of Asia to the north of China on the KWQ, abbreviated as the ANC-KWQ, is also of Chinese origin. What I have observed through my analysis is that Western knowledge of the region to the north of China was extremely limited in the sixteenth century, evidenced by the fact that on the ANC-KWQ, c. 56% of the 102 geographical items (23 of them or c. 23% are of Chinese origin) are missing from the major sixteenth-century European maps, whereas the ANC-KWQ reveals clearly a Ming political condition in the fifteenth century, long before the first Europeans' arrival.

Finally, in Section 3, I have deduced that the era of the ANC-KWQ based on the political condition revealed on this map is from 1419 to right before 1438. However, because the entire KWQ has been deduced as a map drawn in the year 1433 and the town of Tyumen (Item 28) was founded by Taibuga (泰布加) probably sometime between 1405 and 1428, the area north of China must logically be explored in the era from 1428 to 1433. This narrowed period falls nicely in the range of 1428 to 1438. The year 1433 was when the Ming Treasure Fleets returned to China after they completed their seventh (and last) voyage to the Western Ocean. It was also the year when the source map of the entire KWQ was drawn.

The above conclusions are based on analysing the locations and histories of the complete 102 geographical items and the fourteen annotations depicted on the ANC-KWQ and comparing them as best as possible with the six major sixteenth-century European maps discussed in this chapter.

Appendix

The following geographical items are categorized into three different zones. Items preceded by "●" are geographical items which do not appear on any of the six European maps discussed in this chapter. The Chinese characters on the KWQ are almost entirely Chinese traditional characters. Since many readers use the Chinese simplified characters, I include both: all the Chinese characters in the main text and in the Appendix are expressed first with the Chinese traditional characters, then followed by the Chinese simplified characters, separated by a slash "/". When both characters are the same, that expression is presented only once. Chinese pinyin is listed for each item in this Appendix. The sign "@" means "annotation", and all the geographical items and annotations are numbered in sequence in this appendix.

Item #	Name	Pinyin	Etymology	History, Geography and My Comments

Zone I: Western ANC-KWQ (including European Russia, West of Ural Mountains from the Coast of the Arctic Ocean to the River Ural, and Part of Central Asia)

Fig. A7.1. This figure is extracted from Fig. 7.2(a) (public domain); Items 1-35 and annotation @1 are depicted on this Figure.

Item #	Name	Pinyin	Etymology	History, Geography and My Comments
● 1	波的亞	Bo De Ya	Boddia[51]	The place cannot be identified on a modern map.
● 2	非尔馬勒祁亞	Fei Er Ma Le Qi Ya	Feirmalechia[52]	The place cannot be identified on a modern map.
3	各勒利亞	Ge Le Li Ya	Korelia region	各勒利亞 may be Karelia (Korelia, Karelen, Karjala),[53] the land of the Karelian people, which is today split between Finland and Russia. Korelia is depicted as Corelia on the Jenkinson/Ortelius (1562).

4	縛羅得没	Fu Luo De Mei	Vladimir: this item is a repetition of Item 8.	Vladimir[54] is a city and administrative center of Vladimir Oblast (region), western Russia. The founding date of Vladimir is disputed between 990 and 1108. Vladimir was one of the medieval capitals of Russia.[55] Later the city was attacked by Tatars (韃靼人/鞑靼人) several times in the fifteenth century and became a minor local center. But in 1796 it was made a seat of provincial government.[56] On the ANC-KWQ, 縛羅得没/缚罗得没 (Vladimir) is depicted at c. 63°N, 41°E (KWQ coordinates; this latitude corresponds to c. 62.1°N on a modern map, after multiplication by 0.9856; this latitudinal conversion has been explained in my earlier book; the longitudinal coordinate does not need conversion).[57] On a modern map, it is located at 56.1°N, 40.4°E. The longitudinal coordinates are very close to each other, but the latitude of 縛羅得没/缚罗得没 (Vladimir) is almost 6° further north on the ANC-KWQ than on a modern map. The large error may be caused by the less accurate latitudinal reading near the pole (the Earth's magnetic lines deviate more from the geographical longitudes near both the North and South Poles). Vladimir is depicted as Volodemer on the Jenkinson/Ortelius (1562).[58]
5	没失箇辣	Mei Shi Ge La	Meschora: the lowlands near Moscow.	Meschora (没失箇辣) marshes[59] are near Moscow. Meschora is depicted on the Jenkinson/Ortelius (1562).
6	縛羅荅	Fu Luo Da	Vologda[60]	Vologda (縛羅荅/缚罗荅) is a city in, and the administrative center of, Vologda Oblast, Russia. The official founding year is 1147.[61] However, historians and archaeologists believe that the city was founded in the thirteenth century, because Vologda was mentioned in a 1264 agreement between the Novgorod Republic (諾夫哥羅德共和國/诺夫哥罗德共和国) and the Grand Prince of Vladimir (弗拉基米爾大公/弗拉基米尔大公) as an outlying possession of the Novgorod Republic.[62] Vologda is depicted on the Jenkinson/Ortelius (1562).

(*Continued*)

Item #	Name	Pinyin	Etymology	History, Geography and My Comments
● 7	瓦郎尾可	Wa Long Wei Ke	Ualamueico[63]	The place cannot be identified on a modern map.
● 8	縛羅得抹尔	Fu Luo De Mo Er	Vladimir: this item is a repetition of Item 4.	Please also see explanation in Item 4. On the ANC-KWQ, 縛羅得没/缚罗得没 (Vladimir) in Item 4 is depicted at c. 63°N, 41°E (KWQ coordinates; this latitude corresponds to c. 62.1°N on a modern map, after multiplication by 0.9856; this; this latitudinal conversion has been explained in my earlier book; the longitudinal coordinate does not need conversion).[64] On a modern map, Vladimir is located at 56.1°N, 40.4°E. The longitudinal coordinates are very close-to each other, but the latitude of 縛羅得没/缚罗得没 (Vladimir) is almost 6° further north on the ANC-KWQ than on a modern map. However, 縛羅得抹尔/縛羅得抹尔 (Vladimir) in Item 8 is depicted on the ANC-KWQ at c. 61.5°N, 51.5°E (KWQ coordinates; the latitude corresponds to c. 60.6°N on a modern map, after multiplication by 0.9856; the longitudinal coordinate does not need conversion).[65] On a modern map, Vladimir is located at 56.1°N 40.4°E (modern coordinates). The longitudinal coordinates show a difference of about 10°, but the latitude of 縛羅得抹尔/縛羅得抹尔 (Vladimir) is c. 4.5° further north on the ANC-KWQ than on a modern map, a slight improvement over 縛羅得没/缚罗得没 in Item 4. Comparing 縛羅得没/缚罗得没 with 縛羅得抹尔/縛羅得抹尔, it seems that 縛羅得没/缚罗得没 in Item 4 better represents Vladimir than 縛羅得抹尔/縛羅得抹尔 in Item 8. Vladimir is depicted as Volodemer on Jenkinson/Ortelius (1562) only once. Its latitude is at c. 43°N, being too far south from its latitude 56.1°N on a modern map.
9	勿欽	Wu Qin	Vachin[66]	The place cannot be identified on a modern map. Vachin is depicted on the Jenkinson/Ortelius (1562).

10	沒斯箇未突	Mei Si Ge Wei Tu	Moskovia,[67] Moscovia, or Moscovy (region): the name derived from Moscow and the Moskva River (a river that flows through western Russia).[68]	Moscovia (沒斯箇未突/沒斯箇未突)[69] is a historical region in Central Russia (it is, broadly, the various areas in European Russia). In the region lived different Russian folks and tribes for a long period. The name Moscovia emerged in the thirteenth century.[70] *Hence, the ANC-KWQ is not a map before the thirteenth century, unlike the European portion of the KWQ, which can be dated in an era from 1157 to 1166.*[71] Moscovia was also the political and geographical name of the Russian state and the Tsardom of Russia (俄羅斯沙皇國/俄罗斯沙皇国; 1547–1721)[72] in Western sources, together with the name *Russia* from the fifteenth to the beginning of the eighteenth century.[73] "Moskovia" is depicted on the Ortelius (1570; #2).
● 11	迷色录	Mi Se Lu	Miselow[74]	The place cannot be identified on a modern map, but it was in present-day European Russia. The European Russia is the western and most populated part of the Russian Federation. It is geographically situated in present-day Europe, unlike the country's sparsely populated and vastly larger eastern part, Siberia in Asia.
12	蓋大期	Gai Da Qi	Caitachi	The place cannot be identified on a modern map, but it was located on present-day Taman Peninsula of Russia. Caitachi is depicted on the ANC-Ortelius (1570: #5) near the northeast coast of the Black Sea.
13	伯帝兀尔祁	Bo Di Wu Er Qi	Petigorski[75]	Petigorski (伯帝兀爾祁/伯帝兀尔祁) was present-day Pyatigorsk, a city in Stavropol Krai in North Caucasus or Southwestern Russia. The writings of the fourteenth-century Arabian traveller Ibn Battuta presented the earliest known mention of the mineral springs[76] of the place which later was founded a city in 1780. Petigorski is depicted on the Jenkinson/Ortelius (1562).
14	弥色曷尔迷	Mi Se He Er Mi	Cerenisegorni	The place cannot be identified on a modern map, but it was in present-day European Russia. Cerenisegorni is depicted on the Jenkinson/Ortelius (1562).

(*Continued*)

Item #	Name	Pinyin	Etymology	History, Geography and My Comments
15	孟日力里亞	Meng Ri Li Li Ya	Mingrelia[77]	Mingrelia (孟日力裡亞/孟日力里亞) is a historic province in the western part of Georgia, see Fig. A7.2, which was formerly known as Odishi (meaning "once [*odeshi*] this land was ours").[78] The place is primarily inhabited by the Mingrelians, a subgroup of Georgians. In the eleventh to fifteenth centuries, Mingrelia (孟日力裡亞/孟日力里亞) was a part of the United Kingdom of Georgia (1008–1490).[79] The kingdom fell to the Mongol invasions in the thirteenth century but managed to re-assert sovereignty by the 1340s. However, by the end of the fifteenth century Georgia turned into a fractured entity,[80] and in 1466 collapsed into anarchy. Its constituent kingdoms became independent states between 1490 and 1493, and finally into five semi-independent principalities.[81] From the sixteenth century to 1857, the independent Principality of Mingrelia (1557–1857) was under the rule of the House of Dadiani.[82] Mingrelia is bordered by the Black Sea to the west. But on the ANC-KWQ, it is erroneously depicted near and to the northwest of the Caspian Sea. **Fig. A7.2.** Historical Mingrelia in the western part of the modern international borders of Georgia (*Source*: Giorgi Balakhadze, under CC BY-SA 4.0, https://commons.wikimedia.org/wiki/File:Historical_Samegrelo_in_modern_international_borders_of_Georgia.svgn);[83] the Black Sea is at left and the Caspian Sea is far to the right (not shown).

				Mingrelia is not depicted on the European maps in Figs. 7.2(b)–7.2(g). but it is depicted on Abraham Ortelius's atlas Teatrum Orbis Terarum, Antwerp, 1570, map no. 50;[84] the map is the best known of all the sixteenth-century maps of the Ottoman Empire.
● 16	一目國	Yi Mu Guo	Yimou Guo or the Country of One-eyed Men.	一 means "one", 目 "eye" and 國 kingdom. Hence, 一目國 means the "Country of One-eyed Men". According to textual research, the "Country of One-eyed Men" mentioned in Chapter 3 titled "Classic of Regions Beyond the Seas: North (海外北經/海外北经)" in the book *Shan Hai Jing* (山海經/山海经) or *The Classic of Mountains and Seas,* was in the region of the present-day Altai Mountains (阿爾泰山脈/阿尔泰山脉) where Russia, China, Mongolia and Kazakhstan converge. The one-eyed men had only one eye in the centre of their faces. They were also recorded in the works of Herodotus (希羅多德/希罗多德; c. 484–c. 425 B.C.; a Greek historian and geographer),[85] the father of Western historiography, and their active area was in the Altai Mountains. It seems that the "Country of One-eyed Men" recorded in the East and the West coincide. This tribe was the overlord of the Central Asian grasslands in the seventh century B.C. or even earlier. Some people think that each of them wore a one-eyed mask as a witchcraft dress for worshiping. However, this legendary prehistoric kingdom is still depicted anachronously on the ANC-KWQ.

(Continued)

Item #	Name	Pinyin	Etymology	History, Geography and My Comments
				But on the ANC-KWQ, 一目國 (the "Country of One-eyed Men") is depicted very close to the northern side of the Caspian Sea, which does not seem to match the region of the present-day Altai/Altay Mountains (阿爾泰山脈/阿尔泰山脉), see Fig. A7.3. The Altai Mountains lie in the northwest of Mongolia and are a mountain range in Central and Eastern Asia, where Russia, China, Mongolia and Kazakhstan converge. The distance from Mongolia to the Caspian Sea is great, about 3,654 km/2,270.5 mi.  **Fig. A7.3.** Map of the Altai Mountain Range (*Source*: Rudyologist, under CC BY-SA 4.0, https://commons.wikimedia.org/wiki/File:Altai_Mountains.jpg)[86] in the northwest of Mongolia; Russia is to its north, Kazakhstan to its west and China to its south. This "country of one-eyed men" is also explained in a different context as a legendary tribe of one-eyed people of northern Scythia (斯基泰; a kingdom created by the Scythians during the sixth to third centuries B.C. in the Pontic-Caspian steppe shown in Fig. A7.4 .

				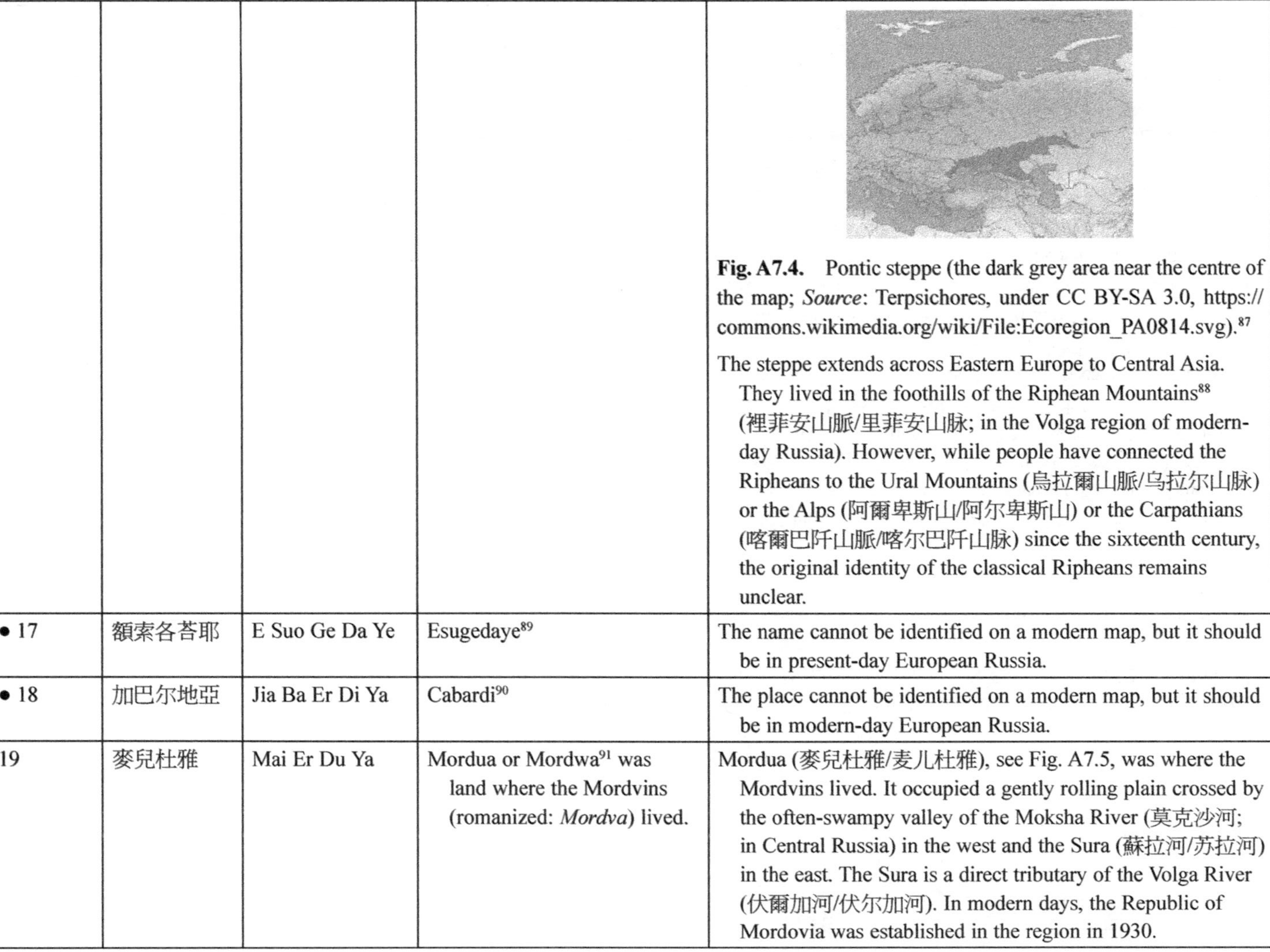 **Fig. A7.4.** Pontic steppe (the dark grey area near the centre of the map; *Source*: Terpsichores, under CC BY-SA 3.0, https://commons.wikimedia.org/wiki/File:Ecoregion_PA0814.svg).[87] The steppe extends across Eastern Europe to Central Asia. They lived in the foothills of the Riphean Mountains[88] (裡菲安山脈/里菲安山脉; in the Volga region of modern-day Russia). However, while people have connected the Ripheans to the Ural Mountains (烏拉爾山脈/乌拉尔山脉) or the Alps (阿爾卑斯山/阿尔卑斯山) or the Carpathians (喀爾巴阡山脈/喀尔巴阡山脉) since the sixteenth century, the original identity of the classical Ripheans remains unclear.
● 17	額索各苔耶	E Suo Ge Da Ye	Esugedaye[89]	The name cannot be identified on a modern map, but it should be in present-day European Russia.
● 18	加巴尔地亞	Jia Ba Er Di Ya	Cabardi[90]	The place cannot be identified on a modern map, but it should be in modern-day European Russia.
19	麥兒杜雅	Mai Er Du Ya	Mordua or Mordwa[91] was land where the Mordvins (romanized: *Mordva*) lived.	Mordua (麥兒杜雅/麦儿杜雅), see Fig. A7.5, was where the Mordvins lived. It occupied a gently rolling plain crossed by the often-swampy valley of the Moksha River (莫克沙河; in Central Russia) in the west and the Sura (蘇拉河/苏拉河) in the east. The Sura is a direct tributary of the Volga River (伏爾加河/伏尔加河). In modern days, the Republic of Mordovia was established in the region in 1930.

(*Continued*)

Item #	Name	Pinyin	Etymology	History, Geography and My Comments
				Fig. A7.5. Modern-day Mordovia (dark area) on the map of Russia (*Source*: File:Map of Russia (2014–2022) — Mordovia.svg: Stasyan117; derivative work: Seryo93, under CC BY-SA 4.0, https://commons.wikimedia.org/wiki/File:Map_of_Russia_(2014%E2%80%932022)_-_Mordovia_(Crimea_disputed).svg).[92] The Mongols conquered vast areas of Eastern Europe in the thirteenth century. They established the Khanate of the Golden Horde (1242–1502),[93] and conquered the Mordvins lands. Mordvins fought against Mongols and later alongside Russians. When the Golden Horde disintegrated in the 1430s (starting from 1419), some Mordvins became subjects of the Khanate of Kazan (喀山汗國/喀山汗国. 1438–1552),[94] whereas others were incorporated into Muscovy.[95] After Ivan IV of Russia[96] annexed the Khanate of Kazan in 1552, the Mordvin lands were seized by the Russian monarchy. Mordua is depicted on the Jenkinson/Ortelius (1562).
20	南斡牙	Nan Wo Ya	Nanwoya[97] or Nagaia[98]; it may be the modern Veliky Ustyug[99] (大烏斯秋格/大乌斯秋格). The town of Veliky Ustyug was first mentioned in a chronicle in 1207.[100]	On the ANC-KWQ, 南斡牙 (Nanwoya or Nagaia) may be the present-day Veliky Ustyug (大烏斯秋格/大乌斯秋格) in Russia, which is a town in Vologda Oblast, Russia. The town is in the northeast of the oblast at the confluence of the Sukhona River (蘇霍納河/苏霍纳河) and the Yug River (尤格河), see Fig. A7.6. Nagaia is depicted on the ANC-Jenkinson/Ortelius (1562).

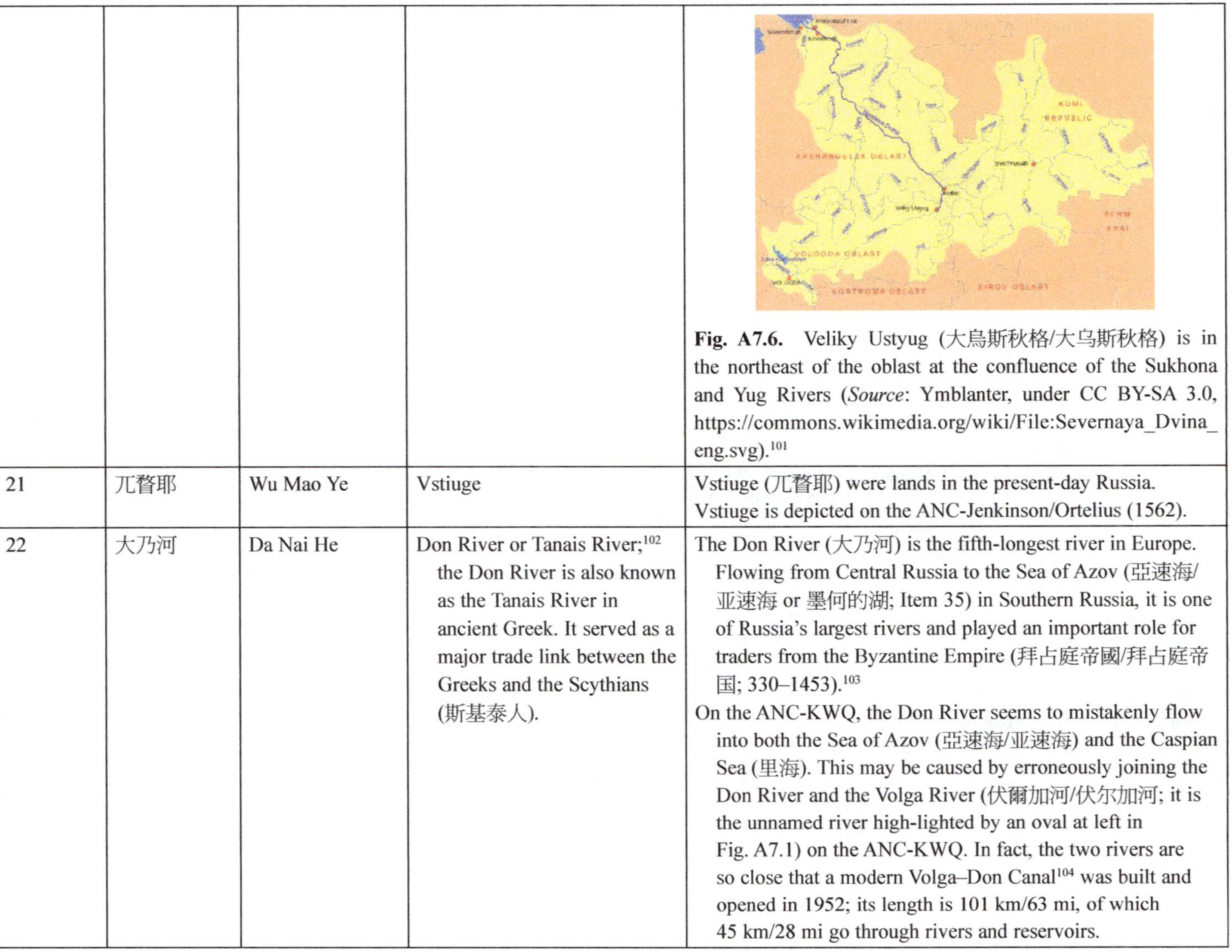

Fig. A7.6. Veliky Ustyug (大烏斯秋格/大乌斯秋格) is in the northeast of the oblast at the confluence of the Sukhona and Yug Rivers (*Source*: Ymblanter, under CC BY-SA 3.0, https://commons.wikimedia.org/wiki/File:Severnaya_Dvina_eng.svg).[101]

21	兀瞀耶	Wu Mao Ye	Vstiuge	Vstiuge (兀瞀耶) were lands in the present-day Russia. Vstiuge is depicted on the ANC-Jenkinson/Ortelius (1562).
22	大乃河	Da Nai He	Don River or Tanais River;[102] the Don River is also known as the Tanais River in ancient Greek. It served as a major trade link between the Greeks and the Scythians (斯基泰人).	The Don River (大乃河) is the fifth-longest river in Europe. Flowing from Central Russia to the Sea of Azov (亞速海/亚速海 or 墨何的湖; Item 35) in Southern Russia, it is one of Russia's largest rivers and played an important role for traders from the Byzantine Empire (拜占庭帝國/拜占庭帝国; 330–1453).[103] On the ANC-KWQ, the Don River seems to mistakenly flow into both the Sea of Azov (亞速海/亚速海) and the Caspian Sea (里海). This may be caused by erroneously joining the Don River and the Volga River (伏爾加河/伏尔加河; it is the unnamed river high-lighted by an oval at left in Fig. A7.1) on the ANC-KWQ. In fact, the two rivers are so close that a modern Volga–Don Canal[104] was built and opened in 1952; its length is 101 km/63 mi, of which 45 km/28 mi go through rivers and reservoirs.

(*Continued*)

Item #	Name	Pinyin	Etymology	History, Geography and My Comments

Fig. A7.7. The map shows the Don River flowing into the Sea of Azov. The Volga River is at the right side of this map, flowing into the Caspian Sea, see Fig. A7.7. Mr. Karl Musser created this map based on USGS data (*Source*: No machine-readable author provided. Kmusser assumed (based on copyright claims), under CC BY-SA 2.5, https://commons.wikimedia.org/wiki/File:Donrivermap.png).[105] The red dot indicates the Kerch Strait (刻赤海峽/刻赤海峡) which connects the Black Sea and the Sea of Azov, separating the Kerch Peninsula (刻赤半島/刻赤半岛) of Crimea (克里米亞/克里米亚) in the west from the Taman Peninsula (塔曼半島/塔曼半岛 of Russia's Krasnodar Krai (克拉斯諾達爾邊疆區/克拉斯诺达尔边疆区) in the east.

Fig. A7.8. Area around the Caspian Sea. The yellow area indicates the (approximate) drainage area; the Volga (伏爾加/伏尔加), Ural (烏拉爾/乌拉尔) and Atrek (阿特雷克) rivers flow into the Caspian Sea (*Source*: Redgeographics, under CC BY-SA 4.0, https://commons.wikimedia.org/wiki/File:CaspianSeaDrainage_v1.png).[106]

				The Don and Volga Rivers are both depicted on the ANC-Ortelius (1570; #1). The Tanais and Volga Rivers are both depicted on the Ortelius (1570; #2). The Volga River is also depicted on the Ortelius (1570; #3), but the Don River is depicted as an unnamed river.
23	亞私大蠟甘	Ya Si Da La Gan	Astrakhan (city)	Astrakhan (亞私大蠟甘/亚私大蜡甘; c. 600–1500) was in the southeast of the European part of Russia in the Caspian Lowland in the lower reaches of the Volga River, see Fig. A7.9. The city developed from a village into one of the main trade and political centres of the Golden Horde (1242–1502; originally a Mongol and later Turkicized khanate established in the thirteenth century and originating as the northwestern sector of the Mongol Empire).[107] Astrakhan was first mentioned by travelers in the early thirteenth century as Xacitarxan.[108] The first mention of the town Astrakhan was recorded in 1333. In the thirteenth and fourteenth centuries, it was one of the main trade and political centers of the Golden Horde.[109] 

Fig. A7.9.   Map of the Volga drainage basin (*Source*: Karl Musser, under CC BY-SA 2.5, https://commons.wikimedia.org/wiki/File:Volgarivermap.png).[110]

(*Continued*)

Item #	Name	Pinyin	Etymology	History, Geography and My Comments
				In 1395, the city was sacked by Timur (帖木兒/帖木儿; 1336–1405; in 1370 Timur gained control of the Western Chagatai Khanate ruled by the second son of Genghis Khan, and his descendants and successors, and became a Turco-Mongol conqueror who founded the Timurid Empire).[111] Astrakhan was rebuilt afterwards and became the capital of the Khanate of Astrakhan (亞私大蠟甘汗國/亚私大蜡甘汗国; 1466–1556; a Tatar rump state of the Golden Horde) in 1459.[112] In 1547, the city was seized by the Kahn of Crimean Sahib I Giray.[113] In 1556, Astrakhan was besieged and burned by Ivan the Terrible[114] (1530–1584; he was the grand prince of Moscow from 1533 to 1547 and the first Tsar of all Russia from 1547 to 1584). In Fig. A7.1, the unnamed river highlighted by the red oval at the left side should be the Volga River. But the ANC-KWQ mistakenly depicts the Volga River as another river to this river's east (see Item 25). Such a mistake does not show up in the other sixteenth century European maps analysed in this chapter. *This is further strong evidence to show that the ANC-KWQ cannot be a direct or adapted copy of these European maps.* Astrakhan is depicted as Astracan on the Ortelius (1570; #2), the Ortelius (1570l #3) and the Jenkinson/Ortelius (1562).
24	葛尔莫奇	Ge Er Mo Qi	Kalmyk was the region where Qalmaq or Calmucchi lived.	Kalmyk is alternatively spelled as Kalmuck or Qalmaq (葛爾莫奇/葛尔莫奇), and Kalmyk Oirat is sometimes called "Russian Oirat" or "Western Mongol". 葛爾莫奇/葛尔莫奇 seems to be a repetition of哥兒墨/哥儿墨 in Item 74. Both items denote where the western Mongols lived. Their ancestors were originally part of the Oirat tribes that lived in an area in Siberia and gradually spread eastwards to western China.

				The unnamed river near Kalmyk, high-lighted by an oval at right in Fig. A7.1, may be the Ural River and the mountains depicted are Ural Mountains. Here on the ANC-KWQ, 葛爾莫奇/葛尔莫奇 (Calmucchi or Qalmaq) is depicted near and to the north of the Caspian Sea, whereas on a 1784 map, Calmucchi is depicted far to the east of the Caspian Sea.[115] Calmucchi/Qalmaq is depicted as Colmack on Jenkinson/Ortelius (1562).[116]
25	勿爾瓦河即亦得	Wu Er Wa He Ji Yi De	Volga River is *Itil*;[117] Volga was formerly referred to as the Itil or *Atil*.	The Turkic peoples living along the Volga River (伏爾加河/伏尔加河) formerly referred to it as *Itil* (亦得) or *Atil* River. In modern Turkic languages, the Volga is known as *Edil* (亦得) in Kazakh.[118] The Volga (see Fig. A7.9) is the longest river in Europe and the longest river to flow into a closed basin. Rising in the Valdai Hills (瓦爾代山/瓦尔代山) about 225 m/738 ft above sea level northwest of Moscow and about 320 km/200 mi southeast of Saint Petersburg (聖彼得堡/圣彼得堡), the Volga heads east first and then turns south, to discharge into the Caspian Sea below Astrakhan (亞私大蠟甘/亚私大蜡甘) at 28 m/92 ft below sea level. Along the way, it bends toward the Don ("the big bend"; see Fig. A7.9) near Volgograd (伏爾加格勒/伏尔加格勒), formerly Stalingrad (斯大林格勒). The ANC-KWQ misidentified a river to the right of the Ural River (烏拉爾河/乌拉尔河; high-lighted by an oval at right in Fig. A7.1) as the Volga River. It is not clear which river this misidentified river may be. The Ural River flows through Russia and Kazakhstan near the continental border between Europe and Asia, as the Volga River. The Ural River originates in the southern Ural Mountains (烏拉爾山脈/乌拉尔山脉) and discharges into the Caspian Sea. It is the third-longest river in Europe after the Volga and the Danube (多瑙河). Before 1775, the Ural was known as the Yaik.

(Continued)

Item #	Name	Pinyin	Etymology	History, Geography and My Comments
				The Yaik River is depicted correctly on the Ortelius (1570; #2), the Ortelius (1570; #3) and the Jenkinson/Ortelius (1562). The Volga River is correctly depicted on the European maps shown in Figs. 7.2(b)–7.2(g). *The misidentification of the Volga River on the ANC-KWQ shows another layer of strong evidence that the ANC-KWQ is not a copy of the European maps shown in Figs. 7.2(b)–7.2(g).*
• 26	黑其泰	Hei Qi Tai	Kara Khitai[119]	The term Kara Khitai (黑其泰) is often translated as the *Black Khitans* in Mongolian, but its original meaning is unclear today.[120] "Black Khitans" (黑契丹) has also been used in Chinese. "Qara" means "black" and it corresponds to the Liao's dynastic color black.[121] The Qara Khitai Empire (1124–1218) was established by Yelü Dashi (耶律大石) who led nomadic Khitans west by way of Mongolia after the collapse of the Liao Dynasty (916–1125). The Jurchens (女真人), once vassals of the Khitans, had allied with the Song Dynasty (960–1279) and overthrown the Liao. Yelü recruited Khitans and other tribes to form an army, and in 1134 captured Balasagun (八剌沙衮/八剌沙衮; an ancient city in modern-day Kyrgyzstan) from the Kara-Khanid Khanate (喀喇汗國/喀喇汗国; 840–1212), which marks the start of the Qara Khitai Empire or Great Liao or Western Kiao (大遼/大辽 or西遼/西辽; 1124–1218) in Central Asia. Khitan is the origin of "Cathay", a foreign name for China. The name *Cathay* originates from the term *Khitan* (契丹), a para-Mongolic nomadic people who ruled the Liao Dynasty in northern China from 916 to 1125, and who later migrated west after they were overthrown by the Jin Dynasty (1115–1234) to form the Qara Khitai (Western Liao or Great Liao; 西遼/西辽 or大遼/大辽; 1124–1218) for another century thereafter.

				The ANC-KWQ still depicted this outdated history on the map.
• 27	亞細亞: 沙尔馬齊亞	Ya Xi Ya: Sa Er Ma Qi Ya	Sarmatia Asiatica	Sarmatia Asiatica (Asiatic Sarmatia; 亞細亞: 沙尔馬齊亞/亚细亚: 沙尔马齐亚), see Fig. A7.10, was the name used in Ptolemy's *Geography* (c. 150) for a part of Sarmatia (a large region which included parts of Europe and Asia.).[122] The other part was Sarmatia Europea or European Sarmatia (歐邏巴: 沙尔馬齊亞/欧逻巴: 沙尔马齐亚) which was situated further west. European Sarmatia largely corresponded to what was later known as Grand Duchy of Lithuania (立陶宛大公國/立陶宛大公国; c. 1236–1795).[123] Sarmatia was present on most maps of the region from the time of Ptolemy until the end of the eighteenth century. Maciej Miechowita (1457–1523) used "Sarmatia" for the Black Sea region and further divided it into Sarmatia Europea which included East Central Europe, and Sarmatia Asiatica. Filippo Ferrari (1551–1626) also differentiated the two Sarmatias.[124] 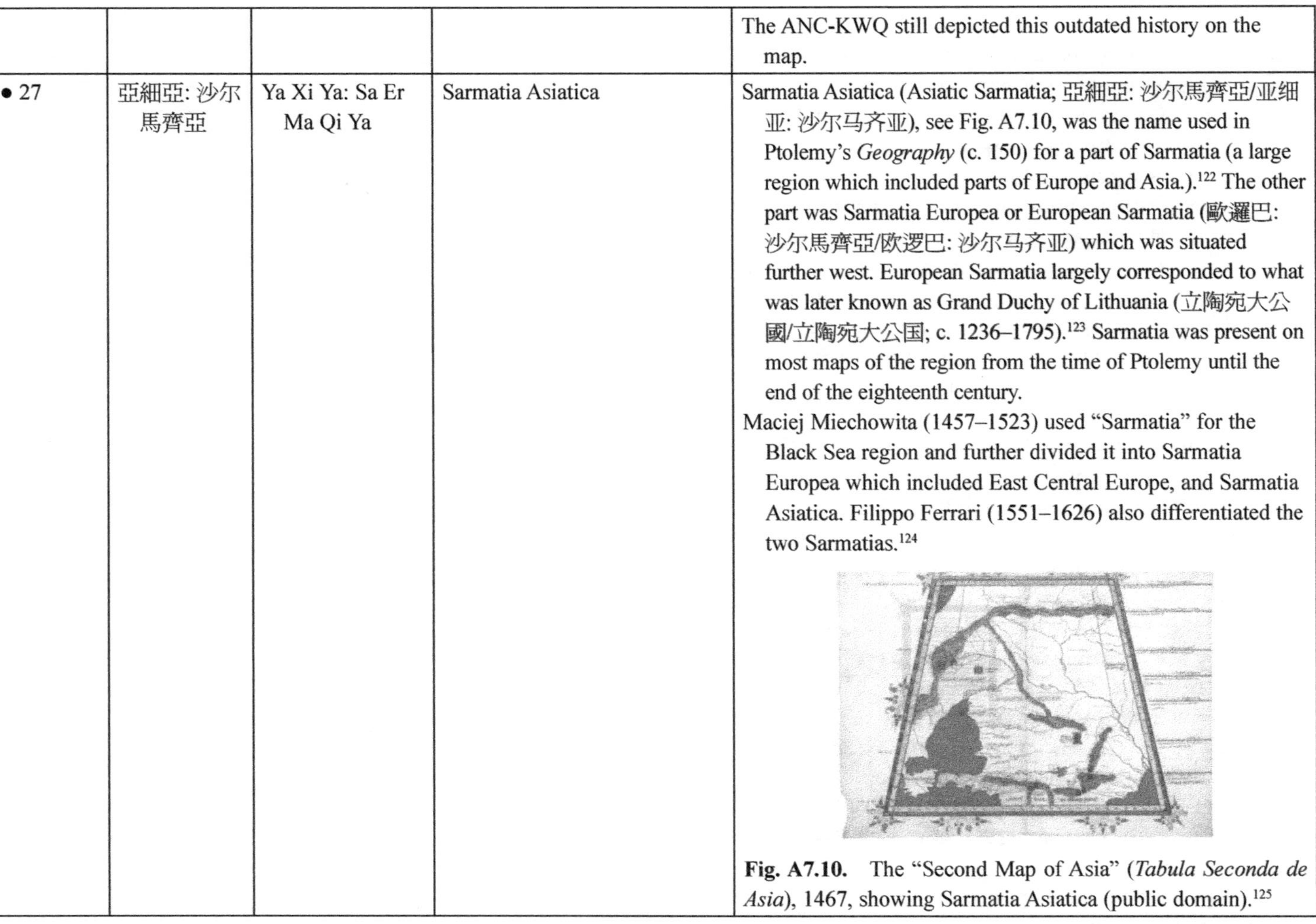 **Fig. A7.10.** The "Second Map of Asia" (*Tabula Seconda de Asia*), 1467, showing Sarmatia Asiatica (public domain).[125]

(*Continued*)

Item #	Name	Pinyin	Etymology	History, Geography and My Comments
				However, on the ANC-KWQ, both Asiatic Sarmatia and European Sarmatia are depicted to the north of the Caspian Sea (里海), which is to the far northeast of the Black Sea (黑海). Why is it so? We need to know some history about the Sarmatians to answer this question. The Sarmatians were members of a people of Iranian origin from the central parts of the Eurasian Steppe. They started migrating westward around the fourth and third centuries B.C. and reached the Volga to the east and bordered the shores of the Black and Caspian seas as well as the Caucasus (高加索; a region between the Black Sea and the Caspian Sea[126] to the south. They formed a large confederation and dominated the Pontic steppe from about the third century B.C. to the fourth century A.D.[127] In later centuries, the Sarmatians were assimilated by other civilization or absorbed into other ethnic groups. Only the Alans (阿蘭人/阿兰人),[128] an ancient and medieval Iranian equestrian nomadic pastoral people of the North Caucasus, who were a Sarmatian sub-tribe, survived in the North Caucasus into the Early Middle Ages (it is typically regarded as lasting from the late fifth or early sixth century to the tenth century). They ultimately gave rise to the modern Ossetic (奧塞梯) ethnic group.[129] The Asiatic Sarmatia and European Sarmatia depicted on the ANC-KWQ may refer to these smaller groups of Sarmatians who lived in the North Caucasus (north of the Caspian Sea) during the Early Middle Ages. Here we see that the ANC-KWQ still depicts very old geographical information in this part of the world. None of the six sixteenth-century European maps shown in Figs. 7.1(b)–7.1(g) depicts Sarmatia Asiatica because it is outdated.

				This gives another layer of evidence showing that the ANC-KWQ is not a direct or adapted copy of these European maps.
● 28	歐邏巴: 沙尔馬齊亞	Ou Luo Ba: Sa Er Ma Qi Ya	Sarmatia Europea	Please see the discussion in Item 27. None of the six sixteenth-century European maps shown in Figs. 7.1(b)–7.1(g) depicts Sarmatia Europea because it is outdated. *This gives another layer of evidence showing that the ANC-KWQ is not a direct or adapted copy of these European maps.*
@ 1	兩沙尔馬齊極寒人衣獸皮不露面只露口眼食馬血風俗朴實犯竊者即殺之	Liang Sha Er Ma Qi Ji Han Ren Yi Shou Pi Bu Lu Mian Zhi Lu Kou Yan Shi Ma Xue Feng Su Pu Shi Feng Qie Zhe Ji Sha Zhi	My translation is as follows: The two Sarmatias are extremely cold. People there wear animal skins as clothes and do not reveal their faces; they only expose their mouths and eyes. They drink horse blood, and their customs are simple: those who steal will be killed. The Sarmatians were a large confederation of ancient Eastern Iranian equestrian nomadic peoples who dominated the Pontic steppe from about the third century B.C. to the fourth century A.D. By the Early Middle Ages, the only people related to the Sarmatians may be the Alans. Please see Item 27 for a detailed discussion.
29	杜亦挐	Du Yi Na	Duina[130] or Dvina	The place was in Russia. But it cannot be identified on a modern map. Duina (杜亦挐) is depicted as Dvina on the Jenkinson/Ortelius (1562).
30	公多辣	Gong Duo La	Kondia or Konda Principality[131]	Kondia/Konda Principality (公多辣), see Fig. A7.11, was the name of a Mansi principality until the late sixteenth century.[132]

(*Continued*)

Item #	Name	Pinyin	Etymology	History, Geography and My Comments
				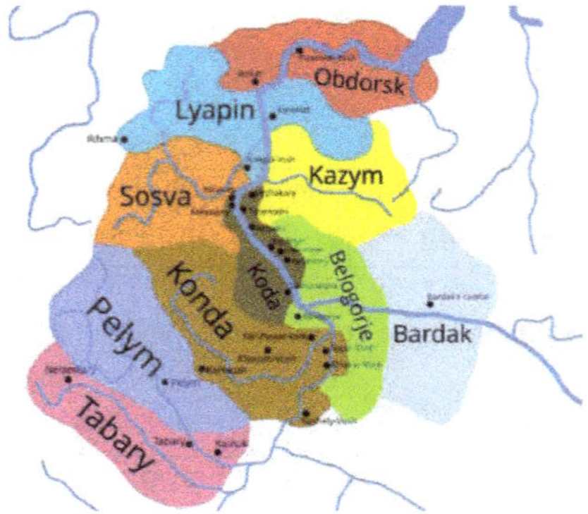 **Fig. A7.11.** Map of the Ugrian Khanty principalities. Kondia or Konda is near the centre of the map (*Source*: Jyrg, under CC BY-SA 4.0, https://commons.wikimedia.org/wiki/File:Map_of_yugra.png).[133] Little is known about when Kondia/Konda was established. But from the beginning, the principality of Koda (1400s–1643), located in Northwestern Siberia, came into conflict with the southern Khanty states of Pelym, Konda, and Tabary. From this history, we know that Konda existed before the 1400s and lasted until the late sixteenth century. Kondia/Konda is depicted as Condora on the Jenkinson/Ortelius (1562) and the Ortelius (1570; #3).
31	莫奕	Mo Yi	Moyeda[134]	The place was in Russia. But it cannot be identified on a modern map. Moyeda is depicted on the Jenkinson/Ortelius (1562).
32	合多蠟	He Duo La	The land around Obdorsk (the former name of Salekhard)[135] was referred to as Obdorsky krai, or Obdoriya (合多蠟/合多蜡). On ancient maps it is depicted as Obdora[136] (formally Obdorsk).	Salekhard, formally Obdorsk (合多蠟/合多蜡), see Fig. A7.12, is a town on the Ob River (鄂畢河/鄂毕河); the name supposedly derives from that). It is an important river port of the Russian Far North and crosses the Arctic Circle (北極圈/北极圈). The town, Salekhard, was established in 1595, but the land around Obdorsk was referred to as Obdorsky krai, or Obdoriya (合多蠟/合多蜡).

				On the ANC-KWQ, 合多蠟/合多蜡 is depicted without crossing the Artic Circle. Obdoriya is depicted as Obdora on the Ortelius (1570: #2), the Ortelius (1570: #3) and the Jenkinson/Ortelius (1562). Fig. A7.12. The location of Salekhard (dark area), formally Obdorsk (*Source*: Uwe Dedering, under CC BY-SA 4.0, https://commons.wikimedia.org/wiki/File:Russia_administrative_location_map.svg).[137]
33	三國	San Guo	Sanguo	三國/三国, literally, means "three kingdoms". This place cannot be identified on a modern map. Sanguo (三國/三国) is depicted as Sim on the Ortelius (1570; #3).[138]
● 34	白海	Bai Hai	White Sea; due to its high latitude (65.5°N) and severe cold climate, for about five months of the year (December to May) the White Sea is covered with ice.[139]	The White Sea (白海) is a southern inlet of the Barents Sea (巴倫支海/巴伦支海) located on the northwest coast of Russia. It is surrounded by Karelia (卡累利阿) to the west, the Kola Peninsula (科拉半島/科拉半岛) to the north, and the Kanin Peninsula (卡寧半島/卡宁半岛) to the northeast, see Fig. A7.13. Residents of Novgorod city[140] (c. ninth century-present; one of the oldest cities in Russia) knew this sea's commercial significance for navigation from at least the eleventh century and rapidly explored it. But the first European ship to arrive in Kholmogory (霍爾莫戈里/霍尔莫戈里; a village port) was the English ship Edward Bonaventure commanded by Richard Chancellor (c. 1521–1556) in 1553.[141] This Arctic village port lies along the Northern Dvina River (北德維納河/北德维纳河), 75 km/47 mi

(*Continued*)

Item #	Name	Pinyin	Etymology	History, Geography and My Comments
				southeast of the city of Arkhangelsk (阿爾漢格爾斯克/阿尔汉格尔斯克; shown in Fig. A7.13), and the village has existed since 1355. It served traders as a river port en route to the White Sea, then grew into a large commercial center in the fifteenth and sixteenth centuries. Between the fifteenth and early eighteenth centuries, the White Sea served as the major trade route in and out of Russia. On the ANC-Mercator (1569), there is Oceanvs Cythicvs qui & Mar Tabin. On the ANC-Ortelius (1570; #1) there is Tazata. On the ANC-Plancius (1594), there is Oceanus Tartaricus (韃靼洋/鞑靼洋). Both the Ortelius (1570: #2) and the Ortelius (1570: #3) depict Septentrio in the Arctic Polar region. On the Jenkinson/Ortelius (1562), the sea is depicted as Mar Septentrionale (北海, meaning "Sea of the North"). 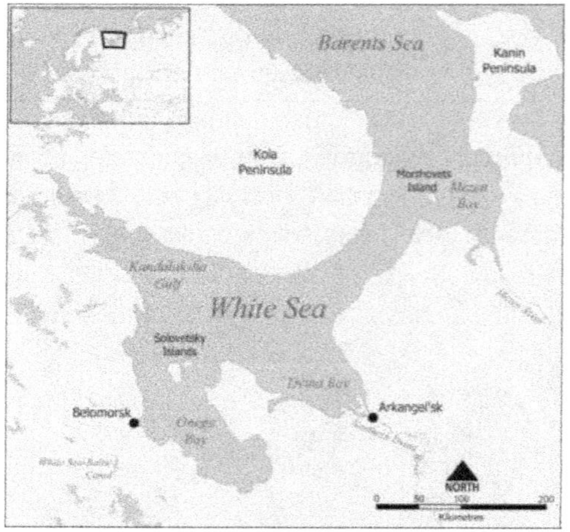

Fig. A7.13.   Map of the White Sea in Russia. Also showing the various bays and gulfs in the sea (*Source*: NormanEinstein, under CC BY-SA 3.0, https://commons.wikimedia.org/wiki/File:White_Sea_map.png).[142]

| 35 | 墨何的湖 | Mo He De Hu | Sea of Azov; Latin: *Palus Maeotis*[143] (墨何的湖) | *Palus Maeotis* (墨何的湖) is the present-day Sea of Azov (亞速海/亚速海). It is an inland shelf sea in Eastern Europe connected to the Black Sea by the narrow (about 4 km/2.5 mi) Strait of Kerch (刻赤海峽), see Fig. A7.7. The Sea of Azov is sometimes regarded as a northern extension of the Black Sea (黑海).[144] The sea is bounded by present-day Russia (俄羅斯/俄罗斯) on the east, and by Ukraine (烏克蘭/乌克兰) on the northwest and southwest, currently under Russian occupation. The Sea of Azov is depicted as Palus Maeotis on the ANC-Jenkinson/Ortelius (1562; #4).[145] |

Zone II: Middle ANC-KWQ (mainly Western Siberia and a portion of present-day Central Asia; Siberia is a region in Northern Asia separated from European Russia by the Ural Mountains in the west and bounded by the Pacific Ocean in the east)

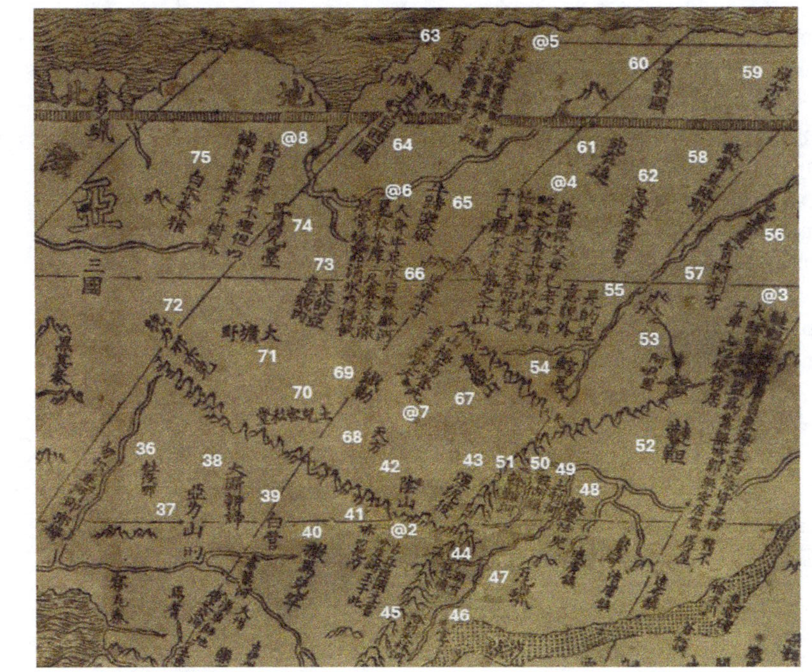

Fig. A7.14.   This figure is extracted from Fig. 7.2(a) (public domain); Items 36-75 and annotations @2-@8 are depicted on this Figure.

(*Continued*)

Item #	Name	Pinyin	Etymology	History, Geography and My Comments
36	杜沒那	Du Mei Na	Chimgi-Tura/City of Chingis, also known as Tyumen;[146] it was derived from the Turkic and Mongol word for "ten thousand" — *tumen*. The town of Tyumen was founded by Taibuga (泰布加) probably sometime between 1405 and 1428,[147] revealing that *the ANC-KWQ is a map not before the era of 1405 to 1428.*	Tyumen (杜沒那; Chimgi-Tura or Chingi-Tura, also known as Tyumen), see Figs. A7.15 and A7.16, was a medieval city of the Siberian Tatars in the twelfth to sixteenth centuries, in Western Siberia. The city was situated just east of the Ural Mountains, with the Tura River (圖拉河/图拉河) crossing the city from northwest to southeast. It was often regarded as the first Siberian city, counting eastward. On the ANC-KWQ, the mountains to the left of Tyumen (杜沒那) may correspond to the Ural Mountains. Then the river depicted as勿爾瓦河/勿尔瓦河即亦得 (Volga was formerly referred to as the Itil or Atil) on the ANC-KWQ should correspond to Ural River which flows into the Caspian Sea. The town of Tyumen was founded by Taibuga probably sometime between 1405 and 1428,[148] and became a capital of the Khanate of Sibir (西伯利亞汗國/西伯利亚汗国; 1468–1598) until the early sixteenth century when its ruler Khan Muhammad chose a new capital named Qasliq, see Fig. A7.15. The area of the Khanate of Sibir had once formed an integral part of the Mongol Empire, and later was controlled by the White Horde (白帳汗國/白帐汗国; 1226–1428) and the Golden Horde (金帳汗國/金帐汗国; 1242–1502). The Khanate of Sibir (西伯利亞汗國/西伯利亚汗国; 1468–1598) was the northernmost Muslim state in recorded history. Its defeat by Yermak Timofeyevich in 1582 marked the beginning of the Russian conquest of Siberia.[149] Tyumen is depicted as Tvmen on the ANC- Jenkinson/Ortelius (1562).

				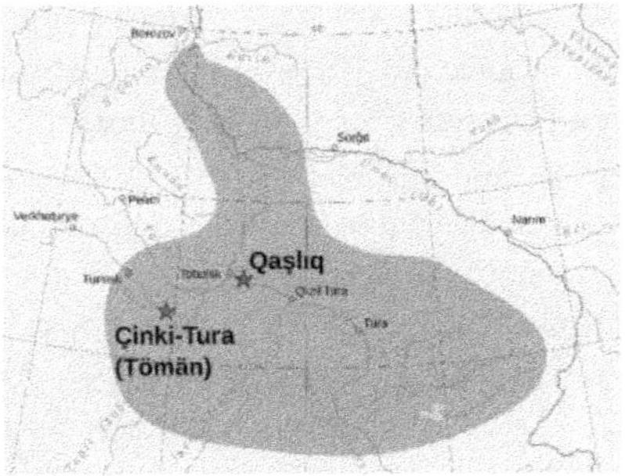  **Fig. A7.15.** Approximate extent of the Khanate of Sibir during the fifteenth and sixteenth centuries (*Source*: Ro4444, under CC BY-SA 4.0, https://commons.wikimedia.org/wiki/File:Siberian_Khanate_map_English_revised.svg).[150]  **Fig. A7.16.** *Tumen* (Chimgi-Tura) on Sigismund von Herberstein's map, published in 1549 (public domain).[151] Tyumen is depicted as Tvmen on the Jenkinson/Ortelius (1562).
● 37	亞力山的	Ya Li Shan De	Alessandria[152]	The place cannot be located on a modern map.
38	大厮耕諦	Da Si Geng Di	Tashkend[153] or Tashkent; Tashkent comes from the Turkic *tash* and *kent*, literally translated as "Stone City" or "City of Stones".[154]	Tashkend/Tashkent (大厮耕諦/大厮耕谛or塔什干) also historically known as Chach and Tashkend, is now the capital and the largest city of Uzbekistan, as well as the most populous city in Central Asia.[155] It is in today's northeastern Uzbekistan, near the border with Kazakhstan.

(*Continued*)

Item #	Name	Pinyin	Etymology	History, Geography and My Comments
				After the eleventh century, the name evolved from Chachkand to Tashkand. The modern spelling of "Tashkent" reflects Russian orthography and 20th century Soviet influence.[156] Tashkend/Tashkent is depicted as "Taskent" the Jenkinson/Ortelius (1562).
39	白營	Bai Ying	Poin[157]	Poin (白營/白营) was a town in modern Keriya (Yutian) County, Xinjiang.[158] 白營 is depicted as Poin on the Ortelius (1570; #3).[159]
40	薩馬兒罕	Sa Ma Er Han	Samarkand; the name comes from Sogdian *samar* "stone, rock" and *kand* "fort, town."[160] Alexander the Great conquered Samarkand in 329 B.C. The city was known as Maracanda by the Greeks.[161]	Samarkand (薩馬兒罕/萨马儿罕), also known as Samarqand, is a city in southeastern Uzbekistan (Fig. A7.17) and among the oldest continuously inhabited cities in Central Asia. Samarkand was one of the largest cities of Central Asia along the Silk Road between China, Persia (波斯) and Europe. In 457–509, Samarkand was part of the Kidarite State (基達利特州/基达利特州)[162] in Central Asia. The Mongols conquered Samarkand in 1220,[163] and it remained part of the Chagatai Khanate (察合台汗國/察合台汗国; one of four Mongol successor realms) until 1370.[164] In 1370, the conqueror Timur (帖木兒/帖木儿), the founder and ruler of the Timurid Empire (帖木兒帝國/帖木儿帝国; 1370–1507), made Samarkand his capital. Ibn Battuta (伊本・白圖泰/伊本・白图泰) visited the city in 1333 and called Samarkand "one of the greatest and finest of cities, and most perfect of them in beauty."[165] In 1501, Samarkand was chosen as the capital of the newly formed Khanate of Bukhara (布哈拉汗國/布哈拉汗国; 1501–1785). From 1599 to 1756, Samarkand was ruled by the Ashtrakhanid (阿什特拉卡尼德) branch[166] of the Khanate of Bukhara.

| • 41 | 亦力把力 | Yi Li Ba Li | Llbalk;[168] "Yilibali" is a homonym of "Yili Bali", which is near today's Yili (伊犁) in Xingjiang (新疆). | The Eastern Chagatai Khanate (東察合台汗國/东察合台汗国; 1347–1570 or 1347–1705;[169] also known as Moghulistan Khanate) was a Mongol and later Turkicised khanate. It moved its capital to Yilibali (亦力把力) starting from 1418, the sixteenth year of the Ming Yongle period).[170]
The Eastern Chagatai Khanate originated from the Chagatai Khanate (察合台汗國/察合台汗国). Chagatai (察合台) is the second son of Genghis Khan (成吉思汗). When Genghis Khan conquered a large area of land, he gave these lands to his four sons, namely the four major Khanates of the Yuan Dynasty: the Golden Horde (金帳汗國/金帐汗国; 1242–1502), the Chagatai Khanate (察合台汗國/察合台汗国; 1227–1570 or 1347–1705),[171] the Wokuotai Khanate (窩闊台汗國/窝阔台汗国; 1251–1309), and the Ilkhanate (伊利汗國/伊利汗国; 1256–1333/1335).[172] |

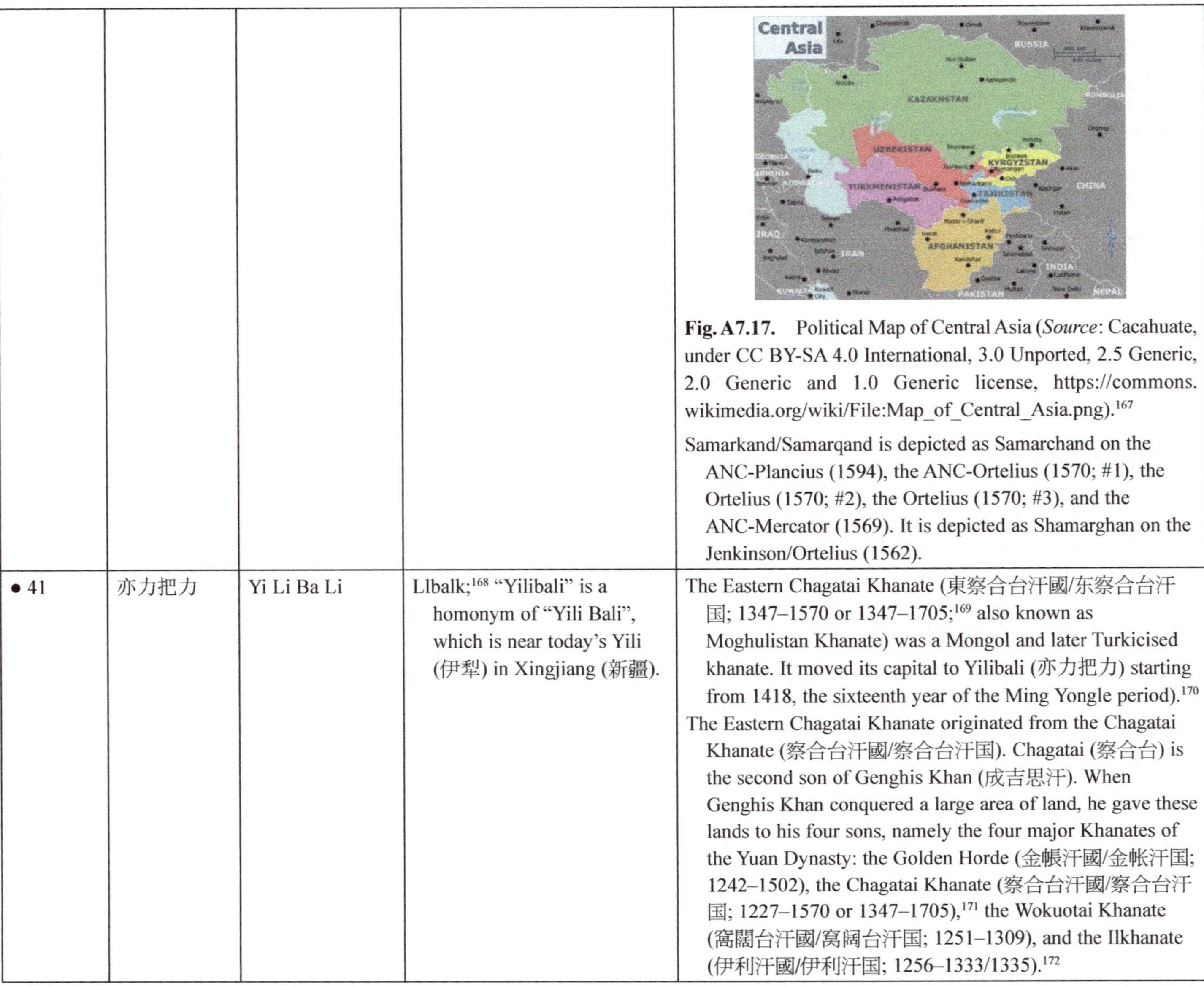

Fig. A7.17.   Political Map of Central Asia (*Source*: Cacahuate, under CC BY-SA 4.0 International, 3.0 Unported, 2.5 Generic, 2.0 Generic and 1.0 Generic license, https://commons.wikimedia.org/wiki/File:Map_of_Central_Asia.png).[167]

Samarkand/Samarqand is depicted as Samarchand on the ANC-Plancius (1594), the ANC-Ortelius (1570; #1), the Ortelius (1570; #2), the Ortelius (1570; #3), and the ANC-Mercator (1569). It is depicted as Shamarghan on the Jenkinson/Ortelius (1562).

(Continued)

Item #	Name	Pinyin	Etymology	History, Geography and My Comments
				The Khanate of the Great Khan (大汗汗國/大汗汗国; 1251–1368), encompassing Mongolia, China, and East Asia, refers to the Yuan Dynasty in China, which was founded by Kublai Khan (忽必烈汗), Genghis Khan's grandson. Kublai completed the conquest of China started by Genghis Khan and became the first Yuan ruler of the entire country. Initially, the rulers of the Chagatai Khanate recognised the supremacy of the Great Khan (Yuan Emperor), but by the reign of Kublai Khan, he no longer obeyed the emperor's orders. During the 1340s, the Chagatai Khanate was divided into two, becoming the East Chagatai Khanate (1347–1570 or 1347–1705) and the West Chagatai Khanate which fell to the Timurid Empire in 1369, and the Khanate ended in 1402. After the fall of the Yuan Dynasty in 1368, the Eastern Chagatai Khanate took the Ming Dynasty as its suzerain, and after that, most of the Ming historians no longer record the country as the Eastern Chagatai Khanate, but referred it by the name of its capital, 亦力把里 or 亦力把力. Therefore, on many maps of the Ming Dynasty, Yilibali is depicted to replace the East Chagatai Khanate. However, Yilibali came to pay tribute less and less after 1465 and finally ended when Turpan Khanate (吐魯番汗國/吐鲁番汗国; also known as the Eastern Moghulistan) conquered Yilibali in the early sixteenth century.[173] Instead, the Turpan City became the capital of the Eastern Chagatai Khanate and the name Yilibali (亦力把力) disappeared from the records of the Ming Dynasty. Yilibali is not depicted on any of the sixteenth century European maps analysed in this chapter. Because after the early sixteenth century the Turpan City became the capital of the Eastern Chagatai Khanate and most of the nomadic tribes in the Yilibali area in the Kazakh Khanate (哈薩克汗國/哈萨克汗国; 1465–1847) and became its components.

@ 2	古丘茲國元嘗分建諸王子于此	Gu Qiu Zi Guo Yuan Chang Fen Jian Zhu Wang Zi Yu Ci	My translation and comment (between square brackets) are as follows:

My translation and comment (between square brackets) are as follows:

This is where the ancient Kingdom of Qui Zi [丘茲 or龟茲、曲先、丘慈、邱慈、屈支、丘茲、拘夷、歸茲、屈茨、庫徹、庫車; Chinese transcriptions of the Han or the Tang imply that *Küchï* was the original form of the country's name] was.[174] During the Yuan Dynasty, several princes have established territories here.

The annotation states that the land was once the homeland of the ancient Buddhist Kingdom of Kucha (庫車/库车) or Kuche (庫車/库车; it was a Uyghur town; second century B.C.–648 A.D.). It was located on the branch of the Silk Road that ran along the northern edge of what is now the Taklamakan Desert[175] (塔克拉瑪乾沙漠/塔克拉玛干沙漠) in the Tarim Basin (塔里木盆地) and south of the Muzat River (穆扎特河). The former area of Kucha now lies in present-day Aksu Prefecture (阿克蘇地區/阿克苏地区), Xinjiang, China.

In 630, Xuanzang (玄奘; 602–664),[176] a well-known Chinese Buddhist monk, scholar, traveler and translator visited Kucha during the Early Tang period. Buddhism was introduced to Kucha before the end of the first century.[177] However, it was not until the fourth century that the kingdom became a major center of Buddhism.[178]

The book titled *Hudud ul-'alam min al-mashriq ila al-maghrib*, written in 982 by an unknown Arab or Persian writer and presented to the ruler of Guzgan (古茲甘; present day northern Afghanistan), states the following regarding the area of Kucha (translation and comment between square brackets by Sheng-Wei Wang):

Kucha is located on the Chinese border and belongs to China, but the indigenous people, Dokuzoguzes [多庫佐古澤斯/多库佐古泽斯], at times are engaged in raids and looting. This city has many advantages.

(*Continued*)

Item #	Name	Pinyin	Etymology	History, Geography and My Comments
				Kucha was conquered by the Tang Dynasty in the seventh century. After the later periods of Tibetan and the Uyghur domination, it became an important center of the Uyghur Kingdom of Qocho (843–fourteenth century; 高昌回鶻/高昌回鹘). In 1209, the Kingdom of Qocho became a vassal of the Mongol Empire,[179] because during the Yuan Dynasty (1271–1368), several princes separately established their territories there, referring to Genghis Khan giving huge lands to his four sons, namely the four major Khanates of the Yuan Dynasty as stated in Item 41. Kucha became part of the Chagatai Khanate. By the 1370s, the Kingdom of Qocho ceased to exist.[180] In 1758, the Qing Dynasty took control of the area, and the Chinese character name "Kuche" (庫車/库车) was made the name of the area.
● 42	陰山	Yin Shan	Yin Mountains: the Yin Mountains are also known by their Chinese name as "Yinshan".	The Yin Mountains (陰山/阴山)[181] are a mountain range stretching across about 1,000 km/ 621.37 mi of northern China. They form the southeastern border of the Gobi Desert (戈壁) and cross the Chinese provinces of Inner Mongolia and Hebei. Among other things, the range is notable for its petroglyphs.[182]
● 43	漠泥彼	Mo Ni Bi	Monibei[183]	This place cannot be identified on a modern map.
● 44	濶百花渡	Kuo Bai Hua Du	Kuobaihuadu[184]	On the ANC-KWQ, 濶百花渡is depicted near the confluence of the Orkhon River (鄂爾渾河/鄂尔浑河 or 哈剌禾林河; see explanation in Item 45) and the Selenge River (色楞格河; a major river in Mongolia and Buryatia, Russia). This place is Sükhbaatar (蘇赫巴托爾/苏赫巴托尔), a city in northern Mongolia on a modern map. Hence, 濶百花渡 (Kuobaihuadu) may be the present-day Sükhbaatar city. The Dzhida River (吉達河/吉达河) is a left tributary of the Selenge River (色楞格河), see Fig. A7.18.

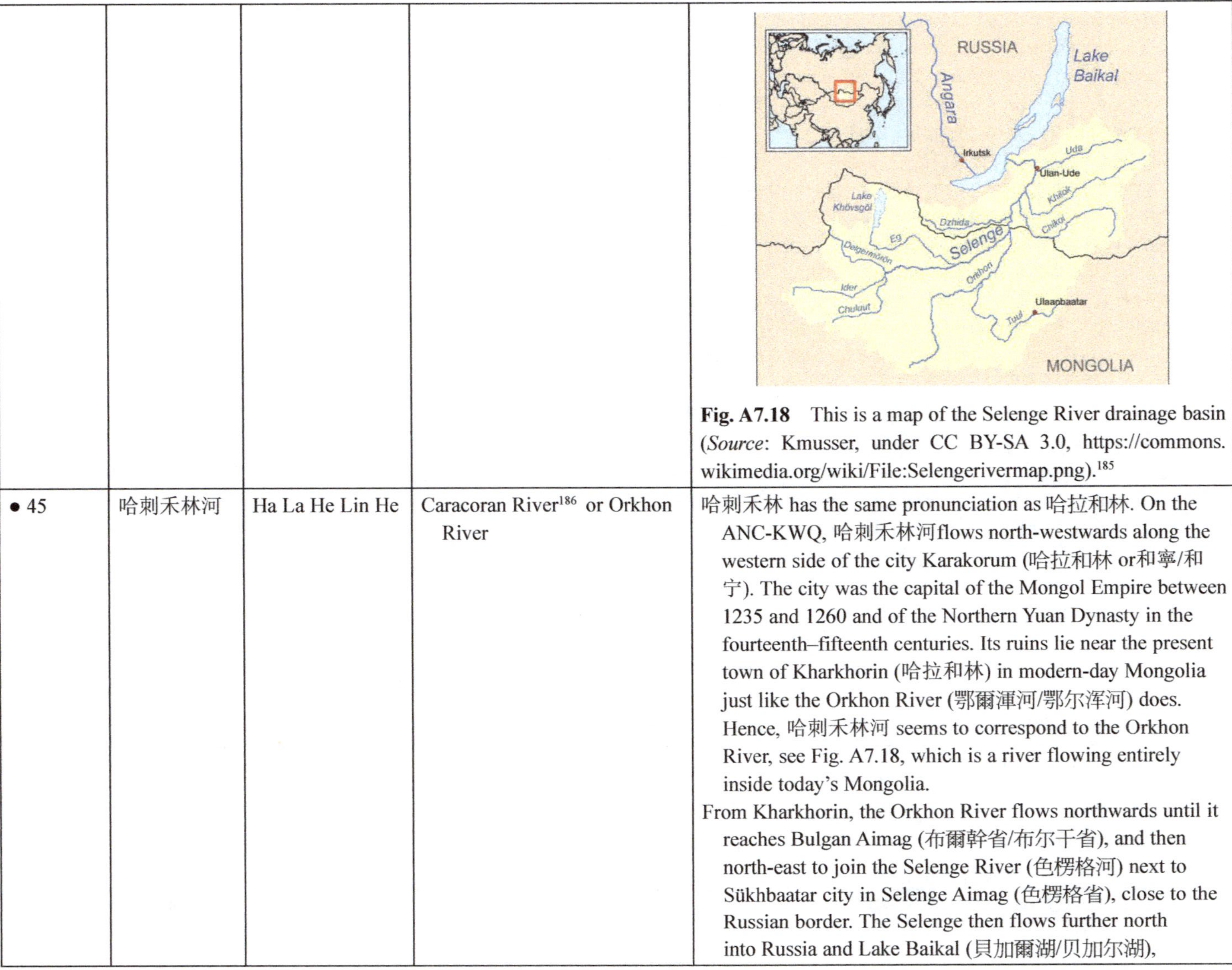

Fig. A7.18   This is a map of the Selenge River drainage basin (*Source*: Kmusser, under CC BY-SA 3.0, https://commons. wikimedia.org/wiki/File:Selengerivermap.png).[185]

● 45	哈剌禾林河	Ha La He Lin He	Caracoran River[186] or Orkhon River	哈剌禾林 has the same pronunciation as 哈拉和林. On the ANC-KWQ, 哈剌禾林河 flows north-westwards along the western side of the city Karakorum (哈拉和林 or和寧/和宁). The city was the capital of the Mongol Empire between 1235 and 1260 and of the Northern Yuan Dynasty in the fourteenth–fifteenth centuries. Its ruins lie near the present town of Kharkhorin (哈拉和林) in modern-day Mongolia just like the Orkhon River (鄂爾渾河/鄂尔浑河) does. Hence, 哈剌禾林河 seems to correspond to the Orkhon River, see Fig. A7.18, which is a river flowing entirely inside today's Mongolia. From Kharkhorin, the Orkhon River flows northwards until it reaches Bulgan Aimag (布爾幹省/布尔干省), and then north-east to join the Selenge River (色楞格河) next to Sükhbaatar city in Selenge Aimag (色楞格省), close to the Russian border. The Selenge then flows further north into Russia and Lake Baikal (貝加爾湖/贝加尔湖),

(*Continued*)

Item #	Name	Pinyin	Etymology	History, Geography and My Comments
				see Fig. A7.18. The Orkhon is the longest river in Mongolia. Major tributaries of the Orkhon River are the Tuul River (圖拉河/图拉河) and Tamir River (塔米爾河/塔米尔河). However, on the KWQ, Lake Baikal is not depicted.
46	和寧	He Ning	Hening/Caracorum/Karakorum	Karakorum (often spelt Qaraqorum or Caracorum; 和寧/和宁 or 哈拉和林) is in the Orkhon Valley of central Mongolia and was the capital of the Mongol Empire[187] from 1235 to 1260. Karakorum was later replaced as the Mongol capital by Daidu (大都; present-day Beijing) and Xanadu/Shangdu (上都; the summer capital of Yuan Dynasty; the city was originally named Kaiping). But the Northern Yuan Dynasty (1368–1635) still used Karakorum as its capital. In the early fifteenth century, various parts of Mongolia fell apart; the two Mongolian confederations of Dada/Tatars/Tarta[188] and Wala/Oirats attacked each other, and Karakorum gradually declined.[189]
●47	瓦剌	Wa La	Wala/Oirats;[190] Oirats, also formerly Eluts and Eleuths (厄魯特) are the westernmost group of the Mongols whose ancestral home was in the Altai region of Siberia, Xinjiang and western Mongolia. The first documented reference to Elut and Yelut was dated in the sixth century.[191]	The rump state after the collapse of the Yuan Dynasty (1271–1368) is known by various names, including the Northern Yuan Dynasty (北元; 1368–1635). However, even the name "Northern Yuan (北元)" is sometimes limited in its usage to referencing only the period between 1368 and 1388. In 1388,[192] the name of the country was renamed as "Mongolia (蒙古本部)". Later, the Mongolian headquarters and the four Oirats (四衛拉特/四卫拉特 or 瓦剌諸部/瓦剌诸部; 瓦剌諸部/瓦剌诸部; 1399–1634) split; Wuliangha (兀良哈), Hami (哈密) and other tribes temporarily surrendered to the Ming Dynasty. The Ming Dynasty called the political realm of Mongolian tribes in Eastern Mongolia as Dada (韃靼/鞑靼; later the region occupied by this

political realm was known in English as "Tartary"; Item 52; 1388–1635) and the confederation of the Mongolian tribes in Western Mongolia as Oirats (瓦剌or瓦蠟/瓦蜡; Alliance of the Four Oirats or the Oirat Confederation; 1399–1634; Item 47).[193]

In the third year (1438) of Zhengtong (正統/正统) in the Ming Dynasty, Tuo Tuo Buhua (脫脫不花), supported by Tuo Huan (脫歡/脱欢) of the Oirat tribe, defeated the Great Khan Adai (大汗阿岱), and unified the Mongolian tribes.[194]

Then, in the fourteenth year (1449) of the Zhengtong (正统) period in the Ming Dynasty, the son of Tuo Huan first invaded the Ming China and captured the Ming Emperor Yingzong Zhu Qizhen (明英宗朱祁鎮/明英宗朱祁镇).[195] In 1453, he declared himself the "Da Yuan Tian Sheng Da Khan (大元田盛大可汗)",[196] but died in 1454, and the Oirats gradually declined afterwards. Mongolia was divided again.

Later, the Dayan Khan (達延汗/达延汗; 1472–1517; from Eastern Mongolia) revived Mongolia and reunified the various ministries in the late fifteenth century (1487).[197] Since then, the Oirats completely lost their power to compete with Eastern Mongolia and gradually moved westward. In 1771, the leader of the Oirats surrendered to the Qing court (清廷), ending the independent history of Oirat.

For a more detailed explanation, please also refer to Subsection 2.2 in the main text of this chapter.

The Oirats are not depicted on any of the sixteenth century's European maps analysed in this chapter, because by then the Oirats already lost their strength to confront the Tartars in the Eastern Mongolian tribes. Only Tartary or Tartaria is depicted on these European maps, but the meaning was very different: *it became a blanket term used in Western European literature and cartography for a vast part of Asia bounded by the Caspian Sea, the Ural Mountains, the Pacific Ocean, and the northern borders of China, India and Persia, at a time when this region was largely unknown to European geographers.*

(Continued)

Item #	Name	Pinyin	Etymology	History, Geography and My Comments
● 48	撒里怯兒	Sa Li Qie Er	Sarikil[198]	Sarikil (薩里川/萨里川、撒里怯兒/撒里怯儿or撒阿里客額兒/撒阿里客额儿) is in the west of the upper reaches of the Kherlen River (克魯倫河/克鲁伦河) in today's Mongolia.[199] Sarikil was the grassland where one of the four palaces of Genghis Khan was located. In the period after the Mongol Empire, Ming Dynasty troops in the Yongle era, who went to subjugate Mongolia, passed through this area.[200]
● 49	土剌河	Tu La He	Tuul River[201]	On the ANC-KWQ, the Tuul River (土剌河; see Fig. A7.18) is depicted to the east of Onon (幹難河/干难河). This is incorrect. The Tuul River is in the Selenge River basin, whereas the Onon River is in the Amur River drainage basin which is to the east of the Selenge River basin and extends to the Pacific Ocean coastal area. The Tuul River flows through the Selenge River into Lake Baikal, see Fig. A7.18. It is part of the Yenisey River system that flows into the Arctic Ocean. But on the ANC-KWQ, the Tuul River mistakenly flows into the Lena River which also flows into the Arctic Ocean. The Battle of Tuul River was a famous battle between the Northern Yuan (1368–1635) and Ming (1368–1644) dynasties. In the first month of the fifth year (1372) of the Hongwu (洪武) period in the Ming Dynasty, Zhu Yuanzhang (朱元璋; reigning from 1368 to 1398) sent troops to conquer Timur (帖木兒/帖木儿), the general of the Northern Yuan Dynasty. The Ming troops defeated the Northern Yuan army in the Tuul River area in present-day Ulaanbaatar (烏蘭巴托/乌兰巴托), Mongolia (see Fig. A7.18), and captured many northern Yuan soldiers.
● 50	殺胡鎮	Sha Hu Zhen	Shahuzhen[202]	The place is recorded in Ming Dynasty's historical record.[203]

| 51 | 幹難河 | Gan Nan He | Gannan River/Onon River[204] is a river in present-day Mongolia and Russia. | The Gannan/Onon (鄂嫩河; old name was幹難河/干难河) is a river in Mongolia and Russia. Figure A7.19 shows the Onon, Shilka (石勒喀河) and Amur (Heilongjiang; 黑龍江/黑龙江; Amur is the world's tenth longest river and the natural border between the Russian Far East and Northeast China (historically Outer and Inner Manchuria). On the ANC-KWQ, the Onon River (幹難河/干难河) is depicted as a river flowing into the Selenge River, which then flows into the Arctic Sea. But this is not correct. The Onon River drainage basin is to the east of and separated from the Selenge River basin. The Onon River originates from an area which constitutes the divide between the Arctic (Tuul River; see Fig. A7.18) and Pacific (Kherlen River and Onon River; see Fig. A7.19) basins and is consequently named "Three River Basins". The upper Onon is one of the areas that are claimed to be the place where Genghis Khan was born and grew up.[205] 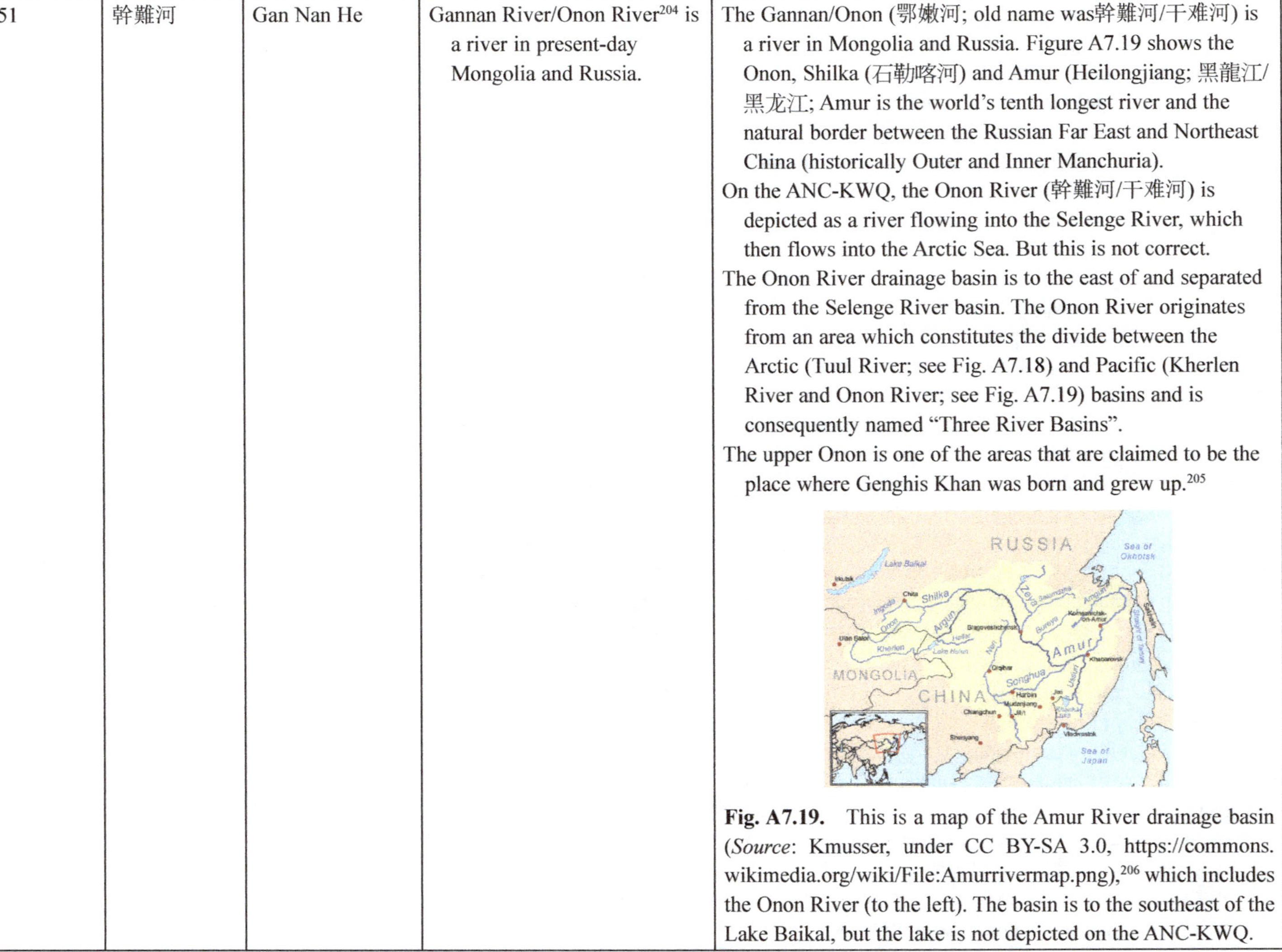 **Fig. A7.19.**   This is a map of the Amur River drainage basin (*Source*: Kmusser, under CC BY-SA 3.0, https://commons.wikimedia.org/wiki/File:Amurrivermap.png),[206] which includes the Onon River (to the left). The basin is to the southeast of the Lake Baikal, but the lake is not depicted on the ANC-KWQ. |

(*Continued*)

Item #	Name	Pinyin	Etymology	History, Geography and My Comments
				The Ming Yongle Emperor (Zhu Di; 朱棣; 1360–1424) named the Gannan (斡難河/干難河) as Xuanming River (玄冥河) in May 1410: "…命都督薛祿祭斡難川山川，賜名玄冥河 (translation by Sheng-Wei Wang: … ordered the governor Xue Lu to offer sacrifices to the mountains and rivers of the Gannan River and named it Xuanming River)."[207] In 1410, the Ming Yongle Emperor led his army to the north and in May, they annihilated the forces of Bunyashir (本雅失里) of the Eastern Mongolia at the Onon River,[208] because Bunyashir had executed the Ming ambassador.[209]
52	韃靼	Da Da	Dada; later the region where the Dada people (Tartars) lived was known in English as "Tartary" (Latin: *Tartaria*). Dada (韃靼/靼韃) was the Ming Dynasty designation for the confederation of the Eastern Mongolian tribes originally ruled by the Northern Yuan Dynasty after the end of the Yuan Dynasty period. To the west are the confederation of the Mongolian Oirat tribes who were nomadic in the northwest of the Mongolian Plateau.[210]	The name of the Dada people/Tartars (韃靼人/韃靼人) already appeared in the Northern and Southern dynasties (420–589). It was the name used by the Turkic (突厥) tribes to call the Shiwei (室韋/室韦) tribes. Shiwei refers to the descendants of the Tribal Khanate Rouran (柔然汗國/柔然汗国; 402–555; a branch of the Xianbei 鮮卑/鲜卑 tribe)[211] in the north of Khitan (契丹) or related tribes of the Mongolian language family. After Rouran (柔然) was defeated by the Turkic people, it was divided into two branches, the north and the south branches. The south branch came to the Laoha River (老哈河) and the Xila Mulun River (西拉沐倫河/西拉沐伦河) in the upper reaches of the Liao River (遼河/辽河; the principal river in southern Northeast China) for hunting; they became the origin of the Khitan (契丹) people; the north branch went to the southern region of the present-day Stanovoy Range (斯塔諾夫山脈/斯塔诺夫山脉) or Outer Khingan Range (外興安嶺/外兴安岭) located in southeastern parts of the Russian Far East;[212] they were called Shiwei and became the ancestor of the Mongols. Shiwei and Khitan came from the same origin, separated by the Greater Khingan Range (大興安嶺/大兴安岭) which is the

watershed between the Nen River (嫩江) and the Songhua River (松花江) systems to the east, and the Amur and its tributaries to the northwest.[213]

In Turkic, "Shiwei" is called "Tatar". In the tenth century, the Tatars became a subordinate of the Liao Dynasty (遼朝/辽朝; 916–1125). Then, after the fall of the Liao Dynasty, their confederation became a vassal state of the Jin Dynasty (金朝 or 大金; 1115–1234). In the twelfth century, their pastures were in the Hulunbuir (呼倫貝爾/呼伦贝尔) area. After the rule of the Khitans in the Liao Dynasty and the Jurchens (女真人) in the Jin Dynasty, most of the Tatars were integrated into the forming of Mongolia and became one of the main sources of the Mongolian nation.

At the beginning of the thirteenth century, different groups of these Mongolian Turkic nomads became part of the troops of the Mongol conqueror Genghis Khan. Afterward, the Mongols and the Turks were mixed with each other. *Therefore, the Mongolian troops that invaded Russia and Hungary were collectively called Tatars by Europeans.* For example, on the ANC-Ortelius (1570; #1) and the Jenkinson/Ortelius (1562), the Latin word *Tartaria* is depicted in European Russia; on the ANC-Plancius (1594), *Tartaria* is depicted from the region of European Russia across the entire Siberia to the Pacific Coast; on the Ortelius (1570; #2) and the Ortelius (1570; #3), *Tartaria* covers the region of European Russia and Western-Central Siberia. The meanings of *Tartaria* on these maps become very different from Dada (韃靼/鞑靼) on the ANC-KWQ.

Other details are given in the discussion of Oirats (瓦蠟/瓦蜡 or 瓦剌; Item 47).

(Continued)

Item #	Name	Pinyin	Etymology	History, Geography and My Comments
@ 3	韃靼地方甚廣自東海至西海皆是種類不一大概習非以盜爲業無城郭無定居架房屋於車上以便移居	Da Dan Di Fang Shen Guang Zi Dong Hai Zhi Xi Hai Jie Shi Zhong Lei Bu Yi Da Gai Xi Fei Yi Dao Wei Ye Wu Cheng Guo Wu Ding Ju Jia Fang Wu Yu Che Shang Yi Bian Yi Ju	My translation and comments (between square brackets) are as follows: The territory of the Tatars is vast, which extends from the eastern sea [referring to the Pacific Ocean; on the KWQ there is 大東洋/大东洋, lit. "Big/Great Eastern Ocean"] to the western sea [this is less obvious; there is 大西洋, lit. "Big/Great Western Ocean" or "Atlantic Ocean" at the western end of the European Continent) in the far west]. There are different tribes. Most of them do not have good habits but steal for living. They do not build cities and have no settlements. They build huts on carriages for easy moving from place to place. The last sentence denotes that the Tatars are nomadic people living in covered wagons; their huts are built on well-wheeled chariots. However, did the Tatars reach the western sea (Atlantic Ocean)? This requires an explanation. In East Asia, it was first seen in Turkic inscriptions and some Chinese records in the Tang Dynasty that there were translated names like "Data, Tatan, Tatar and Dada". But what these names referred to vary with the times: in the Liao and Jin dynasties, they referred to the Tatar tribes in the east of the Mongolian Plateau; in the Ming Dynasty, they referred specifically to the Tatar regime established in the east of the Mongolian Plateau; whereas in Europe, they were first used to refer to the Bulgars, and later became Russia's collective name for all ethnic groups that used Turkic languages in the Eurasian continent. In modern days, they refer to some Turkic and Mongolian peoples and their descendants once ruled by the Golden Horde in Europe.
● 53	阿只里	A Zhi Li	Azhili[214]	This geographical item cannot be identified on a modern map.

● 54	轄戛	Xia Jia	Same as Qirqiz (Kyrgyz), or Kirghiz, or Khirkhiz; same as 乞里吉里(思) (Item 56)	Xia Jia (轄戛/辖戛) was an ethnic minority in the northwestern region during the Five Dynasties (907–960)[215] in Chinese history. It is recorded in *Xin Wu Dai Shi • Si Yi Fu Lu Er* (新五代史 • 四夷附录二)[216] or *New History of the Five Dynasties, Appendix II on the Four Barbarians*: 又其西, 轄戛, 又其北, 單于突厥, 皆與嫗厥 律略同。 My translation and comments (between square brackets) are as follows: In the west, it is the territory of Xia Jia [轄戛/辖戛]; and in the north, it is the territory of the Turkic people [單于/单于was the title of the tribe's leader]. All of them are generally like the Yujuelü [嫗厥律/妪厥律] people. Again, the depiction of轄戛 (Xia Jia) on the ANC-KWQ indicates that the map has kept this very old geographical item without updating it in accordance with the dynastic changes in Chinese history.
55	是的亞意貌外	Shi Di Ya Yi Mao Wai	Scythia extra Imaum[217]	Scythia (是的亞/是的亚) was a kingdom created by the Scythians (是的亞人/是的亚人)[218] during the sixth to third centuries B.C. in the Pontic–Caspian steppe[219] (steppe extending across Eastern Europe to Central Asia). In the time of the Roman Empire, the name Scythia covered the whole of Northern Asia, from the river Volga on the west, to Serica (which is generally taken as referring to North China and can be reached via the overland Silk Road in contrast to the Sinae, who were reached via the maritime routes) on the east, extending to India on the south. Scythia was divided by Mount Imaus (意貌山) into two parts, called respectively Scythia intra Imaum (是的亞意貌內; Item 73), i.e. on the northwestern side of the range, and Scythia extra Imaum (是的亞意貌外; Item 55), on its southeastern side.[220]

(*Continued*)

Item #	Name	Pinyin	Etymology	History, Geography and My Comments
				The Pontic Scythian Kingdom lasted until the third century B.C., when a related Iranic people (伊朗人), the Sarmatians (薩爾馬提亞人/萨尔马提亚人), conquered much of the Pontic Steppe.[221] After this, the final Scythian polities were eventually assimilated by the Roman Empire, Sarmatians, Goths (哥特人), Huns (匈奴), and Early Slavs (早期斯拉夫人) from the first to the sixth centuries A.D, after which the Scythians disappear from history.[222] Scythia extra Imaum is not depicted on any of the European maps analysed in this chapter, but they are depicted on Totius Europae et Asiae Tabula Geographica, Auctore Thoma D. Aucupario. Edita Argentorati, mdxxii/1522.[223]
@ 4	此國俗父母已老子自殺之而食其肉以此爲恤雙親之苦勞而葬之於己腹不忍棄之於山	Ci Guo Su Fu Mu Yi Lao Zi Zi Sha Zhi Er Shi Qi Rou Yi Ci Wei Xu Shuang Qin Zhi Ku Lao Er Zang Zhi Yu Ji Fu Bu Ren Qi Zhi Yu Shan	My translation is as follows: This country has a national custom: when their parents become old, the children would kill them and eat their flesh to commemorate the hard work of their parents and bury them in their own stomachs. They could not bear to abandon their parents on the mountains. According to Greek historian and geographer Herodotus, the Issedones (an ancient people of Central Asia at the end of the trade route leading north-east from Scythia) practiced similar ritual cannibalism of their elderly males. Afterward, it was followed by a ritual feast at which the deceased patriarch's family ate his flesh, gilded his skull, and placed it in a position of honor much like a cult image.[224]

56	乞里吉里(思)	Qi Li Ji Li	Qilijisi;[225] same as Qirqiz (Kyrgyz), or Kirghiz, or Khirkhiz	乞里吉里 should be corrected as乞里吉思 (Kirghiz) which was a formerly nomadic people dwelling chiefly in Kirghizia (Kyrgyzstan).[226] They were a Turkic ethnic group native to Central Asia. In the Yuan Dynasty they were primarily found in the upper reaches of the Yenisey River (葉尼塞河/叶尼塞河).[227] In 1270, they were under the direct rule of the Yuan Dynasty. Later, after some political and military turmoil, with troops stationed and guarded, the Yuan Dynasty restored its rule in 1293.[228] However, the corrupt Yuan Dynasty was eventually destroyed by the peasants' uprising. The Red Turban Rebellions led by Zhu Yuanzhang (朱元璋) succeeded in establishing the Ming Dynasty (1368–1644) to replace the Yuan Dynasty (1271–1368).[229] 乞里吉思 (Kirghiz) existed as a rump state after the collapse of the Yuan Dynasty in 1368 and lasted until its conquest by the Jurchen-led Later Jin Dynasty (后金; 1616–1636) in 1635.[230]
57	育阿利牙	Yu A li Ya	Ugria/Vgori	Ugria/Vgori (育阿利牙) was in present-day Russia. Historically, the Ugrians or Ugors were the linguistic ancestors of the present-day Hungarians, and the Khanty and Mansi peoples of Western Siberia.[231] By 3000 B.C., the Ugrians formed their own group. Ugria/Vgori is depicted as Iovghoria on the Jenkinson/Ortelius (1562). It was the country of "Zlata Baba" or "Golden Old Woman" who was a legendary idol and an alleged item worshipped by the people of Obdora at the mouth of the Ob River (鄂畢河/鄂毕河).[232] The Golden Woman also appears as Zlata Baba on the Jenkinson/Ortelius (1562).
● 58	路骨莫路耶	Lu Gu Mo Lu Ye	Lugumoluye[233]	This place cannot be identified on a modern map. But it should be in present-day Siberia, Russia.
● 59	漠尔拔	Mo Er Ba	Morpa[234]	This place cannot be identified on a modern map. But it should be in present-day Siberia, Russia.

(Continued)

Item #	Name	Pinyin	Etymology	History, Geography and My Comments
● 60	曷剌國	He La Guo	Kingdom of Hela[235]	The land of the Kingdom of Hela (曷剌國/曷剌国) was in northern Siberia and close to the Arctic Ocean (the Chinese called it the North Sea, lit. "北海"), see Fig. A7.20, roughly in the area near today's Taymyr Peninsula (泰梅爾牛島/泰梅尔牛岛). There lived one of the ancient northern Turkic tribes.[236] The Khanates established by them were no longer in existence after the eighth century.[237] **Fig. A7.20.** Location of the Arctic Ocean north of Siberia, Russia. Taymyr Peninsula faces the Laptev Sea (拉普捷夫海) and the Kara Sea (喀拉海; *Source*: NormanEinstein, under CC BY-SA 3.0, https://commons.wikimedia.org/wiki/File:Laptev_Sea_map.png).[238] The climate there was extremely cold, and the land was covered by deep snow. When the weather was warmer, people used plows drawn by men and horses to grow grains. The horses were all mottled, so the country is named after them as the Kingdom of Mottled Horses (駁馬國/驳马国). The Turks called the horses "Hela", hence, the country was also known as the "Kingdom of Hela (曷剌國/曷剌国)". These people were good fishermen and hunters, and they wore [fish and animal] skins as clothes.[239] In the year 650 during the Tang Dynasty, the Kingdom of Hela sent in envoys and became a tributary country of China.[240]

				Later in the Liao Dynasty (遼/辽or大遼/大辽; Great Liao/ Khitan; 916–1125), the *Liao Shi* (遼史/辽史) or *The History of Liao* also records 曷剌. Scholars have identified that "曷剌" mentioned in the Tang Dynasty documents are the Oirad people.[241] The ANC-KWQ depicts this old kingdom which was no longer in existence in the Ming Dynasty.
● 61	宛在達	Wan Zai Da	Wanzaida[242]	This place cannot be identified on a modern map, but it should be in present-day Siberia, Russia.
62	忌塔意西葛	Ji Ta Yi Xi Ge	Kitaisko[243]	This place cannot be identified on a modern map, but it should be in present-day Siberia, Russia. Kitaisko is depicted as Kitaiſko on Asia ex magna orbis terrae descriptione Gerardi Mercatoris desumpta, studio et industria G.M. Iunioris. Edita A° M.D.LXXXVII/1587.[244]
● 63	鬼國	Gui Guo	Ghost Country	The Ghost Country (鬼國/鬼国) is to the west of the Kingdom of Hela (曷剌國/曷剌国; Item 60) and it takes 60 days to get there.[245]
@ 5	其人夜遊晝隱身剝鹿皮爲衣耳目鼻與人同而口在項上噉鹿及蛇	Qi Ren Ye You Zhou Yin Shen Bo Lu Pi Wei Yi Er Mu Bi Yu Ren Tong Er Kou Zai Xiang Shang Dan Lu Ji She	My translation and comment (between square brackets) are as follows: The people [of the Ghost Country] come out at night and hide during the day. They strip off deer skin to make clothes. Their eyes, nose and ears are like the normal people, but their mouths drop to their necks. They eat deer and snakes. It is estimated that this so-called "Ghost Country" is nothing more than a group of uncivilized savages!
64	其尔目西國	Qi Er Mu Xi Guo	Ciremissorum horda[246]	Beyond the Arctic Circle, there was the territory where lived the Cheremis hordes called the Ciremissorum horda.

(Continued)

Item #	Name	Pinyin	Etymology	History, Geography and My Comments
				Russian Cheremisy (Mari, Russian Marytsy, formerly Cheremis),[247] are a Finno-Ugric people in Eastern Europe, who have traditionally lived along the Volga River and Kama River (卡馬河/卡马河) in Russia. Mari is their own name for themselves and Cheremis was the name applied to them by Westerners and pre-Soviet Russians. In the thirteenth century, the Mari fell under the influence of the Golden Horde[248] (金帳汗國/金帐汗国). In 1443 they became subjects of the Kazan Khanate (喀山汗國/喀山汗国).[249] In 1552, the Mari territory was incorporated into Russia[250] with the Russian conquest of Kazan under Ivan the Terrible. *The above history tells us that Ciremissorum horda (其爾目西國/其尔目西国) no longer existed after 1443, revealing that the ANC-KWQ cannot be a map after 1443.* On the ANC-KWQ, the Ciremissorum horda (其爾目西國/其尔目西国) is erroneously depicted near the Arctic Ocean and crosses the Arctic Circle, whereas historical records show that they were mainly in the Volga Valley.[251] Ciremisorum horda is depicted on the ANC-Ortelius (1570: #2) and the ANC-Ortelius (1570: #3).
● 65	牛蹄突厥	Niu Ti Tu Jue	Niuti Turkic people	This item Niuti Turkic (牛蹄突厥) people will be mentioned again later with獸室韋 (Item 100). They were a nomadic tribe that arose in the Altai Mountains in the middle of the sixth century.[252] In the eighth century, under the joint attack of the Tang Dynasty and the Uighurs (維吾爾族/维吾尔族), the country perished. Since then, the country has withdrawn from the stage of history in northern China.[253] Again, the ANC-KWQ is keeping very old geographical and historical information on the map without updating it in accordance with changes in Chinese historical records.

@ 6	人身牛足水曰瓠顱河夏秋冰厚二尺春冬冰澈底常燒器消冰乃得飲	Ren Shen Niu Zu Shui Yue Hu Lu He Xia Qiu Bing Hou Er Chi Chun Dong Bing Che Di Chang Shao Qi Xiao Bing Nai De Yin		My translation and comment (between square brackets) are as follows: The people have a man's body and cow's feet [meaning big feet] and the water is called Hulu River. The river's ice in summer and autumn is two feet thick, and in spring and winter it is frozen to the bottom. People often burn things to melt the ice to drink the water. Since the region where the Niuti Turkic stayed was in northern Siberia and close to the Arctic Ocean, the climate was sub-Arctic. In winter, temperatures could drop to below −50°C (−58°F); record low temperatures can approach −70°C (−94°F).[254]
● 66	黑車子	Hei Che Zi	Heichezi[255]	Heichezi (黑車子/黑车子) was one of the Shiwei tribes (室韋部落/室韦部落).[256] In the tenth century, Heichezi (黑車子/黑车子) was one of the 59 vassal states of the Liao Dynasty (遼朝'辽朝 or大遼/大辽 or 契丹國/契丹国; 916–1125).[257] Detailed discussions of the Shiwei tribes are given in Item 79. The ANC-KWQ is keeping very old geographical information on the map without updating it in accordance with changes in Chinese historical records.
67	意貌山	Yi Mao Shan	Imaus Mountain[258]	Mount Imeon (意貌山) is an ancient name for the Central Asian complex of mountain ranges comprising the present Hindu Kush (興都庫什山脈/兴都库什山脉), Pamir (帕米爾/帕米尔) and Tian Shan (天山), extending from the Zagros Mountains (札格羅斯山脈/札格罗斯山脉) in the southwest to the Altay Mountains (阿爾泰山/阿尔泰山) in the northeast, and linked to the Kunlun (崑崙山/昆仑山), Karakoram (喀喇崑崙山脈/喀喇昆仑山脉) and Himalayas (喜馬拉雅山/喜马拉雅山) to the southeast. The term

(Continued)

Item #	Name	Pinyin	Etymology	History, Geography and My Comments
				"Imaus Mount" was used by Hellenistic-era scholars.[259] The mountain system was crossed by a segment of the Silk Road. A detailed description of the mountainous territory and its people was given in the Armenian geography index *Ashharatsuyts* written by Anania Shirakatsi in the seventh century A.D.[260] Mount Imeon was famous for its lapis lazuli deposits. The Venetian adventurer Marco Polo visited the mines in 1271 during his famous journey to China. He followed the Silk Road to cross the mountains by way of Wakhan (瓦罕; a rugged, mountainous part of the Pamir, Hindu Kush and Karakoram regions of Afghanistan.).[261] Imaus Mons is depicted on the Ortelius (1570; #2) and the Ortelius (1570; #3).
@ 7	山極高登此看星覺大	Shan Ji Gao Deng Ci Kan Xing Jue Da		My translation is as follows: The mountain is very high, and by climbing up there to see the stars, they appear even bigger. It seems that when observers climb high up the mountain, the distances to the stars become shorter, hence, the stars appear bigger to the observers. In fact, scientifically, this effect is negligible (the nearest star is 40 trillion km/$2.48548477 \times 10^{13}$ mi away). But stars do appear brighter when we are higher, because of scattering in the atmosphere.
● 68	天方	Tian Fang	Tianfang; literally, 天 means "sky" and 方 means "direction (方向)". The word Tianfang first appeared in China in the book titled *Jian Yi Jing* (簡易經/简易经) or *A Concise Interpretation of*	In Chinese history, Tianfang had two meanings: (1) it specifically referred to Mecca,[262] the birthplace of Islam.[263] However, recent studies by Lam Yee Din (林貽典/林贻典) and Sheng-Wei Wang (王勝煒/王胜炜) have shown that the Tianfang/Mo-jia referred by Ma Huan (馬歡/马欢; c. 1380s–c. 1460)[264] was Tunisia in North Africa;[265] and (2) it generally referred to Arabic, Persian and other Western Regions (西域) west of the Chinese Central Plains.

				the Book of Changes. The book says that at the time when there was only water and no land, there was only one direction, which was the south. It has the same meaning as the sky; hence it is called Tianfang.	For the second meaning, in general, the "Western Regions" in a narrow sense refers specifically to the east of Congling (蔥嶺/葱岭) and no further west. But in the broad sense, "Western Regions" refers to all areas that can be reached through the Western Regions in the narrow sense, including the areas in Central and Western Asia, the Indian Peninsula, Eastern Europe and Northern Africa. During different Chinese dynasties, the range of the Western Regions also varied. For example, in the Han Dynasty (202 B.C.–9 A.D., 25–220 A.D), the Western Regions mostly referred to many countries and regions west of Yumen Pass (玉門關/玉门关) or Jade Gate[266] and Yang Pass[267] (陽關/阳关) at the southern foot of the Tian Shan Mountains (天山). On the ANC-KWQ, Tianfang (天方) is depicted near and marginally to the southeast of 土兒客私堂/土儿客私堂 (Turkestan, also known as Turkistan; Item 70; also spelled Turkistan which is a historical region in Central Asia). This location corresponds roughly to today's Gansu Province (甘肅省/甘肃省) in China. Both Yumen Pass and Yang Pass are in this province, and the Western Regions or天方 (Tianfang) start from the west of these two passes.
● 69	鐵勒	Tie Le	Tiele (tolos)[268]		The Tiele (Chinese: 鐵勒/铁勒or 敕勒, lit. "People of the Carts"), were a tribal confederation of Turkic ethnic origins[269] and existed from the later part of the fourth century to the middle part of the eighth century.[270] The Tiele people lived to the north of China proper and in Central Asia, emerging after the disintegration of the confederacy of the Xiongnu (匈奴; third century B.C.–first century A.D.). They lived a nomadic life[271] and were good at shooting on horseback. They were fierce, cruel, and very greedy.[272]

(Continued)

Item #	Name	Pinyin	Etymology	History, Geography and My Comments
				They established the Xueyantuo Khanate (薛延陀汗國/薛延陀汗国; 628–723; an ancient Tiele tribe and khaganate in Northeast Asia) in the seventh century and the Uighur Khanate (回鶻汗國/回鹘汗国; Uyghur)[273] in the eighth century. Finally, when the Eastern Turkic Khaganate (東突厥汗國/东突厥汗国) was defeated and absorbed by the Tang Dynasty (618–907, with an interregnum between 690 and 705), Xueyantuo (薛延陀) occupied the territory of the former Turkic Khaganate (突厥汗國/突厥汗国).[274] Shortly after 646, the Uyghur and the rest of the twelve Tiele chiefs arrived at the Chinese court. They were bestowed either with the title of commander-in-chief (都督 *dudu*) or prefect (刺史 *cishi*) under the loose control of the northern protectorate or the "Pacified North" (安北府) in the *jimi* (羈縻/羁縻) system.[275] After that, the Chinese history books no longer use the name of Tiele. Again, the ANC-KWQ keeps a very old geographical item on the map without updating it in accordance with changes in Chinese historical records.
70	土兒客私堂	Tu Er Ke Si Tang	Turkestan (Italianised: *Turchestan*);[276] also spelled as Turkistan.	Turkestan (土兒客私堂/土儿客私堂), lit. "Land of the Turks", was a historical region in Central Asia corresponding to the regions of Transoxiana (lower Central Asia roughly corresponding to modern-day eastern Uzbekistan, western Tajikistan, parts of southern Kazakhstan, parts of Turkmenistan and southern Kyrgyzstan)[277] and East Turkistan (Xinjiang).[278] The history of Turkestan dates to at least the third millennium B.C.[279] The region was a focal point for cultural diffusion, as the Silk Road traversed it.

The entire territory was held at various times by Turkic forces, such as the Göktürks (突厥人; a Turkic people in medieval Inner Asia),[280] until the conquest by Genghis Khan (成吉思汗; c. 1162–1227) and the Mongols[281] in 1220. Figure A7.21 shows how in 1335 the empire fractured into four khanates:[282] the Golden Horde (1242–1502) in Eastern Europe, the Chagatai Khanate (1226–1705) in Central Asia, the Ilkhanate (1256–1335) in Southwest Asia, and the Yuan Dynasty (1271–1368) in East Asia based in modern-day Beijing, and the Yuan emperors held the nominal title of khagan of the empire. The four divisions each pursued their own interests and objectives and fell at different times.[283] The Wokuotai Khanate (窩闊台汗國/窝阔台汗国; 1251–1309) ceased to exist in 1309, hence, it is not depicted on this map.

Among the six sixteenth-century European maps shown in Figs. 7.2(b)–7.2(g), Turkestan/*Turchestan* is depicted as Tvrchestan on the Jenkinson/Ortelius (1562).

Fig. A7.21. Approximate outlines of nations in Asia in 1335 (public domain).[284]

(Continued)

Item #	Name	Pinyin	Etymology	History, Geography and My Comments
● 71	大壙野	Da Kuang Ye	Da Kuoye;[285] Syverma Plateau	Literally, 大壙野/大圹野 means "a vast desolate land" where it is difficult for people to find food and water to survive. Hence, it is a dangerous place for people to stay. 大壙野/大圹野 may refers to the Syverma Plateau (錫維爾馬高原/锡维尔马高原) which is a mountain plateau in Krasnoyarsk Krai (克拉斯諾亞爾斯克邊疆區/克拉斯诺亚尔斯克边疆区), Siberia, Russia. It is a part of the Central Siberian Plateau. The plateau is in a largely uninhabited mountainous area in Siberia, one of the Great Russian Regions (大俄羅斯地區/大俄罗斯地区).
72	貌力那亦兒	Mau Li Na Yi Er	Maurenaher	Maurenaher (貌力那亦兒/貌力那亦儿) may correspond to the region of Transoxiana.[286] Maurenaher is depicted as Mavrenaher on the Ortelius (1570; #3).[287]
73	是的亞意貌內	Shi Di Ya Yi Mao Nei	Scythia intra Imaum[288]	See Item 55 for a detailed explanation. Scythia intra Imaum is not depicted on any of the six European maps shown in Figs. 7.2(b)–7.2(g), but it is depicted on Totius Europae et Asiae Tabula Geographica, Auctore Thoma D. Aucupario. Edita Argentorati, mdxxii/1522.[289]
● 74	哥兒墨	Ge Er Mo	Calmucchi[290]	The Calmucchi (Plural of calmucco) or Kalmyks (哥兒墨/哥儿墨 or 喀爾木克/喀尔木克) were western Mongols, whose ancestors were originally part of the Oirat tribes that lived in an area that extended from Siberia and Dzungaria (準噶爾/准噶尔)[291] which was a geographical subregion in Northwest China that corresponded to the present-day northern half of Xinjiang (also known as Beijiang, lit. "Northern Xinjiang"). They were scattered in detached groups over a vast territory, within the borders of Russia, Mongolia, and western China.[292]

				Their language belongs to the Mongolian branch of the Altaic language family. Since ancient times the Kalmyks have led a nomadic life.[293] The first Kalmyks immigrated to Russia at an uncertain time. Here on the ANC-KWQ, the region where the Calmucchi/Kalmyks (哥兒墨/哥儿墨) stayed is depicted in northern Russia and much closer to the Arctic Ocean in comparison with its location depicted on a 1784 map.[294] Nurlan Kenzheakhmet and Alpamys Abu, on the other hand, propose[295] that 哥兒墨 should be 哥兒黑士 (Ge'erheishi) and refers to Kirgessi (another name for Kazakhs); Kirgessi is depicted on the Jenkinson/Ortelius (1562), but its location is further away from the Arctic Ocean.
@ 8	此國死者不埋但以鐵鏈挂其屍于樹林	Ci Guo Si Zhe Bu Mai Dan Yi Tie Lian Gua Qi Shi Yu Shu Lin		My translation is as follows: The dead people in this country are not buried but hung with iron chains in the woods. This kind of burial may be regarded as a kind of "Sky Burial" as a farewell for their dead. It is like Tibetans who see the vultures as Dakinis, like angels who take souls into the heavens to await reincarnation.[296]
75	白尔米雅	Bai Er Mi Ya	Permia[297]	The Principality of Great Perm or simply Perm, in Latin *Permia* (白爾米雅/白尔米雅; 1323–1505), was a medieval historical region, see Fig. A7.22, in what is now the Perm Krai (彼爾姆邊疆區/彼尔姆边疆区) of the Russian Federation.

(Continued)

Item #	Name	Pinyin	Etymology	History, Geography and My Comments
				 Fig. A7.22. Map of Northern Russia, including Permia; it was drawn by Gerardus Mercator in 1595 (public domain).[298] The same name is likely reflected in the toponym Bjarmaland in Norse sagas since the Viking Age[299] and in geographical accounts until the sixteenth century.[300] The term is usually seen to have referred to the southern shores of the White Sea and the basin of the Northern Dvina River (北德維納河/北德维纳河) as well as some of the surrounding areas. Today, these territories comprise a part of the Arkhangelsk Oblast (阿爾漢格爾斯克州/阿尔汉格尔斯克州) of Russia,[301] as well as the Kola Peninsula (科拉半島/科拉半岛),[302] see Fig. A7.23.

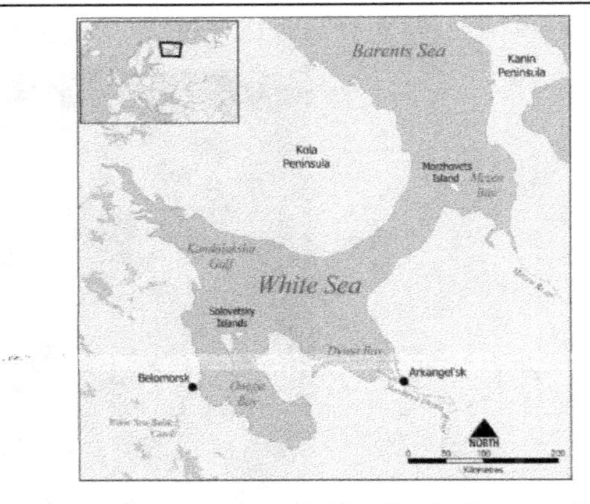

Fig. A7.23. A map of the White Sea in Russia which also shows the various bays and gulfs in the sea; the Northern Dvina River to the right flows into the Dvina Bay in the White Sea (*Source*: NormanEinstein, under CC BY-SA 3.0, https://commons.wikimedia.org/wiki/File:White_Sea_map.png).[303]

Northern Dvina (北德維納河/北德维纳河) is a river in northern Russia flowing into the White Sea; along with the Pechora River (伯朝拉河)[304] to the east, see Fig. A7.24, it drains most of Northwest Russia into the Arctic Ocean. It should not be confused with the Western Dvina (西德維納河/西德维纳河) which is a large river flowing through Belarus (白俄羅斯/白俄罗斯) and Latvia (拉脫維亞/拉脫维亚) into the Gulf of Riga (里加灣/里加湾) of the Baltic Sea (波羅的海/波罗的海).

The Bjarmians cannot be connected directly to any existing group of people living today, but it is likely that they were a separate group of Finnic speakers in the White Sea area.[305]

The Bjarmians declined in the thirteenth century and became tributaries of the Novgorod Republic (諾夫哥羅德共和國/诺夫哥罗德共和国; 1136–1478). During the fourteenth and fifteenth centuries, the Bjarmians were finally assimilated by the Slavs (斯拉夫人).[306]

(Continued)

Item #	Name	Pinyin	Etymology	History, Geography and My Comments
				Permia is depicted on the Jenkinson/Ortelius (1562).[307] 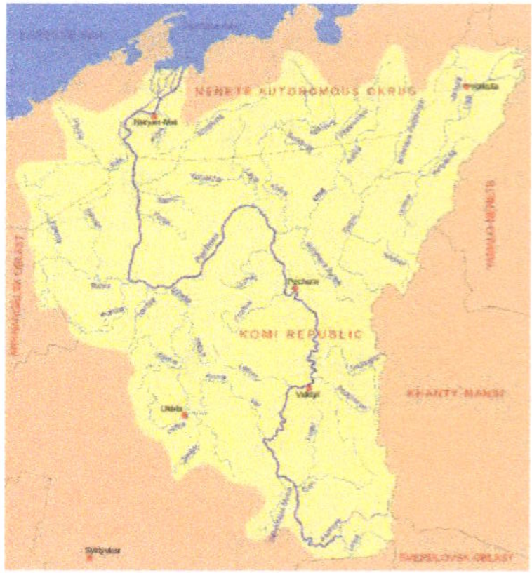 **Fig. A7.24.** Pechora River (伯朝拉河) catchment area and tributaries (*Source*: Pechora.svg: СафроновАВ; derivative work: Geanixx, under CC BY-SA 3.0, https://commons.wikimedia.org/wiki/File:Pechora-en.svg).[308]

Zone III: Eastern Siberia

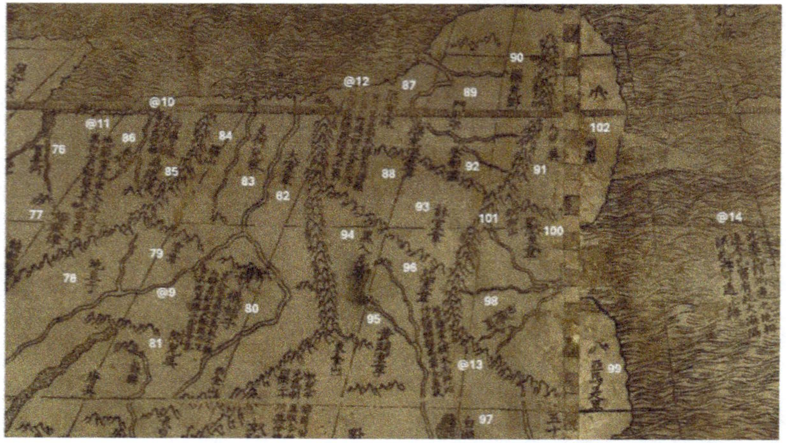

Fig. A7.25. This figure is extracted from Fig. 7.2(a) (public domain); Items 76-102 and annotations @9-@14 are depicted on this Figure.

76	賀喜河	He Xi He	Hexi River[309] may be the Ob River.	Hexi River (賀喜河/贺喜河) may be the Ob River (鄂畢河/鄂毕河 or 阿比河)[310] in Siberia, which flows into the Arctic Ocean. The other two great Siberian rivers that flow into the Arctic Ocean are the Yenisei River 葉尼塞河/叶尼塞河) and the Lena River (勒拿河). In Western Siberia, abutting the Ural Mountains, the huge West Siberian Plain is drained by the Ob River and the Yenisei River and contains wide tracts of swampland, see Fig. A7.26. The European maps in Figs. 7.1(b)–7.1(g) all depict the Ob River.  Fig. A7.26. River routes based on descriptions by James Forsyth's *A History of the Peoples of Siberia*, 1992 (*Source*: Kmusser, under CC BY-SA 3.0, https://commons.wikimedia.org/wiki/File:Siberiariverroutemap.png).[311]
• 77	嫗厥律	Yu Jue Lu	Yujuelü[312] or Wugu	The Wugu (烏古/乌古)[313] were a people roaming the eastern region of modern Mongolia during the Liao (遼/辽; 907–1125) and Jin (金; 1115–1234) periods. Chinese sources also used the transcriptions Wuguli (烏古里/乌古里 or 烏骨里/乌骨里), Wuhuli (烏虎里/乌虎里), Yujue (於厥/于厥 or 羽厥), Yujueli (於厥里/于厥里 or 尉厥里), Yuxieli (於諧里/于諧里), Yuguli (於骨里/于骨里), and Yujuelü (嫗厥律/妪厥律). The eastern neighbours of the Wugu were the Shiwei (室韋/室韦).[314] This is also correctly depicted on the ANC-KWQ in Fig. A7.25.

(*Continued*)

Item #	Name	Pinyin	Etymology	History, Geography and My Comments
				At the end of the Liao Dynasty, the Jurchens (女真人) rose, and the Liao Dynasty fell. Except for a part of the Wugu people who participated in the Western Expedition of Yelu Dashi (耶律大石), most of them surrendered to the Jurchens.[315] At the end of the Jin Dynasty and the beginning of the Yuan Dynasty, the Wugu people gradually merged into Jurchen and Mongolia.[316] Again, the depiction of 嫗厥律 (Yujuelü) on the ANC-KWQ indicates that the KWQ has kept a very old geographical item without updating it according to dynastic changes in Chinese history.
● 78	地豆于	Di Dou Yu	Didouyu[317] or Didougan	The Didouyu (地豆於 or 地豆于)[318] or Didougan (地豆乾/地豆干)[319] was a nomadic tribe during the fifth-sixth century in west Manchuria (滿洲/满洲; a historic region in Northeastern China: ancestral home of the Manchu who were a Tungistic people — meaning "from Tunguska (通古斯)" — of Northeastern China). Their territory lived the Turkic (突厥; Eastern Turk, on their west) people, the Rouran (柔然; on their northwest) people, and the Khitan (契丹; on their southeast) people. The record of *Wei Shu* (魏書/魏书) or *The Book of Wei* states "地豆於國在失韋 (即室韋) 西千餘里", meaning "The country of Didouyu is more than a thousand miles to the west of Shiwei" (translation by Sheng-Wei Wang). It is roughly in today's territory of the West Ujimqin Banner (西烏珠穆沁旗/西乌珠穆沁旗) in Inner Mongolia.[320] Again, the depiction of 地豆于 on the ANC-KWQ indicates that the map has kept a very old geographical item without updating it according to dynastic changes in Chinese history.

| • 79 | 室韋 | Shi Wei | Shiwei | Shiwei (室韋/室韦) were a Mongolic people[321] that inhabited far-eastern Mongolia, northern Inner Mongolia, northern Manchuria and the area near the Okhotsk Sea (鄂霍次克海) coast, see Fig. A7.27.

Shiwei and Khitan (契丹) originated from the same source, with Khingan Range (興安嶺/兴安岭 as the boundary; Khitan in the south and Shiwei in the north. They lived on both sides of the middle and upper reaches of Heilongjiang (黑龍江/黑龙江; Amur River) and the Nen River (嫩江) Basin. See also the explanation of 韃靼/鞑靼 in Item 52.

Shiwei were hunters, mostly catching minks, raising cattle, horses and pigs, eating their meat and wearing their skins. They also grew wheat and millet. They lived in the city in the summer and hunted for water and grass in the winter.

Fig. A7.27.   This is a modern map showing the location of the Sea of Okhotsk (鄂霍次克海). The sea is bordered by Russia and Japan. It also shows parts of the countries and territories that make up the region of Northeast Asia (*Source*: NormanEinstein, under CC BY-SA 3.0, https://commons.wikimedia.org/wiki/File:Sea_of_Okhotsk_map_with_state_labels.png).[322] |

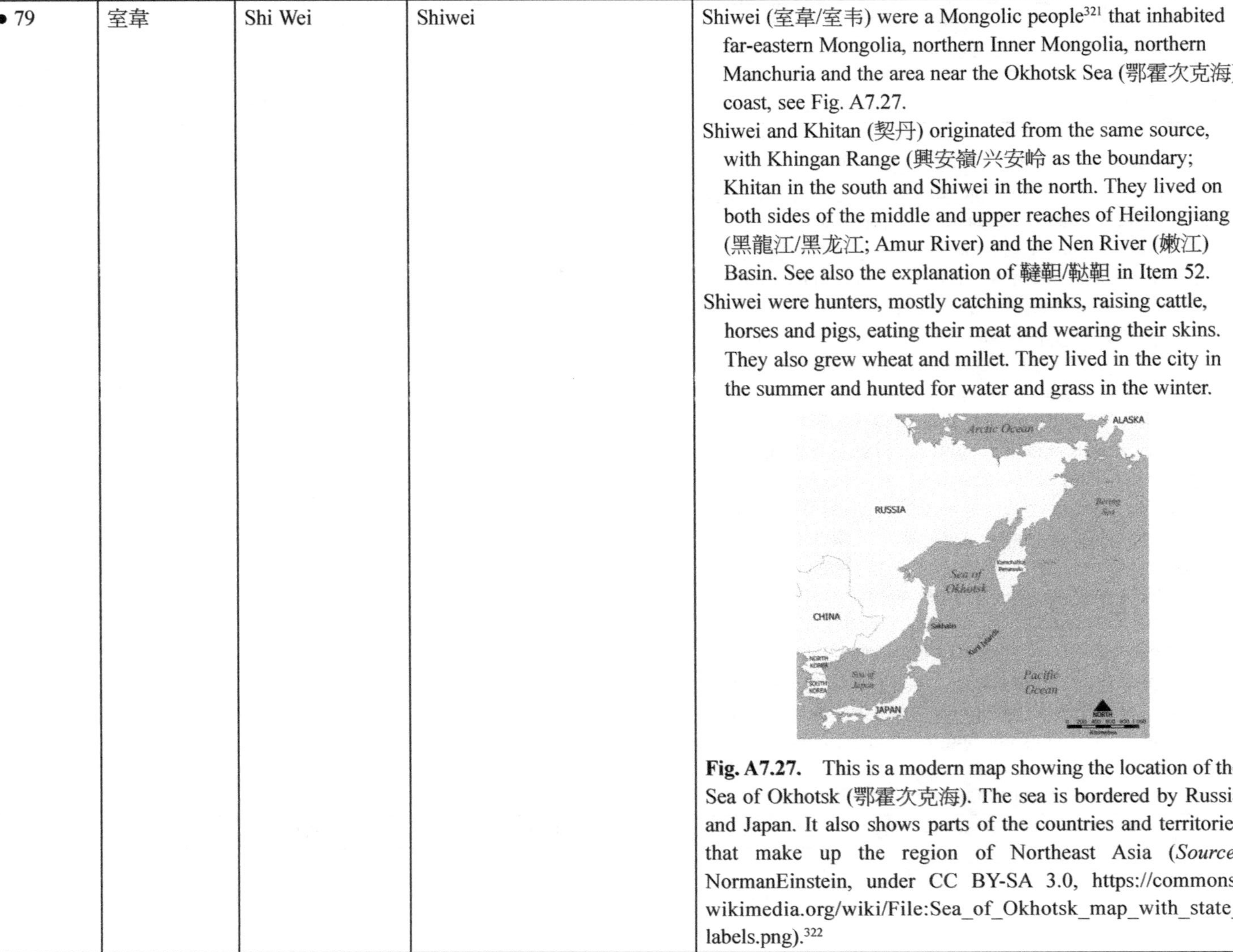

(*Continued*)

Item #	Name	Pinyin	Etymology	History, Geography and My Comments
				From time to time, Shiwei sent envoys to the Northern Zhou Dynasty (北周; 557–581) and Northern Qi Dynasty (北齊/北齐; 550–577) to pay tribute to China. Later, they were divided into five tribes: Nan Shiwei (南室韋/南室韦), Bei Shiwei (北室韋/北室韦), Bo Shiwei (鉢室韋/钵室韦), Shenmoda Shiwei (深末怛室韋/深末怛室韦), and Da Shiwei (大室韋/大室韦). Each of them had different affiliations and slightly different customs and habits.[323] Originally belonging to the Xiongnu (匈奴) and Xianbei (鮮卑/鲜卑), they were attached to the Turks in the Tang Dynasty (618–907, with an interregnum between 690 and 705). After the Turks were replaced by the Huihe (回紇/回纥; the ancestors of the Uighurs in present-day Xinjiang, China,), the Shiwei people surrendered to the Khitan (契丹), later joined the Mongol Empire and became a branch of Mongolia.[324] Shiwei was recorded from the time of the Northern Wei (北魏; 386–534) until the rise of the Mongols under Genghis Khan (c. 1162–1227). In 1206 he formally adopted the title "Genghis Khan" at an assembly,[325] when the names "Mongol" and "Tatar" were applied to all the Shiwei tribes. The ANC-KWQ depicts very old geographical items without updating them to reflect political changes in Chinese history.
● 80	襪結子	Wa Jie Zi	Mierqi or Merkits[326]	Mierqi (襪結子/袜结子 or 轙劫子 or 蔑兒乞/蔑儿乞)[327] was the name of a northern minority nomad tribe living to the southeast of Lake Baikal (貝加爾湖/贝加尔湖) from the tenth to the thirteenth centuries (tenth century–1217).

			In 1217, Genghis Khan[328] (成吉思汗 or 鐵木真/铁木真; the founder and first Great Khan or Emperor of the Mongol Empire, which became the largest contiguous empire in history after his death) sent the Subotai (速波台) of the Wuliangha (兀良哈)[329] tribe to hunt down and kill the Mierqi people, and the Mierqi tribe was officially disbanded. Most of the Mierqi were scattered among various Mongolian and Turkic tribes in Central Asia. Again, the ANC-KWQ depicts 襪結子 (Mierqi/Merkits), retaining outdated geographical information on this map. There is an annotation next to 襪結子 (Mierqi/Merkits); it is taken from *Xin Wu Dai Shi • Si Yi Fu Lu Er* (新五代史 • 四夷附錄二/新五代史 • 四夷附录二)[330] or *New History of the Five Dynasties, Appendix II on the Four Barbarians*. This annotation is translated and explained in @9 as shown below.
@ 9	其人髦首披布为衣不鞍而骑大弓长箭尤善射遇人辄杀而生食其肉其國三面皆室韋	Qi Ren Mao Shou Pi Bu Wei Yi Bu An Er Qi Da Gong Chang Jian You Shan She Yu Ren Zhe Sha Er Sheng Shi Qi Rou Qi Guo San Mian Jie Shi Wei	My translation and comments (between square brackets) are as follows: The Wajiezi [襪結子/袜结子; Item 80] people wear long hair and cover their bodies with cloth. They ride horses without saddles and are particularly good at shooting with a big bow and long arrows. They kill the people [from other tribes] whom they encounter and eat their flesh raw. Their country is surrounded by Shiwei territories on three sides. On the ANC-KWQ, Wajiezi (襪結子/袜结子) faces Shiwei (室韋; Item 80) to its west, Da Shiwei (大室韋/大室韦; Item 82) to its north, and Huangtou Shiwei (黄頭室韋; Item 99) to its east.
● 81	測兒吳	Ce Er Wu	Zeru[331]	This geographical term cannot be identified on a modern map.
● 82	大室韋	Da Shi Wei	Da Shiwei	Da Shiwei was one of the Shiwei tribes mentioned in Item 79. Once again, the KWQ has kept a very old geographical item without updating it according to dynastic changes in Chinese history.

(Continued)

Item #	Name	Pinyin	Etymology	History, Geography and My Comments
● 83	支何兒察	Zhi He Er Cha	Zhiheerca[332]	This geographical item cannot be located on a modern map.
● 84	羅山	Luo Shan	Luo Mountains	This geographical item cannot be identified on a modern map.
● 85	烏洛侯	Wu Luo Hou	Uluohou[333]	The Uluohou (烏洛侯/乌洛侯) people appeared in the Northern Wei Dynasty (北魏; 386–535). The Uluohou people were mainly dispersed in the Tuoba Xianbei (拓拔鮮卑/拓拔鲜卑) homeland of the Greater Khingan Mountains (大興安嶺/大兴安岭) near the upper reaches of the Nen River (嫩江). They merged into Shiwei in the Tang Dynasty (618–907, with an interregnum between 690 and 705).[334] Again, the depiction of 烏洛侯 (Uluohou) on the KWQ indicates that KWQ has kept very old geographical items on this map without updating them according to dynastic changes in Chinese history.
@ 10	土下濕多露氣而寒人尙勇不爲奸竊	Tu Xia Shi Duo Lu Qi Er Han Ren Shang Yong Bu Wei Jian Qie	My translation is as follows: The soil is damp, and the air is dewy. The people are brave and do not act out morally incorrect behaviour or commit stealing. This annotation appears in *Wei Shu • Juan 107 • Lie Chuan Di Ba Shi Ba • Wu Luo Hou* (魏書 • 卷107 • 列傳第八十八 • 烏洛侯/魏书 • 卷107 • 列传第八十八 • 乌洛侯) or *Book of Wei, Juan (Volume) 107, Biography 88, Uluohou.* Because the local climate is cold and the soil is damp, people build houses in the woods in summer at a certain height above the ground; in winter, they build crypt or semi-crypt houses which are called "ground cellars (地窖子 or "Ground Shacks)" today.[335]

| 86 | 兒水 | Mao Shui | Mao Shui[336] | Mao Shui (兒水) may be the present-day Yenisei River (葉尼塞河/叶尼塞河) which is the fifth-longest river system in the world, and the largest to drain into the Arctic Ocean. It is the central one of the three large Siberian rivers that flow into the Arctic Ocean (the other two being the Ob and the Lena), see Fig. A7.26. The Yenisei divides the Western Siberia Plain in the west from the Central Siberia Plateau to the east, see Fig. A7.28.
A significant feature of the Upper Yenisei River is Lake Baikal, the deepest and oldest lake in the world. But Lake Baikal is not depicted on the ANC-KWQ, and only the Upper Yenisei is depicted.
On the ANC-Mercator (1569), the ANC-Ortelius (1570; #1) and the ANC-Plancius (1594), respectively, there is an unnamed river closely related to the Yenisei River.
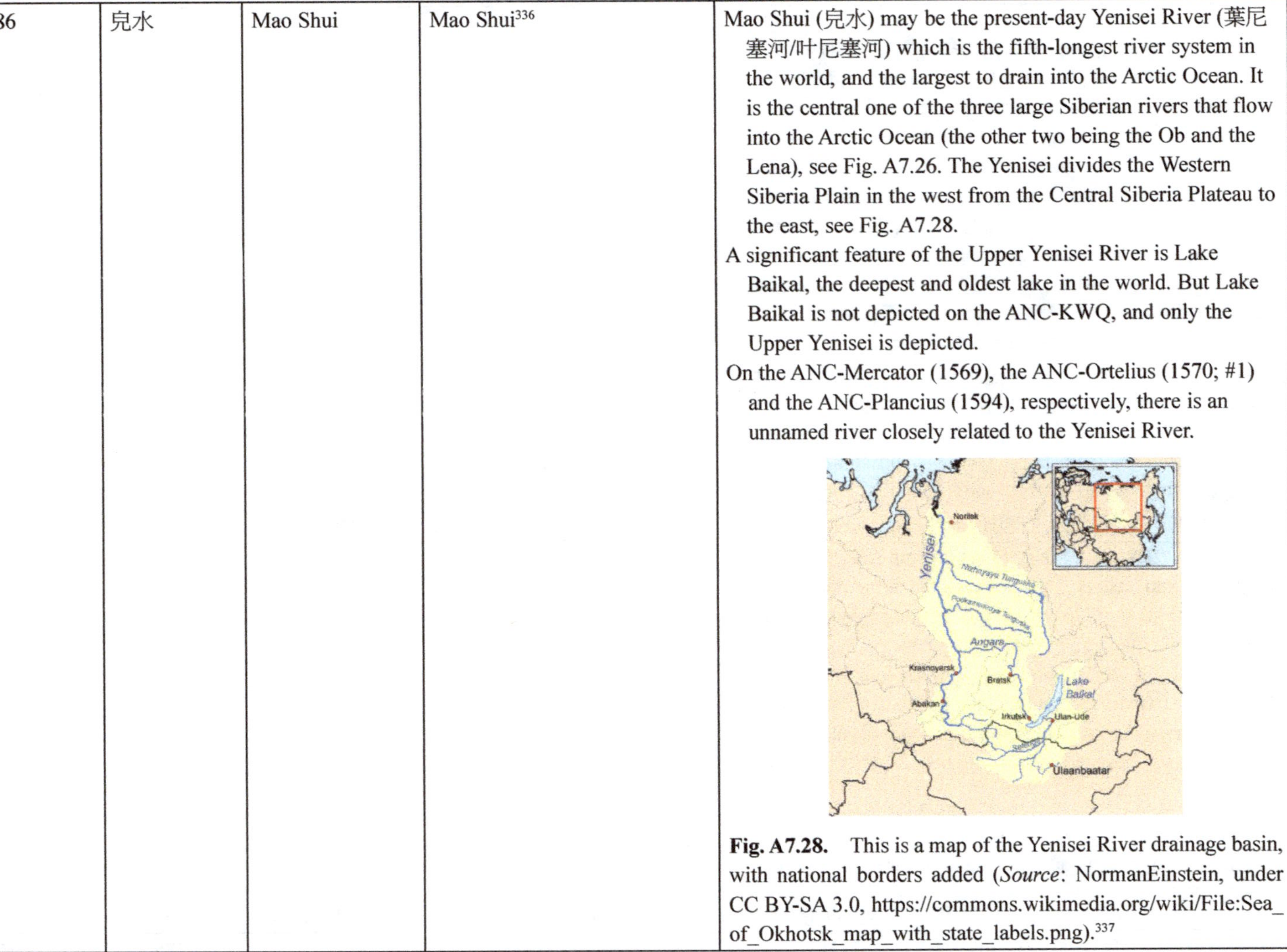
Fig. A7.28.   This is a map of the Yenisei River drainage basin, with national borders added (*Source*: NormanEinstein, under CC BY-SA 3.0, https://commons.wikimedia.org/wiki/File:Sea_of_Okhotsk_map_with_state_labels.png).[337] |

(*Continued*)

Item #	Name	Pinyin	Etymology	History, Geography and My Comments
@ 11	地嚴寒水出大魚又多黑白黃貂鼠其人最勇	Di Yan Han Shui Chu Da Yu You Duo Hei Bai Huang Diao Shu Qi Ren Zui Yong		My translation is as follows: The place is severely cold and rich in big fish. There are also many black, white and yellow minks. The people are most courageous. On the ANC-KWQ, Mao Shui (兒水; Yenisei River) is depicted in Siberia and near the present-day Arctic Ocean. In Siberia, the average temperature in January is almost everywhere below –10°C/14°F, and it goes down to –45°C (–49°F) in the Eastern inland areas. The daily average in July is around freezing (0°C or 32°F) on the northern islands and along the Arctic coast.[338] Hence, this annotation correctly states the climate of Mao Shui (兒水).
• 87	區度寐	Qu Du Mei	Kingdom of Qudumei	The Kingdom of Qudumei (區度寐/区度寐) was in the north of the Outer Khingan Mountains (外興安嶺/外兴安岭) and the eastern part of Siberia.[339] After the Sui Dynasty (隋朝; 581–618) and during the early Tang Dynasty (唐朝; 618–907, with an interregnum between 690 and 705), China set up a Command Post at Shiwei, whose jurisdiction was north of the Outer Khingan Mountains.[340] Some hold that the Qudumei (區度寐/区度寐) people may be the Ewenki people (鄂温克人; a Tungusic people of North Asia).[341] But recent studies[342] show that they may be the Yakuts (雅庫特人/雅库特人) or Sakha (薩哈人/萨哈人) who were the Mongols mainly living in the area of today's Republic of Sakha (薩哈共和國/萨哈共和国) in the Russian Federation (俄羅斯聯邦/俄罗斯联邦), see Fig. A7.29, to the east of the Lena River (勒拿河), see Fig. A7.30, with some extending to several districts of the Krasnoyarsk (克拉斯諾亞爾斯克/克拉斯诺亚尔斯克) region.

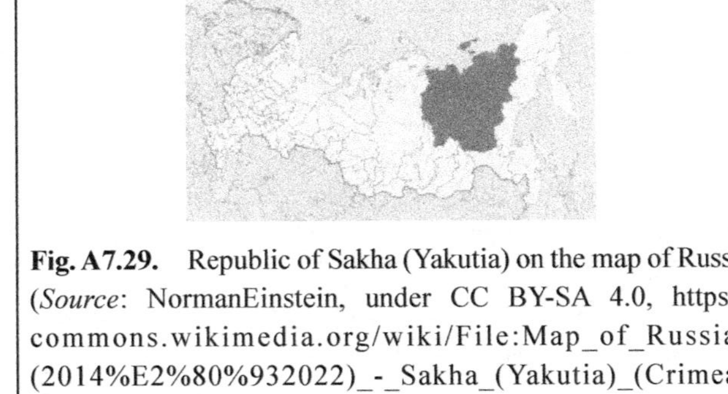

Fig. A7.29.   Republic of Sakha (Yakutia) on the map of Russia (*Source*: NormanEinstein, under CC BY-SA 4.0, https://commons.wikimedia.org/wiki/File:Map_of_Russia_(2014%E2%80%932022)_-_Sakha_(Yakutia)_(Crimea_disputed).svg).[343]

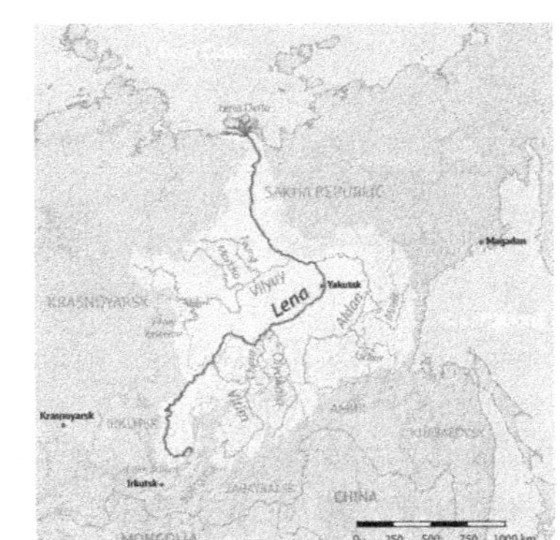

Fig. A7.30.   Map of Siberia's Lena River basin, with major tributaries (*Source*: Shannon1, under CC BY-SA 4.0 International, 3.0 Unported, 2.5 Generic, 2.0 Generic and 1.0 Generic license, https://commons.wikimedia.org/wiki/File:Lena_River_basin.png);[344] Lake Baikal is not depicted on the KWQ.

The ANC-KWQ depicts very old geographical items without updating them to reflect political changes in Chinese history.

(*Continued*)

Item #	Name	Pinyin	Etymology	History, Geography and My Comments
@ 12	其人甚長而衣短不索發皆裹頭居土窟中唯有豬更無諸畜人輕捷一跳三丈餘又能立浮臥浮履水沒腰與陸走不異	Qi Ren Shen Chang Er Yi Duan Bu Suo Fa Jie Guo Tou Ju Tu Ku Zhong Wei You Zhu Geng Wu Zhu Chu Ren Qing Jie Yi Tiao San Zhang Yu You Neng Li Fu Wo Fu Lu Shui Mo Yao Yu Lu Zou Bu Yi	My translation and comment (between square brackets) are as follows: The people here are very tall, but their clothes are short. They do not bundle their hair but all wrap their heads. They live in caves and raise pigs, but no other livestock. Their bodies are nimble and can jump up and down more than 30 feet [1 Chinese *Zhang* (市丈) of length equals 10.94 feet in length; but this description must be an exaggeration]. They can float by standing or by lying flat. They can walk in waist-high water just like walking on land. This annotation is based on endnote 339, in which it also states that the Kingdom of Qudumei, written as "驅度寐" was to the north of Shiwei.
● 88	深末怛室韋	Shen Mo Da Shi Wei	Shenmoda Shiwei[345]	Shenmoda Shiwei (深末怛室韋/深末怛室韦) was one of the Shiwei tribes in the Northern Qi Dynasty (北齊/北齐; 550–577 A.D.) or the Sui Dynasty (581–618). It no longer existed as one of the Shiwei tribes in the Tang Dynasty (618–907, with an interregnum between 690 and 705). The ANC-KWQ depicts very old geographical items without updating them to reflect political changes in Chinese history.
89	門臥尔	Men Wo Er	Mongol[346]; the designation "Mongol" briefly appeared in eighth century records of Tang China to describe a tribe of Shiwei. It resurfaced in the late eleventh century during the Khitan-ruled	Mongols[347] (門臥爾/門臥尔) were members of any of a group of traditionally pastoral peoples of Mongolia, which is a region of eastern Asia east of the Altai Mountains (阿爾泰山脈/阿尔泰山脉). On the ANC-KWQ, 門臥尔 (Mongol) is depicted in the general area where the Shiwei tribes are located. In Item 79, the detailed background of the Shiwei tribes who were a

			Liao Dynasty (916–1125). After the fall of Liao in 1125, the Khamag Mongols became a leading tribe on the Mongolian Plateau. But their wars with the Jurchen-ruled Jin Dynasty and the Tatar confederation had weakened them.	Mongolic people is given. In short, Shiwei was an ancient ethnic group distributed in the upper reaches of the upper Amur/Heilongjiang River Basin and the Nen River Basin from the Northern Wei Dynasty (北魏; 386–535) to the Liao (916–1125) and Jin (1115–1234) dynasties. The Shiwei people had the same origin as the Khitan (契丹) and Xi (奚) peoples living to their south. From Item 79, we know that the Shiwei tribes later joined the Mongol Empire and became a branch of Mongolia.[348] Hence, after entering the twelfth century (in the early Jin Dynasty), there are no records of Shiwei's activities in history books anymore. In the thirteenth century, the word Mongol grew into an umbrella term for a large group of Mongolic-speaking tribes united under the rule of Genghis Khan.[349] This Eastern Siberia portion of the ANC-KWQ has kept very old geographical information even though the various Shiwei tribes had long been incorporated into the federation of the Khitans that founded the Liao Dynasty (遼/辽; 916–1125) that would eventually conquer northern China.[350] Hence, we may interpret the appearance of 門臥尔 (Mongol) on the ANC-KWQ to mean that the map intends to signal that the various Shiwei tribes on the map were all Mongolic people. The true histories of the Chinese Ming Dynasty, the Mongolic rulers and their territories are briefly summarised as follows: The Mongol Empire (1206–1368) was the largest contiguous land empire in history.[351] Originating in present-day Mongolia in East Asia, the Mongol Empire at its height stretched from the Sea of Japan to parts of Eastern Europe, extending northward into parts of the Arctic; eastward and southward into parts of the Indian subcontinent; they also attempted invasions of Southeast Asia and conquered the Iranian Plateau; and westward as far as the Levant (an approximate

(Continued)

Item #	Name	Pinyin	Etymology	History, Geography and My Comments
				historical geographical term referring to a large area in the Eastern Mediterranean region of West Asia and core territory of the Middle East) and the Carpathian Mountains (喀爾巴阡山脈/喀尔巴阡山脉).[352] In 1260, Kublai Khan (忽必烈汗; 1215–1294)[353] became the Great Khan and moved his capital from Helin (和林) north of the Gobi Desert (戈壁) to Yanjing (燕京; the former name of Beijing), which was later renamed Dadu (Great Capital; 大都).[354] In 1271 he founded the Yuan Dynasty (元朝 or 大元; 1271–1368), and in 1279 he subdued the Southern Song (南宋; 1127–1279), bringing the whole of China under his centralized rule.[355] The subsequent Ming Dynasty (明朝; 1368–1644) placed the areas where Mongols lived under the administration of more than 20 garrison posts commanded by Mongolian manorial lords.[356] By the end of the Hongwu Emperor's 30-year reign in 1398, his new dynasty controlled the whole of modern China proper and dominated the northern frontier regions from Hami (哈密; in present-day Xinjiang Province) through Inner Mongolia (內蒙古), and into northern Manchuria (滿洲/满洲) which was the region of northeastern China that now covers the provinces of Heilongjiang (黑龍江/黑龙江), Jilin (吉林), and Liaoning (遼寧/辽宁). Some geographers include northeastern Inner Mongolia as well.[357] In the early fifteenth century, the Oirat and Tatar Mongols living west and north of the Gobi Desert pledged their allegiance to the Ming Empire.

				Then the Yongle Emperor (reigned 1402–24) subjugated Nam Viet (南越), personally campaigned against the reorganizing Mongols in the north, and sent large naval expeditions overseas, chiefly under the eunuch Admiral Zheng He (鄭和海軍上將/郑和海军上将), to demand tribute from rulers as far away as Africa and beyond.[358] He also returned the empire's capital to Beijing, giving that city its present-day name.[359] *The ANC-KWQ depicts the region of the Mongols in the far Eastern Siberia corner and denotes many regions occupied by the various Shiwei tribes, reflecting a very old and outdated political condition in far Eastern Siberia.* On the ANC-Mercator (1569), "Mongul" is depicted in the northeastern Siberia corner. On the ANC-Ortelius (1570; #1), "Mongol" is depicted in the northeastern Siberia corner. On the ANC-Plancius (1594), "Mongul" is depicted in Central Siberia.
● 90	羅荒野	Luo Huang Ye	Luo Huang Ye	On ancient Chinese maps, Siberia (西伯利亞/西伯利亚) was called "羅荒野/罗荒野" which is a vast and sparsely populated geographical region constituting all North Asia, from the Ural Mountains in the west to the Pacific Ocean in the east. The famous Lake Baikal (貝加爾湖/贝加尔湖) is in Siberia. But the ANC-KWQ only depicts 羅荒野 (Siberia) as being in the far corner of Northeastern Asia, see Fig. A7.25. Siberia was inhabited long before the Stone Age. During the last ice age, the climate was cold, so the water level was relatively low. The current Bering Strait (白令海峽/白令海峡) used to have a land bridge connecting Siberia and Alaska. *One theory proposes that most of the present-day Native American population came to the Americas from Siberia via this land bridge.*[360]

(*Continued*)

Item #	Name	Pinyin	Etymology	History, Geography and My Comments
				Siberia was also the cradle of many powerful nations in ancient times. The Xiongnu (匈奴人; third century B.C–first century A.D.), the Xianbei (鮮卑人/鲜卑人; third century B.C.–third century A.D.), the Turkic (突厥人; sixth century–744),[361] the Dada/Tartars (韃靼人/鞑靼人; 732–1635),[362] the Mongols (蒙古人; eighth century–fourteenth century), the Jurchen (女真人; eighth century–eighteenth century) and other nationalities all rose from Siberia. The Russian conquest of Siberia took place in the sixteenth, seventeenth, and eighteenth centuries when the Khanate of Sibir (西伯利亞汗國/西伯利亚汗国; an autonomous Khanate formed after the breakup of Golden Horde) became a loose political structure of vassalages that were being undermined by the activities of Russian explorers.[363]
91	白兒吳	Bi Er Wu	Bargu[364]	This geographical item cannot be located on a modern map. Bargu is depicted on the ANC-Mercator (1569) and the ANC-Plancius (1594); on the Ortelius (1570; #2) and the Ortelius (1570; #3), Bargu is depicted as "Bargv."
92	登都國	Deng Du Guo	Tenduc[365]	In *The Travels of Marco Polo, Book 1*,[366] Chapter 59, titled "Concerning the Province of Tenduc, and the Descendants of Prester John", it is stated that "Tenduc is a province which lies towards the east, and contains numerous towns and villages; among which is the chief city, also called TENDUC. The king of the province is of the lineage of Prester John, George by name; …". However, the Tenduc Province mentioned by Marco Polo was in Eastern Suiyuan;[367] but Suiyuan (綏遠/绥远) was in northwestern China, not Siberia. You wonder: did Marco Polo visit Siberia? Polo never produced a map, but his family drew several maps to the Far East based on his writing. These collections of maps were signed by Polo's daughters.[368] The maps not only

contain his journey but also sea routes to Japan, Siberia's Kamchatka Peninsula (堪察加牛島/堪察加牛岛), the Bering Strait (白令海峽) and even to the coastlines of Alaska, centuries before the rediscovery of the Americas by Europeans.[369]

What is equally confusing is that Tenduc had the third location in history as shown in Fig. A7.31.

Even 450 years after the destruction of the Western Xia/Xi Xia/Great Xia/Tangut Empire (西夏 or大夏 or 1038–1227), the "Kingdom of Tenduc or Tangut" was still shown on some European maps as China's northwestern neighbour, see Fig. A7.31, not in eastern Siberia. The Tamgut (黨項/党項) people were a Sino-Tibetan people who founded and inhabited the Western Xia Dynasty.

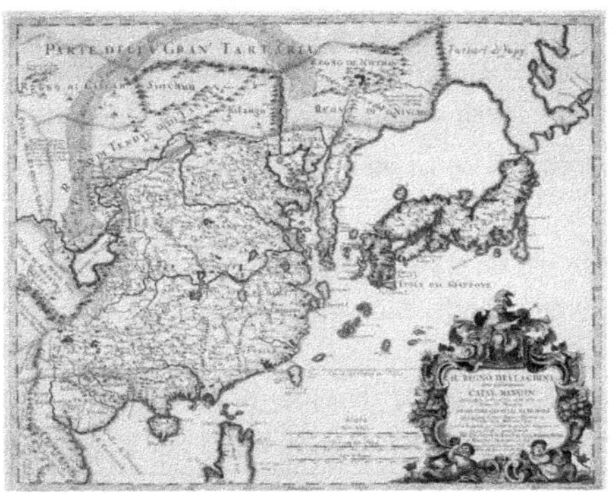

Fig. A7.31. The map shows that 450 years after the destruction of the Western Xia/Xi Xia/Great Xia/Tangut (西夏 or 大夏) Empire (1038–1227) or the "Kingdom of Tenduc or Tangut" is still depicted on some European maps as China's northwestern neighbour; here is such as this 1682 map of China by Giacomo Cantelli: Il regno della China detto presentemente Catay e Mangin diuiso sopra le carte piu esatte nelle sue principali prouincie ("The Kingdom of China, presently called Catay and Mangin, divided into its principal provinces on a most precise map"; public domain).[370]

(*Continued*)

Item #	Name	Pinyin	Etymology	History, Geography and My Comments
				On the ANC-KWQ, Tenduc (登都國/登都国) seems to be erroneously depicted in Eastern Siberia. Tenduc is also erroneously depicted on the ANC-Mercator (1569), ANC-Ortelius (1570; #1) and ANC-Plancius (1594) in similar way.
● 93	鉢室韋	Bo Shi Wei	Bo Shiwei[371]	Bo Shiwei (鉢室韋/钵室韦) was one of the Shiwei tribes. It had customs like those of Northern Shiwei (北室韋/北室韦).[372] Then ANC-KWQ depicts very old geographical items without updating them to reflect political changes in Chinese history.
● 94	黑入瓦牙	Hei Ru Wa Ya	Hegiuuaia[373]	This geographical item cannot be located on a modern map.
● 95	黃頭室韋	Huang Tou Shi Wei	Huangtou Shiwei[374]	Huangtou Shiwei (黃頭室韋/黄头室韦) was one of the Shiwei tribes. The name was changed to Huangtou Jurchen (黃頭女真/黄头女真) in the Jin Dynasty (1115–1234).[375] The external features of these people included a beard and their hair being golden yellow, most of their pupils being green, only a few being golden (yellow). These were typical Caucasian (白種人) characteristics.[376] Once again, the ANC-KWQ keeps an outdated name Huangtou Shiwei (黃頭室韋/黄头室韦) on the map without updating it into Huangtou Jurchen (黃頭女真/黄头女真).
● 96	北室韋	Bei Shi Wei	Northern Shiwei[377]	Bei Shiwei or Northern Shiwei (北室韋/北室韦) was one of the Shiwei tribes. It was named after its location being in the north of Nan Shiwei or Southern Shiwei (南室韋/南室韦). But Southern Shiwei is not depicted on the ANC-KWQ, because, after the Tang Dynasty, it had decomposed into many branch tribes such as the Huangtou Shiwei (黃頭室韋/黄头室韦; Item 95).[378]

@ 13	地多積雪人騎木而行以防坑陷捕貂爲業衣魚皮	Di Duo Ji Xue Ren Qi Mu Er Xing Yi Fang Keng Xian Bu Diao Wei Ye Yi Yu Pi		My translation is as follows: The land is full of accumulated snow and people ride on wood to avoid falling into pits and traps. They catch minks for living and wear fish skins as clothes. Bei Shiwei or Northern Shiwei (北室韋/北室韦; Item 96) was in today's Far East of Russia, which is the coldest region in Russia with an average high temperature of only 3°C. The climate is predominantly frosty cold and in the winter months the temperatures are below zero.
● 97	白湖	Bai Hu	White Lake; today's Lake Khanka (Chinese: 興凱湖/兴凯湖)	白 means "white" and 湖 "lake". On the ANC-KWQ, 白湖 located at 48.5°N, 141.5°E (KWQ latitudinal coordinate has been converted to modern coordinate), is today's Lake Khanka (興凱湖/兴凯湖; modern coordinates are 45°N, 132.4°E; it is the largest lake in the Russian Far East), see Fig. A7.32. **Fig. A7.32.** The red dot denotes where Lake Khanka is located (*Source*: Das steinerne Herz, under CC BY-SA 3.0, https://commons.wikimedia.org/wiki/File:Russia_Primorsky_Krai_location_map.svg).[379] The northern part of the lake is in Chinese territory (in Heilongjiang Province). The lake is pear-shaped, with an extension in its northern part. Lake Khanka is a freshwater lake on the border between Primorsky Krai (濱海邊疆區/滨海边疆区), Russia and Heilongjiang Province (黑龍江省/黑龙江省) in Northeast China. The lake is also included on UNESCO's "World Biosphere Reserves".[380]

(*Continued*)

Item #	Name	Pinyin	Etymology	History, Geography and My Comments
				In ancient Chinese books, "white fish" refers mostly to crawfish with cocked mouth. The reason is probably its large body, white color, and delicious taste. The "skeleton fish" in Lake Khanka fits this description and is also often called "white fish" locally, hence, the name of the lake.[381] The English name of the fish is Culter Alburnus or Topmouth Culter (翹嘴鰲/翹嘴鮊).
● 98	包得河	Bau De He	Baode River[382] is the Muling/ Muren River.	To know which river 包得河 (lit. "Baode River") may be, Figs. A7.33 and 7.34 may give some clues. **Fig. A7.33.** A map shows the Amur River/Heilongjiang (黑龍江/黑龙江; lit. "Black Dragon River"; Amur River) watersheds (*Source*: Kmusser, under CC BY-SA 3.0, https://commons.wikimedia.org/wiki/File:Amurrivermap.png).[383] The Amur/Heilongjiang is the world's tenth longest river, forming the border between the Russian Far East and Northeastern China (Inner Manchuria). The Songhua River (松花江) is the longest tributary of the Amur and the other three major affluents are the Shilka River (石勒喀河), the Argun River (阿爾貢河/阿尔贡河), and the Ussuri River (烏蘇里江/乌苏里江). The Nen River (嫩江) meets its longest tributary of the Second (Di'er) Songhua River (第二松花江/西流松花江) near Da'an to form the Songhua River. The Second (Di'er) Songhua River flows along the left side of Lake Khanka.

On the ANC-KWQ, to the left of Lake Khanka, we can see two unnamed rivers meet. The one on the left may be identified as the Nen River, and the one on the right may be the Songhua River, but its flowing direction is north to south, unlike the Songhua River from west to northeast. The lower portion of the unnamed (Nen) river may be the Second (Di'er) Songhua River, but it is erroneously depicted as flowing into the Japan Sea.

On the ANC-KWQ, to the north of the Baode River, there are also two unnamed rivers: the one close to Shou Shiwei (獸室韋/兽室韦; Item 100) may be the Amur River/Heilongjiang. The one just to its south may be the Songhua River, but it is totally disconnected from the Second (Di'er) Songhua River by a huge mountain range. This leaves Baode River (包得河) a possible candidate for the Ussuri River (烏蘇里江/乌苏里江), but it is depicted as very short river on the ANC-KWQ. or its tributary. It is blocked by a huge mountain range and does not flow along the east bank of Lake Khanka/White Lake (興凱湖/兴凯湖 or 白湖; 97). Then, near White Lake there depicts two unnamed rivers which may correspond to the two tributaries of the Ussuri River as follows:

The Russian map in Fig. A7.34 shows the Ussuri River (Yccypu) with two tributaries: (1) Muling/Muren (Russian: Мулинхэ; 穆稜河) is a river in Northeast China; it is also a left tributary of the Ussuri; and (2) Sungacha (Russian: Сунгача; 松阿察河) is the boundary river between the People's Republic of China and the Russian Federation, and is a tributary of the Ussuri River, originating from Lake Khanka.

(*Continued*)

Item #	Name	Pinyin	Etymology	History, Geography and My Comments
				Fig. A7.34. A detailed map of the Ussuri River with two tributaries and Lake Khanka (*Source*: СафроновAB at ru.wikipedia, under CC BY-SA 3.0, https://commons.wikimedia.org/wiki/File:Ussuri.png).[384]\n\nIn short, on the ANC-KWQ, the rivers in the Amur River/ Heilongjiang watersheds are poorly depicted. The reason may be due to difficulties in surveying these rivers in the mountainous regions of the Khingan Range (興安嶺/兴安岭) in Northeast China.
● 99	亞馬是里		Iammasceli[385]	This geographical item cannot be located on a modern map.
● 100	獸室韋	Shou Shi Wei	Shou Shiwei[386]	Shou Shiwei (獸室韋/兽室韦) was one of the Shiwei tribes, but it has been rarely mentioned in historical records. However, during the period of the Five Dynasties and Ten Kingdoms[387] (五代十國/五代十国; an era of political upheaval and division from 907 to 979; the five dynastic states quickly succeeded one another in the Central Plains, and more than a dozen concurrent dynastic states, collectively known as the Ten Kingdoms, were established elsewhere, mainly in South China), a Han man named Hu Qiao (胡嶠) experienced a wandering life in the Kingdom of Liao (遼/辽 or 大遼/大辽 or 大契丹國/大契丹国; Liao Dynasty; 916–1125) for seven years (916–923). After returning to the Central Plains, he compiled what he saw

and heard there into a book titled *Xian Lu Ji* (陷虜記/陷虏记) or *The Story of the Captive*.[388]

In that book, Hu Qiao introduced in detail the specific information of the Kingdom of Liao and its neighbouring countries. Among them, there were more than a dozen countries and tribes in the north of the Kingdom of Liao, and there were often frictions among them. These countries and tribes included Shou Shiwei (獸室韋/兽室韦), Shiwei (室韋/室韦), Huangtou Shiwei (黃頭室韋/黄头室韦), Heichezi (黑車子/黑车子; a branch of Shiwei), Niuti Turkic (牛蹄突厥) and many more.

The book also records a story that the Kingdom of Liao sent an expedition to explore Siberia. They went all the way north and deep into Siberia, but in the end the group chose to return, after which the interactions between ancient China and Siberia were cut off.

The expedition sent by the Liao Dynasty first arrived in Hulun Buir (呼倫貝爾/呼伦贝尔), and then arrived in the Dog Country (狗國/狗国), which was the northernmost known territory of the Kingdom of Liao. Afterwards, a group of people climbed over a big mountain, went to Outer Khingan Range (外興安嶺/外兴安岭), and formally set foot in Siberia. The group continued to go north, passing 43 cities and dozens of tribes along the way. People there had very different lifestyles. They lived a nomadic life. Due to backwardness in technology and poor living conditions, they could only build houses with tall arbor (喬木/乔木) trees and the tops of the houses were covered with bark, which shows the poor living conditions in Siberia at that time.

When heading north, the expedition also encountered tribes speaking unheard languages and many rare birds and animals they had never been seen before. At this time, the expedition had already set foot in the untouched virgin forest surrounded by tigers and leopards, which are very dangerous, so the expedition had to stop there and turn back.

(*Continued*)

Item #	Name	Pinyin	Etymology	History, Geography and My Comments
				The expedition spent nearly a year on this trip, and they told people what they saw and heard during their trip in detail, which opened people's eyes. According to the description of the expedition, the group reached as far as 60 degrees of north latitude. This was another great geographical discovery in ancient China after the exploration of the Western Regions by Zhang Qian (張騫/张骞; unknown birth year–114 B.C.) back in the Han Dynasty (202 B.C.–9 A.D.; 25–220 A.D.; 9–23 A.D.: Xin); the tribes which were gathered in Siberia then were called the Xiongnu (匈奴; third century B.C.–first century A.D.) by the people in the Central Plains. By the time of the Yuan Dynasty, the Mongols not only expanded their territory westward to Eastern Europe, but also annexed land further north, reaching as far as the Arctic Circle, and included all the areas south of the Arctic Ocean into their territory. During this ancient period, people in the Central Plains further deepened their knowledge of Siberia. Chinese astronomer Guo Shoujing (郭守敬, 1231–1314/1316)[389] established 27 observatories across the country of the Yuan Dynasty. His northernmost observation station was located near Beihai (北海) at 67 degrees of north latitude (this is the KWQ latitude; after multiplying it by 0,9856 gives 66 °N in modern latitude). 北海 is part of the present-day Arctic Sea. it is depicted in Fig. A7.25 to the immediate north of annotation @11 and right on the Arctic Circle encircles the North Pole. The reading of 66 °N is very close to the present Arctic Circle encircles the North Pole at c. 66.57°N.
● 101	胡布山	Hu Bu Shan	Hubu Mountain[390]	It is known that in the Tang Dynasty, the distribution of the Shiwei became increasingly widespread, reaching more than twenty tribes.[391] Among them, *Bei Shi* (北史) or *The Northern History* states that the Bo Shiwei (鉢室韋/钵室韦) lived by Hubu Mountain ("依胡布山而住").[392]

• 102	狗國	Gou Guo	Dog Nation[393]	Dog Nation (狗國/狗国) here refers to the area from the confluence of the Ussuri River River (烏蘇里江/乌苏里江) into the Heilongjiang (黑龍江/黑龙江; Amur River) from where it flows into the sea. It was named after dogs, because in winter dogs were used for transportation. In the Yuan Dynasty, a dog station was set up.[394]

According to *Xin Wu Dai Shǐ • Juan Qī Shí San* (新五代史 • 卷七十三) or *New History of the Five Dynasties, Juan (Volume) 71,* written in the Northern Song Dynasty (北宋; 960–1127), there is a description about the Dog Nation as follows:

又北, 狗國, 人身狗首, 長毛不衣, 手搏猛獸, 語爲犬 噑, 其妻爲人, 能漢語, 生男爲狗, 女爲人, 自相婚嫁, 穴居食生, 而妻女人食。

My translation and comments (between square brackets) are as follows:

Further to the north there is the Dog Nation, where men wear a dog mask over their head, grow long hair and wear no clothes. They hand fight against fierce beasts and howl like dogs. Their wives are normal women and can speak Chinese. When the wives give birth to a boy, the boy will also dress like a dog [and do hunting work like his father], while a girl will be a normal woman like her mother [doing household work]. They have free marriage] and live in cave-like igloos. Their men eat raw food [this is understandable because men go out hunting for days or even months without returning home. Under the conditions of endless ice and snow, it is impossible to find branches for fire. The harsh natural conditions in far northeastern Asia force them to be flexible about food intake in the wild. In the Arctic and sub-Arctic regions, food will not spoil at low

(Continued)

Item #	Name	Pinyin	Etymology	History, Geography and My Comments
				temperatures such as tens of degrees below zero degree Celsius. So, it is possible for men to get used to eating raw food], and women can eat normally cooked food [at home]. We know that in ancient days some of these men and women from the far northeastern Asia might have migrated to today's Alaska through the current Bering Strait which used to have a land bridge connecting Siberia and Alaska.[395]  **Fig. A7.35.** Map of the Inuit Circumpolar Council of Eskimo peoples, showing the Yupik (Yu'ik, Siberian Yupik) and Inuit (Iñupiat, Inuvialuit, Nunavut, Nunavik, Nunatsiavut, Greenlandic Inuit) in the Arctic and sub-Arctic regions (*Source*: Kmusser, under CC BY-SA 3.0, https://commons.wikimedia.org/wiki/File:Inuit_conf_map.png);[396] the Arctic Circle encircles the North Pole at c. 66.57°N. The Arctic region sits inside the Arctic Circle and the subarctic region lies just south of it. Earth's arctic and subarctic regions are extremely cold, icy areas of land and sea that receive almost no sunlight during their long, dark winters. Temperatures rarely rise above freezing.

When we look at the history of the Yupik and Inuit (Inuk in singular form), we find that early on they lived on islands and lands (referring to areas such as the Aleutian Islands, northern Alaska, and the Mackenzie Delta in northwestern Canada) near the Arctic Circle and the Bering Sea. The domestication and utilization of dogs is a very important part of their life. They are mainly engaged in land or sea hunting supplemented by fishing. Prey is the main source of livelihood. Animal fur is used for clothing, oily fat is used for lighting and cooking, and bones and teeth are used as tools and weapons. The men hunt and build semi-underground igloos with snow blocks to live in, and the women make leather goods and do sewing. Due to the cold temperature, these dwellings only have a small opening for entry and exit, see Fig. A7.36. Their language is close to East Asian languages. These people are short in comparison with the Whites, and they have yellow skin and black hair. Such features are quite consistent with the Mongolian race: in general, Mongolian nomads are shorter and skinnier compared with sedentary people due to nutrition and environmental differences. Usually, nomad men are around 166–170 cm in height, and women are around 155–160 cm in height.[397]

Fig. A7.36. A nearly complete, medium-sized igloo, with excavation under the door and the exterior unfinished.[398]

(*Continued*)

Item #	Name	Pinyin	Etymology	History, Geography and My Comments
@ 14	此處古謂兩邊之地相連今已審 有此大海隔開北海可通北海	Ci Chu Gu Wei Liang Bian Zhi Di Xiang Lian Jin Yi Shen You Ci Da Hai Ge Kai Bei Hai Ke Tong Bei Hai		My translation and comments (between square brackets) are as follows: In ancient times, it was said that the lands on both sides were connected. Now it is known that there is this sea [the present-day Bering Sea separated from the North Sea by the Bering Strait] separating the North Sea [Arctic Ocean] and that one can reach the North Sea. Although from at least 1562, European geographers thought that there was a Strait of Anian between Asia and North America, the first European to enter the sea separating the Pacific Ocean from the Arctic Oceans is the Danish-born Russian navigator Vitus Bering in 1728.[399] Hence, the strait bears his name. In 1732, Mikhail Gvozdev crossed it for the first time, from Asia to America. As late as 1741, a Russian expedition led by Vitus Bering, along with George Steller, made the first "discovery" of Alaska, landing near what today is Kayak Island.[400] How was Matteo Ricci in 1602 able to depict this strait between Alaska and Russia's Far East before the eighteenth century, if he did not use a Chinese source map?

Endnotes

[1]Ronnie Po-Chia Hsia. *Matteo Ricci and the Catholic Mission to China, 1583–1610: A Short History with Documents (Passages: Key Moments in History)*. Indianapolis, USA: Hackett Publishing Company, Inc., 2016.

[2]Sheng-Wei Wang. *Chinese Global Exploration in the Pre-Columbian Era: Evidence from an Ancient World Map*. Singapore: World Scientific Publishing Co., 2023. Sheng-Wei Wang. "Chapter nine: Secret Revealed by Ancient Maps." *The Last Journey of the San Bao Eunuch, Admiral Zheng He*, Hong Kong, China: Proverse Hong Kong, 2019, pp. 290–301.

[3]Sheng-Wei Wang. *Chinese Global Exploration in the Pre-Columbian Era: Evidence from an Ancient World Map, op. cit.*, p. 115.

[4]Sheng-Wei Wang. *Chinese Global Exploration in the Pre-Columbian Era: Evidence from an Ancient World Map, op. cit.*, p. 116.

[5]Sheng-Wei Wang. *Chinese Global Exploration in the Pre-Columbian Era: Evidence from an Ancient World Map, op. cit.*, p. 118.

[6]"File:Kunyu Wanguo Quantu (坤輿萬國全圖).jpg." *Wikimedia Commons: The Free Encyclopedia*, Wikimedia Foundation, 11 July 2024, upload.wikimedia.org/wikipedia/commons/7/71/Kunyu_Wanguo_Quantu_(坤輿萬國全圖).jpg

[7]Sheng-Wei Wang. *Chinese Global Exploration in the Pre-Columbian Era: Evidence from an Ancient World Map, op. cit.*, p. 128.

[8]Sheng-Wei Wang. "A Chinese-Based World Map Depicts Europe Between 1157–1166." *Chinese Global Exploration in the Pre-Columbian Era: Evidence from an Ancient World Map, op. cit.*, pp. 71–140.

[9]"File:Kunyu Wanguo Quantu (坤輿萬國全圖).jpg." *Wikimedia Commons*, 11 July 2024, upload.wikimedia.org/wikipedia/commons/7/71/Kunyu_Wanguo_Quantu_(坤輿萬國全圖).jpg

[10]Sheng-Wei Wang. *Chinese Global Exploration in the Pre-Columbian Era: Evidence from an Ancient World Map, op, cit.*, p. 119.

[11]"File:Kunyu Wanguo Quantu by Matteo Ricci Plate 1-3.jpg." *Wikimedia Commons*, 13 June 2023. upload.wikimedia.org/wikipedia/commons/b/b8/Kunyu_Wanguo_Quantu_by_Matteo_Ricci_Plate_1-3.jpg

[12]"File:Mercator 1569 world map composite.jpg." *Wikimedia Commons*, 26 Nov. 2016, upload.wikimedia.org/wikipedia/commons/4/4b/Mercator_1569_world_map_composite.jpg

[13]"File:OrteliusWorldMap1570.jpg." *Wikimedia Commons*, 12 July 2022, upload.wikimedia.org/wikipedia/commons/e/e2/OrteliusWorldMap1570.jpg

[14]"File:1594 double hemisphere world map by Petrus Plancius.jpg." *Wikimedia Commons*, 2 Sept. 2022, upload.wikimedia.org/wikipedia/commons/0/0d/1594_double_hemisphere_world_map_by_Petrus_Plancius.jpg

[15]"File:1570 Ortelius Map of Asia (first edition) - Geographicus - AsiaeNovaDescriptio-ortelius.jpg." *Wikimedia Commons*, 22 Mar. 2011, upload.wikimedia.org/wikipedia/commons/a/a3/1570_Ortelius_Map_of_Asia_(first_edition)_-_Geographicus_-_AsiaeNovaDescriptio-ortelius.jpg

[16]"File:Tartariae sive Magni Chami Regni ẗypus. LOC 83690086.jpg." *Wikimedia Commons*, 5 June 2022, upload.wikimedia.org/wikipedia/commons/6/67/Tartariae_sive_Magni_Chami_Regni_ẗypus._LOC_83690086.jpg

[17]"File:1562-anthony jenkinson.2.jpg." *Wikimedia Commons*, 18 May 2024, upload.wikimedia.org/wikipedia/commons/0/08/1562-anthony_jenkinson.2.jpg

[18]Anthony Jenkinson, *et al. Early Voyages and Travels to Russia and Persia by Anthony Jenkinson and other Englishmen, Volumes I-II: With some Account of the First Intercourse of ... Caspian Sea (Hakluyt Society, First Series)*. London, UK: Hakluyt Society, 2020.

[19]M. C. Elliot. "The Limits of Tartary: Manchuria in imperial and national geographies." *The Journal of Asian Studies*, vol. 59, no. 3, 2000, pp. 603–646.

[20]E. Denison Ross. "The Orkhon Inscriptions: Being a Translation of Professor Vilhelm Thomsen's Final Danish Rendering." *Bulletin of the School of Oriental Studies, University of London*, vol. 5, no. 4, 1930, pp. 861–76. doi:10.1017/S0041977X00090558

[20]Library of Congress. "The Russian Discovery of Siberia." www.loc.gov. Library of Congress, 2025, https://www.loc.gov/collections/meeting-of-frontiers/articles-and-essays/exploration/russian-discovery-of-siberia/

[22]Benson Bobrick. *East of the Sun: The Epic Conquest and Tragic History of Siberia*. Montpelier VT, USA: Russian Information Services, Inc., 2014.

[23]C. Michael Hogan. "Silk Road, North China - Ancient Trackway in China." www.megalithic.co.uk, Megalithic Portal, 2025, https://www.megalithic.co.uk/article.php?sid=18006

[24]Saglar Bougdaeva. "The Yelu language of war and peace: A revised oirad translation of the Altai Runic Inscriptions (6th–9th centuries)." *Central Asiatic Journal*, vol. 66, no. 1–2, 2024, pp. 27–46.

[25]Frederick W. Mote and Denis Twitchett, eds. *The Cambridge History of China, Vol. 7: The Ming Dynasty, 1368-1644, Part 1*, Cambridge, UK: Cambridge University Press, 1988.

[26]Henry H. Howorth. *History of the Mongols, Vol. 1: From the 9th to the 19th Century; The Mongols Proper and the Kalmuks*. London, UK: Forgotten Books, 2018.

[27]Wang Yongqiang (王永強). *Zhong Guo Shao Shu Min Zu Wen Hua Shi Tu Dian • Bei Fang Juan* (中国少数民族文化史图典•北方卷) or *Illustrated Dictionary of the Cultural History of China's Ethnic Minorities: Volume of the Northern Region*, Guangxi, China: Guangxi Education Publishing House (广西教育出版社), 1999, p. 308.

[28]Cao Yongnian (曹永年). *Meng Gu Min Zu Tong Shi • Di San Juan* (蒙古民族通史 • 第三卷) or *General History of the Mongolian Nation, Volume Three*, Inner Mongolia, China: Inner Mongolia University Press (內蒙古大學出版社), 2002, p. 116.

[29]Liu Qingyang (劉清陽). *Ming Qing She Hui Ge Yao Ji • Shang Juan •· Shi Shi Pian* (明清社會歌謠集 • 上卷 • 時事篇) or *Collection of Social Ballads of the Ming and Qing Dynasties, Volume 1 of Two, Current Affairs*, Morrisville, NC, USA: Lulu Press, Inc., 2015, p. 59.

[30]Ph. de Heer. *The Care-Taker Emperor: Aspects of the Imperial Institution in Fifteenth-Century China as Reflected in the Political History of the Reign of Chu Chi'i-Yü*. Leiden, Netherlands: Brill Academic Publishers, 1986; Cao Yongnian (曹永年). *Meng Gu Min Zu Tong Shi • Di San Juan* (蒙古民族通史 • 第三卷) or *General History of the Mongolian Nation, Volume Three*, Inner Mongolia, China: Inner Mongolia University Press (內蒙古大學出版社), 2002, p. 116.

[31]Chahryar Adle, *et al. History of Civilization in Central Asia, Volume V*. New Delhi, India: Motilal Banarsidass, 2008.

[32]Library of Congress. "The Russian Discovery of Siberia." *Op. cit.*

[33]Liu Ying Sheng Zhu. *Chagatai Khanate History*. Shanghai, China: Shanghai Ancient Books Publishing House, 1991. The Chagatai Khanate split into Eastern Chagatai Khanate (Moghulistan) and Western Chagatai Khanate in 1347. The Timurid Empire replaced the original Western Chagatai Khanate and continued the Chagatai Khanate. From 1363, the Chagatais progressively lost Transoxiana to the Timurids. The reduced realm came to be known as Moghulistan, which lasted until the late fifteenth century, when it broke off into the main Turpan Khanate (吐魯番汗國/吐鲁番汗国; 1487–1570) and the branch Yarkant Khanate (葉爾羌汗國/叶尔羌汗国; 1514–c. 1705). In 1570, the Yarkant Khanate made an Eastern

Campaign which led to the destruction of the Turpan Khanate and ended the Eastern Chagatai Khanate's main branch. The Yarkant Khanate was eventually conquered by the Dzungar Khanate (準噶爾汗國/准噶尔汗国) in 1705.

[34]Hu Wenkang (胡文康). *Xinjiang Guangji* (新疆广记) or *Xinjiang Records*, Xinjiang, China: Xinjiang People's Publishing House (新疆人民出版社), 2005, p. 40.

[35]Frank McLynn. *Genghis Khan: His Conquests, His Empire, His Legacy.* Boston, MA, USA: Da Capo Press, 2016.

[36]Xu lianli (徐良利). *Yi Er Han Guo Shi Yan Jiu* (伊儿汗国史研究) or *Studies on the History of the Ilkhanate.* Beijing, China: People's Publishing House (人民出版社), 2009.

[37]Ronald Findlay and Mats Lundahl. "The First Globalization Episode: The Creation of the Mongol Empire, or the Economics of Chinggis Khan." In Göran Therborn and Habibul Khondker, eds. *Asia and Europe in Globalization.* Leiden, Netherlands: Brill, 2006, pp. 13–54.

[38]*Ibid.*

[39]Chen Gaohua (陳高華). "Ili Baliq 亦力把里." *Zhongguo Da Baike Quanshu* (中國大百科全書) or *Encyclopedia of China, Vol. 3.* Bejing/Shanghai, China: The Encyclopedia of China Publishing House, 1991, p. 1400.

[40]*Ibid.*

[41]Charles J. Halperin. *Russia and the Golden Horde: The Mongol Impact on Medieval Russian History.* Bloomington, IN, USA: Indiana University Press, 1987.

[42]István Vásáry. "The Crimean Khanate and the Great Horde (1440s–1500s): A Fight for Primacy." In Meinolf Arens and Denise Klein, eds. *Das frühneuzeitliche Krimkhanat (16.–18. Jahrhundert) zwischen Orient und Okzident.* Wiesbaden, Germany: Harrassowitz, 2012, pp. 13–26; George Vernadsky. *The Mongols and Russia.* New Haven, CT, USA: Yale University Press, 1953; Janet Martin. *Medieval Russia: 980–1584.* Cambridge, UK: Cambridge University Press, 2007.

[43]Donald Rayfield. *A Seditious and Sinister Tribe: The Crimean Tatars and Their Khanate.* London, UK: Reaktion Books, 2024.

[44]Janet Martin. *Op. cit.*

[45]*Ibid.*

[46]Michael Khodarkovsky. *Russia's Steppe Frontier: The Making of a Colonial Empire, 1500–1800.* Bloomington, IN, USA: Indiana University Press, 1955.

[47]Sheng-Wei Wang. "Chapter 1: Chinese Explored Cape Breton Island Long before the Europeans." *Chinese Global Exploration in the Pre-Columbian Era: Evidence from an Ancient World Map, op. cit.,* pp. 7–8.

[48]*Ibid.*

[49]Siu-Leung Lee (李兆良). "(坤輿萬國全圖)成圖斷代." *Kunyu Wanguo Quantu Jie Mi • Ming Dai Ce Hui Shi Jie* (坤輿萬國全圖解密:明代測繪世界) or *Deciphering the Kunyu Wanguo Quantu, A Chinese World Map: Ming Chinese Mapped the World Before Columbus, op, cit.,* p. 51.

[50]Sheng-Wei Wang. *Chinese Global Exploration in the Pre-Columbian Era: Evidence from an Ancient World Map, Op. cit.*

[51]Huang Shijian (黄时鉴) and Gong Yingyan (龚缨晏). Li Ma Dou Shi Jie Di Tu Yan Jiu (利玛窦世界地图研究) or *Research on Matteo Ricci's World Map,* Shanghai, China: Shanghai Chinese Classics Publishing House (上海古籍出版社), 2004, p. 196.

[52]Huang Shijian (黄时鉴) and Gong Yingyan (龚缨晏). *Op. cit.,* p. 196.

[53]Nickolay Knipovitch. Sudhir Raghav, trans. "Korelia." In *Encyclopaedic Dictionary of Brockhaus and Efron.* http://heninen.net/, Andrew Heninen, 2024, http://heninen.net/korela/korelija_e.htm

[54]Janet Martin. *Op. cit.*

[55]*Ibid.*

[56]*Ibid.*

[57]Sheng-Wei Wang. *Chinese Global Exploration in the Pre-Columbian Era: Evidence from an Ancient World Map, op. cit.*, pp. 5–6, 239–240.

[58]Nurlan Kenzheakhmet and Alpamys Abu. "Some Medieval and Post-Golden Horde's Towns of the Itil (Volga) and Syr-Darya Basins According to the Arabic and Chinese Maps." *Golden Horde Review*, vol. 9, no. 3, 2021, p. 623. https://www.researchgate.net/publication/354971343_Some_Medieval_and_Post-Golden_Horde's_Towns_of_the_Itil_Volga_and_Syr-Darya_Basins_According_to_the_Arabic_and_Chinese_Maps.

[59]E. A. Kashina and A.V. Yemelyanov. "Bone bird figurines of the Meschora Lowlands in the Final Stone Age // Problems of prehistoric and medieval archaeology of the Oka river basin. (in Russian)." https://www.researchgate.net/, ResearchGate, 2025, www.researchgate.net/publication/264733103_Bone_bird_figurines_of_the_Meschora_Lowlands_in_the_Final_Stone_Age_Problems_of_prehistoric_and_medieval_archaeology_of_the_Oka_river_basin_in_Russian

[60]Huang Shijian (黄时鉴) and Gong Yingyan (龚缨晏). *Op. cit.*, p. 207.

[61]Foundation of an administrative centre. "Foundation of the city of Vologda – 1147." http://vologda-oblast.ru/, Vologda Oblast Government, 2025, https://web.archive.org/web/20110722202517/http://vologda-oblast.ru/main.asp?V=440&LNG=ENG

[62]Gerol d Ivanovich Vzdornov. *Vologda*, Open Library, 1972, p. 7.

[63]Huang Shijian (黄时鉴) and Gong Yingyan (龚缨晏). *Op. cit.*, p. 186.

[64]Sheng-Wei Wang. *Chinese Global Exploration in the Pre-Columbian Era: Evidence from an Ancient World Map, op. cit.*, pp. 5–6, 239–240.

[65]*Ibid.*

[66]Huang Shijian (黄时鉴) and Gong Yingyan (龚缨晏). *Op. cit.*, p. 187.

[67]Huang Shijian (黄时鉴) and Gong Yingyan (龚缨晏). *Op. cit.*, p. 193.

[68]Max Vasmer. "Москва." In O. N Trubachyov and B. O. Larin, eds. *Этимологический словарь русского языка [Russisches etymologisches Wörterbuch] (in Russian) (2nd ed.)*. Moscow, USSR: Progress, 1986–1987.

[69]Michael T. Florinsky. *Russia: A History and an Interpretation in Two Volumes.* London, UK: MacMillan Co., 1955.

[70]*Ibid.*

[71]Sheng-Wei Wang. "Chapter 3: A Chinese-Based World Map Depicts Europe Between 1157 and 1166." *Chinese Global Exploration in the Pre-Columbian Era: Evidence from an Ancient World Map, Op. cit.*, pp. 71–140.

[72]George Vernadsky. *The tsardom of Moscow, 1547–1682.* New Haven, CT, USA: Yale University Press, 1969.

[73]Michael T. Florinsky. *Op. cit.*

[74]Huang Shijian (黄时鉴) and Gong Yingyan (龚缨晏). *Op. cit.*, p. 199.

[75]Huang Shijian (黄时鉴) and Gong Yingyan (龚缨晏). *Op. cit.*, p. 193.

[76]Ibn Battutah. *The Travels of Ibn Battutah.* London, UK: Picador, 2003, p. 124.

[77]Huang Shijian (黄时鉴) and Gong Yingyan (龚缨晏). *Op. cit.*, p. 197.

[78]Tsarévitch Wakhoucht. Marie-Félicité Brosset, ed. ღეოღრაჴიული აღწერა საქართვჽლოჲსა. *Description géographique de la Géorgie [Geographic description of Georgia] (in Georgian and French)*, S.-Pétersbourg, Russia: A la typographie de l'Académie Impériale des Sciences, 1842, pp. 392–395.

[79]Ronald Grigor Suny. *The Making of the Georgian Nation*. Bloomington, IN, USA: Indiana University Press, 1994.

[80]*Ibid.*

[81]Donald Rayfield. *Edge of Empires: A History of Georgi*. London, UK: Reaktion Books, 2013, p. 162.

[82]*Ibid.*

[83]"File:Historical Samegrelo in modern international borders of Georgia.svg." *Wikimedia Commons*, 8 Oct. 2020, upload.wikimedia.org/wikipedia/commons/thumb/d/da/Historical_Samegrelo_in_modern_international_borders_of_Georgia.svg/942px-Historical_Samegrelo_in_modern_international_borders_of_Georgia.svg.png

[84]File:Abraham Ortelius - Tvrcici imperii descriptio.jpg." *Wikimedia Commons*, 6 Oct. 2010, upload.wikimedia.org/wikipedia/commons/9/94/Abraham_Ortelius_-_Tvrcici_imperii_descriptio.jpg

[85]Herodotus. David Grene, trans. *Herodotus: The History*. Chicago, IL, USA: The University of Chicago Press, 1988.

[86]"File:Altai Mountains.jpg." *Wikimedia Commons*, 3 Aug. 2023, upload.wikimedia.org/wikipedia/commons/f/f1/Altai_Mountains.jpg

[87]"File:Ecoregion PA0814.svg." *Wikimedia Commons*, 15 Sept. 2020, upload.wikimedia.org/wikipedia/commons/thumb/9/9e/Ecoregion_PA0814.svg/2630px-Ecoregion_PA0814.svg.png

[88]Nurlan Kenzheakhmet and Alpamys Abu. *Op. cit.*, p. 622. doi: 10.22378/2313-6197.2021-9-3.611–653.

[89]Nurlan Kenzheakhmet and Alpamys Abu. *Op. cit.*, p. 623. doi: 10.22378/2313-6197.2021-9-3.611–653.

[90]Huang Shijian (黄时鉴) and Gong Yingyan (龚缨晏). *Op. cit.*, p. 190.

[91]Huang Shijian (黄时鉴) and Gong Yingyan (龚缨晏). *Op. cit.*, p. 201.

[92]"File:Map of Russia - Mordovia (Crimea disputed).svg." *Wikimedia Commons*, 16 Jan. 2023, upload.wikimedia.org/wikipedia/commons/thumb/0/09/Map_of_Russia_%282014%E2%80%932022%29_-_Mordovia_%28Crimea_disputed%29.svg/1541px-Map_of_Russia_%282014%E2%80%932022%29_-_Mordovia_%28Crimea_disputed%29.svg.png

[93]Marie Favereau. *The Horde: How the Mongols Changed the World*. Cambridge, MASS, USA: Belknap Press: An Imprint of Harvard University Press, 2022; Charles J. Halperin. *Op. cit.*

[94]Viacheslav Shpakovsky and David Nicolle. Gerry Embleton, Illustrator. *Armies of the Volga Bulgars & Khanate of Kazan: 9th–16th centuries (Men-at-Arms, 491)*. Oxford, UK: Osprey Publishing, 2013.

[95]Robert Crummey. *Formation of Muscovy 1304–1613 (Longman History of Russia)*. Boston, MASS, USA: Addison-Wesley Longman Ltd., 1987.

[96]Paul Bushkovitch. *A Concise History of Russia*. Cambridge, UK: Cambridge University Press, 2011.

[97]Huang Shijian (黄时鉴) and Gong Yingyan (龚缨晏). *Op. cit.*, p. 198.

[98]Nurlan Kenzheakhmet and Alpamys Abu. *Op. cit.*, p. 620. https://www.researchgate.net/publication/354971343_Some_Medieval_and_Post-Golden_Horde's_Towns_of_the_Itil_Volga_and_Syr-Darya_Basins_According_to_the_Arabic_and_Chinese_Maps

[99]Ibid.

[100]Георгий Михайлович Лаппо, ed. Энциклопедия Города России (俄罗斯城市百科全书), Moscow, Russia: ДРОФА, 2003, p. 64. ISBN 5-7107-7399-9.

[101]"File:Severnaya Dvina eng.svg." *Wikimedia Commons*, 31 Oct. 2020, upload.wikimedia.org/wikipedia/commons/thumb/2/24/Severnaya_Dvina_eng.svg/2560px-Severnaya_Dvina_eng.svg.png

[102]Huang Shijian (黄时鉴) and Gong Yingyan (龚缨晏). *Op. cit.*, p. 185.

[103]Aleksandr Aleksandrovich Vasil'ev. *History of the Byzantine Empire, 324–1453*. Madison, WI, USA: University of Wisconsin Press, 2012.

[104]Unknown author. *The Lenin Volga-Don Canal*. Moscow, USSR: Foreign Languages Publishing House, 1956.

[105]"File:Donrivermap.png." *Wikimedia Commons*, 12 May 2022, upload.wikimedia.org/wikipedia/commons/0/0c/Donrivermap.png

[106]"File:CaspianSeaDrainage v1.png." *Wikimedia Commons*, 3 Dec. 2021, upload.wikimedia.org/wikipedia/commons/e/ee/CaspianSeaDrainage_v1.png

[107]Charles J. Halperin. *Op. cit.*

[108]Institution of the Tatar Encyclopaedia. "Хажитархан" *Tatar Encyclopaedia (in Tatar)*. Kazan, Tatarstan, Russia: the Republic of Tatarstan Academy of Sciences, 2002.

[109]UC Berkeley Professional Development Provider. "Lands of the Golden Horde & the Chagatai: 1332–1333." https://orias.berkeley.edu/, Berkeley Orias, 2025, https://orias.berkeley.edu/resources-teachers/travels-ibn-battuta/journey/lands-golden-horde-chagatai-1332-1333

[110]"File:Volgarivermap.png." *Wikimedia Commons*, 23 July 2021, upload.wikimedia.org/wikipedia/commons/7/76/Volgarivermap.png

[111]Icon Group International. *Astrakhan: Webster's Timeline History, 1240–2007*. Las Vegas, NV, USA: ICON Group International, Inc., 2010.

[112]*Ibid.*

[113]*Ibid.*

[114]*Ibid.*

[115]The Barry Lawrence Ruderman Map Collection. "La Tartaria Indipendente…il Paese de Calmuchi quello degli Usbeks, e il Turkesan…1784." https://exhibits.stanford.edu/, Stanford Libraires, 2025, https://exhibits.stanford.edu/ruderman/catalog/jt613xx3971

[116]Nurlan Kenzheakhmet and Alpamys Abu. *Op. cit.*, p. 619. https://www.researchgate.net/publication/354971343_Some_Medieval_and_Post-Golden_Horde's_Towns_of_the_Itil_Volga_and_Syr-Darya_Basins_According_to_the_Arabic_and_Chinese_Maps

[117]Huang Shijian (黄时鉴) and Gong Yingyan (龚缨晏). *Op. cit.*, p. 187.

[118]R. G. Akhmetyanov. *Brief Historical and Etymological Dictionary of the Tatar Language*. Kazan, Tatarstan, Russia: Tatar Book Publishers, 2001. p. 76.

[119]Nurlan Kenzheakhmet and Alpamys Abu. *Op. cit.*, p. 621. https://www.researchgate.net/publication/354971343_Some_Medieval_and_Post-Golden_Horde's_Towns_of_the_Itil_Volga_and_Syr-Darya_Basins_According_to_the_Arabic_and_Chinese_Maps

[120]Michal Biran. *The Empire of the Qara Khitai in Eurasian History: Between China and the Islamic World*. Cambridge Studies in Islamic Civilization. Cambridge, UK: Cambridge University Press, 2005.

[121]Yuan Julian Chen. "Legitimation Discourse and the Theory of the Five Elements in Imperial China." *Journal of Song-Yuan Studies*, vol. 44, no. 44, 2014, pp. 325–364. https://www.researchgate.net/publication/297647263_Legitimation_Discourse_and_the_Theory_of_the_Five_Elements_in_Imperial_China

[122]UNRV Roman History. "Sarmatia." www.unrv.com, United Nations of Roma Victrix (UNRV), 2024, www.unrv.com/provinces/sarmatia.php

[123]Norman Davies. *Litva: The Rise and Fall of the Grand Duchy of Lithuania: A Selection from Vanished Kingdoms*. London, UK: Penguin Books, 2013.

[124]Howell Lloyd, et. al., eds. *European Political Thought 1450-1700: Religion, Law and Philosophy*. New Haven, CT, USA: Yale University Press, 2008.

[125]"File:Ptolemy Cosmographia 1467 - Central Russia and Sarmatia.jpg." *Wikimedia Commons*, 23 Oct. 2021, upload.wikimedia.org/wikipedia/commons/f/f1/Ptolemy_Cosmographia_1467_-_Central_Russia_and_Sarmatia.jpg

[126]Richard Brzezinski and Mariusz Mielczarek. Gerry Embleton, illustrator. *The Sarmatians 600 BC–AD 450*. Oxford, UK: Osprey Publishing, 2002.

[127]*Ibid.*

[128]James B. Minahan. *One Europe, Many Nations: A Historical Dictionary of European National Groups*. Santa Barbara, CA, USA: Greenwood Publishing Group, 2000, p. 518; Agustí Alemany. *Sources on the Alans: A Critical Compilation*. Leiden, Netherlands: Brill, 2000.

[129]Tatiana Mastyugina and Lev Perepelkin. Irina Zviagelskaia and Vitalii Viacheslavovich Naumkin, eds. *An Ethnic History of Russia: Pre-revolutionary Times to the Present*. Manhattan, NY, USA: Praege Publishers, 1996, p. 80.

[130]Huang Shijian (黄时鉴) and Gong Yingyan (龚缨晏). *Op. cit.*, p. 192.

[131]*Ibid.*

[132]Rein Taagepera. *The Finno-Ugric Republics and the Russian State*. Oxfordshire, UK: Routledge, 2013.

[133]"File:Map of yugra.png." *Wikimedia Commons*, 19 Nov. 2021, upload.wikimedia.org/wikipedia/commons/b/b7/Map_of_yugra.png

[134]*Ibid.*

[135]John Gibson. "Correct Map of Europe from the Best Authorities (1758)." https://dl.mospace.umsystem.edu/, University of Missouri Digital Libraries, 2025, https://dl.mospace.umsystem.edu/mu/islandora/object/mu:110880/print_object

[136]Nurlan Kenzheakhmet and Alpamys Abu. *Op. cit.*, p. 620. https://www.researchgate.net/publication/354971343_Some_Medieval_and_Post-Golden_Horde's_Towns_of_the_Itil_Volga_and_Syr-Darya_Basins_According_to_the_Arabic_and_Chinese_Maps

[137]"File:Russia administrative location map.svg." *Wikimedia Commons*, 19 Oct. 2022, upload.wikimedia.org/wikipedia/commons/thumb/e/e1/Russia_administrative_location_map.svg/2362px-Russia_administrative_location_map.svg.png

[138]Nurlan Kenzheakhmet and Alpamys Abu. *Op. cit.*, p. 622. doi: 10.22378/2313-6197.2021-9-3.611–653.

[139]Nocolay Usov. "Kartesh D1: White Sea." www.st.nmfs.noaa.gov, METABASE Explorer: The Marine Ecological Time Series database, 2025, https://www.st.nmfs.noaa.gov/copepod/time-series/ru-10101/

[140]Valentin L. Yanin. "The Archaeology of Novgorod." *Scientific American,* Special Issue, c. 1994, pp. 120–127.

[141]G. Patrick March. "3: Ivan IV and the Muscovite Drang nach Osten." *Eastern Destiny: Russia in Asia and the North Pacific,* Westport, CT, USA: Praeger Publishers, 1996, p. 26.

[142]"File:White Sea map.png." *Wikimedia Commons*, 27 Mar. 2021, upload.wikimedia.org/wikipedia/commons/b/b3/White_Sea_map.png

[143]Roger Bacon. John Henry Bridges, ed. *The Opus Majus of Roger Bacon, Vol. 1*, Cambridge, UK: Cambridge University Press, 2010, p. 357.

[144]WorldAtlas. "Sea Of Azov." www.worldatlas.com, WorldAtlas, 2025, www.worldatlas.com/seas/sea-of-azov.html

[145]Nurlan Kenzheakhmet and Alpamys Abu. *Op, cit.*, p. 623. https://www.researchgate.net/publication/354971343_Some_Medieval_and_Post-Golden_Horde's_Towns_of_the_Itil_Volga_and_Syr-Darya_Basins_According_to_the_Arabic_and_Chinese_Maps

[146]Huang Shijian (黄时鉴) and Gong Yingyan (龚缨晏). *Op. cit.*, p. 192.

[147]Миллер Г. Ф. Глава первая. События древнейших времён до русского владычества // История Сибири — М.-Л.: АН СССР, 1937. — Т. 1. — С. 189–194.

[148]*Ibid.*

532 *Chinese Explored Asia Long Before the Europeans*

[149]W. Bruce Lincoln. *The Conquest of a Continent: Siberia and the Russians.* New York, NY, USA: Random House, 1993.

[150]"File:Siberian Khanate map English revised.svg." *Wikimedia Commons*, 3 Nov. 2022, upload.wikimedia. org/wikipedia/commons/thumb/9/9c/Siberian_Khanate_map_English_revised.svg/2560px-Siberian_ Khanate_map_English_revised.svg.png

[151]"File:Herberstein-Moscovia-NE.png." *Wikimedia Commons*, 25 Jan. 2023, upload.wikimedia.org/ wikipedia/commons/b/b0/Herberstein-Moscovia-NE.png

[152]Huang Shijian (黄时鉴) and Gong Yingyan (龚缨晏). *Op. cit.*, p. 194.

[153]Huang Shijian (黄时鉴) and Gong Yingyan (龚缨晏). *Op. cit.*, p. 185.

[154]Edward C. Sachau. *Alberuni's India: an Account of the Religion, Philosophy, Literature, Geography, Chronology, Astronomy, Customs, Laws and Astrology of India about AD 1030, vol. 1.* London, UK: Kegan Paul, Trench, Trübner & Co., 1910. p. 298.

[155]Toshkent Shahar Statistika Boshqarmasi. "Toshkent Shahri Raqamlarda." https://toshstat.uz/, O'zbekiston Respublikasi Davlat statistika qo'mitasining Elektron interaktiv xizmatlar portaliga o'tish, 2025, https:// toshstat.uz/uz/

[156]Karel Balas, *et al. Tashkent: A Modernist Capital.* Milan, Italy: Rizzoli, 2024.

[157]Nurlan Kenzheakhmet and Alpamys Abu. *Op. cit.*, p. 621. https://www.researchgate.net/ publication/354971343_Some_Medieval_and_Post-Golden_Horde's_Towns_of_the_Itil_Volga_and_Syr-Darya_Basins_According_to_the_Arabic_and_Chinese_Maps

[158]*Ibid.*

[159]*Ibid.*

[160]Adrian Room. *Placenames of the World: Origins and Meanings of the Names for 6,600 Countries, Cities, Territories, Natural Features and Historic Sites*, London, UK: McFarland, 2006, p. 330.

[161]Leon D. Seltzer. *The Columbia Lippincott Gazetteer of the World.* New York, NY, USA: Columbia University Press, 1952.

[162]Étienne de la Vaissière. James Ward, trans. *Sogdian Traders: A History*, Leiden, Netherlands: Brill Academic Publishers, 2005, pp. 108–111.

[163]Liu Ying Sheng Zhu. *Op. cit.*

[164]*Ibid.*

[165]Ibn Battutah, *The Travels of Ibn Battuta*, London, UK: Picador, 2002, p. 143.

[166]Thomas Welsford. *Four Types of Loyalty in Early Modern Central Asia: The Tqy-Tmrid Takeover of Greater M War al-Nahr, 1598-1605.* Leiden, Netherlands: Brill, 2012.

[167]"File:Map of Central Asia.png." *Wikimedia Commons*, 12 Aug. 2023, upload.wikimedia.org/wikipedia/ commons/6/68/Map_of_Central_Asia.png

[168]Huang Shijian (黄时鉴) and Gong Yingyan (龚缨晏). *Op. cit.*, p. 191.

[169]The Chagatai Khanate split into Eastern Chagatai Khanate and Western Chagatai Khanate in 1347. The Timurid Empire replaced the original Western Chagatai Khanate and continued the Chagatai Khanate. From 1363, the Chagatais progressively lost Transoxiana to the Timurids. The reduced realm came to be known as Moghulistan, which lasted until the late fifteenth century, when it broke off into the main Turpan Khanate (吐魯番汗國/吐鲁番汗国; 1487–1570) and the branch Yarkant Khanate (葉爾羌汗國/叶尔羌汗国; 1514–c. 1705). In 1570, the Yarkant Khanate made an Eastern Campaign which led to the destruction of the Turpan Khanate and ended the Eastern Chagatai Khanate's main branch. The Yarkant Khanate was eventually conquered by the Dzungar Khanate (準噶爾汗國/准噶尔汗国) in 1705.

[170]Daily Kanji (每日汉字). "明朝的'亦力把里'是哪？成吉思汗子孙的封地，在今天的新疆 (Where is 'Yilibaoli' of the Ming Dynasty? The fiefdom of Genghis Khan's descendants is in today's Xinjiang)！" www.sohu.com, Daily Kanji (每日汉字), 2025, https://www.sohu.com/a/354675740_276265

[171]Frank McLynn. *Genghis Khan: His Conquests, His Empire, His Legacy.* Boston, MA, USA: Da Capo Press, 2016.

[172]Xu lianli (徐良利) · *Yi Er Han Guo Shi Yan Jiu* (伊儿汗国史研究) or *Studies on the History of the Ilkhanate.* Beijing, China: People's Publishing House (人民出版社). 2009.

[173]*Ibid.*

[174]John Edward Hill. *Through the Jade Gate - China to Rome. A Study of the Silk Routes 1st to 2nd Centuries CE. Vol. I*, North Charleston, SC, USA: CreateSpace, 2015, pp. 121–125,

[175]Charles Blackmore. *Crossing the Desert of Death: Through the Fearsome Taklamakan.* London, UK: John Murray (Publishing House), 1677.

[176]Sally Hovey Wriggins. *The Silk Road Journey With Xuanzang.* Boulder, CO, USA: Westview Press, 2003.

[177]Mariko Namba Walter. *Tokharian Buddhism in Kucha: Buddhism of Indo-European Centum speakers in Chinese Turkestan before the 10th century C.E. (Sino-Platonic papers).* Philadelphia, PA, USA: Dept. of Asian and Middle Eastern Studies, University of Pennsylvania, 1998.

[178]*Ibid.*

[179]George Qingzhi Zhao. *Marriage as Political Strategy and Cultural Expression: Mongolian Royal Marriages from World Empire to Yuan Dynasty.* Lausanne, Switzerland: Peter Lang Inc., International Academic Publishers, 2008.

[180]*Ibid.*

[181]Zhang Lihan (張立漢), ed. Zhong Guo Shan He Quan Shu • Shang [M] (中國山河全書 • 上 [M])or *Encyclopedia of Mountains and Rivers of China, Part 1 [M].* Qingdao, Shandong, China: Qingdao Publishing House (青島出版社), 2005.

[182]Claudia Moser and Cecelia Feldman, eds. *Locating the Sacred: Theoretical Approaches to the Emplacement of Religion.* Havertown, PA, USA: Oxbow Books, 2014.

[183]Huang Shijian (黄时鉴) and Gong Yingyan (龚缨晏). *Op. cit.,* p. 205.

[184]Huang Shijian (黄时鉴) and Gong Yingyan (龚缨晏). *Op. cit.,* p. 206.

[185]"File:Selengerivermap.png." *Wikimedia Commons,* 8 June 2021, upload.wikimedia.org/wikipedia/commons/0/07/Selengerivermap.png

[186]Huang Shijian (黄时鉴) and Gong Yingyan (龚缨晏). *Op. cit.,* p. 199.

[187]David O. Morgan. *The Mongols.* Hoboken, NJ, USA: Wiley-Blackwell, 2007.

[188]Danielle Ross. *Tatar Empire: Kazan's Muslims and the Making of Imperial Russia.* Bloomington, IN, USA: ⌐Indiana University Press, 2020.

[189]Owen Lattimore. *The Desert Road to Turkestan.* New York, NY, USA: Kodansha USA, 1995.

[190]Huang Shijian (黄时鉴) and Gong Yingyan (龚缨晏). *Op. cit.,* p. 186.

[191]Saglar Bougdaeva. "The Yelu Language of War and Peace: A Revised Oirad Translation of the Altai Runic Inscriptions (6th–9th centuries)." *Central Asiatic Journal,* vol. 66, no. 1–2, 2024, pp. 27–46.

[192]Luc Kwanten. *Imperial Nomads: A History of Central Asia, 500–1500.* Philadelphia, PA, USA: University of Pennsylvania Press, 1979.

[193]Wang Yongqiang (王永强). *Zhong Guo Shao Shu Min Zu Wen Hua Shi Tu Dian • Bei Fang Juan* (中国少数民族文化史图典 • 北方卷) or *Illustrated Dictionary of the Cultural History of China's Ethnic Minorities, Volume of the Northern Region,* Guangxi, China: Guangxi Education Publishing House (广西教育出版社), 1999, p. 308.

[194]Cao Yongnian (曹永年). *Meng Gu Min Zu Tong Shi • Di San Juan* (蒙古民族通史 • 第三卷) or *General History of the Mongolian Nation, Volume Three*, Inner Mongolia, China: Inner Mongolia University Press (內蒙古大學出版社), 2002, p. 116.

[195]Liu Qingyang (劉清陽). *Ming Qing She Hui Ge Yao Ji • Shang Juan • Shi Shi Pian* (明清社會歌謠集 • 上卷 • 時事篇) or *Collection of Social Ballads of the Ming and Qing Dynasties, Volume 1, Current Affairs*, Morrisville, NC, USA: Lulu Press, Inc., 2015, p. 59.

[196]Ph. de Heer. *Op. cit.*, p. 99.

[197]Jack Weatherford (傑克 • 魏澤福). Huang Zhongxian (黃中憲), trans. *Cheng Ji Si Han De Nu Er Men* (成吉思汗的女兒們) or *The Secret History of the Mongol Queens: How the Daughters of Genghis Khan Rescued His Empire*. Chinese Taipei: China Times Publishing Co. (時報出版), 2018.

[198]Huang Shijian (黃时鉴) and Gong Yingyan (龔缨晏). *Op. cit.*, p. 206.

[199]Gao Wende (高文德), editor in chief. *Zhong Guo Shao Shu Min Zu Shi Da Ci Dian* (中国少数民族史大辞典) or *Dictionary of Chinese Minority History*, Jilin, China: Jilin Education Publishing House (吉林教育出版社), 1995, p. 2007.

[200]Compiled by Ming Dynasty of China (中國明朝編修). *Da Ming Tai Zhong Wen Huang Di Shi Lu • Juan Yi Bai Wu Shi Er* (大明太宗文皇帝實錄 • 卷一百五十二) or *Records of Taizong – Emperor Wen – of the Ming Dynasty, Juan (Volume) 152*; an electronic version can be found at https://ctext.org/wiki.pl?if=gb&chapter=136709

[201]Huang Shijian (黃时鉴) and Gong Yingyan (龔缨晏). *Op. cit.*, p. 184.

[202]Huang Shijian (黃时鉴) and Gong Yingyan (龔缨晏). *Op. cit.*, p. 201.

[203]Yang Shiqi (杨士奇 ; 1365–1444), *et al. Da Ming Tai Zhong Wen Huang Di Shi Lu • Juan Yi Bai Wu Shi Er* (大明太宗文皇帝實錄 • 卷一百五十二) or *Records of Taizong – Emperor Wen – of the Ming Dynasty, Juan (Volume) 152*; an electronic version can be found at https://ctext.org/wiki.pl?if=gb&chapter=136709

[204]Huang Shijian (黃时鉴) and Gong Yingyan (龔缨晏). *Op. cit.*, p. 205.

[205]Author unknown. Christopher P. Atwood, trans. *The Secret History of the Mongols*. London, UK: Penguin Classics, 2023.

[206]"File:Amurrivermap.png." *Wikimedia Commons*, 7 Aug. 2023, upload.wikimedia.org/wikipedia/commons/f/fa/Amurrivermap.png

[207]Yang Shiqi (杨士奇 ; 1365–1444), *et al. Da Ming Tai Zhong Wen Huang Di Shi Lu • Juan Yi Bai Si* (大明 太宗文皇帝實錄 • 卷一百四) or *Records of Taizong - Emperor Wen - of the Ming Dynasty, Vol. 104*; an electronic version can be found at https://ctext.org/wiki.

[208]Hok-lam Chan. "The Chien-wen, Yung-lo, Hung-hsi, and Hsüan-te reigns, 1399–1435." *The Cambridge History of China, Volume 7: The Ming Dynasty, 1368–1644, Part 1*, Cambridge, UK: Cambridge University Press, 1998; Morris Rossabi. "The Ming and Inner Asia." *The Cambridge History of China, Volume 8: The Ming Dynasty, 1398–1644, Part 2*. Cambridge, UK: Cambridge University Press, 1998.

[209]Hok-lam Chan. *Op. cit.*

[210]Wang Yongqiang (王永強). *Zhong Guo Shao Shu Min Zu Wen Hua Shi Tu Dian • Bei Fang Juan* (中国少数民族文化史图典 • 北方卷) or *Illustrated Dictionary of the Cultural History of China's Ethnic Minorities, Juan (Volume) of the Northern Region*, Guangxi, China: Guangxi Education Publishing House (广西教育出版社), 1999, p. 308.

[211]Zhang Jiuhe (张久和). *Yuan Menggu Ren De Li Shi: Shiwei–Dadan Yan Jiu* (原蒙古人的历史: 室韦--达怛研究) or *History of the Original Mongols: Research on Shiwei-Dadan*, Beijing, China: Higher Education Press (高等教育出版社), 1998, pp. 27–28.

[212]The mountains in the south bank of Heilongjiang and the west of the Nen River are called the Greater Khingan Range; and the mountains in the east of the Nen River are called the Lesser Khingan Range; the

mountains in the north of Heilongjiang and located in Russia are called the Outer Khingan Range, and it is called the Stanovoy Range in Russia.

[213]*Ibid.*

[214]Huang Shijian (黄时鉴) and Gong Yingyan (龚缨晏). *Op. cit.*, p. 197.

[215]Xiu Ouyang. Richard L. Davis, trans. *Historical records of the Five Dynasties*. New York, NY, USA: Columbia University Press, 2004.

[216]Ouyang Ciu (歐陽修; 1007–1072). *Xin Wu Dai Shi • Si Yi Fu Lu Er* (新五代史 • 四夷附录二) or *New History of the Five Dynasties, Appendix II on the Four Barbarians*; the electronic version of this article can be found at https://www.gushiwen.cn/guwen/bookv.aspx?id=46653FD803893E4FEDAE6D488B5D3FED

[217]Huang Shijian (黄时鉴) and Gong Yingyan (龚缨晏). *Op. cit.*, p. 198.

[218]The British Museum. "Introducing the Scythians." www.britishmuseum.org, The British Museum, 2025, https://www.britishmuseum.org/blog/introducing-scythians

[219]Harry Thurston Peck. *Harper's Dictionary of Classical Literature and Antiquities*. Lanham, MD, USA: Cooper Square Publishing Llc, 1963; Harry Thurston Peck. "Scythia." www.gtp.gr, Greek Travel Pages. 2025, www.gtp.gr/LocInfo.asp?infoid=49&IncludeWide=1&code=RRU0SK&PrimeCode=RRU0SK&Level=4&PrimeLevel=4&LocId=15498&lng=2

[220]*Ibid.*

[221]F. R. Grahame, Creator. *The Archer and the Steppe, or, The Empires of Scythia: a History of Russia and Tartary, From the Earliest Ages Till the Fall of the Mongul Power in Europe, in the Middle of the Sixteenth Century*. Open Library: Legare Street Press, 2021.

[222]*Ibid.*

[223]Nurlan Kenzheakhmet and Alpamys Abu. *Op. cit.*, p. 621. https://www.researchgate.net/publication/354971343_Some_Medieval_and_Post-Golden_Horde's_Towns_of_the_Itil_Volga_and_Syr-Darya_Basins_According_to_the_Arabic_and_Chinese_Maps

[224]Herodotus. David Grene, trans. *The History*, New York, NY, USA: University of Chicago Press, 1988.

[225]Huang Shijian (黄时鉴) and Gong Yingyan (龚缨晏). *Op. cit.*, p. 186.

[226]Han Rulin (韩儒林). "元代的吉利吉思及其邻近诸部 (The Kyrgyz and Their Neighbouring Tribes in the Yuan Dynasty)." *Journal of Chinese Historical Studies* (中国史研究), vol. 1, 1979.

[227]*Ibid.*

[228]*Ibid.*

[229]Frederick W. Mote. "The rise of the Ming dynasty, 1330–1367." In Frederick W. Mote and Denis Twitchett, eds. *The Cambridge History of China, Volume 7: The Ming Dynasty, 1368–1644, Part 1*, Cambridge, UK: Cambridge University Press, 1988, pp. 11–57.

[230]Han Rulin (韩儒林). *Op. cit.*

[231]Denis Sinor, ed. *The Cambridge history of early Inner Asia*. Cambridge, UK: Cambridge University Press, 1990, pp. 230–232; András Róna-Tas. Nicholas Bodoczky, trans. *Hungarians and Europe in the Early Middle Ages: An Introduction to Early Hungarian History*, Budapest, Hungary: Central European Univ. Press, 1999.

[232]Nurlan Kenzheakhmet and Alpamys Abu. *Op. cit.*, p. 619. https://www.researchgate.net/publication/354971343_Some_Medieval_and_Post-Golden_Horde's_Towns_of_the_Itil_Volga_and_Syr-Darya_Basins_According_to_the_Arabic_and_Chinese_Maps

[233]Huang Shijian (黄时鉴) and Gong Yingyan (龚缨晏). *Op. cit.*, p. 204.

[234]Huang Shijian (黄时鉴) and Gong Yingyan (龚缨晏). *Op. cit.*, p. 205.

[235]Huang Shijian (黄时鉴) and Gong Yingyan (龚缨晏). *Op. cit.*, p. 198.

[236]Ma Duanlin (馬端臨; 1254–1323). *Wen Xian Tong Kao • Juan 348* (文献通考 • 卷348) or *Comprehensive Examination of Literature, Juan (Volume) 348*; An electronic version can be found at https://ctext.org/datawiki.pl?if=gb&res=807291&remap=gb

[237]*Ibid.*

[238]"File:Laptev Sea map.png." *Wikimedia Commons*, 27 Nov. 2021, upload.wikimedia.org/wikipedia/commons/f/f0/Laptev_Sea_map.png

[239]Ma Duanlin (馬端臨; 1254–1323). *Op. cit.*

[240]*Ibid.*

[241]BTG. "Hela in the History of Liao ((辽史)中的曷剌)." https://zhuanlan.zhihu.com/, Zhihu, 2025, https://zhuanlan.zhihu.com/p/273519454

[242]Huang Shijian (黄时鉴) and Gong Yingyan (龚缨晏). *Op. cit.*, p. 197.

[243]Nurlan Kenzheakhmet and Alpamys Abu. *Op. cit.*, p. 621. doi: 10.22378/2313-6197.2021-9-3.611–653

[244]*Ibid.*

[245]Wang Pu (王溥; 922–982）*Tang Hui Yao • Juan Yi Bai* (唐会要 • 卷一百) or *Tang Huiyao, Juan (Volume) 100*; an electronic version can be found at https://ctext.org/wiki.pl?if=gb&chapter=411923

[246]Huang Shijian (黄时鉴) and Gong Yingyan (龚缨晏). *Op. cit.*, p. 194.

[247]Peter Heather. *Goths and Romans 332-489*. Oxford, UK: Clarendon Press, 1994; Walter Goffart. *The Narrators of Barbarian History*, Notre Dame, IN, USA: University of Notre Dame Press, 2005.

[248]Marie Favereau. *The Horde: How the Mongols Changed the World*. Cambridge, MA, USA: Belknap Press: An Imprint of Harvard University Press; 2022.

[249]Viacheslav Shpakovsky and David Nicolle. *Armies of the Volga Bulgars & Khanate of Kazan: 9th–16th centuries*. Oxford, UK: Osprey Publishing, 2013.

[250]Martin Sixsmith. *Russia: A 1000-Year Chronicle of the Wild East*. New York, NY, USA: Harry N. Abrams, 2013.

[251]*Ibid.*

[252]Xue Zongzheng (薛宗正). *Tu Jue Shi* (突厥史) or *History of the Turks*. Beijing, China: China Social Sciences Press (中國社會科學出版社), 1992.

[253]*Ibid.*

[254]Robert R. Dickson, *et al.*, eds. *Arctic-Subarctic Ocean Fluxes: Defining the Role of the Northern Seas in Climate*. New York, NY, USA: Springer, 2008.

[255]Huang Shijian (黄时鉴) and Gong Yingyan (龚缨晏). *Op. cit.*, p. 203.

[256]Ouyang Xiu (歐陽修; 1007–1072). *Xin Wu Dai Shi • Juan Qi Shi San • Si Yi Fu Lu Di Er* (新五代史 • 卷七十三 • 四夷附录第二) or *New History of the Five Dynasties, Juan (Volume) 73, Appendix of the Four Barbarians, No. 2*; an electronic version can be found at http://www.guoxue.com/shibu/24shi/Newwudai/xwd_073.htm; Gao Wende (高文德), editor in chief. *Zhong Guo Shao Shu Min Zu Shi Da Ci Dian* (中国少数民族史大辞典) or *Dictionary of Chinese Minority History*, Jilin, China: Jilin Education Publishing House (吉林教育出版社), 1995.

[257]Gao Wende (高文德), editor in chief. *Op. cit.*

[258]Huang Shijian (黄时鉴) and Gong Yingyan (龚缨晏). *Op. cit.*, p. 205.

[259]Bruce P. Lenman and Hilary Marsden, eds. *Chambers Dictionary of World History*. London, UK: Chambers Harrap Pub Ltd, 2005.

[260]Anania Shirakatsi. Introduction, Translation and Commentary by Robert H. Hewsen. *The Geography of Ananias of Sirak (Asxarhacoyc): The Long and the Short Recensions*. Wiesbaden. Germany: Reichert Verlag, 1992; Suren Eremian. "Reconstructed map of Central Asia from 'Ashharatsuyts'." www.kroraina.com, Suren Eremian, 2025, https://www.kroraina.com/armen_ca/map_casia_b.jpg

[261]Marco Polo. Ronald Latham, trans. *The Travels of Marco Polo*. London, UK: Penguin Classics, 1958.

[262]Zhāng Tingyu (張廷玉; 1672–1755), *et al. Míng Shi • Xi Yu Chuan Si • Tian Fang* (明史 • 西域傳四 • 天方) or *History of Ming • Biography of the Western Regions IV, Tianfang*; an electronic version can be found at https://ctext.org/wiki.pl?if=gb&res=410835

[263]Muhammad Asad. *The Road to Mecca*. London, UK: The Book Foundation, 2014.

[264]Ma Huan accompanied Admiral Zheng He to sail to the Western Ocean during the fourth, sixth and seventh voyages. He is the author of the book titled *Ying Ya Sheng Lan* (瀛涯胜览) or *The Overall Survey of the Ocean's Shores* which he started writing in 1416 and completed in 1451. The book describes his personal experiences and what he had learned during his voyages. He knew several classical Chinese and Buddhist texts and learned Arabic well enough to translate into Chinese; Ma Huan. J.V.G. Mills, trans. *Ying-yai Sheng-lan: The Overall Survey of the Ocean's Shores* (1433). London, UK: Hakluyt Society, 1970.

[265]Sheng-Wei Wang. "Chapter Two: Ma Huan's *Tianfang*/Mo-jia was ancient Tunisia in North Africa." *The Last Journey of the Sam Bao Eunuch, Admiral Zheng He*, Hong Kong, China: Proverse Hong Kong, 2019, pp. 53–90; Lam Yee Din (林贻典). "The Country of Tian Fang is in Tunisia." *Zheng He Research Updates* (郑和研究动态), vol. 2 (total 24 volumes), 2014.

[266]UNESCO World Heritage Centre. "Silk Roads: the Routes Network of Chang'an-Tianshan Corridor." https://whc.unesco.org/, UNESCO World Heritage Convention, 2024, https://whc.unesco.org/en/list/1442/multiple=1&unique_number=1985

[267]Ibid.

[268]Huang Shijian (黄时鉴) and Gong Yingyan (龚缨晏). *Op. cit.*, p. 209.

[269]Xue Zongzheng (薛宗正). *Op. cit.*

[270]*Ibid.*

[271]*Ibid.*

[272]Wei Zheng (魏徵; 580–643), *et al. Sui Shu • Juan 84* (隋書 • 卷 84) or *Book of Sui, Vol. 84*; an electronic version can be found at https://ctext.org/wiki.pl?if=gb&res=386407

[273]Lin Gan (林幹). *Tu Jue Yu Hui He Shi* (突厥與回紇史) or *History of the Turks and the Uighurs*. Hohhot, Inner Mongolia, China: Inner Mongolia People's Publishing House (內蒙古人民出版社), 2007.

[274]Ouyang Xiu (歐陽修 ; 1007–1072) and Song Qi (宋祁; 998–1061) *Xin Tang Shu • Hui Gu Chuan* (新唐書 • 回鶻傳) or *New Book of Tang, Biography of the Uighurs*. An electronic version can be found at https://ctext.org/wiki.pl?if=gb&res=182378

[275]Ulrich Theobald. "Tiele 鐵勒, Tölös." www.chinaknowledge.de, Ulrich Theobald, 2025, http://www.chinaknowledge.de/History/Altera/toeloeshs.html

[276]Huang Shijian (黄时鉴) and Gong Yingyan (龚缨晏). *Op. cit.*, p. 184.

[277]Jaswant Singh Jaswant Singh. *Travels in Transoxiana: In the Lands over the Hindu Kush and Cross the Amu Darya*. Delhi, India: Rupa & Co, 2006.

[278]Sheila Blair. *The Monumental Inscriptions from Early Islamic Iran and Transoxiana*, Leiden, Netherlands: Brill, 1992, p. 115.

[279]William Henry Swinton. *The Heart of Asia: A History of Russian Turkestan and the Central Asian Khanates from the Earliest Times*. London, UK: Forgotten Books, 2012.

[280]Arzu Emel Altınkılıç. "Göktürk giyim kuşamının plastik sanatlarda değerlendirilmesi." *Journal of Social and Humanities Sciences Research*, 2020, pp. 1101–1110.

[281]John Man. *The Mongol Empire: Genghis Khan, His Heirs and the Founding of Modern China*. London, UK: Congi, 2016.

[282]David Christian. *A History of Russia, Central Asia and Mongolia, Vol. 1: Inner Eurasia from Prehistory to the Mongol Empire*. Hoboken, NJ, USA: Wiley-Blackwell, 1998.

[283]*Ibid.*

[284]"File:Asia in 1335.svg." *Wikimedia Commons*, 27 Dec. 2022, upload.wikimedia.org/wikipedia/commons/thumb/b/b6/Asia_in_1335.svg/2560px-Asia_in_1335.svg.png

[285]Huang Shijian (黄时鉴) and Gong Yingyan (龚缨晏). *Op. cit.*, p. 185.

[286]Nurlan Kenzheakhmet and Alpamys Abu. *Op. cit.*, p. 621. https://www.researchgate.net/publication/354971343_Some_Medieval_and_Post-Golden_Horde's_Towns_of_the_Itil_Volga_and_Syr-Darya_Basins_According_to_the_Arabic_and_Chinese_Maps

[287]*Ibid.*

[288]Huang Shijian (黄时鉴) and Gong Yingyan (龚缨晏). *Op. cit.*, p. 198.

[289]Nurlan Kenzheakhmet and Alpamys Abu. *Op. cit.*, pp. 620–621. https://www.researchgate.net/publication/354971343_Some_Medieval_and_Post-Golden_Horde's_Towns_of_the_Itil_Volga_and_Syr-Darya_Basins_According_to_the_Arabic_and_Chinese_Maps

[290]Huang Shijian (黄时鉴) and Gong Yingyan (龚缨晏). *Op. cit.*, p. 200.

[291]Carlo Tagliavini, *et al.* "Calmucchi." www.treccani.it, Treccani, 2025, https://www.treccani.it/enciclopedia/calmucchi_%28Enciclopedia-Italiana%29/

[292]*Ibid.*

[293]Carlo Tagliavini, *et al. Op. cit.*

[294]The Barry Lawrence Ruderman Map Collection. "La Tartaria Indipendente…il Paese de Calmuchi quello degli Usbeks, e il Turkesan…1784." https://exhibits.stanford.edu/, Stanford Libraries, 2025, https://exhibits.stanford.edu/ruderman/catalog/jt613xx3971

[295]Nurlan Kenzheakhmet and Alpamys Abu. *Op. cit.*, p. 619. doi: 10.22378/2313-6197.2021-9-3.611–653

[296]Matthew Carney. "Tibetan Sky Burials: Stumbling upon the secret ritual of feeding the dead to vulture 'angels'." www.abc.net.au, Australian Broadcasting Corporation, 2025, https://www.abc.net.au/news/2017-09-16/stumbling-upon-ritual-of-feeding-the-dead-to-vulture-angels/89483322

[297]Huang Shijian (黄时鉴) and Gong Yingyan (龚缨晏). *Op. cit.*, p. 189.

[298]"File:Permia map.jpg." *Wikimedia Commons*, 24 July 2024, upload.wikimedia.org/wikipedia/commons/e/eb/Permia_map.jpg

[299]Joonas Ahola, *et al*, eds. *Fibula, Fabula, Fact - The Viking Age in Finland.* Helsinki, Finland: Suomen Kirjallisuuden Seura, 2018.

[300]"File:Bjarmaland,Carta Marina.jpg." *Wikimedia Commons*, 25 Nov. 2023, upload.wikimedia.org/wikipedia/commons/0/0b/Bjarmaland%2CCarta_Marina.jpg

[301]"File:Murmansk in Russia.svg." *Wikimedia Commons*, 16 Oct. 2020, upload.wikimedia.org/wikipedia/commons/thumb/6/6d/Murmansk_in_Russia.svg/2560px-Murmansk_in_Russia.svg.png

[302]"File:Kola peninsula.png." *Wikimedia Commons*, 9 Mar. 2022, upload.wikimedia.org/wikipedia/commons/e/e6/Kola_peninsula.png

[303]"File:White Sea map.png." *Wikimedia Commons*, 27 March 2021, upload.wikimedia.org/wikipedia/commons/b/b3/White_Sea_map.png

[304]"File:Pechora-en.svg." *Wikimedia Commons*, 19 Jan. 2024, upload.wikimedia.org/wikipedia/commons/thumb/f/f8/Pechora-en.svg/1892px-Pechora-en.svg.png

[305]Joonas Ahola, *et al.*, eds. *Op, cit.*

[306]Leontev Aleksandr. *Bjarmaland northern cradle of Russia.* Moscow, Russia: Eksmo, 2004.

[307]Nurlan Kenzheakhmet and Alpamys Abu. *Op. cit.*, p. 620. https://www.researchgate.net/publication/354971343_Some_Medieval_and_Post-Golden_Horde's_Towns_of_the_Itil_Volga_and_Syr-Darya_Basins_According_to_the_Arabic_and_Chinese_Maps

[308]"File:Pechora-en.svg." *Wikimedia Commons*, 13 Oct. 2020, upload.wikimedia.org/wikipedia/commons/thumb/f/f8/Pechora-en.svg/1892px-Pechora-en.svg.png

[309]Huang Shijian (黄时鉴) and Gong Yingyan (龚缨晏). *Op. cit.*, p. 204.

[310]Yong Wu Long Qian (用勿龍潛). "A New Discussion on the Territory of the Ming Dynasty, Serial 08, Chapter 8: Hexi River (新论明朝疆域（连载 08） 第八篇：贺喜河)." https://zhuanlan.zhihu.com/, Cian Hu Zai Hao Guo Xue Gong Zhong Shu Fu (玄乎哉 昊國學宮中樞府), 2025, https://zhuanlan.zhihu.com/p/643147157

[311]"File:Siberiariverroutemap.png." *Wikimedia Commons*, 31 Oct. 2020, upload.wikimedia.org/wikipedia/commons/1/1f/Siberiariverroutemap.png

[312]Huang Shijian (黄时鉴) and Gong Yingyan (龚缨晏). *Op. cit.*, p. 206.

[313]Gao Wende (高文德), ed. *Op. cit.*, p. 332.

[314]*Ibid.*

[315]*Ibid.*

[316]*Ibid.*

[317]Huang Shijian (黄时鉴) and Gong Yingyan (龚缨晏). *Op. cit.*, p. 190.

[318]Wei Shou (魏收; 506–572). *Wei Shu • Juan 100 • Lie Chuan Di Ba Shi Ba • Di Dou Yu Tiao* (魏書 • 卷 100 • 列傳第八十八 • 地豆于條) or *Book of Wei, Juan (Volume) 100, Biography 88, Didouyu*; an electronic version can be found at https://ctext.org/wiki.pl?if=gb&res=801592

[319]Li Yanshou (李延寿; Year of birth and death unknown). *Bei Shi Juan 94 • Lie Chuan Di Ba Shi Er • Di Dou Gan Tiao* (北史卷 94 • 列傳第八十二 • 地豆干條) or *Northern History • Juan (Volume) 94 • Biography 82, Didougan*; an electronic version can be found at https://ctext.org/wiki.pl?if=gb&res=21033&remap=gb

[320]*Ibid.*

[321]Zhang Jiuhe (张久和). *Yuan Meng Gu Ren De Li Shi • Shi Wei-Da Dan Yan Ju* (原蒙古人的历史: 室韦-达怛研究) or *History of the Original Mongols: Research on Shiwei-Dadan*, Beijing, China: High Education Press (高等教育出版社), 1998, pp. 27–28.

[322]"File:Sea of Okhotsk map with state labels.png." *Wikimedia Commons*, 6 Mar. 2024, upload.wikimedia.org/wikipedia/commons/9/92/Sea_of_Okhotsk_map_with_state_labels.png

[323]Lu Yier (陆一2). "Ancient Peoples of the World: The Shiwei Tribe ((世界古代民族) 室韦族.)" http://www.360doc.com/, 360doc.com, 2025, http://www.360doc.com/content/17/1103/12/8716899_700548576.shtml

[324]Wei Zheng (魏徵; 580–643). *Sui Shu • Juan 84 • Qi Dan* (隋書 • 卷 84 • 契丹) or *Book of Sui, Juan (Volume) 84, Khitan*; an electronic version can be found at https://ctext.org/wiki.pl?if=gb&res=386407

[325]Frank McLynn. *Genghis Khan: His Conquests, His Empire, His Legacy*. Boston, MA, USA: Da Capo Press, 2016.

[326]Gao Wende (高文德), ed. *Op. cit.*, p. 2442.

[327]Ouyang Xiu (歐陽修; 1007–1072). *Xin Wu Dai Shi • Juan Qi Shi San • Si Yi Fu Lu Di Er* (新五代史 • 卷七十三 • 四夷附录第二) or *New History of the Five Dynasties, Juan (Volume) 73, Appendix of the Four Barbarians, No. 2*; an electronic version can be found at http://www.guoxue.com/shibu/24shi/Newwudai/xwd_073.htm

[328]Frank McLynn. *Genghis Khan: His Conquests, His Empire, His Legacy*. Boston, MA, USA: Da Capo Press, 2016.

[329]Staff Writer. "Brief Introduction to Duoyan Sanwei (简述朵颜三卫)." www.163.com, NetEase, 2025, https://www.163.com/dy/article/H4TP2FF60552Q4X0.html

[330]Richard L. Davis, trans. *Historical Records of the Five Dynasties*. New York, USA.: Columbia University Press, 2004.

[331]Huang Shijian (黄时鉴) and Gong Yingyan (龚缨晏). *Op. cit.*, p. 204.

[332]Huang Shijian (黄时鉴) and Gong Yingyan (龚缨晏). *Op. cit.*, p. 186.

[333]Huang Shijian (黄时鉴) and Gong Yingyan (龚缨晏). *Op. cit.*, p. 200.

[334]Wei Shou (魏收; 506–572). *Wei Shu • Juan 107 • Lie Chuan Di Ba Shi Ba • Wu Luo Hou* (魏書 • 卷 107 • 列傳第八十八 • 烏洛侯) or *Book of Wei, Juan (Volume) 107, Biography 88, Wu Luo Hou*; an electronic version can be found at https://ctext.org/wiki.pl?if=gb&res=801592

[335]*Ibid.*

[336]Huang Shijian (黄时鉴) and Gong Yingyan (龚缨晏). *Op. cit.*, p. 192.

[337]"File:Sea of Okhotsk map with state labels.png." *Wikimedia Commons*, 6 Mar. 2024, upload.wikimedia. org/wikipedia/commons/9/92/Sea_of_Okhotsk_map_with_state_labels.png

[338]World Climate Guide. "Weather and climate in Siberia." www.climatestotravel.com, Climates to Travel, 2025, https://www.climatestotravel.com/climate/siberia

[339]Du You (杜佑; 735–812). *Tong Dian Juan Er Bai* (通典 • 卷二百) or *An Encyclopaedia of Statecraft, Juan (Volume) 200*; an electronic version can be found at https://ctext.org/tongdian/zhs

[340]Zheng Tianting (郑天挺). *Zhong Guo Li Shi Da Ci Dian Xu Ben* (中国历史大辞典音序本) or *Comprehensive Dictionary of Chinese History, Chinese Pinyin Sequence Edition - 2007 Edition*, Shanghai, China: Shanghai Lexicographical Publishing House (上海辞书出版社), p. 2347.

[341]*Ibid.*

[342]National Library of Medicine. "Genetic evidence suggests a sense of family, parity and conquest in the Xiongnu Iron Age nomads of Mongolia." www.PublMed.gov, National Library of Medicine, 2025, pubmed. ncbi.nlm.nih.gov/32734383/; https://www.360kuai.com/pc/938bae6ce162bb459?cota=3&kuai_so=1&sign= 360_3574ab00&refer_scene=so_4

[343]"File:Map of Russia - Sakha (Yakutia) (Crimea disputed).svg." *Wikimedia Commons*, 13 Dec. 2022, upload.wikimedia.org/wikipedia/commons/thumb/3/3e/Map_of_Russia_(2014–2022)_-_Sakha_(Yakutia)_ (Crimea_disputed).svg/1541px-Map_of_Russia_(2014–2022)_-_Sakha_(Yakutia)_(Crimea_disputed).svg. png

[344]"File:Lena River basin.png." *Wikimedia Commons*, 18 Oct. 2021, upload.wikimedia.org/wikipedia/ commons/f/f7/Lena_River_basin.png

[345]Huang Shijian (黄时鉴) and Gong Yingyan (龚缨晏). *Op. cit.*, p. 202.

[346]Huang Shijian (黄时鉴) and Gong Yingyan (龚缨晏). *Op. cit.*, p. 192.

[347]Nicholas Morton. *The Mongol Storm: Making and Breaking Empires in the Medieval Near East*. New York, NY, USA: Basic Books, 2022.

[348]Wei Zheng (魏徵; 580–643). *Sui Shu • Juan 84 • Qi Dan* (隋書 • 卷 84 • 契丹) or *Book of Sui, Juan (Volume) 84, Khitan*; an electronic version can be found at https://ctext.org/wiki.pl?if=gb&res=386407

[349]Jack Weatherford. *Genghis Khan and the Making of the Modern World*. New York, NY, USA: Crown, 2005.

[350]Ruixi Zhu, *et al. A Social History of Middle-Period China: The Song, Liao, Western Xia and Jin Dynasties*. Cambridge, UK: Cambridge University Press, 2017.

[351]"File:Mongol Empire map 2.gif." *Wikimedia Commons*, 20 Dec. 2022, commons.wikimedia.org/wiki/ File:Mongol_Empire_map_2.gif

[352]*Ibid.*

[353]Jonathan Clements. *Brief History of Khubilai Khan*. London, UK: Robinson, 2019.

[354]*Ibid.*

[355]*Ibid.*

[356]Jeff Hays. "Mongols after the End of the Mongol Empire." https://factsanddetails.com/, Facts and Details Project, 2025, https://factsanddetails.com/about.html

[357]Wu Han. Noah Sullivan, trans. *Biography of Zhu Yuanzhang: Emperor Taizu of Ming*. San Diego, CA, USA: Daybreak Studios, 2024.

[358]Sheng-Wei Wang. *The Last Journey of the San Bao Eunuch, Admiral Zheng He. Op. cit.*; Louise Levathes. *When China Ruled The Seas: The Treasure Fleet of the Dragon Throne 1405–1433*. New York, NY, USA: Oxford University Press, 1994.

[359]Shih-shan Henry Tsai. *Perpetual Happiness: The Ming Emperor Yongle*. Seattle, Wash., USA: University of Washington Press, 2011.

[360]Jennifer Raff. *Origin: A Generic History of the Americas*. New York, NY, USA: Twelve, 2022.

[361]Xue Zongzheng (薛宗正). *Tu Jue Shi* (突厥史) or *History of the Turks. Op. cit.*

[362]Li Jiaxing (利家興) and Zhuang Yanling (莊延齡). "A Millennium History of Tatar and the Tatar Research in European Academic Circles from the 18th to 19th Centuries (韃靼千年史與 18–19 世紀歐洲學界的 '韃靼' 研究)." *Global History Review* (全球史評論), 2023.

[363]Martin Sixsmith. *Russia: A 1000-Year Chronicle of the Wild East*. New York, NY, USA: Abrams Press, 2012.

[364]Huang Shijian (黄时鉴) and Gong Yingyan (龚缨晏). *Op. cit.*, p. 189.

[365]Huang Shijian (黄时鉴) and Gong Yingyan (龚缨晏). *Op. cit.*, p. 204.

[366]Marco Polo. Henry Yule, trans. Henri Cordier, ed. and annotated. *The Travels of Marco Polo*. London, UK.: John Murray, 1920; an electronic version can be found at https://en.wikisource.org/wiki/The_Travels_of_Marco_Polo/Book_1/Chapter_59

[367]Rev Claude L. Pickens Jr. "What Marco Polo Called Tenduc - Eastern Suiyuan. The mighty gates of Kweisui.." https://hpcbristol.net/, University of Bristol, 2025, https://hpcbristol.net/visual/Hv47-162

[368]Elmastudio. "Did Marco Polo go to Alaska?" www.medievalists.net, medievalists.net, 2025, medievalists.net/2014/09/marco-polo-go-alaska/

[369]*Ibid.*

[370]"File:CEM-36-Regno-della-China-2355.jpg." *Wikimedia Commons*, 20 Aug. 2024, upload.wikimedia.org/wikipedia/commons/b/b3/CEM-36-Regno-della-China-2355.jpg

[371]Huang Shijian (黄时鉴) and Gong Yingyan (龚缨晏). *Op. cit.*, p. 204.

[372]Gao Wende (高文德), editor in chief. *Op. cit.*

[373]Huang Shijian (黄时鉴) and Gong Yingyan (龚缨晏). *Op. cit.*, p. 203.

[374]Huang Shijian (黄时鉴) and Gong Yingyan (龚缨晏). *Op. cit.*, pp. 202–203.

[375]Ye Longli. (葉隆禮'). *Qi Dan Guo Zhi • Juan Er Shi Liu • Zhu Fan Ji • Huang Tou Nu Zhen* (契丹國志 • 卷二十六 • 諸藩記 • 黃頭女真) or History of the *Khitans, Volume 26, Records of the Vassals, Huangtou Jurchen*; an electronic version can be found at https://ctext.org/wiki.pl?if=gb&res=959804&remap=gb

[376]*Ibid.*

[377]Huang Shijian (黄时鉴) and Gong Yingyan (龚缨晏). *Op. cit.*, p. 188.

[378]Gao Wende (高文德), editor in chief. *Op. cit.*

[379]"File:Russia Primorsky Krai location map.svg." *Wikimedia Commons*, 29 Oct. 2020, upload.wikimedia.org/wikipedia/commons/thumb/3/34/Russia_Primorsky_Krai_location_map.svg/2075px-Russia_Primorsky_Krai_location_map.svg.png

[380]Photo by Sun Hui. "Amazing Khanka Lake in China's Heilongjiang." www.china.org.cn, China.org.cn, 2025, http://www.china.org.cn/travel/2013-08/30/content_29871943_2.htm

[381]Xiao Ning (萧宁). "What kind of fish is Xingkai Lake's whitefish (兴凯湖白鱼到底是什么鱼)?" www.shidicn.com/, Wet Land Protection (湿地保护), 2025, http://www.shidicn.com/sf_46C31847DF1E4127A5DB6E41AF8C522A_151_hexiaoning.html

[382]Huang Shijian (黄时鉴) and Gong Yingyan (龚缨晏). *Op. cit.*, p. 189.

[383]"File:Amurrivermap.png." *Wikimedia Commons*, 7 Aug. 2023, upload.wikimedia.org/wikipedia/commons/f/fa/Amurrivermap.png

[384]"File:Ussuri.png." *Wikimedia Commons*, 27 Nov. 2021, upload.wikimedia.org/wikipedia/commons/8/87/Ussuri.png

[385]Huang Shijian (黄时鉴) and Gong Yingyan (龚缨晏). *Op. cit.*, p. 195.

[386]Huang Shijian (黄时鉴) and Gong Yingyan (龚缨晏). *Op. cit.*, p. 208.

[387]Peter Lorge. *Five Dynasties and Ten Kingdoms*. Hong Kong, China: The Chinese University of Hong Kong Press, 2011.

[388]Hu Qiao (胡嶠; birth and death years unknown but his book was written in the tenth century of the Later Han Dynasty). *Xian Lu Ji* (陷虏记) or *The Story of the Captive*; an electronic version can be found at https://ctext.org/datawiki.pl?if=en&res=515733&remap=gb

[389]Li Baoxiang (李保祥) and Li Jing (李靖). *Guo Shoujing* (郭守敬) or *Guo Shoujing*. Shijiazhuang, Hebei, China: Huashan literature and Art Publishing House (花山文藝出版社), 2011.

[390]Huang Shijian (黄时鉴) and Gong Yingyan (龚缨晏). *Op. cit.*, p. 198.

[391]Lu Yier (陆一2). *Op. cit.*

[392]Gao Wende (高文德), editor in chief. *Op. cit.*, p. 2007.

[393]Li Shanchang (李善長; 1314–1390). *Yuan Shi • Ben Ji Di Shi San • Shi Zu Shi* (元史•本紀第十三•世祖十) or *History of the Yuan Dynasty, Main History 3, The Tenth Emperor Shizu;* an electronic version can be found at https://ctext.org/wiki.pl?if=gb&chapter=109653&remap=gb

[394]*Ibid.*

[395]Jennifer Raff. *Origin: A Generic History of the Americas. Op. cit.*

[396]"File:Inuit conf map.png." *Wikimedia Commons*, 4 Oct. 2022, upload.wikimedia.org/wikipedia/commons/3/3a/Inuit_conf_map.png

[397]Orgil. "Are Mongolians Tall & Big? What Makes Them So?" www.silkroadmongolia.com, Silk Road Mongolia, 2025, silkroadmongolia.com/are-mongolians-tall-big/#:~:text=While Mongolians are in fact slightly taller and,background or a lineage of physically gifted wrestlers.

[398]"File:Igloo.jpg." *Wikimedia Commons*, 6 Jan. 2022, commons.wikimedia.org/wiki/File:Igloo.jpg

[399]Evgenii G. Kushnarev, *et al. Bering's Search for the Strait: The First Kamchatka Expedition 1725–1730.* Portland, OR, USA: Oregon Historical Society Press,1990.

[400]National Park Service (gov). "Bering Expedition Landing Site National Historic Landmark." www.nps.gov, National Park Service (gov), https://www.nps.gov/places/bering-expedition-landing-site.htm

Index

Great Qing/大清, 8, 143
Great Russian Regions/大俄羅斯地區/大俄罗斯
  地区, 494
Great Wall/萬里長城/万里长城, 130
Great Western Ocean/大西洋, 66, 337, 482
Greater Caucasus/大高加索地區/大高加索地区,
  361, 406
Greater Khingan Mountains/大興安嶺/大兴安岭,
  148, 480, 504
ground cellars/Ground Shacks/地窨子, 504
Gu Shi Ji/Kojiki/Records of Ancient Matters/(古事
  記/古事记), 22, 56
Guade/瓦得爾/瓦得尔, 396
Guang Yu Tu/Enlarged Terrestrial Atlas/(廣輿圖/
  广舆图), 306–307
Guang Yu Tu • Yudi Zongtu/General Geographical
  Map in Enlarged Terrestrial Atlas/(廣輿圖 •
  輿地總圖/广舆图 • 舆地总图), 307
Guangdong/廣東/广东, 134, 194
Guangxi/廣西/广西, 134, 196
Guanzhong/關中/关中, 160
Guard Battalions/衛所/卫所, 105, 135, 151
Guazhou/瓜州, 165
Guilin/桂林, 195
Guizhou/貴州/贵州, 134, 195, 197
Gujarat/吳茶蠟/吳茶蜡, 298, 300–301, 334
Gulf Cooperation Council, 371
Gulf of Aden/亞丁灣/亚丁湾, 361, 369
Gulf of Aqaba/亞喀巴灣/亚喀巴湾, 361
Gulf of Oman/阿曼灣/阿曼湾, 361, 369
Gulf of Riga/里加灣/里加湾, 497
Gulf of Siam/暹羅灣/暹罗湾, 242
Gulf of Suez/蘇伊士灣/苏伊士湾, 361
Gulf of Thailand/泰國灣/泰国湾, 242
Gulf of Tonkin/北部灣/北部湾, 196
Gumal River/古瑪爾河/古玛尔河, 317
Guo Shoujing/郭守敬, 520
Guyuan Inspection Commission/固原巡檢司/固原
  巡检司, 173
Guyuan Li/固原里, 173
Guyuan/固原, 170, 173
Guzarate/梧作剌得, 334
Guzgan/古兹甘, 473
Gyeonggi Province/*Gyeonggi-do*/京畿道, 108,
  116

Gyeongsang Province/*Gyeongsang-do*/慶尙道/庆
  尙道, 110, 114
Gyōki-zu/行基圖/行基図, 11

H

Hachijō-kojima, 65
Haeju/海州, 114
Haijin/sea ban/海禁, xiii, 233, 293
Hailaer River/Yumuchuan/海拉爾河/海拉尔河/榆
  木川, 150
Hainan Island/海南島/海南岛, 204
Hainei Huayi Tu/Map of China and the Barbarian
  Countries within the Seas/(海內華夷圖/
  海内华夷图), 131
*Hainei Nan Jing/The Classic of Regions within the
  Seas: The South*/(海內南經/海内南经), 195
Haixi Jurchens/海西女真, 143
Haixi/海西, 151
Hakodate, 14
Hakuseki, Arai, 63
Hamel, Hendrik, 84
Hamgil Province/Hamgyong, 104
Hamgil/咸吉, 110
Hamgyong Province/*Hamgyŏng-do*/咸鏡道/咸镜
  道, 108, 115
Hamgyong, 115
Hami Wei/哈密衛/哈密卫, 136, 167
Hami/Kumul/Camul/哈密, 167, 510
Hamili/哈密力, 167
Han Dynasty, 5, 18, 162
Han River/漢水/汉水, 175, 191
Han Shu/Book of Han/(漢書/汉书), 321
Handan City/邯鄲市/邯郸市, 179
Handong Wei/罕東衛/罕东卫, 136, 184
Handong Zuowei/罕東左衛/罕东左卫, 136, 164
Hangzhou/杭州, 187
Hankou/漢口/汉口, 191
Hanthawaddy Kingdom/漢達瓦底王國/汉达瓦底
  王国, 241
Hanyang/漢陽/汉阳, 191
Hanzhong/漢中/汉中, 175
Harbin City/哈爾濱市/哈尔滨市, 142
Harima Province/播磨國/播磨国, 31
Hebei Province/河北省, 134, 306
Hehemo/赫曷抹, 377